Electrocatalysis

Electrocatalysis

Editor:

Mulmudi Hemant Kumar, Ph.D.

ArclerPress

www.arclerpress.com

Electrocatalysis

Editor: Mulmudi Hemant Kumar, Ph.D.

708 3rd Avenue, 6th Floor
New York
NY 10017
United States of America
www.arclerpress.com

ISBN: 978-1-68094-418-1

Library of Congress Control Number: 2016951128

About the Editor

Mulmudi Hemant Kumar, Ph.D.

Hemant obtained his PhD from Nanyang Technological University, Singapore in 2014. His interests span from material synthesis via solution processed methods to it application in energy harvesting devices such as photovoltaics and water splitting. He is currently working as a research fellow at the Australian national university on silicon/perovskite solar cells.

Contents

List of Contributors

Carlos F. Zinola
Electrochemical Engineering Group, School of Sciences and School of Engineering, Fundamental Electrochemistry Laboratory, J. R. Rodó No. 1843, UDELAR, C.P.11100 Montevideo, Uruguay

Maria E. Martins
Electrocatalysis Laboratory, INIFTA, UNLP, C.C. 16, Suc. 4, 1900 La Plata, Argentina

Elena Pastor Tejera
Departemento de Química-Física, P.O. Box 38207 ULL, La Laguna, Spain

Newton Pimenta Neves Jr.
Laboratório de Hidrogênio, Instituto de Física "Gleb Wataghin", Universidade Estadual de Campinas (UNICAMP), Campinas, SP, Brazil

Minhua Shao
Department of Chemical and Biomolecular Engineering, The Hong Kong University of Science and Technology, Clear Water Bay, Kowloon, Hong Kong

Nemanja danilovic
Materials Science Division, Argonne National Laboratory, Argonne, Illinois 60439, USA

ram subbaraman
Materials Science Division, Argonne National Laboratory, Argonne, Illinois 60439, USA

dusan strmcnik
Materials Science Division, Argonne National Laboratory, Argonne, Illinois 60439, USA

Vojislav r. Stamenkovic
Materials Science Division, Argonne National Laboratory, Argonne, Illinois 60439, USA

nenad m. Markovic
Materials Science Division, Argonne National Laboratory, Argonne, Illinois 60439, USA

Andrzej Wieckowski
Department of Chemistry, University of Illinois at Urbana-Champaign, Urbana, IL 61801, USA

Matthew Neurock
Departments of Chemical Engineering and Chemistry, University of Virginia, Charlottesville, VA 22904-4741, USA

Ye Wang
Shanghai Key Laboratory for Molecular Catalysis and Innovative Materials, Collaborative Innovation Center of Chemistry for Energy Materials and Department of Chemistry, Fudan University, Shanghai 200433, China,

Shouzhong Zou
Department of Chemistry, American University, Washington, DC 20016, USA

Wen-Bin Cai
Shanghai Key Laboratory for Molecular Catalysis and Innovative Materials, Collaborative Innovation Center of Chemistry for Energy Materials and Department of Chemistry, Fudan University, Shanghai 200433, China,

Hui Meng
Department of Physics and Siyuan Laboratory, College of Science and Engineering, Jinan University, Guangzhou 510632, China

Dongrong Zeng
Department of Physics and Siyuan Laboratory, College of Science and Engineering, Jinan University, Guangzhou 510632, China

Fangyan Xie
Instrumental Analysis & Research Center, Sun Yat-sen University, Guangzhou 510275, China

Athanasios Papaderakis
Physical Chemistry Laboratory, Department of Chemistry, Aristotle University of Thessaloniki, Thessaloniki, Greece
Centre for Research and Technology Hellas, Chemical Process and Energy Resources Institute, Thessaloniki, Greece

Nikolaos Pliatsikas
Department of Physics, Aristotle University of Thessaloniki, Thessaloniki, Greece

Chara Prochaska
Physical Chemistry Laboratory, Department of Chemistry, Aristotle University of Thessaloniki, Thessaloniki, Greece

Kalliopi M. Papazisi
Centre for Research and Technology Hellas, Chemical Process and Energy Resources Institute, Thessaloniki, Greece

Stella P. Balomenou
Centre for Research and Technology Hellas, Chemical Process and Energy Resources Institute, Thessaloniki, Greece

Dimitrios Tsiplakides
Physical Chemistry Laboratory, Department of Chemistry, Aristotle University of Thessaloniki, Thessaloniki, Greece
Centre for Research and Technology Hellas, Chemical Process and Energy Resources Institute, Thessaloniki, Greece

Panagiotis Patsalas
Department of Physics, Aristotle University of Thessaloniki, Thessaloniki, Greece

Sotiris Sotiropoulos
Physical Chemistry Laboratory, Department of Chemistry, Aristotle University of Thessaloniki, Thessaloniki, Greece

Ksenija Đ. Popović
ICTM – Institute of Electrochemistry, University of Belgrade, Njegoševa 12, P. O. Box 473, 11000 Belgrade, Serbia

Jelena D. Lović
ICTM – Institute of Electrochemistry, University of Belgrade, Njegoševa 12, P. O. Box 473, 11000 Belgrade, Serbia

Nagappan Ramaswamy
Laboratory for Electrochemical Advanced Power, Department of Chemistry and Chemical Biology, Center of Renewable Energy Technology (NUCRET), Northeastern University, 360 Huntington Avenue, Boston, MA 02115, USA

Sanjeev Mukerjee
Laboratory for Electrochemical Advanced Power, Department of Chemistry and Chemical Biology, Center of Renewable Energy Technology (NUCRET), Northeastern University, 360 Huntington Avenue, Boston, MA 02115, USA

I. Istadi, Ardian D. Yudhistira
Laboratory of Chemical and Process Engineering, Department of Chemical Engineering, Diponegoro University, Jln. Prof. Soedarto, Kampus Undip Tembalang, Semarang 50275, Indonesia

Didi D. Anggoro
Laboratory of Chemical and Process Engineering, Department of Chemical Engineering, Diponegoro University, Jln. Prof. Soedarto, Kampus Undip Tembalang, Semarang 50275, Indonesia

Luqman Buchori
Laboratory of Chemical and Process Engineering, Department of Chemical Engineering, Diponegoro University, Jln. Prof. Soedarto, Kampus Undip Tembalang, Semarang 50275, Indonesia

Periasamy Arun Prakash
Department of Chemical Engineering and Biotechnology, National Taipei University of Technology, No.1, Section 3, Chung-Hsiao East Road, Taipei 106, Taiwan (ROC)

Umasankar Yogeswaran
Department of Chemical Engineering and Biotechnology, National Taipei University of Technology, No.1, Section 3, Chung-Hsiao East Road, Taipei 106, Taiwan (ROC)

Shen-Ming Chen
Department of Chemical Engineering and Biotechnology, National Taipei University of Technology, No.1, Section 3, Chung-Hsiao East Road, Taipei 106, Taiwan (ROC)

Hartmut WendtIBerl Institut TU Darmstadt, Wendt Ingenieure, Darmstaedter Str. 65, D-64807 Dieburg, Germany

Estevam V. Spinacé
Instituto de Pesquisas Energéticas e Nucleares, Av. Prof. Lineu Prestes, 2242, 05508-900 São Paulo - SP, Brasil

Almir Oliveira Neto
Instituto de Pesquisas Energéticas e Nucleares, Av. Prof. Lineu Prestes, 2242, 05508-900 São Paulo - SP, Brasil

Marcelo Linardi
Instituto de Pesquisas Energéticas e Nucleares, Av. Prof. Lineu Prestes, 2242, 05508-900 São Paulo - SP, Brasil

James Highfield
Heterogeneous Catalysis, Institute of Chemical & Engineering Sciences (ICES, A * Star), 1 Pesek Road, Jurong Island, 627833, Singapore

Aleksandar R. Žerađanin
Department of Interface Chemistry and Surface Engineering, Max-Planck-Institut für Eisenforschung, Max-Planck-Straße 1, 40237 Düsseldorf, Germany

Kai Wan, Zhi-Peng Yu
Key Laboratory on Fuel Cell Technology of Guangdong Province, School of Chemistry and Chemical Engineering, South China University of Technology, Guangzhou 510641, China

Zhen-Xing Liang
Key Laboratory on Fuel Cell Technology of Guangdong Province, School of Chemistry and Chemical Engineering, South China University of Technology, Guangzhou 510641, China

Jakob G. Howalt
Department of Energy Conversion and Storage, Technical University of Denmark, DK-4000 Roskilde, Denmark
Center for Atomic-scale Materials Design, Department of Physics, Technical University of Denmark, DK-2800 Kgs. Lyngby, Denmark

Tejs Vegge
Center for Atomic-scale Materials Design, Department of Physics, Technical University of Denmark, DK-2800 Kgs. Lyngby, Denmark

Preface

Electrocatalysis is a branch of electrochemistry which has become a necessity for industrial electrochemical processes. Though, most of the applications have run for academic purposes but not for technical uses in the industry. These days, the application of new concepts of electrocatalysis for industrial electrochemical processes has appeared as a necessity and not only attracting attention for chemists but for engineers. Consequently, industrial electrocatalytic processes have only been presented in the literature from the chemical engineering point of view with few further electrochemical explanations and mainly during the late seventies.

This book entitled Electrocatalysis focuses on the various aspects, right from fundamentals to it applications in fuel cells, Hydrogen evolution reaction (HER), electro-oxidation, oxygen reduction, sensors and solar fuels. The purpose of first chapter is to provide fundamentals and applications of electrocatalysis. Second chapter presents electrocatalysis in fuel cells. The aim of third chapter is to deal with electrocatalysis of the hydrogen evolution reaction (HER) in acid and alkaline media. Fourth chapter deals with contrast and synergy between electrocatalysis and heterogeneous catalysis. Fifth chapter is aimed at presenting an overview of the advances in the study of reaction mechanisms and electrocatalytic materials for ethanol oxidation reaction (EOR) with a focus on Pt- and Pd-based catalysts. Sixth chapter covers the latest advances in Pd-based catalysts for fuel cells. Seventh chapter focuses on ternary Pt-Ru-Ni catalytic layers for methanol electrooxidation prepared by electrodeposition and galvanic replacement. Eighth chapter deals with the effects influencing the overall formic acid oxidation at Platinum–Bismuth catalysts. In ninth chapter, a systematic study of oxygen reduction reaction (ORR) reaction mechanisms on Pt and non-Pt surfaces were presented. Electrocatalysis system for biodiesel synthesis from palm oil over dielectric-barrier discharge plasma reactor has been investigated in tenth chapter. Eleventh chapter provides a brief review on direct electrochemistry of catalase for electrochemical sensors. The aim of twelfth chapter is to deal with electrocatalysis and electrocatalysts for low temperature fuel cells. Thirteenth chapter reviews advances in photocatalysis and photoelectrocatalysis that exploit solar (photonic) primary energy in relevant endergonic processes, viz., H2 generation by water splitting, bio-oxygenate photoreforming, and artificial photosynthesis (CO2 reduction). Fourteenth chapter explores on the impact of spatial distribution of morphological patterns on the efficiency of electrocatalytic gas evolving reactions. Fifteenth chapter discusses about polyaniline-derived ordered mesoporous carbon as an efficient electrocatalyst for oxygen reduction reaction. In last chapter, the presence of oxygen species, e.g., resulting from a dehydrogenation reaction of residual water from a solid electrolyte or an aqueous electrolyte, has been investigated to understand the implications it has on the catalytic properties for electrochemical ammonia production.

Editor

Mulmudi Hemant Kumar, Ph.D.

INTRODUCTION

An electrocatalyst is a catalyst that participates in electrochemical reactions. Catalyst materials modify and increase the rate of chemical reactions without being consumed in the process. Electrocatalysts are a specific form of catalysts that function at electrode surfaces or may be the electrode surface itself. An electrocatalyst can be heterogeneous such as a platinum surface or nanoparticles, or homogeneous like a coordination complex or enzyme. The electrocatalyst assists in transferring electrons between the electrode and reactants, and/or facilitates an intermediate chemical transformation described by an overall half-reaction.

Half-reaction	$E°$ (V)
$H_2(g) \rightleftarrows 2H^+ + 2e^-$	≡ 0
$O_2(g) + 4H^+ + 4e^- \rightleftarrows 2H_2O$	+1.23

There are multiple ways for many transformations to occur. For example, hydrogen and oxygen can be combined to form water through a free-radical mechanism commonly referred to as combustion. Useful energy can be obtained from the thermal heat of this reaction through an internal combustion engine with an upper efficiency of 60% (for compression ratio of 10 and specific heat ratio of 1.4) based on the Otto thermodynamic cycle. It is also possible to combine the hydrogen and oxygen through redox mechanism as in the case of a fuel cell. In this process, the reaction is broken into two half-reactions which occur at separate electrodes. In this situation the reactant's energy is directly converted to electricity.

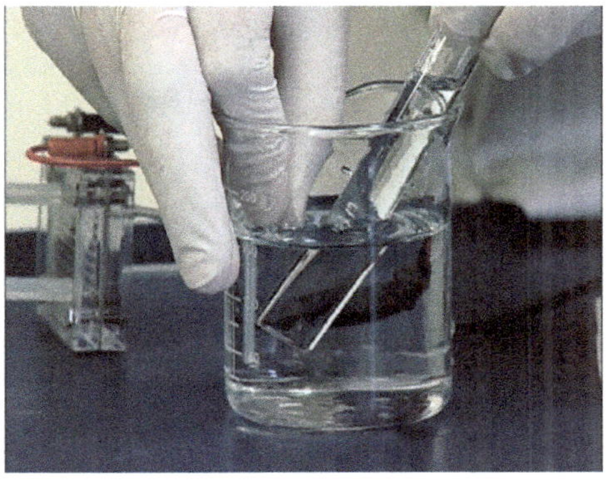

This process is not governed by the same thermodynamic cycles as combustion engines, it is governed by the total energy available to do work as described by the Gibbs free energy. In the case of this reaction, that limit is 83% efficient at 298K. This half-reaction pair and many others don't achieve their theoretical limit in practical application due to lack of an effective electrocatalyst.

One of the greatest drawbacks to galvanic cells, like fuel cells and various forms of electrolytic cells, is that they can suffer from high activation barriers. The energy diverted to overcome these activation barriers is transformed into heat. In most exothermic combustion reactions this heat would simply propagate the reaction catalytically. In a redox reaction, this heat is a useless byproduct lost to the system. The extra energy required to overcome kinetic barriers is usually described in terms of low faradaic efficiency and high over potentials. In the example above, each of the two electrodes and its associated half-cell would require its own specialized electrocatalyst.

Half-reactions involving multiple steps, multiple electron transfers, and the evolution or consumption of gases in their overall chemical transformations, will often have considerable kinetic barriers. There is often more than one possible reaction at the surface of an electrode. For example, during the electrolysis of water, the anode can oxidize water through a two electron process to hydrogen peroxide or a four electron process to oxygen. The presence of an electrocatalyst could facilitate either of the reaction pathways. Like other catalysts, an electrocatalyst lowers the activation energy for a reaction without altering the reaction equilibrium. Electrocatalysts go a step further than other catalysts by lowering the excess energy consumed by a redox reaction's activation barriers.

What is Catalysis?

Catalysis is the increase in the rate of a chemical reaction due to the participation of an additional substance called a catalyst. With a catalyst, reactions occur faster because they require less activation energy. They are not consumed in the catalyzed reaction, catalysts can continue to act repeatedly. Often only tiny amounts are required in principle.

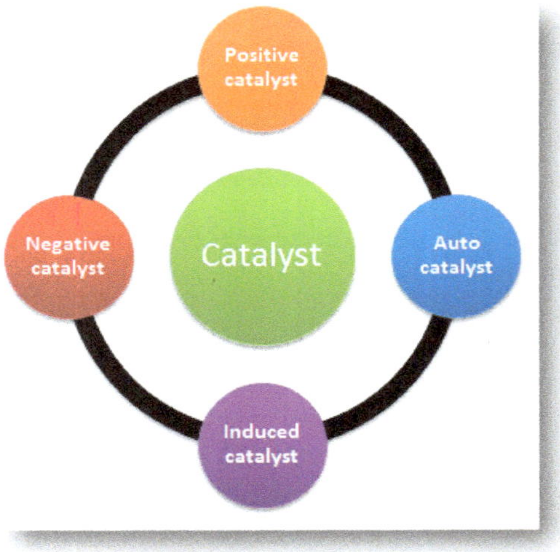

Technical perspective

In the presence of a catalyst, less free energy is required to reach the transition state, but the total free energy from reactants to products does not change. A catalyst may participate in multiple chemical transformations. The effect of a catalyst may vary due to the presence of other substances known as inhibitors or poisons (which reduce the catalytic activity) or promoters (which increase the activity). The opposite of a catalyst, a substance that reduces the rate of a reaction, is an inhibitor.

Catalyzed reactions have a lower activation energy (rate-limiting free energy of activation) than the corresponding unanalyzed reaction, resulting in a higher reaction rate at the same temperature and for the same reactant concentrations. However, the detailed mechanics of catalysis is complex. Catalysts may affect the reaction environment favorably [how?], or bind to the reagents to polarize bonds, e.g. acid catalysts for reactions of carbonyl compounds, or form specific intermediates that are not produced naturally, such as osmate esters in osmium tetroxide-catalyzed dihydroxylation of alkenes, or cause dissociation of reagents to reactive forms, such as chemisorbed hydrogen in catalytic hydrogenation.

Kinetically, catalytic reactions are typical chemical reactions; i.e. the reaction rate depends on the frequency of contact of the reactants in the rate-determining step. Usually, the catalyst participates in this slowest step, and rates are limited by amount of catalyst and its "activity". In heterogeneous catalysis, the diffusion of reagents to the surface and diffusion of products from the surface can be rate determining. A nanomaterial-based catalyst is an example of a heterogeneous catalyst. Analogous events associated with substrate binding and product dissociation apply to homogeneous catalysts.

Although catalysts are not consumed by the reaction itself, they may be inhibited, deactivated, or destroyed by secondary processes. In heterogeneous catalysis, typical secondary processes include coking where the catalyst becomes covered by polymeric side products. Additionally, heterogeneous catalysts can dissolve into the solution in a solid–liquid system or sublimate in a solid–gas system.

Overview of catalysts

The production of most industrially important chemicals involves catalysis. Similarly, most biochemically significant processes are catalyzed. Research into catalysis is a major field in applied science and involves many areas of chemistry, notably organometallic chemistry and materials science. Catalysis is relevant to many aspects of environmental science, e.g. the catalytic converter in automobiles and the dynamics of the ozone hole. Catalytic reactions are preferred in environmentally friendly green chemistry due to the reduced amount of waste generated, as opposed to stoichiometric reactions in which all reactants are consumed and more side products are formed. Many transition metals and transition metal complexes are used in catalysis as well. Catalysts called enzymes are important in biology.

A catalyst works by providing an alternative reaction pathway to the reaction product. The rate of the reaction is increased as this alternative route has a lower activation energy than the reaction route not mediated by the catalyst. The disproportionation of hydrogen peroxide creates water and oxygen, as shown below.

$$2\ H_2O_2 \rightarrow 2\ H_2O + O_2$$

This reaction is preferable in the sense that the reaction products are more stable than the starting material, though the unanalyzed reaction is slow. In fact, the decomposition of hydrogen peroxide is so slow that hydrogen peroxide solutions are commercially available. This reaction is strongly affected by catalysts such as manganese dioxide, or the enzyme peroxidase in organisms. Upon the addition of a small amount of manganese dioxide, the hydrogen peroxide reacts rapidly. This effect is readily seen by the effervescence of oxygen. The manganese dioxide is not consumed in the reaction, and thus may be recovered unchanged, and re-used indefinitely. Accordingly, manganese dioxide catalysis this reaction.

The Concept of Electrocatalysis

In chemistry, the concept of a catalyst is defined as a substance that can alter the velocity of a certain chemical reaction without any chemical change. This definition can be directly extrapolated for electrocatalysis. An electrocatalyst is an electrode material that interacts with some certain species during a Faradaic reaction but still remain unaltered. Since electrode reactions are heterogeneous, so electrocatalysts are usually heterogeneous catalysts, which means that the reactions take place on the surfaces of catalysts, and there exist adsoprtion/desoprtion steps on the surfaces of electrocatalysts. In order to compare the catalytic activity of different electrode materials, one can compare the current density at a constant over potential, or measure the over potential at a constant current density. A good electrocatalyst should show high current density at low over potential.

The concept of electrocatalysis is rather recent, and has arisen since the advent of research into fiel cells as a practical means of energy conversion. The performance of such devices depends, at the present stage of technology, almost entirely on problems concerning electrode reaction rates rather than, for instance, mass-transport processes. Typical $H_2/0_2$ fuel cells; working at low temperatures in acid electrolyte, me limited by the intrinsic rate of the oxygen electrode. To date, platinum is the most effective material for both hydrogen oxidation and oxygen reduction in acid medium. There is, however, a considerable difference in the relative rates at a given over potential that can be maintained on a platinum electrode for these two processes. Under ordinary conditions, the hydrogen oxidation rate is approximately 10^7 times more rapid than that of oxygen reduction. "His difference results in an effective loss in efficiency, due to cathode polarization, of perhaps 30 percent, and requires the use of platinum metal catalysts in high loadings. This slow rate of the oxygen-reduction reaction is well brown, and is manifested in practice by the low open-circuit potential of the cathode which corresponds to rate control by an alternative process, the electrode potential being determined by platinum oxidation (4), or impurity oxidation (5), or by some parasitic redox process.

The hydrogen electrode process on platinum has one of the fastest specific rate constants known (apart from metal-metal ion reactions of reactive metals and some redox processes) in aqueous solutions, whereas oxygen reduction on platinum must by comparison be considered a very slow process, though it is by no means the slowest reaction that has been observed. * is the wide range of rate constants that me encountered for the same process - for instance, some 10 orders of magnitude for the hydrogen electrode reaction (6), and at least 5 orders of magnitude for oxygen evolution on oxidized platinum (7). laboratory has shown that rate constants for oxygen reduction on oxide-free noble metals in acid electrolyte vary by approximately 7 orders of magnitude. Wide rate variations for other types of reactions, for instance hydrocarbon oxidation, have been noted. In this paper, past work on electrocatalysis will be briefly reviewed and correlated with some recent work on the electrocatalytic properties of Group VIII and Ib metals and alloys for the oxygen reduction reaction in acid solution. Some of the consequences of this study concerning the fuel cell oxygen electrode are discussed. In particular, the electrocatalytic mechanisms of the hydrogen and oxygen electrodes will be explored in detail, to determine the likelihood of future increase in 0_2 electrode exchange currents.

The Mechanism of Electrocatalysis

The specific reactions has given some insight into the factors controlling rates on different electrode substrates. The reactions most broadly studied for this purpose have been –

1. Hydrogen evolution in acid solution
2. Oxygen evolution on noble metal oxides in acid solution
3. Certain hydrocarbon reactions, for example, ethylene oxidation

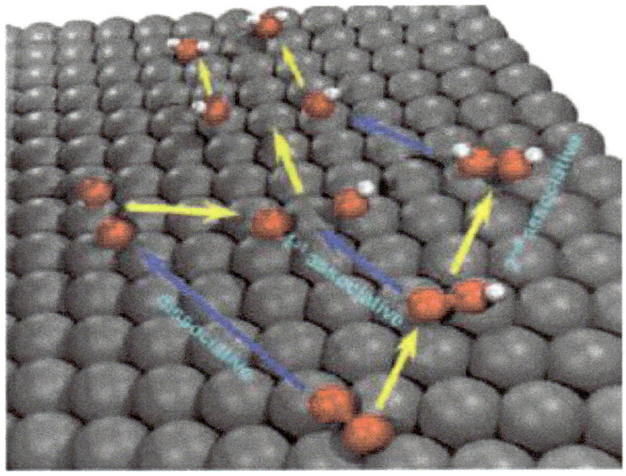

In comparing reaction rates for a specific process on different metals, it is important to first ascertain that two necessary conditions are met:

1. The process in question must be the same on the group of metals - that is, the rate-determining step must be the same in each case, and the reaction path, at least as far as the rate-determining step, must also be the same.
2. The process has been compared on the different metals under the same conditions - that is, with the same surface preparation in solutions of the highest possible purity, so that poisoning of the electrode surface by adsorption of capillary-active matter is avoided,

Electrocatalyst Surfaces

Specific characteristics of solid electrocatalysts, such as surface heterogeneity, degree of dispersion, and surface dynamics, play an important role in determining the optimization of the electrocatalytic process. These characteristics are considered in electrocatalyst design aiming to produce a material with a high density number of active centers on a large electrode surface area that should be easily accessible for reactants. From a practical standpoint, cheap, handily machinable, and environmentally friendly materials are desirable. All these requirements have to be evaluated for long-term efficient operation. Preliminary indications of electrocatalytic potentialities can be derived from no stationary process data, although a full evaluation is required at technical scales to vouch for real applications.

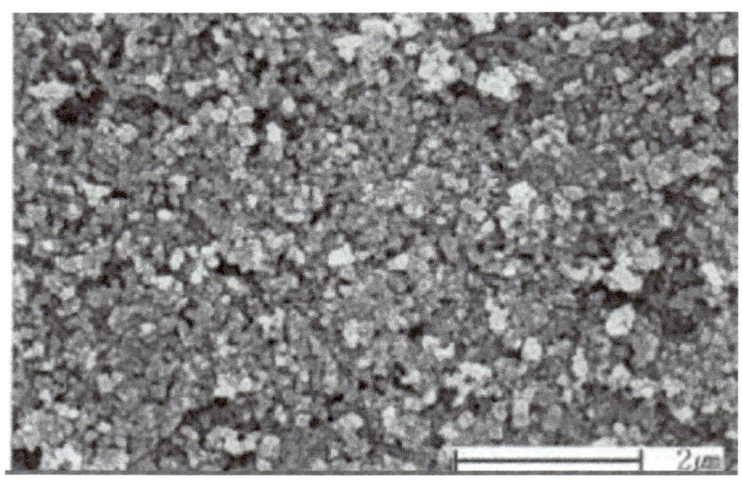

Electrocatalysts can be used at low, intermediate, and high dispersion conditions. The distribution of crystallographic planes at the surface can be varied from that corresponding to a smooth, single-crystal surface to those of rough polycrystalline particles dispersed in a conducting substrate. The number, density, and energy distribution of active sites available at the electrode surface depends on the surface roughness structure of solid electrocatalysts. The concept of roughness is related to the geometry and distribution of reacting centers, the size and morphology of dispersed electrocatalyst particles, and their aspect ratios. Capillary effects appear through porosity, the facile surface diffusion of species interacting with surface atoms, and relaxation of the proper solid surface. Handling this set of properties becomes essential for producing efficient electrocatalysts.

Metal Single-Crystal Surfaces

Single-crystal metal electrodes can be prepared by different techniques and are utilized as reference solid surfaces in which the kinetics of electrochemical reactions undergoes on a well-defined crystallography. Details on the preparation and use of metal single-crystal electrodes in electrocatalysis can be obtained elsewhere. At room temperature, the surface order of real single crystals is usually restricted to atomically smooth terraces of the order of some hundreds of square nanometers. Under different applied potentials, the interactions of metal single-crystal surfaces with OH species and water produce superstructures due to surface reconstruction. The use of well-defined single-crystal surfaces represents an important step for understanding the influence of surface crystallography on the kinetics of heterogeneous electrocatalysis, for determining the structure of adlayers at the atomic and molecular levels, and for evaluating adsorbate—substrate interactions and substrate dynamics.

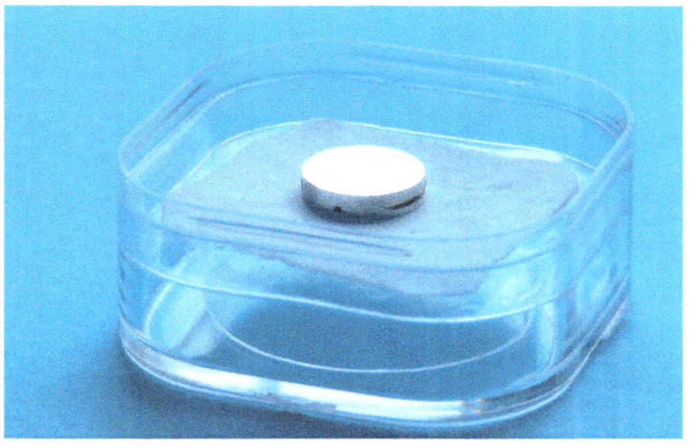

CHAPTER 1

ELECTROCATALYSIS: FUNDAMENTALS AND APPLICATIONS

Carlos F. Zinola, Maria E. Martins, Elena Pastor Tejera, and Newton Pimenta Neves Jr,

Electrocatalysis is a special field in Electrochemistry that has gained a special growth after the late eighties due to the application of new hybrid techniques. However, most of the applications have run for academic purposes but not for technical uses in the industry. Nowadays, the application of new concepts of electrocatalysis for industrial electrochemical processes has appeared as a necessity and not only attracting attention for chemists but for engineers. Therefore, industrial electrocatalytic processes have only been presented in the literature from the chemical engineering point of view with few further electrochemical explanations and mainly during the late seventies. The design and preparation of an electrocatalyst (electrodes for technological uses) are based on new concepts such as a controlled surface roughness, atomic topographic profiles, defined catalytic centre sites, atomic rearrangements, and phase transitions in the course of the electrochemical reactions.

Maybe the best interpretation of electrocatalysis is due to A. J. Appleby as an electrochemical reaction with an adsorbed species, either as reactant and/or product, which can change the kinetics of the reaction and in some cases also the mechanism. Although the old conception of electrocatalysis is mainly related to an experimental field of electrochemistry when using platinum-type metals, the advances in theory and modeling are main keys for a successful interpretation of the data.

The perspective and limitations of electrocatalysis in the laboratory together with their theoretical implications are also presented to give the readers a clearer view of the problems inside electrocatalytic reactions. This special issue also gathers the experiences of different authors who have worked on the principles of electrocatalysis with relevancy to experimental situations of particular importance in electrochemical engineering. This is the case of the work by T. Fahidy in the use of the Bayer's rule for an electrochemical reactor in order to avoid the possibility of reactor breakdown, reactor's safety, and environmental considerations well beyond purely scientific quantities.

One of the most important features of electrocatalysis arises from the different electrode kinetics using different crystallographic orientations for the electrode surface. A cathodic treatment develops (111) stepped planes similarly to the electrofacetting performed after applying the square wave program shown with X-ray diffraction patterns. These arrangements are well explained in the paper by C. Zinola, using combined techniques and particularly studying the methanol electrooxidation reaction. In spite of being most of the electrocatalytic reactions studied in the scientific community only for rather simple species, some of them involve low-weight amino acids, showing changes in the adsorbed configurations and chiral responses when the nature of the electrode surface is changed, as shown by K. Watanabe. Chiral behavior was clearly observed as an oxidation current difference between the enantiomers of alanine, aspartic acid, and glutamic acid.

The fast and contemporary advances in electrocatalysis have been already treated in other books but only considering the progress in the application of surface science and ultrahigh vacuum techniques to electrochemistry with no extrapolation to the possible industrial applications of these findings. Some combined techniques, such as differential electrochemical mass spectrometry, show the possibility of the continuous analysis of an electrode reaction with the detection of volatile or gaseous products in the course of the electrochemical reaction as shown by M. J. Lázaro for carbon dioxide electroreduction. Differential electrochemical mass spectrometry studies showed that the carbon support degradation, the distribution of

Citation: Carlos F. Zinola, Maria E. Martins, Elena Pastor Tejera, and Newton Pimenta Neves, Jr., "Electrocatalysis: Fundamentals and Applications," International Journal of Electrochemistry, vol. 2012, Article ID 874687, 2 pages, 2012. doi:10.1155/2012/874687.

products, and the catalytic activity toward the carbon dioxide electroreduction depend significantly on the surface chemistry of the carbon support.

There are various applications of electrocatalysis for technological electrochemical reactions, organic electrosynthesis, galvanoplasty, electrode sensors, fuel cells, batteries preparations, and so forth Some of them are presented here in order to give an insight into the possibilities of this science. Thus, the avoiding of interferences in the case of hydrogen peroxide electroreduction has been gained using alloys such as palladium-gold electrodes as reported by R. Mussarlieva with application for glucose biosensors. The surface topography of the electrode modified with Pd + Au mixed in proportions 90 : 10, exhibiting optimal combination of sensitivity and linear dynamic range towards hydrogen peroxide electrochemical reduction.

However, it seems that the main application is the case of fuel cells and one of the scopes is the possibility of long-lasting experiments without contamination or poisoning of the electrode surfaces. A. S. Aricó studied the continuous use of IrO_2/Pt electrocatalysts in fuel cells with good results that are also presented in this special issue. The IrO_2/Pt composition of 14/86 showed the highest performance for water electrolysis and the lowest one as fuel cell. Polymer electrolyte fuel cells are one of the most used devices for low-to-medium power density sizes. However, platinum electrocatalysts are often changed to platinum/ruthenium alloys because of the application of the electrocatalytic Langmuir-Hinshelwood surface mechanism. In this issue the use of dendrimers during the synthesis of platinum/ruthenium electrocatalysts for polymer electrolyte fuel cells shows interesting results that are strongly dependent on the physical characteristics of the alloy as presented by E. Pastor. After the activation of the catalysts by heating at 350°C, the real importance of the use of these encapsulating molecules and the effect of the generation of the dendrimer become visible. Not only the use of alloys of combined electrodeposited species offers electrocatalytic differences for fuel cell application, but also the preparation of Au core-Pd shell/C electrodes as shown by X. Liu for a direct borohydride-hydrogen peroxide fuel cell. The results show that the Au-Pd/C catalyst with core-shell structure exhibits much higher catalytic activity for the direct oxidation of $NaBH_4$ than pure Au/C catalyst, i.e. a direct borohydride-hydrogen peroxide fuel cell, in which the Au-Pd/C with core-shell structure is used as the anode catalyst and the Au/C as the cathode catalyst.

New advances relevant to electrocatalysis such as spectroscopic or hybrid techniques with new theoretical considerations act as a final point from them it is possible to evolve to a new projection of the electrocatalytic reaction itself. One of the purposes of this special issue is to present the field of electrocatalysis as an interfacial approach to a dynamic and permanently changing electronic-to-ionic conductor interphase with real technical applications.

CHAPTER 2

ELECTROCATALYSIS IN FUEL CELLS

Minhua Shao

INTRODUCTION

Low temperature fuel cells are expected to come into widespread commercial use in the areas of transportation and stationary and portable power generation, and thus will help solve energy shortage and environmental issues. Despite their great promise, commercialization has been hindered by lower-than-predicted efficiencies and the high cost of the electrocatalysts in the electrodes. The sluggish kinetics of the oxygen reduction reaction (ORR) are one of the main reasons for the high overpotential in a hydrogen proton exchange membrane fuel cell (PEMFC). The introduction of Mirai, the first mass-produced fuel cell vehicles (FCVs), by Toyota in Japan in 2014, and in North America in the following year, has accelerated the development of FCVs by other automotive companies. For instance, Honda and Hyundai announced the mass production of their own FCVs in 2016 and 2017, respectively. The current sale price of a new Mirai is about $57,000 US. One of the main reasons for the high sale price is the high Pt loading in the fuel cell stacks, especially at the cathode electrode, where the ORR occurs. The Pt loading at the anode, where the hydrogen oxidation reaction (HOR) occurs, can be reduced to as low as 0.05 mg cm^{-2} due to the extremely high reaction rate on Pt surfaces, while a much higher Pt loading ($\geq$0.2 mg cm^{-2}) at the cathode is required using Pt or Pt alloys as the ORR electrocatalysts in order to achieve a desirable cell performance. Pt is a costly and scarce metal. Thus, reducing its loading or even completely replacing it with abundant and cheap materials would be advantageous to lower the cost of FCVs. Recent research efforts have been focused on developing advanced Pt alloys, core-shell structures, shape-controlled nanocrystals and non-precious-metal (NPM) catalysts [1,2,3,4].

In addition to ORR activity, one must also consider the durability of the electrode during fuel cell operation in the harsh environment [5]. The life of a fuel cell stack in a FCV has to last at least 10 years in order to compete with the conventional combustion engine. It has been confirmed that the fuel cell performance gradually declined during operation [5]. The main reasons for the degradation of the catalyst layer are the dissolution of the Pt and the corrosion of the carbon support. As a consequence, both catalysts and supports that are more stable than Pt nanoparticles and carbon black may be needed to meet the durability requirements.

Promising supports include alternative carbon supports, carbides and oxides [6]. In other types of low temperature fuel cells, for instance direct alcohol fuel cells (DAFCs), the slow fuel oxidation reactions and fast performance decay caused by poisonous CO species adsorbed on catalyst surfaces are the other major contributions to their low performances [1]. Thus, the development of more active catalysts with higher tolerance to CO poisoning is required for high-performance and long-life DAFCs. The most promising DAFCs include DMFC, DEFC and DFAFC which use methanol, ethanol and formic acid as fuel, respectively.

THIS SPECIAL ISSUE

This Special Issue aims to cover recent progress and trends in designing, synthesizing, characterizing and evaluating advanced electrocatalysts and supports for both ORR and small organic molecule oxidation reactions, as well as theoretical understanding in fuel cell reactions. I am honored to be the Guest Editor for this Issue, which includes 34 high quality papers. The nine reviews together with 25 original research papers cover a very broad spectrum of electrocatalysis in fuel cells. The diversities of these contributions are illustrated in Figure 1.

Citation: Minhua Shao, Electrocatalysis in Fuel Cells, doi:10.3390/catal5042115.

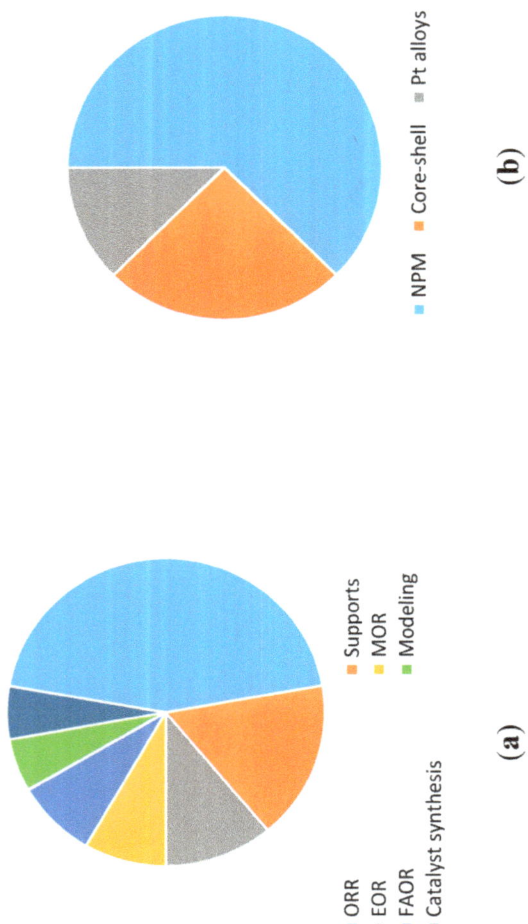

Figure 1. The distribution of the 34 contributions in this Special Issue (**a**) and the 16 contributions in oxygen reduction reaction; (**b**) in different research areas. ORR: oxygen reduction reaction; EOR: ethanol oxidation reaction; MOR: methanol oxidation reaction; FAOR: formic acid oxidation; NPM: non-precious metal.

From Figure 1a, it is obvious that the ORR is still the most important topic in fuel cell electrocatalysis. There are 16 papers that cover some important development of ORR electrocatalysts including NPM, core-shell and Pt alloy catalysts. As shown in Figure 1b, a significant effort is focused on completely replacing Pt in the ORR catalysts by developing novel NPM materials. Wei *et al.* summarized the recent progress in design and synthesis of metal-free nitrogen-doped carbon materials, including nitrogen-doped carbon nanotubes (NCNTs) and nitrogen-doped graphene (NG) for ORR in both acidic and alkaline media [7]. The Fe and N-doped carbon composites (Fe-N-C) are the most promising candidates, having comparable activity to Pt/C [4]. Liu *et al.* reviewed the progress made in the past five years in the areas of Fe-N-C electrocatalysts for ORR and understanding the possible active sites in this type of catalyst [8]. The Fe-N-Cs prepared from Fe-doped zeolitic imidazolate frameworks (ZIFs) are among the most active ones in catalyzing the ORR [9]. Barkholtz *et al.* optimized the synthesis and post-treatment protocols of ZIF-based Fe-N-C nanocomposites, as well as the membrane electrode assembly (MEA) fabrication process, and achieved an impressive fuel cell performance of 221.9 mA cm^{-2} at 0.8 V [10]. Armel *et al.* [11] emphasized the importance of the morphology control of Fe-N-C on ORR activity by adjusting the crystal size of ZIF-8, milling speed and heating mode. With the smallest ZIF-8 crystal size (100 nm), the best H_2/O_2 fuel cell performance of 900 mA cm^{-2} at 0.5 V was obtained, which was double the value obtained with previous synthesis protocol [11]. Zhang and Chen developed a novel method to prepare Fe-N-C catalysts by using a cationic surfactant cetyltrimethylammonium bromide (CTAB) as the template and the negatively charged persulfate ions as the oxidative agent to stimulate the aniline polymerization, resulting in a unique one-dimensional (1D) semi-tubular structure of PANI [12]. On the other hand, SBA-15 was used as the template by Wan *et al.* in the synthesis of nitrogen-doped ordered mesoporous carbon [13]. Qiao *et al.* found that P, N dual-doped reduced graphene oxide synthesized by pyrolyzing a mixture of graphite oxide and diammonium hydrogen phosphate was very active for ORR [14]. Non-noble metal oxides and chalcogenides are also promising catalysts for ORR. Some interesting works on the synthesis and evaluation of Ti-Nb oxides [15], CoS [16] and $FeSe_2$ [17] were also included in this Special Issue.

Core-shell structures consisting of a cheaper core and an atomic thin Pt shell have attracted great attention due to their extremely high Pt utilization and activity enhancement from the core materials [18]. Hu *et al.* designed a core-shell catalyst with a nitride (PdNiN) alloy core and a Pt monolayer. Its stability was dramatically enhanced compared with that of the previously reported structure with a pure Pd core [19]. Inoue *et al.* developed a new method without using any surfactant to synthesize clean Pd nanoparticles as the core for Pt monolayer deposition [20]. On the other hand, Caballero-Manrique *et al.*used Cu nanoparticles as the sacrificing template to prepare core-shell catalysts consisting of Pt and Pt-Ru shells [21]. Amra*et al.* tried to explain the strain and ligand effects from the $Pd_{1-x}Cu_x$ alloy core on the Pt monolayer based on the first principles density functional theory (DFT) calculations [22].

Since the discovery of Pt alloys as superior ORR catalysts for fuel cells in the 1980s, they have been considered as the second generation fuel cell catalysts after pure Pt. Pt alloys not only significantly reduce the Pt loading, but also enhance the catalytic activity and stability in comparison with Pt. Shen *et al.* gave a nice review on some of the recent approaches in developing Pt alloy electrocatalysts for the ORR [23]. The particle size effect of Pt alloys on the fuel cell performance and decay rate is very important and has not been systematically studied. Gummalla *et al.* compared the initial performances and decay trends of Pt_3Co/C cathodes in PEMFCs with three different particle sizes (4.9 nm, 8.1 nm, and 14.8 nm), but with the identical Pt loading [24]. The initial mass activity of the 4.9 nm Pt_3Co-based electrode was the highest, as well as the performance decay rate. The impact of PEMFC operating conditions, including upper potential, relative humidity, and temperature, on the alloy catalyst decay trends were also carefully studied for the first time.

Some non-carbon-based materials have been explored as alternative supports for ORR catalysts [6]. Lori and Elbaz summarized the latest studies on ceramic supports including carbides [25], oxides, nitrides, borides, and some composite materials [26]. Alternative carbon supports including carbon nanotubes, ordered mesoporous carbon, and colloid imprinted carbon were reviewed by Banham *et al.* [27]. The importance of carbon wall thickness was highlighted. Functionalized graphitic supports with pyrene carboxylic acid also showed superior durability [28].

The performance of fuel cell catalysts is certainly dependent on the preparation methods. Holade *et al.* [29] and Job *et al.*[30] summarized the recent advances in the preparation of carbon-supported nanocatalysts based on colloidal methods. The correlation between the structure of the catalysts and their activities and the effects from the synthesis methods were discussed. In addition, the fuel cell performance is also strongly influenced by the composition and fabrication protocols of MEA. A semi-empirical model was developed by Myles *et al.* to understand the performance of the cell as a function of the ratio of Nafion ionomer to carbon support (I/C ratio) in high temperature PEMFCs [31].

There are three papers in this issue focusing on Pt-based nanocomposites for the methanol oxidation reaction (MOR). Pt/C-Mn_xO_{1+x} [32] and $H_3PMo_{12}O_{40}$-Pt/reduced graphene oxide [33] were found to have better performance than Pt/C and PtRu/C due to the synergetic effects between Pt nanoparticles and hybrid supports. Wu *et al.* synthesized a conductive copolymer based on indole-6-carboxylic acid and 3,4-ethylenedioxythiophene (EDOT) as the support for Pt particles [34]. This nanocomposite also showed good activity for MOR. Chen *et al.* prepared a multi-component nanoporous PtRuCuW electrocatalyst by chemical and mechanical dealloying [35]. The unique ligament/channel nanoporous structure showed an enhanced activity for MOR compared to PtRu/C.

The advances in the study of reaction mechanisms and electrocatalytic materials (mainly Pt- and Pd-based catalysts) for the ethanol oxidation reaction (EOR) were reviewed by Wang *et al.* [36]. PdW alloys [37] and hollow PdCu nanocubes [38], as well as PtMn alloys [39], showed some improvement over pure Pd or Pt toward EOR.

Meng *et al.* presented a comprehensive review on the Pd-based electrocataysts' formic acid oxidation reaction (FAOR), MOR, EOR and ORR [40]. The high activity of Pd-based materials toward FAOR was also supported by a couple of original research papers included in this issue [41,42].

Finally, I would like to thank Keith Hohn, Editor-in-Chief, Mary Fan, Managing Editor, and the staff of the Catalysts Editorial Office for their great support during the preparation for this Special Issue. I also thank all the authors for their great contributions and referees for their time reviewing the manuscripts. I believe these excellent papers collected in this Special Issue will make significant contributions to the electrocatalysis community.

REFERENCES

1. *Electrocatalysis in Fuel Cells: A Non- and Low- Platinum Approach*; Shao, M., Ed.; Springer: London, UK, 2013.
2. Nie, Y.; Li, L.; Wei, Z. Recent advancements in Pt and Pt-free catalysts for oxygen reduction reaction. *Chem. Soc. Rev.* **2015**, *44*, 2168–2201.
3. Wang, Y.-J.; Zhao, N.; Fang, B.; Li, H.; Bi, X.T.; Wang, H. Carbon-supported Pt-based alloy electrocatalysts for the oxygen reduction reaction in polymer electrolyte membrane fuel cells: Particle size, shape, and composition manipulation and their impact to activity. *Chem. Rev.* **2015**, *115*, 3433–3467.
4. Jaouen, F.; Proietti, E.; Lefèvre, M.; Chenitz, R.; Dodelet, J.-P.; Wu, G.; Chung, H.T.; Johnston, C.M.; Zelenay, P. Recent advances in non-precious metal catalysis for oxygen-reduction reaction in polymer electrolyte fuel cells. *Energy Environ. Sci.* **2011**, *4*, 114–130.
5. Borup, R.; Meyers, J.; Pivovar, B.; Kim, Y.S.; Mukundan, R.; Garland, N.; Myers, D.; Wilson, M.; Garzon, F.; Wood, D. Scientific aspects of polymer electrolyte fuel cell durability and degradation. *Chem. Rev.* **2007**, *107*, 3904–3951.
6. Luo, Y.; Alonso-Vante, N. The effect of support on advanced Pt-based cathodes towards the oxygen reduction reaction. State of the art. *Electrochim. Acta* **2015**, *179*, 108–118.
7. Wei, Q.; Tong, X.; Zhang, G.; Qiao, J.; Gong, Q.; Sun, S. Nitrogen-doped carbon nanotube and graphene materials for oxygen reduction reactions. *Catalysts* **2015**, *5*, 1574–1602.
8. Liu, J.; Li, E.; Ruan, M.; Song, P.; Xu, W. Recent progress on Fe/N/C electrocatalysts for the oxygen reduction reaction in fuel cells. *Catalysts* **2015**, *5*, 1167–1192.
9. Proietti, E.; Jaouen, F.; Lefèvre, M.; Larouche, N.; Tian, J.; Herranz, J.; Dodelet, J.-P. Iron-based cathode catalyst with enhanced power density in polymer electrolyte membrane fuel cells. *Nat. Commun.* **2011**, *2*, 416.
10. Barkholtz, H.; Chong, L.; Kaiser, Z.; Xu, T.; Liu, D.-J. Highly active non-PGM catalysts prepared from metal organic frameworks. *Catalysts* **2015**, *5*, 955–965.
11. Armel, V.; Hannauer, J.; Jaouen, F. Effect of zif-8 crystal size on the O_2 electro-reduction performance of pyrolyzed Fe–N–C catalysts. *Catalysts* **2015**, *5*, 1333–1351.
12. Zhang, S.; Chen, S. Surfactant-template preparation of polyaniline semi-tubes for oxygen reduction. *Catalysts* **2015**, *5*, 1202–1210.
13. Wan, K.; Yu, Z.-P.; Liang, Z.-X. Polyaniline-derived ordered mesoporous carbon as an efficient electrocatalyst for oxygen reduction reaction. *Catalysts* **2015**, *5*, 1034–1045.
14. Qiao, X.; Liao, S.; You, C.; Chen, R. Phosphorus and nitrogen dual doped and simultaneously reduced graphene oxide with high surface area as efficient metal-free electrocatalyst for oxygen reduction. *Catalysts* **2015**, *5*, 981–991.
15. Ishihara, A.; Tamura, Y.; Chisaka, M.; Ohgi, Y.; Kohno, Y.; Matsuzawa, K.; Mitsushima, S.; Ota, K.-i. Titanium-niobium oxides as non-noble metal cathodes for polymer electrolyte fuel cells. *Catalysts* **2015**, *5*, 1289–1303.
16. Zhong, H.; Xi, J.; Tang, P.; Li, D.; Feng, Y. Positive effect of heat treatment on carbon-supported CoS nanocatalysts for oxygen reduction reaction. *Catalysts* **2015**, *5*, 1211–1220.
17. Zheng, Q.; Cheng, X.; Li, H. Microwave synthesis of high activity $FeSe_2$/C catalyst toward oxygen reduction reaction. *Catalysts* **2015**, *5*, 1079–1091.
18. Adzic, R.R.; Zhang, J.; Sasaki, K.; Vukmirovic, M.B.; Shao, M.; Wang, J.X.; Nilekar, A.U.; Mavrikakis, M.; Valerio, J.A.; Uribe, F. Platinum monolayer fuel cell electrocatalysts. *Top. Catal.* **2007**, *46*, 249–262.

19. Hu, J.; Kuttiyiel, K.; Sasaki, K.; Su, D.; Yang, T.-H.; Park, G.-G.; Zhang, C.; Chen, G.; Adzic, R. Pt monolayer shell on nitrided alloy core—a path to highly stable oxygen reduction catalyst. *Catalysts* **2015**, *5*, 1321–1332.

20. Inoue, H.; Sakai, R.; Kuwahara, T.; Chiku, M.; Higuchi, E. Simple preparation of Pd core nanoparticles for Pd core/Pt shell catalyst and evaluation of activity and durability for oxygen reduction reaction. *Catalysts* **2015**, *5*, 1375–1387.

21. Caballero-Manrique, G.; Brillas, E.; Centellas, F.; Garrido, J.; Rodríguez, R.; Cabot, P.-L. Electrochemical oxidation of the carbon support to synthesize Pt(Cu) and Pt-Ru(Cu) core-shell electrocatalysts for low-temperature fuel cells.*Catalysts* **2015**, *5*, 815–837.

22. Peles, A.; Shao, M.; Protsailo, L. Pt monolayer electrocatalyst for oxygen reduction reaction on Pd-Cu alloy: First-principles investigation. *Catalysts* **2015**, *5*, 1193–1201.

23. Shan, S.; Wu, J.; Kang, N.; Cronk, H.; Zhao, Y.; Zhao, W.; Skeete, Z.; Joseph, P.; Trimm, B.; Luo, J.; *et al.* Nanoscale alloying in electrocatalysts. *Catalysts* **2015**, *5*, 1465–1478.

24. Gummalla, M.; Ball, S.; Condit, D.; Rasouli, S.; Yu, K.; Ferreira, P.; Myers, D.; Yang, Z. Effect of particle size and operating conditions on Pt_3Co PEMFC cathode catalyst durability. *Catalysts* **2015**, *5*, 926–948.

25. Chiwata, M.; Kakinuma, K.; Wakisaka, M.; Uchida, M.; Deki, S.; Watanabe, M.; Uchida, H. Oxygen reduction reaction activity and durability of Pt catalysts supported on titanium carbide. *Catalysts* **2015**, *5*, 966–980.

26. Lori, O.; Elbaz, L. Advances in ceramic supports for polymer electrolyte fuel cells. *Catalysts* **2015**, *5*, 1445–1464.

27. Banham, D.; Feng, F.; Fürstenhaupt, T.; Pei, K.; Ye, S.; Birss, V. Novel mesoporous carbon supports for PEMFC catalysts. *Catalysts* **2015**, *5*, 1046–1067.

28. Negro, E.; Stassi, A.; Baglio, V.; Aricò, A.; Koper, G. Electrocatalytic activity and durability of Pt-decorated non-covalently functionalized graphitic structures. *Catalysts* **2015**, *5*, 1622–1635.

29. Holade, Y.; Sahin, N.; Servat, K.; Napporn, T.; Kokoh, K. Recent advances in carbon supported metal nanoparticles preparation for oxygen reduction reaction in low temperature fuel cells. *Catalysts* **2015**, *5*, 310–348.

30. Job, N.; Lambert, S.; Zubiaur, A.; Cao, C.; Pirard, J.-P. Design of Pt/carbon xerogel catalysts for PEM fuel cells. *Catalysts* **2015**, *5*, 40–57.

31. Myles, T.D.; Kim, S.; Maric, R.; Mustain, W.E. Application of a coated film catalyst layer model to a high temperature polymer electrolyte membrane fuel cell with low catalyst loading produced by reactive spray deposition technology.*Catalysts* **2015**, *5*, 1673–1691.

32. Videla, A.; Osmieri, L.; Esfahani, R.; Zeng, J.; Francia, C.; Specchia, S. The use of C-MnO_2 as hybrid precursor support for a Pt/C-Mn_xO_{1+x} catalyst with enhanced activity for the methanol oxidation reaction (mor). *Catalysts* **2015**, *5*, 1399–1416.

33. Wang, X.; Zhang, X.; He, X.; Ma, A.; Le, L.; Lin, S. Facile electrodeposition of flower-like PMo_{12}-Pt/RGO composite with enhanced electrocatalytic activity towards methanol oxidation. *Catalysts* **2015**, *5*, 1275–1288.

34. Wu, T.-Y.; Kuo, C.-W.; Chen, Y.-L.; Chang, J.-K. Copolymers based on indole-6-carboxylic acid and 3,4-ethylenedioxythiophene as platinum catalyst support for methanol oxidation. *Catalysts* **2015**, *5*, 1657–1672.

35. Chen, X.; Wang, H.; Wang, Y.; Bai, Q.; Gao, Y.; Zhang, Z. Synthesis and electrocatalytic performance of multi-component nanoporous PtRuCuW alloy for direct methanol fuel cells. *Catalysts* **2015**, *5*, 1003–1015.

36. Wang, Y.; Zou, S.; Cai, W.-B. Recent advances on electro-oxidation of ethanol on Pt- and Pd-based catalysts: From reaction mechanisms to catalytic materials. *Catalysts* **2015**, *5*, 1507–1534.

37. Liu, Q.; Liu, M.; Li, Q.; Xu, Q. Preparation and electrocatalytic characteristics of PdW/C catalyst for ethanol oxidation.*Catalysts* **2015**, *5*, 1068–1078.

38. Bai, Z.; Huang, R.; Niu, L.; Zhang, Q.; Yang, L.; Zhang, J. A facile synthesis of hollow palladium/copper alloy nanocubes supported on N-doped graphene for ethanol electrooxidation catalyst. *Catalysts* **2015**, *5*, 747–758.

39. Ghavidel, M.; Easton, E. Improving the ethanol oxidation activity of Pt-Mn alloys through the use of additives during deposition. *Catalysts* **2015**, *5*, 1016–1033.

40. Meng, H.; Zeng, D.; Xie, F. Recent development of Pd-based electrocatalysts for proton exchange membrane fuel cells.*Catalysts* **2015**, *5*, 1221–1274.

41. Wang, L.-L.; Cao, X.-L.; Wang, Y.-J.; Li, Q.-X. Sb surface modification of Pd by mimetic underpotential deposition for formic acid oxidation. *Catalysts* **2015**, *5*, 1388–1398.

42. Qiu, X.; Zhang, H.; Dai, Y.; Zhang, F.; Wu, P.; Wu, P.; Tang, Y. Sacrificial template-based synthesis of unified hollow porous palladium nanospheres for formic acid electro-oxidation. *Catalysts* **2015**, *5*, 992–1002.

CHAPTER 3

ELECTROCATALYSIS OF THE HER IN ACID AND ALKALINE MEDIA

Nemanja Danilovic, Ram Subbaraman, Dusan Strmcnik, Vojislav R. Stamenkovic and Nenad M. Markovic

INTRODUCTION

Electrocatalysis lies at the heart of a spectrum of chemical transformations that occur at electrochemical interfaces. In the near future, it will be the key driving force for technological innovations that are urgently needed to address the delivery of reliable, affordable, and environmentally friendly energy. There is no rigorous definition of the term "electrocatalysis", but in the broadest sense, it is the study of electrode reactions where the kinetics have a strong dependence on the nature of the electrode material. Not surprisingly, virtually every electrochemical reaction in which chemical bonds are made or formed is electrocatalytic, and the kinetics varies by many orders of magnitude for different electrode materials. This is true even for the simplest electrochemical reaction in which chemical bonds are broken, such as the hydrogen evolution reaction (HER), which, for decades, has served as a model reaction for probing how the nature of the electrode material (and/or structure of the double layer3–6) may affect electrochemical transformation of protons (acid electrolytes) or water (basic electrolytes) into molecular hydrogen ($(2H^+ [2H_2O] + 2e^- \leftrightarrows H_2 + [2OH^-])$. Traditionally, the mechanism of the HER is assumed to proceed by an initial formation of hydrogen intermediates (denoted hereafter as Had) which, depending on the pH of solution, are formed via charge-induced discharge of protons or water $(H^+ [H_2O] + e^- \leftrightarrows H_{ad} [+ OH^-])$. This, so-called Volmer step, is then followed either by either a Tafel chemical recombination step $(2H_{ad} \leftrightarrows H_2)$ or by the transfer of a second electron in a Heyrovsky step $(H^+ [H_2O] + H_{ad} + e^- \rightleftarrows H_2 + + [+ OH^-])$. Analytical, experimental, and computation analyses of this reaction have, for the most part, been closely tied to the concept of the volcano plot. In electrocatalysis, such plots, which generally express the rate of an electrocatalytic reaction as a function of more fundamental properties of the catalyst, known as descriptors, date to the early HER rate expressions derived by Parsons and Gerischer. Parsons and Gerischer used of the hydrogen adsorption free energy (determined by the strength of adsorption of H_{ad} on the catalyst surface) as a descriptor for the trends in HER rates on different catalysts. This early analytical volcano was followed by experimentally derived volcanoes, developed by Trasatti and extending the earlier work of Krishtalik, in which the measured rates were correlated with measurements of the surface work function and the bulk heats of hydride formation. More recently, a combination of fundamental surface rate modeling, similar to the techniques employed by Parsons, with atom-specific free energy data obtained from Density Functional Theory calculations, have permitted the development of HER volcanoes in terms of descriptors (metal–hydrogen bonding) that may be rapidly calculated on both metals and metal alloys. It should be noted that some authors, however, have identified some challenges in generating such volcano plots from experimental results, particularly when dealing with metals that are likely to form oxides or hydrides under HER conditions. Given that the intrinsic activity of metal catalysts in alkaline medium are ≈2–3 orders of magnitude lower than in acid solutions, the question arises as to what type of relationships between activity in the HER and the nature of electrode material might be expect at high pH values. This has been addressed recently in several published papers and will be summarized in the remaining part of this paper.

EXPERIMENTAL

The Pt, Ru, Ir, Au, Ag and Cu electrodes were prepared by radio frequency (RF) annealing at ≈800–1100 °C in a 3 % H_2–Ar gas mixture, while the Ti electrode was polished and chemically cleaned. The samples were transferred into an electrochemical cell (made from Teflon) with the surface protected with a drop

Citation: Nemanja danilovic, ram subbaraman, dusan strmcnik, vojislav r. Stamenkovic and nenad m. Markovic, Electrocatalysis of the HER in acid and alkaline media, doi: 10.2298/JSC131118136D.

of DI water and immersed under potential control at 0.05 V vs. the reversible hydrogen electrode (RHE) in 0.1 M KOH and 0.1 M $HClO_4$. $Ni(OH)_2$ -modified metal electrodes (ca. 30–40 % coverage) were prepared by chemical deposition, wherein, the pristine metal electrodes were immersed and equilibrated in 0.01–0.1 M $NiCl_2$ solutions for 2–12 h. For assessing the HER activity, a sweep rate of 50 mV s -1 was employed, while the rotation rate was 1600 rpm. Only the first negative going HER scan was used for activity determination, where the current densities were obtained based on the geometric surface area. Experiments were controlled using an Autolab PGSTAT 302N potentiostat with IR compensation. The used gases were of research grade (5N) Ar.

RESULTS AND DISCUSSIONS

HER Activities in Acid Electrolytes

As in the past, Fig. 1 shows that if activities for the HER of the coinage metals (IB metals: Au, Cu, Ag), the platinum group (or PGM: Pt, Ir, Ru) metals and the valve metal (Ti) in 0.1M HClO4, are plotted as a function of the M–Had binding energy (the M–H_{ad} binding energy values were taken from the literature9), a volcano relationship is established with Ir and Pt at the apex of the volcano curve. Further analysis of Fig. 1 reveals that the IB group elements are positioned on the ascending slope of the volcano with order in activity increasing from Au < Cu < Ag, which is different from the order predicted by DFT (Cu > > Au > Ag). The exact cause of this discrepancy is currently unknown, but considering that Cu is "less noble" than Au and Ag, it is reasonable to assume that the experimentally measured activity of Cu could be influenced by adsorbed spectator species, which may affect the availability of active sites required for adsorption of the hydrogen active intermediates. Fig. 1 also shows that the elements that interact strongly with H_{ad} (such as Ru and Ti) are positioned on the descending slope of the volcano, supporting previous suggestions that the M–Had binding energy can be used as a descriptor for the HER. Not in passing, given that recent analysis has demonstrated that neither Ru nor Ti are bare metals in the HER region, it is suggested that, in fact, experimentally it is very difficult (impossible) to determine unambiguously solely based on the metal–Had energetics what would be the correct position of these two elements in the observed volcano relationship. This is most likely also true for the HER in alkaline solutions, when the rates of the reaction are much slower than in acidic environments. Interestingly, with one recent exception, there are no fundamental studies that have focused on deriving relationships between the nature of metals and activity of the HER at high pH values. Furthermore, there are still debates as to why the rate of the HER is 2 to 3 orders of magnitude lower at pH 13 than at pH 1 and why the reaction is sensitive to the structure of surface atoms in alkaline media but largely insensitive in acidic environments. Finally, the question arises as to whether the HER on metal surfaces in alkaline solutions can approach the activity at low pH values. In the next two sections, focus is first directed on the differences and similarities between the volcano-type relationships in alkaline and acidic environments and, then, this knowledge is used to demonstrate that it is indeed possible to improve the activities of the HER at high pH values, simply by controlling the rates of the water dissociation step, i.e., the Volmer step.

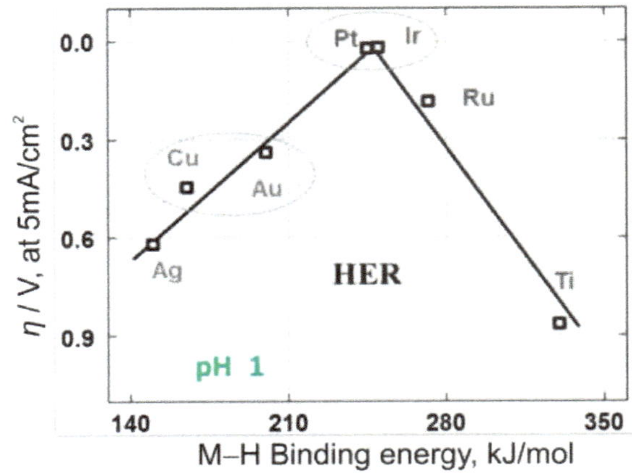

Figure 1. A volcano plot, measured in 0.1 M $HClO_4$ (pH 1), of several metals denoting their HER activity, overpotential (η) at 5 mA cm^{-2} , as a function of their calculated M–H binding energy. At the apex of the plot are the noble metals Pt and Ir, which have nearly identical activities, while on either side are metals that either bind hydrogen too weakly (Ag) or too strongly (Ti). Note that the order of activity of the coinage metals is Au > Cu > Ag, which follows the trend in the M–H binding energies.

HER Activities in Alkaline Electrolytes

Although the HER exhibits a volcano-like behavior in alkaline solutions (Fig. 2), there are three notable differences with change in pH: i) the activities in alkaline solutions are lower than in acid solutions (in line with previous observation that intrinsic activity of metal catalysts in alkaline medium are ≈2–3 orders of magnitude lower than in acid solutions); ii) significantly, in contrast to acid solutions, in alkaline solution Ir is more active than Pt (>55 mV), indicating that there might be substantial difference in the reaction mechanisms between the HER at low and high pH values; iii) the trend in the activity between IB group metals in alkaline solution increases from Ag < Au < Cu, an order of activity that is different from the one observed in acidic environments (Ag < Cu < Au). Clearly, these features cannot be explained solely based on the free energy of hydrogen adsorption since this descriptor should be independent of the pH of the solution. In previous reports, it was argued that the main difference between the HER in acid and alkaline electrolytes lies in the source of the hydrogen; while in acid solution, active intermediates are formed from protons, in alkaline solutions an additional activation process involving the dissociation of water is required to generate the Had. It was recently shown that this is an important difference and that evolution of H_2 from H_2O in alkaline solutions requires materials that are equally effective for water dissociation as well as Had recombination steps. In line with literature data, it was concluded that Cu is more active for the HER in alkaline solution than Au because the former is a more effective water dissociation catalyst; hence, in alkaline solutions, Cu provides a higher rate for generation of hydrogen intermediates from water. The differences observed between Pt and Ir surfaces could be explained in a similar way. Considering that there is no difference in the energetics for the Tafel step on Pt and Ir (the same activities are observed in acid solutions), the higher activity of Ir compared to Pt in alkaline solution can only be due to improved dissociation of water on the Ir surface. Therefore, in order to construct the activity trends of the HER in alkaline solutions, two rather than one descriptor has to be taken into account; namely, synergy between the energy required to dissociate the water molecules (M–water energetics) and adsorption/recombination of hydrogen intermediates (M–H_{ad} energetics).

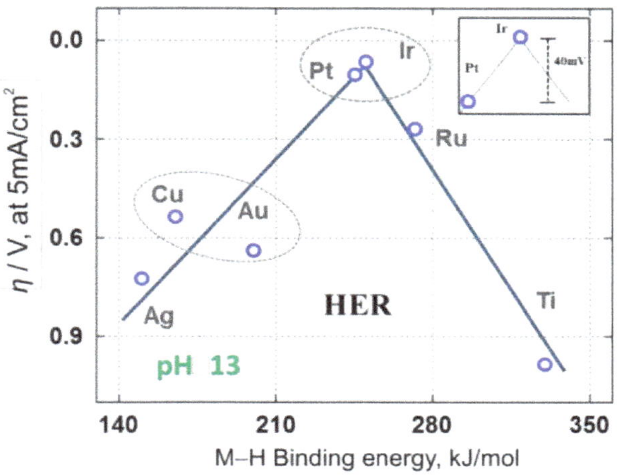

Figure 2. A volcano plot, measured in 0.1 M KOH (pH 13), of several metals denoting their HER activity, overpotential (η) at 5 mA cm^{-2}, as a function of their calculated M–H binding energy. At the apex of the plot are noble metals Pt and Ir, while on either side are metals that either bind hydrogen too weakly (Ag) or too strongly (Ti). Note the distinctions between the volcano plot in acid electrolytes. The order of activity of the coinage metals is Cu > Au > Ag, which no longer follows the trend in their M–H binding energy and the difference in overpotential between Ir and Pt is significantly larger (≈55–60 mV) compared to acid electrolyte values (≈1–5 mV).

HER Activities on Metals Modified With A "Water Dissociation Catalyst" In Alkaline Electrolytes

In order to confirm the importance of both the descriptors in a complex transformation of H_2O to H_2 in alkaline solution, the metal electrodes were modified by $Ni(OH)_2$ clusters, which are known to be extremely active in the water dissociation step. Thus, the activity curves in Fig. 3a and b for such modified systems were compared with the activities for the bare (unmodified) surfaces in the corresponding alkaline and acid environments. A close inspection of Fig. 3 unambiguously reveals that the activities for the $Ni(OH)_2$–M surfaces in alkaline solutions approach the activities in acid solution for the pure metals, i.e., Pt = Ir and Au > Cu > Ag. Based on these result, it is proposed that the rate of proton production from water is significantly enhanced on the Ni(OH)–M surfaces in alkaline environments so that the rate of the overall reaction is again affected predominantly by the M–Had binding energy, as is the case for acidic solutions. It should be remembered, however, that the M–H_{ad} energies alone cannot serve as the physicochemical descriptor for the HER on metal surfaces in alkaline environments. In other words, the sine qua non for the HER in alkaline medium is clearly a synergy between the effectiveness of the catalyst to break water molecules (predominantly on more oxophilic sites, such as defects or hydroxyoxides) and to efficiently collect and recombine hydrogen intermediates on more noble metal sites. Based on this supposition, it is reasonable to suggest that in alkaline solutions, the more defected Pt(110) is more active than the more "perfect" Pt(111) simply because the rate of the HER is proportional to the density of surface defect sites. Along the same lines, metals covered by $Ni(OH)_2$ are more active than bare M surfaces because in a b functional mechanism, the edges of $Ni(OH)_2$ clusters promote the dissociation of water while the produced intermediates are adsorbed and recombined on M sites (Fig. 3).

CONCLUSIONS

It is proposed that: i) due to coverage by spectator species, even in the HER potential region, it is still questionable if it would be possible to establish experimentally a true relationships between M–Had energetics and catalytic activity – therefore, with the exception of Pt, Ir and Au, the experimentally established positions of the other metal catalysts in the observed volcano relations are uncertain over the entire pH range; ii) while in acidic solutions the M–H_{upd} binding energy controls the activity trends (supported by DFT calculations), the trends in activity in alkaline solutions are controlled by a delicate balance between two descriptors: the M–H_{ad} interaction as well as the energetics required to dissociate water molecules; iii) the importance of the second descriptor was confirmed by the introduction of bifunctional catalysts, such as M modified by Ni(OH); e.g., while the latter serves to enhance catalytic decomposition of water, the metal sites are required for the collection and recombination of the produced hydrogen intermediates.

Acknowledgments

This work was supported by the Office of Science, Office of Basic Energy Sciences, Division of Materials Science, U.S. Department of Energy, under contract DE-AC0-06CH11357.

REFERENCES

1. N. M. Markovic, P. N. Ross, Surf. Sci. Rep. 45 (2002) 117
2. R. R. Adzic, F. Feddrix, B. Z. Nikolic, E. Yeager, J. Electroanal. Chem. 341 (1992) 287
3. R. Subbaraman, D. Tripkovic, D. Strmcnik, K.-C. Chang, M. Uchimura, A. P. Paulikas, V. Stamenkovic, N. M. Markovic, Science 334 (2011) 1256
4. D. Strmcnik, M. Escudero-Escribano, K. Kodama, V. R. Stamenkovic, A. Cuesta, N. M. Marković, Nat. Chem. 2 (2010) 880
5. D. Strmcnik, K. Kodama, D. van der Vliet, J. Greeley, V. R. Stamenkovic, N. M. Marković, Nat. Chem. 1 (2009) 466
6. D. Strmcnik, D. F. van der Vliet, K.-C. Chang, V. Komanicky, K. Kodama, H. You, V. R. Stamenkovic, N. M. Marković, J. Phys. Chem Lett. 2 (2011) 2733
7. J. O' M. Bockris, E. C. Potter, J. Electrochem. Soc. 99 (1952) 169
8. A. Lasia, in Fuel Cell Electrocatalysis, W. Vielstich, A. Lamm, H. A. Gasteiger, Eds., Wiley, Chichester, 2003, pp. 416–440
9. S. Trasatti, J. Electroanal. Chem. Interf. Electrochem. 39

(1972) 163

10. R. Parsons, Trans. Faraday Soc. 54 (1958) 1053
11. H. Gerischer, Bull. Soc. Chim. Belg. 67 (1958) 506
12. L. I. Krishtalik, Electrokhimiya 2 (1966) 616
13. J. Greeley, T. F. Jaramillo, J. Bonde, I. B. Chorkendorff, J. K. Norskov, Nat. Mat. 5 (2006) 909
14. J. Greeley, J. K. Norskov, L. A. Kibler, A. M. El-Aziz, D. M. Kolb, ChemPhysChem 7 (2006) 1032
15. W. Schmickler, S. Trasatti, J. Electrochem. Soc. 153 (2006) L31.
16. E. Skúlason, G. S. Karlberg, J. Rossmeisl, T. Bligaard, J. Greeley, H. Jónsson, J. K. Nørskov, Phys. Chem. Chem. Phys. 9 (2007) 3241
17. E. Skúlason, V. Tripkovic, M. E. Björketun, S. Gudmundsdóttir, G. Karlberg, J. Rossmeisl, T. Bligaard, H. Jónsson, J. K. Nørskov, J. Phys. Chem., C 114 (2010) 18182
18. M. M. Jaksic, Electrochim. Acta 29 (1984) 1539
19. J. Greeley N. M., Markovic, Energy Environ. Sci. 5 (2012) 9246
20. N. Danilovic, R. S. D. Strmcnik, K. C. Chang, A. P. Paulikas, V. R Stamenkovic, N.M. Markovic, Angew. Chem. 124 (2012) 12663
21. R. Subbaraman, D. Tripkovic, K.-C. Chang, D. Strmcnik, A. P. Paulikas, P. Hirunsit, M. Chan, J. Greeley, V. Stamenkovic, N. M. Markovic, Nat. Mat. 11 (2012) 550
22. N. M. Markovic, B. N. Grgur, P. N. Ross, J. Phys. Chem., B 101 (1997) 5405
23. N. M. Markovic, S. T. Sarraf, H. A. Gasteiger, P. N. Ross, J. Chem. Soc., Faraday Trans. 92 (1996) 3719
24. D. Strmcnik, M. Uchimura, C. Wang, R. Subbaraman, N. Danilovic, D. van der Vliet, A. P. Paulikas, V. R. Stamenkovic, N. M. Markovic, Nat. Chem. 5 (2013) 300.

CHAPTER 4

CONTRAST AND SYNERGY BETWEEN ELECTROCATALYSIS AND HETEROGENEOUS CATALYSIS

Andrzej Wieckowski and Matthew Neurock

INTRODUCTION

Electrocatalysis and heterogeneous catalysis are closely related in that they involve well-controlled sequences of elementary bond-breaking and making processes and share many common mechanistic principles in the transformation of molecules over supported metal and metal oxide catalysts. While there are many areas of synergy between the two, including the materials that are used and the available reaction pathways and mechanisms, there are also well-established differences [1–4]. Heterogeneous catalysis has often celebrated more detailed insights into reaction mechanisms than electrocatalysis due to the advances in spectroscopy and theory of the gas/solid interface as compared to the more complex aqueous/solid interface in electrocatalysis. As such electrocatalysis has often followed from the mechanistic advances derived from gas phase heterogeneous catalysis. Many of the current efforts in heterogeneous catalysis, however, are focused on energy conversion strategies involving catalytic transformations which proceed at the fluid/solid interface and, as a consequence, are now closely following the leads from electrocatalysis. A number of common mechanistic principles and features are beginning to emerge between the two fields. Understanding the synergies as well as the differences between catalysis and electrocatalysis should thus enable advances in the science and application for both areas. Herein, we compare and contrast some of the fundamental mechanistic constructs as well as the practical applications for electrocatalysis and heterogeneous catalysis. More specifically, we focus on metal catalyzed oxidation processes.

Citation: Andrzej Wieckowski and Matthew Neurock, "Contrast and Synergy between Electrocatalysis and Heterogeneous Catalysis," Advances in Physical Chemistry, vol. 2011, Article ID 907129, 18 pages, 2011. doi:10.1155/2011/907129.

GENERAL COMPARISONS BETWEEN CATALYSIS AND ELECTROCATALYSIS

At the macroscopic level, many of the catalytic materials that are used in catalysis and electrocatalysis are very similar in that they involve supported metal particles, where the interaction between the metal and support is critical to catalyst performance as well as catalyst stability. The metal or metal oxide/support interface can result in sites with unique structural or electronic characteristics, nove bifunctional sites, and/or sites that promote proton and electron transfer. The nature and the strength of the bonds between the metal and the support control the stability of these materials and their resilience to harsh reaction environments.

The characterization of the electronic and atomic structure of the metal and the support in both catalysis as well as electrocatalysis is typically carried out through the use of extended X-ray absorption spectroscopy (XAFS), electron microscopy, X-ray (XPS) and ultraviolet photoelectron spectroscopy (UPS). In addition, many of the most active metals used in electrocatalysis are very often the same as those used in heterogeneous catalysis. For example, Pt and other group VIII metals are known to be very active in the electrocatalytic oxidation of alcohols and the reduction of oxygen in fuel cells, automotive exhaust catalysis and hydrogenolysis, and hydrogenation catalysis in the conversion of petroleum and renewable resources. This is predominantly the result of the well-established Sabatier's Principle which suggests that the metals in middle of the periodic table demonstrate an optimal metal-adsorbate bond strengths necessary to balance surface reaction steps and product desorption steps [5–9].

In addition to similarities, there are well-established differences between traditional gas phase heterogeneous catalysis and electrocatalysis. Perhaps the greatest difference between the two relates to the unique reaction environments in which they are carried out. The gas phase catalytic environment is far less complex than that of the electrified water/metal interface for electrocatalytic systems, thus allowing for more detailed spectroscopic characterization of the working surface intermediates, application of ultrahigh vacuum experiments as well as direct comparisons with theoretical simulations on model surfaces. The presence of solution, ions, charged interfaces, complex surface potentials, and electric fields present in electrocatalytic systems can all act to significantly change the surface chemistry and catalysis that occurs in these environments. These interfaces tend to significantly promote polar reactions and direct heterolytic bond activation steps which would otherwise be unstable and not occur in gas phase catalytic systems. The electrochemical environment, however, is typically much harsher and deleterious to catalyst stability than that found in gas phase catalysis. The dissolution of the metal and the support are thus important concerns for electrocatalytic processes as these steps are enhanced under electrochemical conditions. In addition, the presence of electrolyte often enhances or impedes catalytic kinetics and within certain potential regions can result in poisoning of the surface.

While there are important differences between electrocatalysis and catalysis that result from the presence of solution, counter ions, and electric fields, Nørskov [8–12], Anderson [13–18], and others [1, 6] have been able to model the electrochemical systems by simply carrying out gas phase calculations on well-defined model clusters and surfaces and adding in the critical features that influence the surface chemistry such as local water molecules as well as the influence of potential. This is an important step in that one has the ability to not only understand but begin to tune the reaction chemistry. Understanding the similarities and differences of the molecular transformations that occur at ultrahigh vacuum conditions and electrochemical conditions will undoubtedly drive advances in the development of catalytic and electrocatalytic materials and processes.

In addition to the scientific issues, there are also a number of important technological differences in the "infrastructure" that supports the heterogeneous and electrocatalysis communities. Most of electrocatalysis appears to be centered around fuel cells and more specifically proton exchange membrane (PEM) fuel cells carry out the oxidation of hydrogen, oxygenates, or hydrocarbon molecules to CO_2 and the reduction oxygen to water [19]. This is in clear contrast to heterogeneous catalysis, which spans a wide and diverse range of different molecules with very rich chemistry and stands behind extensive chemical, automotive, petroleum, and pharmaceutical industries. In addition, there have been significant research and development investments in heterogeneous catalysis from the government as well as industry. Methane reforming, methane combustion, ammonia synthesis NO_x conversion, and Fischer Tropsch synthesis, for example, are very large-scale processes which have no analogue in electrocatalysis. Furthermore, the future directions for PEM fuel cell catalysis is very specific with a strong focused effort on resolving the issues related to durability and maintaining catalytic activity for many years. This is quite different than the shorter lifetimes involved in most heterogeneous catalytic processes with the exception of automotive emission catalysis. In electrocatalysis, catalyst must be able to withstand the harsh operating conditions and operate effectively over the lifetime of the vehicle. Catalyst loss and deactivation tend to be quite severe in electrocatalysis due to the presence of solution, ions, and electric fields which not only lead to catalyst poisoning but also catalyst dissolution. This significantly limits the choice of catalytic materials to specific supported metals/alloys, metal oxides, and other stable inorganic materials such as chalcogenides or carbides. This is a very narrow range of possibilities as compared to what is typically practiced in the gas phase heterogeneous catalysis community. The long term durability, aggressive solution conditions (both high and low pH), as well as the cost tend to prevent other avenues available to gas phase catalysis to be applied in the field of electrocatalysis.

Despite the complexity and the challenges of the electrochemical environment and the differences outlined above, many of the fundamental constructs that govern gas phase catalysis are also integral to electrocatalysis. There have been a number of pioneering developments in spectroscopy, kinetic analyses, theory, and synthesis that have occurred over the past few decades that have clearly shown how traditional concepts from heterogeneous catalysis apply directly to electrocatalysis. This includes the elucidation of nature of the active site, competitive adsorption phenomena, the influence of alloys, promoters and poisons, structure sensitivity, surface oxidation state, particle size effects, and metal support interactions. A schematic representation of the complex metal solution interface that would exist in either the catalytic or electrocatalytic oxidation of glucose over a carbon-supported PtRu alloy cluster is shown in Figure 1. In a review that is now over ten years old, Jarvi and Stuve [20] elegantly described the direct link between some of the fundamental principles that control heterogeneous catalysis and electrocatalysis. This included the specific accounting of active sites, the identification of reaction intermediates in elucidating reaction mechanisms, and the role of poisons and modifiers and their influence on catalytic kinetics and aging. The authors nicely showed that the kinetics for catalysis and electrocatalysis were essentially the same and shared a common framework.

The complexity of the aqueous/metal interface and the inability to spectroscopically resolve molecular intermediates at this interface limited early electrocatalytic studies to measurements of macroscopic kinetics with little understanding of the elementary molecular transformations, the active sites or the influence the atomic and the electronic structure. This put electrocatalysis at a distinct disadvantage over traditional gas phase heterogeneous catalysis. The tremendous breakthroughs that have occurred over the past two decades in the ability to characterize the atomic structure of the working surface and reaction intermediates on the surface within the electrochemical environments have made it possible to begin to discuss elementary mechanisms and the influence of specific structural and composition parameters. Breakthroughs in broad-band spectroscopy [21], in situ electrochemical NMR [22–24], EXAFS [25–29], and surface-enhanced infrared spectroscopy [30,31] are helping to significantly advance our understanding of electrocatalysis and in some cases begin to rival that in gas phase catalysis. These methods now allow for the direct insights into the nature of the active sites as well as surface intermediates under actual working conditions. Such insight has led to an exponential growth in the literature in the identification of active sites, reaction mechanisms, and rigorous structure-property relationships. Despite these important advances, the molecular resolution of intermediates under working conditions still presents a significant challenge and as such has been limited to only a few different intermediates and systems. In addition, these methods provide information on model systems that lack the complexity of the actual catalytic environment and often provide only a limited understanding.

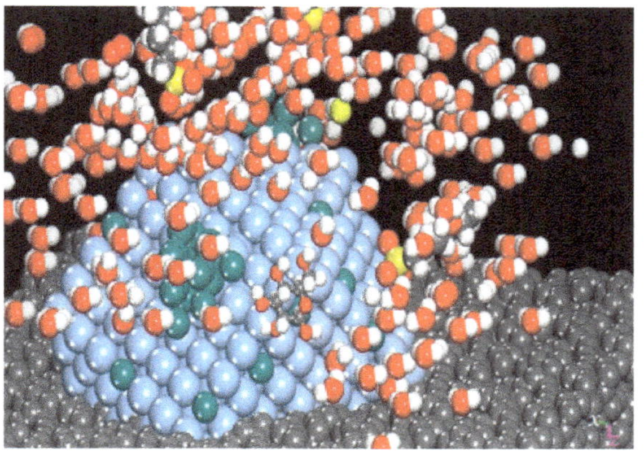

Figure 1: Schematic representation of the complex aqueous/metal interface involved in the catalytic and electrocatalytic oxidation of glucose over a PtRu alloy particle supported on carbon in the presence of electrolyte.

In addition to these advances in spectroscopy, the past two decades have witnessed exponential increases in computational power and tremendous advances in theory and simulation methods. The development of density functional theory along with higher level ab initio wave function methods, and novel embedding methods, for example, has revolutionized our understanding adsorbate bonding and reactivity on well-defined surfaces and organometallic clusters. Due to limited computational resources, methods, and knowledge, most of these initial theoretical studies were focused on modeling ideal single-crystal surfaces under vacuum conditions. The insights and confidence gained from these initial efforts together with further increases in computational methods and nurturing from the experimental electrocatalysis community have helped to "seed" the exponential growth that has occurred in the development and the application of theory in electrocatalysis over the past ten years. Many of the initial developments were based on important insights into the electronic factors that controlled electrocatalytic reactivity. Andserson pioneered the development of reaction center model [13,15, 18, 32–35], whereas Nørkov and colleagues [11, 12] developed a simple but elegant method that directly relates gas phase surface reaction energies to reaction energies at applied potentials. Schmickler et al. [36–38] developed a model Hamiltonian that appropriately captures bond-breaking and bond-making processes that occur over metal surfaces in electrochemical systems by combining fundamental electron transfer and solvent reorganization principles derived from Marcus theory, Newns Anderson theory on surface reactivity, and a tight binding theory.

These initial efforts were subsequently followed by ab initio-based simulation methods to follow chemistry within the aqueous metal interface and the direct relationship to electrocatalysis at applied potentials. There are now a number of rather sophisticated models that include the presence of solution, electrochemical potentials, applied fields, and actual electrolyte in modeling the electrocatalysis. Filhol, Taylor, and Neurock used explicit electrolyte or charge to establish the double layer at the surface [39–43]. The charged surfaces were then referenced to vacuum in order to establish the working potential. Otani et al. used DFT to describe the water metal interface and coupled this with an effective screening medium to represent to polarizable continuum [44,45]. Jionnouchi and Anderson developed a similar approach by combined density functional theory and modified Poisson-Boltzmann theory [46]. Rossmeisl et al. [47] used explicit protons at the water/metal interface to establish the double-layer interface. While the models by Neurock, Otani, Anderson, and Nørskov differ in how they treat the double layer, they are providing more rigorous solutions to the electrochemical transformations that occur on electrode surfaces. It is important, however, to note that these approaches are at best semiquantitative due to limitations of fundamental accuracy of the quantum mechanical methods, modeling electron transfer reactions and simulating long time dynamics, the full reaction environment, or the millions of configurations needed for accurate statistical treatments. Vapor phase density functional calculations of bond energies and activation barriers are typically within the range of 0.1–0.2 eV accuracies but can have outliers [2, 48]. The simulation of electrochemical systems would at best only be 0.3 V. Despite these issues, theory has plaid and will likely continue to play a very valuable role in understanding and establishing trends.

The discussions that follow will look to theory only to provide insights rather than quantitative predictions. All of the simulations reported were carried out gradient corrected periodic DFT calculations using the Vienna Ab Initio Simulation Program (VASP) [49, 50] with four-layer metal slabs in the presence of solution, where the bottom two layers in the metal were held fixed to the bulk lattice positions of the metal. The coordinates of the metal atoms in the top two layers along with all of the atoms in the adsorbates as well as in the solution layer were fully optimized. Transition states were isolated using the nudged elastic band method with climbing [51,52] followed by the dimer method [53]. The specific details are reported in the previous papers [39–43].

Rather than continue to discuss the obvious connections between electrocatalysis and gas phase heterogeneous catalysis, it is perhaps more interesting to discuss the growing efforts for carrying out heterogeneous catalysis in solution and new and emerging results that connect heterogeneous catalysis to well-established principles and phenomena in electrocatalysis. There has been an exponential growth over the past few years in carrying out heterogeneous catalytic reactions in solvents or aqueous media. This has been the result of the significant efforts to convert biomass into chemicals and fuels [54–59]. The carbohydrates that result from the breakdown of biomass are soluble in aqueous media and in addition can be catalytically converted at lower temperatures and much milder conditions than traditional gas phase processes. Similarly, many of the processes used in the selective hydrogenation and selective oxidation of fine chemical and pharmaceutical intermediates are also carried out in aqueous or solvent media that operate at lower temperatures to control both chemical as well as enantiomeric selectivities. Many of the catalyst performance and durability issues found in these systems have strong parallels to those found in electrocatalysis. Despite the similarities, there have been very few attempts to connect or compare the two.

SPECIFIC COMPARISONS BETWEEN CATALYSIS AND ELECTROCATALYSIS: EXAMPLE SYSTEMS

Metal Support Interface

Before discussing specific chemistry, we will first focus the metal-support interface in aqueous phase heterogeneous catalysis and electrocatalysis. While the lower temperatures used in aqueous phase catalysis help to control the reaction selectivity, the presence of water often leads to the hydrolysis of metal support bonds which can significantly limit the supports that can be used due to issues related to metal sintering and dissolution. Much of the initial work in the area of conversion of biorenewables was carried out over traditional transition metal catalysts (Pd, Ru, Pt, and Ni) and their supports including SiO_2, Al_2O_3, TiO_2, and high surface area carbons in order to identify active and selective materials [54–57, 60]. There was little emphasis on the fundamental surface chemistry that occurred in the solution phase or the stability of these materials. Maris et al. [61, 62] were some of the first to identify the potential issues related to metal-support interactions under aqueous phase catalytic conditions. They showed that while Ru supported on SiO_2 leads to 100% selective hydrogenation of glucose to the sugar alcohol, the catalyst used was inherently unstable in aqueous media and resulted in the significant metal particle growth. Subsequent studies by Ketchie et al. [63] used in situ X-ray absorption spectroscopy to monitor the oxidation state of the Ru and follow the metal stability over a range of traditional supports including SiO_2, γ-Al_2O_3, carbon, and TiO_2. Significant sintering of Ru on both SiO_2 and Al_2O_3 supports occurred at the mild conditions associated with the conversion of biorenewables. Both the high surface area activated carbon and titania, on the other hand, were found to be stable supports for aqueous phase catalysis over a range of operating conditions. They showed that different carbons may behave differently and that great care must be taken to elucidate the nature of the metal-support interactions under actual process conditions. Most of the recent studies on the conversion of renewables as well as the hydrogenation of pharmaceutical intermediates are now carried out on activated high surface area carbon due to the stability issues.

Interestingly, many of these issues were resolved in the electrocatalysis community many years earlier as metal dissolution is one of the key issues that limit fuel cell durability. Carbon has been the preferred support throughout electrocatalysis as a result of the stability, durability, conductivity, and reactivity of the metal/carbon interface. More recent efforts have demonstrated that the introduction of titania can help to stabilize the metal/support interactions at the cathode for oxygen reduction.

Despite the advances in both catalysis and electrocatalysis towards stabilizing the metal/support interface, it is clear that this is an important area which will require mechanistic insights into the fundamental processes that lead to dissolution and loss of metal and practical advances to solve the issues of durability.

Oxidation of CO

Catalytic Oxidation of CO

A second example in which heterogeneous catalysis and electrocatalysis are related is the recent discoveries concerning the unique catalytic activity of supported metal particles in the presence of an aqueous medium. Perhaps most evident is the work that has been carried out over supported Au. Up until 1987, gold was considered to be inert and inactive for catalysis. In a pioneering discovery, Haruta et al. [64, 65] demonstrated that nanometer-sized Au particles supported on TiO_2 were highly active for low-temperature CO oxidation in the gas phase. This work led to a tremendous number of follow-up studies aimed at understanding the mechanism by which this reaction proceeds and demonstrates the unique behavior of nanometer and subnanometer-sized Au particles in catalyzing a range of different reactions over TiO_2 as well as other supports [66, 67]. A number of possible explanations for the unique reactivity of Au have been presented in the literature [67] including quantum-size effects [68], increased coordinatively unsaturated edge and corner sites [69], the presence of cationic or anionic Au centers [70, 71], and unique sites at the Au/TiO_2 interface [72]. While the mechanism is still openly debated, much of the literature suggests that sites along the interface are responsible for the high catalytic activity. Some have speculated that the Au-Ti site pairs that result at the Au/TiO_2 interface stabilize the adsorption and activation of O_2 or the formation of hydroxyl intermediates both of which can catalyze these reactions [67]. The later idea is supported by the fact that the introduction of small amounts of water or base significantly promotes CO oxidation over Au/TiO_2 [73, 74]. Catalysis on these active 2–4 nm-sized Au clusters is thought to be quite different than that found in the electrocatalysis, as the former is carried out in the gas phase at low temperatures over small Au clusters without an applied potential or the presence of promoters. In addition, O_2 is the active oxidant in gas phase heterogeneous catalysis, whereas activated water or hydroxyl intermediates are thought to oxidize CO in electrocatalysis.

Kim et al. [75] and Sanchez-Castillo et al. [76] and Ketchie et al. [77, 78] later demonstrated that bulk Au as well as Au nanotubes were also very active in catalyzing CO oxidation if the reactions were carried out in an aqueous media. The rates over bulk Au were found to be over an order of magnitude higher in water than those in gas phase. While CO oxidation can proceed readily in the gas phase for nanometer clusters of Au on TiO_2, water is necessary for carrying the reactions out over bulk Au as well as Au supported on carbon.

These same aqueous phase CO oxidation rates were increased by up to 50 times upon increasing the pH of solution from neutral conditions (pH = 7) to basic conditions (pH = 14) and by over an order of magnitude from those in neutral solutions when the reactions were carried out in the presence of 0.5% H_2O_2 [76].

Dumesic et al. suggested that mechanism for CO oxidation in water may proceed via the formation of surface hydroxyl intermediates that can catalyze CO oxidation similar to the classic bifunctional mechanism suggested in electrocatalysis [79–82] and shown previously from theoretical calculations by Desai and Neurock [83, 84]. In this mechanism, water is activated on Ru sites within a PtRu surface alloy, whereas

CO binds/blocks the Pt sites as is shown in Figure 2. CO oxidation proceeds via a nucleophilic attack of the OH (bound to Ru) on a neighboring CO (bound to Pt) coupled with the heterolytic splitting of O–H to form CO_2 as well as an electron and a proton. This mechanism would also help to explain the significant promotional effects that occur when the reaction is carried out in the presence of base both catalytically [76, 78] as well as electrocatalytically [85,86].

Figure 2: DFT-calculated transition state for the electrocatalytic oxidation of CO on Pt (green spheres) by hydroxyls formed via the dissociation of water over Ru (yellow spheres). The resulting electron is transferred to the metal whereas the proton shuttles away from the surface via the water network (copyright Science [83]).

Electrocatalytic Oxidation of CO over Au

The anodic oxidation of CO over Au/C was actually established in 1965, over two decades before the pioneering work of Haruta [87, 88]. The reaction readily proceeds over single crystal as well as polycrystalline Au electrodes at low temperature in alkaline media at potentials which are 0.5 V lower than those found with Pt which is used in most fuel cells. While CO oxidation readily occurs over single-crystal electrodes in acidic media, the rate is significantly enhanced when carried out in alkaline media. CO oxidation proceeds over Au(111) and Au(110) at potentials of 0.1 and 0.2 V RHE, respectively, which is 0.5 V lower than those recorded in acid media [88]. Similar differences for CO oxidation in acidic and alkaline media also exist for polycrystalline Au. The reaction is thought to proceed via the coupling of CO* and OH* in a mechanism that is similar to that presented above in Figure 2. While the reaction proceeds in acidic media, the activation of water to form OH* on Au at lower potentials is difficult. The reaction is significantly faster in alkaline media as a result of increased formation of OH* at lower potentials and the stronger adsorption of CO at these lower potentials. The higher coverages of CO* and OH* in alkaline media further enhance the rate of reaction at the lower potentials. This shift in potential results in an increase in electron density at the metal surface which further enhances backdonation. Rodriguez et al. [88, 89] suggested that the increased CO adsorption further enhances the adsorption of OH and self-promotes the reaction.

Nearly all of the early electrocatalytic studies were carried out over bulk polycrystalline Au electrodes. The exceptional findings by Haruta et al. [64, 65] and the suggestions of the unique interface for nanoparticles of Au supported on TiO_2 prompted Hayden et al [85, 86] to examine the electrocatalytic oxidation of CO over nanoparticles of Au on TiO_2. They demonstrated considerable enhancements in the electrocatalytic rates of CO oxidation for Au nanoparticles supported on TiO_2 even when the reaction is carried out in acidic media. The enhancement was attributed to the substrate-induced reactivity of Au as discussed below [85, 86].

Particle Size Effects on CO Oxidation

The results from gas phase catalysis carried out over supported Au clusters indicate that the reaction is structure sensitive where the highest catalytic activity occurs for 2–4 nm sized particles supported on TiO_2 [66,67]. The results in aqueous media, however, are not as clear. Ketchie et al. show a significant increase in the activity in moving from 42 nm down to 5 nm [77]. The activity on 2-3 nm-sized clusters, however, was more difficult to quantify as a result of difficulties in the synthesis of monodisperse particles [64, 65, 67]. Under electrochemical conditions, Hayden et al. [85, 86] showed that while polycrystalline Au electrodes readily oxidize CO to CO_2, they require overpotentials greater than 0.7 V as AuO forms and inhibits the reaction. They demonstrated significant structure sensitivity in the reaction, whereby 3 nm Au particles were much more active even at potentials as low as 0.3 V. Au particles smaller than 2.5 nm showed a significant decrease in activity as did particles larger than 3.5 nm. They speculate that the loss in activity below 3 nm is the result of quantum size effects rather than from irreversibly adsorbed oxygen. There appears to be clear similarities for particle size effects in comparing the results for CO oxidation under electrocatalytic conditions with those found for Au/TiO_2 in the gas phase heterogeneous catalysis. The reactions over Au/C in solution, however, were less conclusive.

CO Oxidation Mechanisms

As was suggested earlier, the heterogeneous catalytic and the electrocatalytic oxidation of CO may proceed by common mechanisms or at least common features in the mechanism. We compare the mechanistic ideas for the catalytic and electrocatalytic oxidation of CO over both TiO_2 and carbon supports.

(1) Catalytic and Electrocatalytic Oxidation of CO over TiO_2

The unique reactivity of nanometer-sized Au particles on TiO_2 used in both catalysis as well as electrocatalysis is thought to be dictated by sites at the Au/TiO_2 interface. We have recently shown that the Ti cations in direct proximity to the adsorbed Au become positively charged as a result of local charge transfer from Au to local Ti^{5+} cations as is shown in Figure 3 [90]. This increase in charge along with the direct involvement of both Au and Ti^{5+} atoms stabilize the bidentate adsorption of O_2. This increases the O_2 adsorption strength by nearly 60 kJ/mol and promotes its activation at these dual interface sites. This subsequently catalyzes the reaction between O_2 coadsorbed CO. While O_2 is not present in the electrocatalytic oxidation of CO, water can adsorb and activate at these same perimeter sites. DFT calculations indicate that the adsorption of water at this same charged interfacial Ti^{5+} site is −104 kJ/mol, whereas water on a Ti^{5+} site removed from Au and on a Au site near the support are significantly weaker at −77 kJ/mol and −8 kJ/mol, respectively. These Au-Ti perimeter sites likely attenuate the activation of water. The resulting intermediate (either activated water or surface hydroxyl groups) can then react with coadsorbed CO to form CO_2. This is consistent with the results from Hayden et al. [85, 86] who showed a direct relationship between electrocatalytic activity and the perimeter of the particle and was able to rule out the influence of low coordinate Au atoms and quantum size effects. The differences between the gas phase reactions and the electrocatalytic reactions may simply be the nature of the active oxidant that forms. In the presence of water, the oxidant involved in catalysis and electrocatalysis is likely the same.

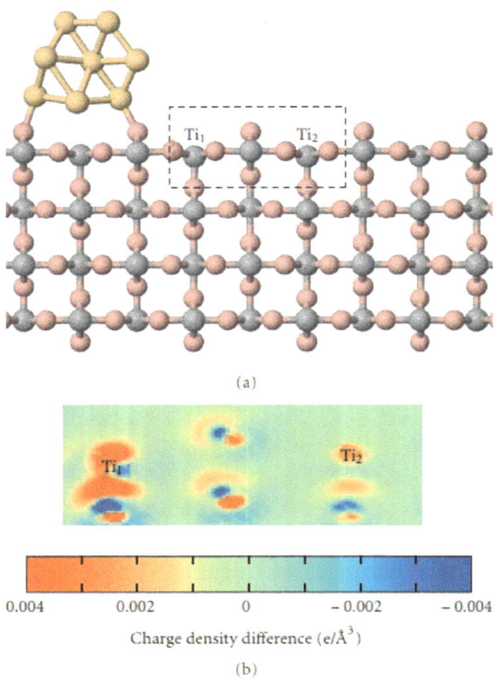

Figure 3: Charge density difference that results at Ti^{5+} centers of TiO_2 adjacent to Au upon the adsorption of Au nanorods or clusters. (Copyright Science [90]).

(2) Au/C The Effects of Water and Base

The heterogeneous oxidation of CO over bulk Au/C in water appears to proceed via a mechanism that involves the nucleophilic addition of OH groups present in solution or on the surface formed under reaction conditions to adsorbed CO*. This results in the formation of CO_2 and either adsorbed hydrogen or a proton and an electron.

$$CO^* + OH^* \longrightarrow COOH^* \longrightarrow CO_2(aq) + H^* + {}^* \quad (1)$$

$$CO^* + OH^* \longrightarrow COOH^* \longrightarrow CO_2(aq) + H^+ + e^- + {}^* \quad (2)$$

The latter is identical to the mechanism typically proposed in electrocatalysis. The results for the heterogeneous CO oxidation in water over Au/C indicate that the reaction can proceed solely by the presence of H_2O_2 and water without oxygen [76]. In the presence of oxygen, small amounts of H_2 are formed possibly as the result of the water gas shift reaction [76]. Both of these results strongly suggest that OH groups are formed and directly participate in the oxidation mechanism [76]. Similar reactions occur in the electrocatalytic oxidation of CO over Au/C in both acid as well as alkaline media.

In order to model the oxidation of CO over bulk Au in solution, we have carried out detailed density functional theory calculations with CO and OH or H_2O adsorbed on a Au(111) surface immersed in water [91]. The resulting activation barrier for the reaction (2) above was calculated to be negligible, whereas the overall reaction energy was calculated to be exothermic by −231 kJ/mol. Various structures along the reaction coordinate are shown in Figure 4. The reaction proceeds via the formation of the OC–OH bond combined with the heterolytic splitting of the O–H bond to form an H_3O^+ intermediate which subsequently undergoes a second proton transfer to another water molecule. The overall mechanism for the heterogeneous catalyzed oxidation of CO may proceed via the formation of local electrochemical cell or circuit, where CO is oxidized by OH resulting in the formation of proton and an electron which subsequently catalyzes the reduction of O_2 to the active OH intermediates or to H_2O_2. This is discussed in somewhat more detail in Section 3.3.2(1). These same ideas on the oxidation of CO can readily be extended to methanol as well as other alcohols.

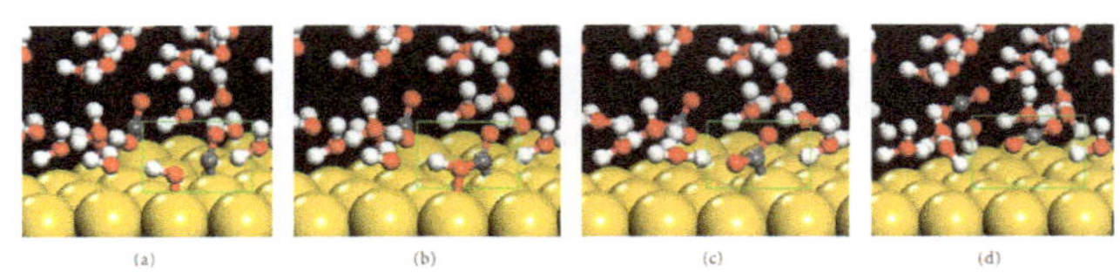

Figure 4: DFT-calculated reaction coordinate for the oxidation of CO via adsorbed OH* over Au. The reaction proceeds via: (a) the coadsorption of CO* and OH*, (b) the nucleophilic attack of OH* on CO* to form the HO*–*CO surface intermediate, (c) cleavage of O–H bond in the transition state, and (d) formation of CO_2 and H_3O^+products. The active CO* and OH* surface species are highlighted in green as they proceed through the reaction.

Alcohol Oxidation

The increasing demand to shift from petroleum-based fuels and chemical to those derived from biomass has significantly increased efforts in both the electrocatalytic and catalytic oxidation of carbohydrates feedstocks. Significant efforts have been focused on the selective oxidation of methanol, ethanol, glycerol, and other C_2–C_6polyols into chemical intermediates via heterogeneous catalysis [92–98] as well as the total oxidation of these fuels to electrical energy via electrocatalysis [99–107]. In the next few sections, we compare some of the similarities and differences involved in the catalytic and electrocatalytic oxidation of these alcohols in acidic as well as alkaline media.

Alcohol Oxidation in Acidic Media

(1) Catalytic

The catalytic oxidation of glycerol as well as ethanol is rather low when carried out in neutral solution over Pd and

Pt with reported turnover frequencies of 0.05 and 0.06 s^{-1}, respectively [98]. The products that form over these metals are predominantly intermediate aldehydes and ketones that result from dehydrogenation with only 25% selectivity to form the acid product. Glycerol oxidation, for example, leads to the formation of glyceraldehyde as well as dihydroxyacetone without further oxidation to C_1 or C_2 acids or CO_2. Supported Au clusters under neutral and acidic conditions were found to be completely inactive [98]. The dehydrogenation reactivity that results over Pd and Pt can proceed via C–H and O–H activation over the metal or by adsorbed oxygen or hydroxyl intermediates that form upon O_2 dissociation or by the subsequent reaction of adsorbed oxygen with water. DFT-calculated activation barriers suggest that O–H activation of ethanol occurs via the reaction of adsorbed ethanol with OH*, whereas the C–H activation of ethanol preferentially occurs via metal sites on the Pt and Pd surfaces [98]. Both reactions, however, are limited on Pt and Pd, as the high surface coverages on these metals result in barriers for both C–H and O–O activation that are significantly greater than 100 kJ/mol. The oxidation of alcohols over Au does not proceed over Au alone under neutral or acidic conditions as gold cannot activate water, O_2, or alcohol [98].

(2) Electrocatalytic

The oxidation of methanol, ethanol, glycerol, and other alcohols occurs in acidic media but requires significantly higher overpotentials than reactions carried out in alkaline media [79, 100, 108–119]. The oxidation of methanol proceeds over most transition metals through a sequence of elementary C–H and O–H bond activation steps which occur on the metal, and ultimately result in the formation of CO [79, 109, 110]. Through the combination of cyclic voltammetry, chronoamperometry, and DFT studies, we demonstrated that the methanol decomposition occurs via a dual path mechanism [1, 120]. At potentials below 0.35 V, the mechanism proceeds predominantly through a sequence of C–H activation steps to form the hydroxyl methylene (CHOH*) intermediate that subsequent breaks the O–H and C–H bonds to form CO*. At potentials above 0.35 V, the O–H bond of methanol can also be activated, resulting in the formation of formaldehyde which can desorb or continue on to form CO. The onset potential was found to be a function of the Pt surface structure. The dual paths for the oxidation of methanol to CO over Pt(111) along with their corresponding reaction energies calculated at 0.5 V RHE are shown in Figure 5 [1, 120]. While CO is an intermediate to CO_2, it readily builds up on the surface and poisons more active metals such as Pt. As such, Pt is typically alloyed with a more oxophilic metal such as Ru to promote the adsorption and dissociation of water thus creating bifunctional sites on the surface [79, 109, 110]. The OH groups that result interrupt the CO adlayer and readily oxidize CO to CO_2. The addition of Ru also helps to weaken the Pt–CO bond thus enhancing CO desorption. The addition of Ru to Pt lowers the overpotential for CO oxidation by ~0.25 V [80–82, 121] as it prevents CO poisoning. The oxidation of ethanol, glycerol and other larger alcohols in acidic media result in dehydrogenation which forms the corresponding aldehyde [114–117]. The subsequent activation of the C–C bond, however, is very difficult over typical metals such as Pt, Pd, or Au, and as such, very limited CO_2 is formed. The dehydrogenation routes are identical to those presented above for the catalytic oxidation of the same alcohols. The higher potentials used electrochemically can activate water at higher potentials, and thus result in the formation of surface hydroxyl intermediates that can subsequently oxidize the aldehyde and ketone intermediates [115]. This leads to the formation of acids as well as carboxylate intermediates which inhibit the surface under acidic conditions and prevent the formation of CH_x intermediates on closed-packed crystal surfaces. Feliu et al. have shown that steps on specific Pt surfaces Pt(554) and Pt(110) can begin to enhance C–C bond breaking and CO_2 formation, but rate is still very limited [111, 114].

The electrooxidation of ethanol over Pt in acidic media has two major limitations which prevent its viability as was discussed by Lia et al. [115]. The first relates to the fact that reaction predominantly produces acetate and acetic acid intermediates, thus resulting in only 2 and 4 electrons, respectively, which are only very minor contributions to total possible current. Both are thus unwanted side products for fuel cell applications. The second limitation is that the path to CO_2 is rather difficult in that it requires the activation of C–C bond as well as the oxidation of both the CH_x and CO intermediates that form. Both of these intermediates tend to inhibit or poison metal surfaces at lower potentials [48].

The characteristic difference between the catalytic oxidation and electrocatalytic oxidation of alcohols in acidic media lies in the generation of the active surface intermediate. In heterogeneous catalysis, the metal plays an important role in activating oxygen. The surface oxygen can directly activate the alcohol or adsorbed water to generate hydroxyl intermediates that aid in activating and oxidizing the alcohol. Under electrocatalytic conditions, the O–H and C–H bonds of the alcohol can be activated on the metal directly as a result of higher potentials or via adsorbed hydroxyl intermediates that form by the activation of water.

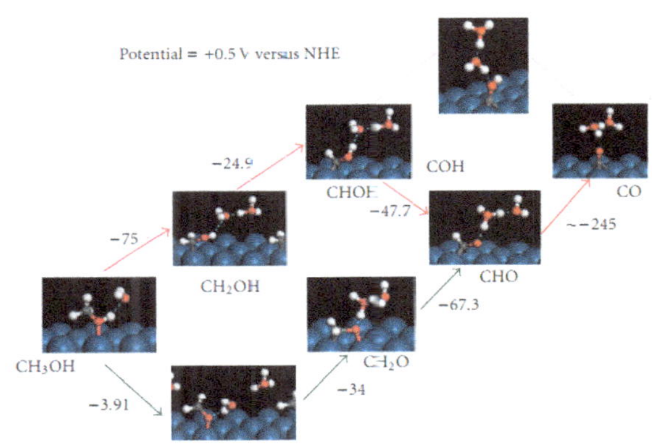

Figure 5: DFT-calculated potential-dependent reaction paths for the oxidation of methanol to CO over Pt(111) [1]. At 0.5 V, both experiments and theory point to the onset of dual paths. The primary path, available over a wide range potentials, is shown in red. It proceeds via a sequence of C–H bond activation steps ultimately forming the hydroxyl methylene intermediate (CHOH) before activating the final O–H bond to form CO. The minor path begins between 0.4–0.5 V RHE proceeds via the initial activation of O–H bond of methanol to form a surface methoxy intermediate which subsequently reacts to form formaldehyde consistent with previous speculations [122].

Alcohol Oxidation in Alkaline Media

(1) Catalytic

The catalytic oxidation of polyols over Pd, Pt, and Au is much more favorable when carried out in alkaline media [78, 98, 123]. The TOF increases by over an order of magnitude on Pt and over two orders of magnitude on Pd upon the addition of a 2/1 ratio of NaOH to glycerol [98]. Remarkably, the TOF over gold in base is over $6\,s^{-1}$, whereas the rate over Au in acidic media is negligible [98]. In addition to these increases in TOF, there are also significant improvements in the overall selectivity to form the acid from the alcohol. For glycerol, the selectivity to glyceric acid was found to be 83% over Pd, 70%–78% over Pt, and 67% over Au [98]. The predominant side products are glycolic and tartronic acid which are shown in the paths outlined in Figure 6. In the presence of base, the second terminal hydroxyl group can be oxidized to form tartronic acid, but this only occurs over Pt and Pd. The oxidation over Au results only in the monofunctional glyceric acid product. In addition to the selective oxygen addition, Pd and Pt can also promote C–C activation resulting in the formation of glycolic, oxalic, lactic, formic, and acetic acids. These C–C activation paths appear to coincide with the formation of hydrogen peroxide [123]. The activation of the C–C bonds have been speculated to occur either through a retroaldol reaction which would be catalyzed by the OH^- base, or via oxidation catalyzed by the hydrogen peroxide that forms [78, 123].

Through detailed labeling studies along with DFT simulations, we established a plausible mechanism for the oxidation of alcohols over Au in basic media [98]. The mechanism also helps to explain the unique promotional effects of OH on Au. We discuss here the energetics involved in the reaction over the model Au(111) surface. The first step involves the dehydrogenation of the alcohol to form the corresponding aldehyde. This can proceed via the activation of the O–H and C–H bonds of the alcohol by the metal, adsorbed oxygen, or adsorbed OH intermediates. As one might expect, Au atoms alone cannot activate the O–H bond of the alcohol. The calculated barrier to activate ethanol to ethoxy over Au(111) in the presence of solution was calculated to be 204 kJ/mol (the transition state is shown in Figure 7(a)). The O–H bonds are much more readily activated by the weakly adsorbed $OH^{\delta-}$ intermediates via a mechanism which involves a proton abstraction by the $OH^{\delta-}$ surface intermediate (the transition state is shown in Figure 7(b)). The inability of Au(111) to activate the O–H and C–H bonds is well established, as bulk Au is quite noble. The binding energy of $OH^{*\delta-}$ on Au(111) in water is only −216 kJ/mol (versus −274 kJ/mol on Pt(111)), thus making it quite basic. The weak interaction promotes its ability to readily abstract a proton from a neighboring O–H on the alcohol. Similarly, the activation of the C–H bond of the ethoxy intermediate to form acetaldehyde does not occur over Au(111) alone (see the transition state in Figure 7(c)) but instead proceeds by the reaction of $OH^{*\delta-}$ with the adsorbed ethoxy intermediate with a barrier of only 12 kJ/mol (see the transition state in Figure 7(d)). The aldehyde that results is a key reaction intermediate to form the acid in alkaline media over Au.

Most of the initial studies in the aqueous phase catalytic oxidation of alcohols assumed that O_2 was responsible for carrying out the oxidation. By carrying out the reaction with ^{18}O-labeled O_2 and H_2O, however, we showed that only labeled water found its way into the resulting acid product that forms [98]. DFT results for Au in an aqueous medium showed that this reaction proceeds via a classic nucleophilic attack of $OH^{*\delta-}$ on the adsorbed aldehyde quite similar to thought found for the oxidation of CO. The barrier to form the geminal diol intermediate shown in Figure 8(a) was only 5 kJ/mol when carried out over Au and 42 kJ/mol for reactions in the solution phase Figure 8(b). The final C–H activation of the germinal diol to form the acid can proceed over the Au itself (21 kJ/mol) or via reaction with adsorbed $OH^{*\delta-}$ (29 kJ/mol). A schematic representation of the mechanism which involves the unique reactivity of OH^- intermediates at the aqueous/Au interface for the selective oxidation of the alcohol to the acid is depicted in Figure 9, whereas the corresponding potential energy surface is shown in Figure 10. This oxidation path proceeds without the activation or incorporation of oxygen from O_2.

While O_2 is not directly involved in any of the steps depicted in the mechanism for alcohol oxidation shown in Figure 10, it is critical as the reaction does not occur without it. Oxygen must somehow be intimately coupled with the overall catalytic process which requires the balance of charge and the regeneration of OH^-. As the result of theoretical calculations, we showed that oxygen is necessary to remove the electrons that are generated as result of the oxidation of the alcohol. The measured rate for the oxidation reaction is 6.1 turnovers per second per site [98]. As the steps outlined in the cycle presented in Figure 10 consume 4 hydroxyl ions, they generate 4 electrons per every turnover.

$$RCH_2OH + 4OH^- \longrightarrow RCOOH + 3H_2O + 4e^- \qquad (3)$$

In order for the reaction to occur catalytically, the electrons that are produced per turnover must be consumed. Each oxygen molecule can effectively remove 4 electrons via the oxygen reduction reaction (ORR).

$$O_2 + 2H_2O + 4e^- \longrightarrow 4OH^- \qquad (4)$$

The oxygen reduction reaction is known to occur quite readily over single crystal Au electrodes. DFT calculations were used to determine the reaction energies and the activation barriers for the most relevant steps for ORR. The results reported in Table 1 indicate that the direct dissociation of O_2 (reaction 5) does not take place over Au as is well established experimentally. Oxygen instead is reduced by the protons from water and the electrons in the metal to form a peroxo intermediate along with OH^- (reaction 6). The peroxo intermediate is subsequently reduced to hydrogen peroxide (reaction 7) which can subsequently dissociate and result in the formation $2OH^-(aq)$. The intermediate formation of hydrogen peroxide is consistent with observed experimental results and suggestions that peroxide is responsible for C–C bond breaking and formation of shorter acids.

This overall ORR cycle removes the 4 electrons from the metal produced via the oxidation, and thus allows the reaction to continue catalytically. The cycle also regenerates $OH^-(aq)$. The overall catalytic reaction then involves the direct coupling of both alcohol oxidation and oxygen reduction cycles in one system. This can thus be considered a local, short-circuited, electrochemical cell, where the oxidation and reduction occur simultaneously at the metal/surface interface. The role of

O_2 then is to simply to remove the electrons from the metal at a rate fast enough to maintain the optimal surface potential of the local electrochemical cell.

Table 1: DFT-calculated reaction energies and activation barriers for possible steps involved in the reduction of O_2 to OH over Au(111) in aqueous media [98].

	ΔE_{rxn} (kJ/mol)	ΔE^* (kJ/mol)	Reaction
$O_2^* + * \rightarrow O^* + O^*$	41	105	(5)
$O_2^* + H_2O^* \rightarrow OOH^* + OH^*$	−4	16	(6)
$OOH^* + * \rightarrow OH^* + O^*$	−56	83	(7)
$OOH^* + H_2O^* \rightarrow HOOH^* + OH^*$	37	48	(8)
$HOOH^* + * \rightarrow OH^* + OH^*$	−86	71	(9)

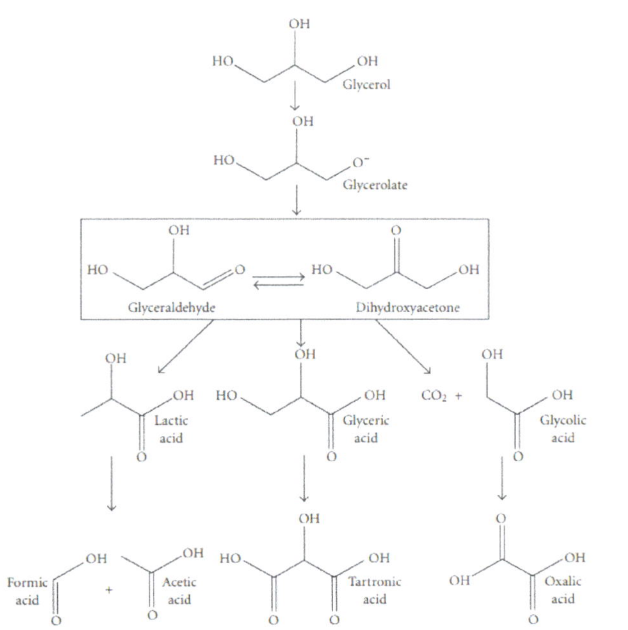

Figure 6: Reaction pathways speculated in the catalytic and electrocatalytic oxidation of glycerol over Au and Pt. Arrows in red depict paths observed over Au and Pt whereas the paths in black are only observed over Pt. Adapted from [123]

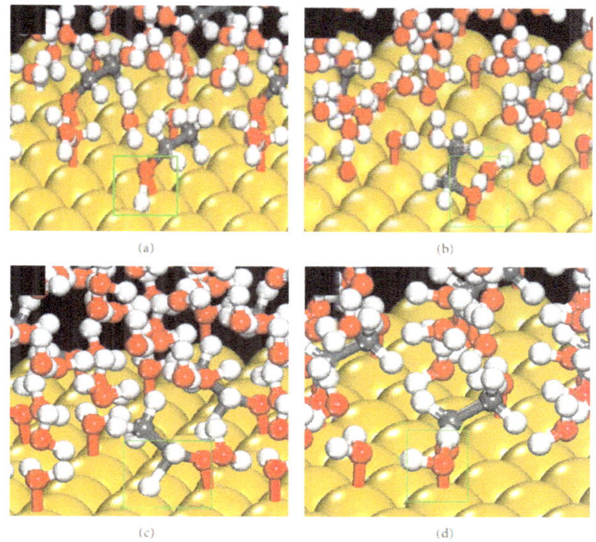

Figure 7: DFT-calculated transition states for the activation of the O–H bond of ethanol adsorbed to Au by (a) metal surface atom or (b) a coadsorbed hydroxyl intermediate on Au(111). The activation of the C–H bonds of the adsorbed ethoxy are activated similarly via (c) Au atoms in the Au(111) surface or (d) via the bound OH groups on the Au(111) surface. The active bond breaking and forming sites are highlighted in green.

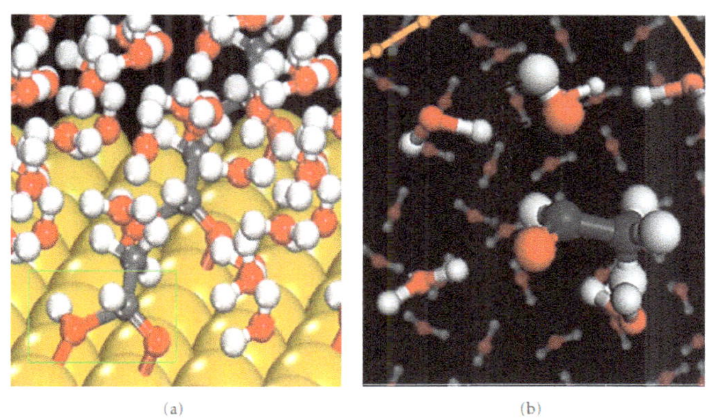

Figure 8: DFT-calculated transition states and their activation barriers for the oxidation of acetaldehyde on (a) the Au(111) surface or (b) in the presence of solution. The active bond breaking and forming sites are highlighted in green.

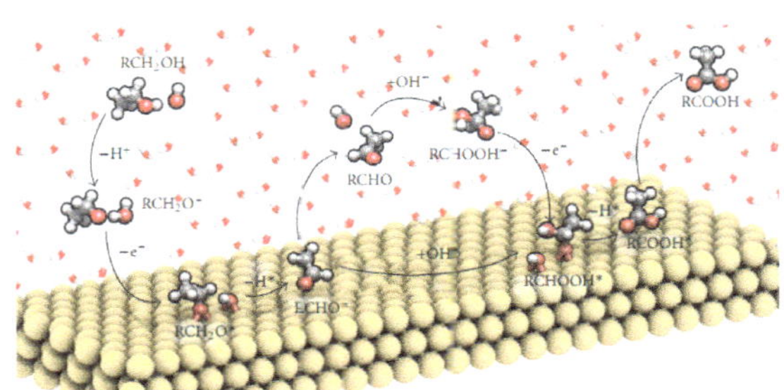

Figure 9: Schematic representaion of the reaction paths for ethanol and other polyol oxidation to acids over Au and Pt in alkaline media based on DFT results and isotopic kinetic labeling experiments. (Copyright Science [98]).

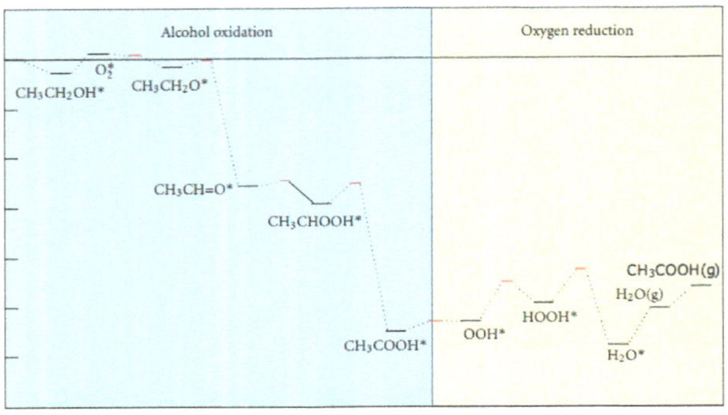

Figure 10: DFT-calculated potential energy surface for the catalytic oxidation of ethanol to acetic acid over Au(111) in the presence of base (OH^-) and in water. Both the OH^- as well as the metal are involved in the mechanism. In order establish an overall cycle, O_2 is reduced via the electrons that from as a result of the oxidation of the alcohol as well as by the reaction protons that form upon the activation of water. Red lines refer to transition states, whereas the black lines refer to reaction intermediates.

(2) Electrocatalytic

The development and application of carbonate as well as anion-exchange membrane electrolytes have significantly renewed interest in the development of alkaline-based direct alcohol fuel cells [100]. Many of the reaction intermediates, products, and paths discussed above for the catalytic oxidation of alcohols in alkaline media have also been identified or speculated to take part in the electrocatalytic oxidation of these same alcohols. As such, the mechanistic insights established from catalytic oxidation should be important in understanding the chemistry and the mechanisms that control electrocatalytic oxidation. Similarly, the detailed insights established from electrochemical methods such as cyclic voltammetry should provide new insights into heterogeneous catalytic oxidation.

The oxidation of methanol, ethanol, as well as other polyols in alkaline media is thought to proceed either by $4e^-$ or $6e^-$ processes thus resulting in the formation of either formate and carbonate intermediates by reactions (5), or (6) [100]

$$RCH_2OH+5OH^- \longrightarrow HCOO^-+4H_2O+4e^- \quad (5)$$

$$RCH_2OH+8OH^- \longrightarrow CO_3^{2-}+6H_2O+6e^- \quad (6)$$

For methanol, the later reaction is speculated to occur through the formation of CO. In the presence of base, surface OH^- groups readily catalyze the oxidation of methanol to CO_2 or carbonate as shown in reactions (5) and (6), respectively. At higher potentials, however, OH binds very strongly to the metal surface and inhibits the reaction. The selectivity to formate or carbonate depends upon the metal, metal surface structure, potential pH, and alcohol concentration present during the reaction. We discuss here the difference in the electrocatalytic oxidation over Pt and Au and compare the changes that result upon changing the pH and concentration.

While the oxidation of methanol as well as other alcohols can be carried out in acid media as was discussed above, the activity is rather low. The rates and currents increase significantly with increasing the pH of solution. The electrocatalytic oxidation of methanol as well as other alcohols over Au is quite high in alkaline media [100,101, 104–108, 110, 118, 124–126]. The higher rates and current densities found in alkaline media are the result of the high reactivity of OH^- anions that are weakly bound to the Au substrate at moderate potentials. The weakly bound hydroxyls are quite basic and will readily activate C–H and O–H bonds (on adsorbed species) as was presented above in the catalytic oxidation of alcohols over Au in basic media. The electrocatalytic activity over different Au substrates including polycrystalline Au, Au(111), and Au(210) all show increases in activity with increases in the solution pH [101]. The adsorption of OH^- anions onto Au is thought to significantly enhance the reactivity of the surface over a range of potentials. Electrocatalytic oxidation can, therefore, proceed at much lower potentials where Au is not oxidized.

The anodic oxidation of methanol begins to occur with the adsorption of OH^-. In general, there are two very different regimes that result for methanol oxidation over Au [101]. In the first regime, methanol is actively oxidized by adsorbed OH^- intermediates on Au at potentials as low as 0.6 V RHE. The second regime which occurs at much more positive potentials and results in the formation of Au–O monolayers which are much less active for carrying out oxidation.

Chen and Lipkowski [127] and Yan et al. [107] provided more details in terms of the changes that take place at the surface as a result of these changes in potential and suggest that there are actually three regimes that result upon increasing the potential. The initial oxidation of methanol and other alcohols proceeds quite readily from 0.07–0.13 V versus Hg/HgO. The current increases linearly with the adsorption of OH^- in this regime. There appears to be a weak amount of charge transfer between the OH^- and the metal surface. The second regime appears between 0.13–0.3 V versus Hg/HgO and results in a higher degree of charge transfer of OH and the Au sites in the surface. Regime 3 which appears at potentials greater than 0.3 V Hg/HgO is the inactive monolayer Au oxide that results. The oxidation of methanol is thought to proceed via the weakly held OH^- intermediates found in both regimes 1 and 2. The more strongly held oxygen intermediates which begin to form in regime 2 are less active. Regime 3 leads to the formation of monolayer Au oxide coverages and thus the loss of the active OH^- sites. For methanol oxidation, the reaction is 0.75 order in methanol and 0.55 order in OH [107]. There is a clear negative shift in the potential and an increase in the current density with an increase in the concentration of weakly adsorbed $OH^{\delta-}$ and methanol coverage.

Ethanol Oxidation over Au

Ethanol oxidation is similar to methanol oxidation in that there is a significant shift to lower potentials and increased current that results upon increasing pH. Ethanol oxidation proceeds at a potential of 0.6 V versus RHE in alkaline solution which is 0.3 V lower than that reported in acid [115]. In addition there is a shift of −0.1 V in the maximum from 1.35 V to 1.25 V. The current density in alkaline media is found to be over an order of magnitude higher than that in neutral or acid conditions [115]. Regardless of the potential or pH there is negligible C–C bond breaking that occurs on Au, and as such, there is very little to no CO_2 formation. The mechanism is thought to be very similar to the path outlined in Figure 10 and presented above for the catalytic oxidation of ethanol. The main path involves the oxidation of ethanol to acetaldehyde through the activation of the acidic O–H and C–H bonds by weakly adsorbed hydroxyl intermediates on the Au surface. Acetaldehyde subsequently reacts with weakly held $OH^{\delta-}$ intermediates via simple nucleophilic attack depending of $OH^{\delta-}$ on the C=O bond of the aldehyde to form the geminal diol intermediate which can further react with OH^- to form acetic acid or adsorbed acetate in alkaline media.

Extension to Polyols

These same ideas have been further extended to larger polyols such as glycerol and glucose in alkaline media [104–106, 125]. Glycerol, for example, reacts via a $4e^-$ transfer process to form glyceric acid on both Au as well as Pt electrodes. At higher potentials, the C–C bond of glyceric acid can further oxidize via a two electron transfer step to form glycolic and formic acids [125]. While this occurs over both Au and Pt, the selectivity to the over-oxidized glycolic state is much higher on Au. This was thought to be due to the fact that Pt deactivates by the formation of a surface oxide intermediate at much lower potential (0.9 V) than that found on Au (~1.3 V) [125]. This results in much lower oxidation potentials and a much narrower window for the subsequent oxidation to proceed. Glyceric acid as well as glycolic acid can both undergo further oxidation on Pt to form tartronic as well as oxalic acid, respectively. The subsequent OH^- addition steps, however, only appear to proceed over Pt as neither of the di-substituted products appear over Au. The results reported by Kwon and Koper [125] for the electrocatalytic oxidation of glycerol discussed here are very similar to those reported by Ketchie et al. [78] for the catalytic oxidation of glycerol. Both show that C–C bond breaking steps can readily take place in the presence of OH^- to form glycolic, formic, and oxalic acids on Au as well as on Pt. Both also reveal that Au will only oxidize one of the terminal $-CH_2OH$ bonds of the polyol, whereas Pt can oxidize both. It is very likely that the detailed mechanism for the formation of glyceric acid established for catalytic systems holds also for the electrocatalytic oxidation. The elementary steps involved in C–C bond breaking are not known for either the catalytic or the electrocatalytic paths. The catalytic routes have been speculated to occur via oxidation with hydrogen peroxide or by base catalyzed paths. There is no evidence, however, for the formation of hydrogen peroxide electrocatalytically and more likely the reaction proceeds solely via the reaction with base.

The similarities and differences between the catalytic and electrocatalytic oxidation of alcohols is very informative. A closer analysis between the two suggests that the two are nearly equivalent with the exception that oxidation of the alcohol which occurs at the anode is decoupled from the oxygen reduction which occurs at the cathode. The two electrodes communicate via charge and ion transfer. In the heterogeneous catalytic system, the alcohol is oxidized where the electrons are directly used at the same aqueous metal interface to reduce O_2. The rate of oxidation is thus directed by the rate at which O_2 can be reduced by electrons. This is directly analogous to that which happens in the PEM alcohol fuel cells.

CO and alcohol oxidation make up just two reaction systems. One can readily draw analogies to other catalytic reaction systems carried out in aqueous media. Desai and Neurock for example carried out first principles DFT calculations to show that an aqueous medium could readily facilitate the catalytic hydrogenation of adsorbed oxygenates [83, 84, 128]. They showed that adsorbed hydrogen could undergo an electron transfer coupled with proton transfer at the metal interface to form a local proton in the form of a hydronium ion that could readily transfer through solution and attach itself with the negatively charged adsorbed surface intermediate. The solution phase here acts as a cocatalyst. As such, the solution provides for a short-circuited electrocatalytic cell which allows for electron and proton generation transfer and facile recombination. The ability carry out this heterolytic proton transfer process is controlled by the work function or "electron affinity" of the metal. Metals such as Pd and Pt should have the ability to convert adsorbed hydrogen into protons that reside in solution and electrons which remain in the metal. This was demonstrated by both Wagner and Moylan [129] and Kizhakevariam and Struve [130] who both showed that adsorbed hydrogen on ideal Pt substrates in the presence of water in UHV (at low temperature) forms hydronium ions in solution near the surface. These results are consistent with the results from theory. More generally the results suggest that the presence of protic media along with high work function metals can carry out hydrogenation reactions much more efficiently through proton coupled electron transfer processes that mimic electrocatalysis.

SUMMARY

While fundamental information concerning reaction mechanisms, active sights and catalytic kinetics gleaned from in situ spectroscopy, detailed theoretical simulations and rigorous kinetic studies for gas phase heterogeneous catalysis has helped guide the development of electrocatalytic systems, the complexity of the reaction environment has often precluded more in-depth or quantitative analyses. The tremendous advances in spectroscopy along with theory, that have taken place over the past few decades, however, have allowed for more detailed resolution of the molecular transformations that occur in electrocatalytic systems along with a more detailed following of the nature of the active centers and their environment. There appears to be an important and growing trend where this knowledge and guiding principles from electrocatalysis are being used to guide heterogeneous the complex aqueous and solvent-based catalytic processes. The knowledge of the complex electrified interface in electrocatalysis bears a number of common similarities to the aqueous/metal interface for catalytic reactions carried out the presence of solution. More detailed fundamental studies which attempt to rigorously compare heterogeneous catalysis in solution with electrocatalysis will continue and will likely be crucial in establishing the links between the two.

Acknowledgments

A. Wieckowski gratefully acknowledges support from the National Science Foundation under Grant no. NSF CHE06-51083 A. Wieckowski and ARO under Grant no. W911NF-08-1-0309. M. Neurock gratefully acknowledges support from the Office of Basic Energy Sciences under Award no. ERKCC61 for the work on metal/solution interfaces. (This work is part of the Fluid Interface Reactions, Structures and Transport (FIRST) Center, an Energy Frontier Research Center funded by the U.S. Department of Energy, Office of Science, Office of Basic Energy Sciences), U.S. Department of Energy, Division of Chemical Sciences, Office of Basic Energy Sciences (DE-FG02-07ER15894) for the work on oxygen reduction, the National Science Foundation Research Center for Biorenewable Chemicals (EEC-0813570), and the National Science Foundation PIRE (NSF OISE-0730277) for the work on alcohol oxidation. M. Neurock would also kindly acknowledge the computational time at the Environmental Molecular Science Laboratory, a national scientific user facility sponsored by the Department of Energy's Office of Biological and Environmental Research and located at Pacific Northwest National Laboratory, the National Center for Computational Sciences at Oak Ridge National Laboratory and the National Energy Research Scientific Computing Center, which is supported by the Office of Science of the U.S. Department of Energy which was used to carry out the work. Lastly, the authers would like to thank Professor Robert J. Davis, David Hibbitts, and Craig Plaisance for helpful discussions.

REFERENCES

1. D. Cao, G. Q. Lu, A. Wieckowski, S. A. Wasileski, and M. Neurock, "Mechanisms of methanol decomposition on platinum: a combined experimental and ab initio approach," Journal of Physical Chemistry B, vol. 109, no. 23, pp. 11622–11633, 2005.
2. M. J. Janik, S. A. Wasileski, C. D. Taylor, and M. Neurock, "First principles simulation of the active sites and reaction environment in electrocatalysis," in Fuel Cell Catalysis: A Surface Science Approach, M. Koper, Ed., John Wiley and Sons, 2008.
3. M. Koper, Ed., Fuel Cell Catalysis: A Surface Science Approach, John Wiley and Sons, 2008.
4. A. Wieckowski and J. K. Nørskov, Fuel Cell Science: Theory, Fundamentals, and Bio-Catalysis, John Wiley and Sons, 2010.
5. R. A. van Santen and M. Neurock, "Theory of surface chemical reactivity," in Handbook of Catalysis, H. K. G. Ertl and J. Weitcamp, Eds., pp. 942–958, Springer, 1997.
6. R. A. van Santen and M. Neurock, Molecular Heterogeneous Catalysis: A Mechanistic and Computational Approach, VCH-Wiley, 2006.
7. R. A. van Santen, M. Neurock, and S. G. Shetty, "Reactivity theory of transition-metal surfaces: a brønsted-evans- polanyi linear activation energy-free-energy analysis," Chemical Reviews, vol. 110, no. 4, pp. 2005–2048, 2010.
8. T. Bligaard and J. K. Nørskov, "Ligand effects in heterogeneous catalysis and electrochemistry,"Electroch imica Acta, vol. 52, no. 18, pp. 5512–5516, 2007.
9. J. Greeley, T. F. Jaramillo, J. Bonde, I. B. Chorkendorff, and J. K. Nørskov, "Computational high-throughput screening of electrocatalytic materials for hydrogen evolution," Nature Materials, vol. 5, no. 11, pp. 909–913, 2006.
10. P. Liu, A. Logadottir, and J. K. Nørskov, "Modeling the electro-oxidation of CO and H_2/CO on Pt, Ru, PtRu and Pt_3Sn," Electrochimica Acta, vol. 48, no. 25-26, pp. 3731–3742, 2003.
11. J. K. Nørskov, J. Rossmeisl, A. Logadottir et al., "Origin of the overpotential for oxygen reduction at a fuel-cell cathode," Journal of Physical Chemistry B, vol. 108, no. 46, pp. 17886–17892, 2004.
12. J. Rossmeisl, J. K. Nørskov, C. D. Taylor, M. J. Janik, and M. Neurock, "Calculated phase diagrams for the electrochemical oxidation and reduction of water over Pt(111)," Journal of Physical Chemistry B, vol. 110, no. 43, pp. 21833–21839, 2006.
13. A. B. Anderson, "O_2 reduction and CO oxidation at the Pt-electrolyte interface. The role of H_2O and OH adsorption bond strengths," Electrochimica Acta, vol. 47, no. 22-23, pp. 3759–3763, 2002.
14. A. B. Anderson and Y. Cai, "Calculation of the Tafel plot for H_2 oxidation on Pt(100) from potential-dependent activation energies," Journal of Physical Chemistry B, vol. 108, no. 52, pp. 19917–19920, 2004.
15. A. B. Anderson, J. Roques, S. Mukerjee, V. S. Murthi, N. M. Markovic, and V. Stamenkovic, "Activation energies for oxygen reduction on platinum alloys: theory and experiment," Journal of Physical Chemistry B, vol. 109, no. 3, pp. 1198–1203, 2005.
16. J. Roques and A. B. Anderson, "Electrode potential-dependent stages in OH_{ads} formation on the Pt_3Cr alloy (111) surface," Journal of the Electrochemical Society, vol. 151, no. 11, pp. E340–E347, 2004.
17. J. Roques, A. B. Anderson, V. S. Murthi, and S. Mukerjee, "Potential shift for OH(ads) formation on the Pt skin on Pt_3Co(111) electrodes in acid theory and experiment," Journal of the Electrochemical Society, vol. 152, no. 6, pp. E193–E199, 2005.
18. R. A. Sidik and A. B. Anderson, "Density functional theory study of O_2 electroreduction when bonded to a Pt dual site," Journal of Electroanalytical Chemistry, vol. 528, no. 1-2, pp. 69–76, 2002.
19. A. Lamm, H. Gasteiger, and W. Vielstich, Eds., Fuel Cell Handbook, Vol. 2, Electrocatalysis, Wiley-VCH, 2003.
20. T. D. Jarvi and E. M. Stuve, "Fundamental aspects of vacuum and electrocatalytic reactions of methanol and formic acid on platinum surfaces," in Electrocatalysis, J. Lipkowski and P. N. Ross, Eds., Wiley-VCH, 1998.
21. G. Q. Lu, A. Lagutchev, D. D. Dlott, and A. Wieckowski, "Quantitative vibrational sum-frequency generation spectroscopy of thin layer electrochemistry: CO on a Pt electrode," Surface Science, vol. 585, no. 1-2, pp. 3–16, 2005.
22. P. K. Babu, H. S. Kim, S. T. Kuk et al., "Activation of nanoparticle Pt-Ru fuel cell catalysts by heat treatment: A195Pt NMR and electrochemical study," Journal of Physical Chemistry B, vol. 109, no. 36, pp. 17192–17196, 2005.
23. H. Yano, J. Inukai, H. Uchida et al., "Particle-size effect of nanoscale platinum catalysts in oxygen reduction reaction: an electrochemical and $_{195}$Pt EC-NMR study," Physical Chemistry Chemical Physics, vol. 8, no. 42, pp. 4932–4939, 2006.
24. T. Kobayashi, P. K. Babu, J. H. Chung, E. Oldfield, and A. Wieckowski, "Coverage dependence of CO surface diffusion on Pt nanoparticles: an EC-NMR study," Journal of Physical Chemistry C, vol. 111, no.

19, pp. 7078–7083, 2007.

25. S. Maniguet, R. J. Mathew, and A. E. Russell, “EXAFS of carbon monoxide oxidation on supported Pt fuel cell electrocatalysts,” Journal of Physical Chemistry B, vol. 104, no. 9, pp. 1998–2004, 2000.
26. J. McBreen and S. Mukerjee, “In situ X-ray absorption studies of a Pt-Ru electrocatalyst,” Journal of the Electrochemical Society, vol. 142, no. 10, pp. 3399–3404, 1995.
27. F. J. Scott, C. Roth, and D. E. Ramaker, “Kinetics of CO poisoning in simulated reformate and effect of Ru island morphology on PtRu fuel cell catalysts as determined by operando X-ray absorption near edge spectroscopy,” Journal of Physical Chemistry C, vol. 111, no. 30, pp. 11403–11413, 2007.
28. F. J. Scott, S. Mukerjee, and D. E. Ramaker, “CO coverage/oxidation correlated with PtRu electrocatalyst particle morphology in 0.3 M methanol by in situ XAS,” Journal of the Electrochemical Society, vol. 154, no. 5, pp. A396–A406, 2007.
29. C. Roth, N. Benker, T. Buhrmester et al., “Determination of O[H] and CO coverage and adsorption sites on PtRu electrodes in an operating PEM fuel cell,” Journal of the American Chemical Society, vol. 127, no. 42, pp. 14607–14615, 2005.
30. G. Samjeske, A. Miki, S. Ye, and M. Osawa, “Mechanistic study of electrocatalytic oxidation of formic acid at platinum in acidic solution by time-resolved surface-enhanced infrared absorption spectroscopy,”Journal of Physical Chemistry B, vol. 110, no. 33, pp. 16559–16566, 2006.
31. G. Samjeske, A. Miki, and M. Osawa, “Electrocatalytic oxidation of formaldehyde on platinum under galvanostatic and potential sweep conditions studied by time-resolved surface-enhanced infrared spectroscopy,” Journal of Physical Chemistry C, vol. 111, no. 41, pp. 15074–15083, 2007.
32. A. B. Anderson, “Theory at the electrochemical interface: reversible potentials and potential-dependent activation energies,” Electrochimica Acta, vol. 48, no. 25-26, pp. 3743–3749, 2003.
33. A. B. Anderson, Y. Cai, R. A. Sidik, and D. B. Kang, “Advancements in the local reaction center electron transfer theory and the transition state structure in the first step of oxygen reduction over platinum,”Journal of Electroanalytical Chemistry, vol. 580, no. 1, pp. 17–22, 2005.
34. A. B. Anderson, N. M. Neshev, R. A. Sidik, and P. Shiller, “Mechanism for the electrooxidation of water to OH and O bonded to platinum: quantum chemical theory,” Electrochimica Acta, vol. 47, no. 18, pp. 2999–3008, 2002.
35. Y. Cai and A. B. Anderson, “The reversible hydrogen electrode: potential-dependent activation energies over platinum from quantum theory,” Journal of Physical Chemistry B, vol. 108, no. 28, pp. 9829–9833, 2004.
36. E. Santos, A. Lundin, K. Potting, P. Quaino, and W. Schmickler, “Model for the electrocatalysis of hydrogen evolution,” Physical Review B, vol. 79, no. 23, Article ID 235436, 2009.
37. E. Santos and W. Schmickler, “Electronic interactions decreasing the activation barrier for the hydrogen electro-oxidation reaction,” Electrochimica Acta, vol. 53, no. 21, pp. 6149–6156, 2008.
38. F. Wilhelm, W. Schmickler, R. R. Nazmutdinov, and E. Spohr, “A model for proton transfer to metal electrodes,” Journal of Physical Chemistry C, vol. 112, no. 29, pp. 10814–10826, 2008.
39. J. S. Filhol and M. Neurock, “Elucidation of the electrochemical activation of water over Pd by first principles,” Angewandte Chemie, vol. 45, no. 3, pp. 403–406, 2006.
40. C. Taylor, R. G. Kelly, and M. Neurock, “First-principles calculations of the electrochemical reactions of water at an immersed Ni(111)/H_2O interface,” Journal of the Electrochemical Society, vol. 153, no. 12, pp. E207–E214, 2006.
41. C. D. Taylor and M. Neurock, “Theoretical insights into the structure and reactivity of the aqueous/metal interface,” Current Opinion in Solid State and Materials Science, vol. 9, no. 1-2, pp. 49–65, 2005.
42. C. D. Taylor, M. Neurock, and J. R. Scully, “First-principles investigation of the fundamental corrosion properties of a model Cu_{38} nanoparticle and the (111), (113) surfaces,” Journal of the Electrochemical Society, vol. 155, no. 8, pp. C407–C414, 2008.
43. C. D. Taylor, S. A. Wasileski, J. S. Filhol, and M. Neurock, “First principles reaction modeling of the electrochemical interface: consideration and calculation of a tunable surface potential from atomic and electronic structure,” Physical Review B, vol. 73, no. 16, Article ID 165402, pp. 1–16, 2006.
44. M. Otani, I. Hamada, O. Sugino, Y. Morikawa, Y. Okamoto, and T. Ikeshoji, “Structure of the water/platinum interface—a first principles simulation under bias potential,” Physical Chemistry Chemical Physics, vol. 10, no. 25, pp. 3609–3612, 2008.
45. M. Otani, I. Hamada, O. Sugino, Y. Morikawa, Y. Okamoto, and T. Ikeshoji, “Electrode dynamics from first principles,” Journal of the Physical Society of Japan, vol. 77, no. 2, Article ID 024802, 2008.
46. R. Jinnouchi and A. B. Anderson, “Electronic structure calculations of liquid-solid interfaces: combination of density functional theory and modified Poisson-Boltzmann theory,” Physical Review B, vol. 77, no. 24, Article ID 245417, 2008.
47. J. Rossmeisl, E. Skulason, M. E. Bjorketun, V. Tripkovic, and J. K. Nørskov, “Modeling the electrified solid-liquid interface,” Chemical Physics Letters, vol. 466, no. 1–3, pp. 68–71, 2008.
48. M. Neurock, W. Vielstich, H. A. Gasteiger, and H. Yokokawa, “First principles modeling for the electrooxidation of small molecules,” in Handbook of Fuel Cells, W. Vielstich, H. A. Gasteiger, and H. Yokokawa, Eds., John Wiley and Sons, 2009.
49. G. Kresse and J. Furthmuller, “Efficiency of ab-initio total energy calculations for metals and semiconductors

using a plane-wave basis set," Computational Materials Science, vol. 6, no. 1, pp. 15–50, 1996.

50. G. Kresse and J. Furthmuller, "Efficient iterative schemes for ab initio total-energy calculations using a plane-wave basis set," Physical Review B, vol. 54, no. 16, pp. 11169–11186, 1996. ·
51. G. Henkelman and H. Jónsson, "Improved tangent estimate in the nudged elastic band method for finding minimum energy paths and saddle points," Journal of Chemical Physics, vol. 113, no. 22, pp. 9978–9985, 2000.
52. G. Henkelman, B. P. Uberuaga, and H. Jónsson, "Climbing image nudged elastic band method for finding saddle points and minimum energy paths," Journal of Chemical Physics, vol. 113, no. 22, pp. 9901–9904, 2000.
53. G. Henkelman and H. Jónsson, "A dimer method for finding saddle points on high dimensional potential surfaces using only first derivatives," Journal of Chemical Physics, vol. 111, no. 15, pp. 7010–7022, 1999. ·
54. R. R. Davda, J. W. Shabaker, G. W. Huber, R. D. Cortright, and J. A. Dumesic, "Aqueous-phase reforming of ethylene glycol on silica-supported metal catalysts," Applied Catalysis B, vol. 43, no. 1, pp. 13–26, 2003. ·
55. P. Gallezot, N. Nicolaus, G. Fleche, and A. Perrard, "Glucose hydrogenation on ruthenium catalysts in a trickle-bed reactor," Journal of Catalysis, vol. 180, no. 1, pp. 51–55, 1998.
56. D. K. Sohounloue, C. Montassier, and J. Barbier, "Catalytic hydrogenolysis of sorbitol," Reaction Kinetics and Catalysis Letters, vol. 22, no. 3-4, pp. 391–397, 1983.
57. M. B. Valenzuela, C. W. Jones, and P. K. Agrawal, "Batch aqueous-phase reforming of woody biomass,"Energy and Fuels, vol. 20, no. 4, pp. 1744–1752, 2006.
58. N. Dimitratos, C. Messi, F. Porta, L. Prati, and A. Villa, "Investigation on the behaviour of Pt(0)/carbon and Pt(0),Au(0)/carbon catalysts employed in the oxidation of glycerol with molecular oxygen in water,"Journal of Molecular Catalysis A, vol. 256, no. 1-2, pp. 21–28, 2006.
59. R. D. Cortright, R. R. Davda, and J. A. Dumesic, "Hydrogen from catalytic reforming of biomass-derived hydrocarbons in liquid water," Nature, vol. 418, no. 6901, pp. 964–967, 2002.
60. J. W. Shabaker, G. W. Huber, R. R. Davda, R. D. Cortright, and J. A. Dumesic, "Aqueous-phase reforming of ethylene glycol over supported platinum catalysts," Catalysis Letters, vol. 88, no. 1-2, pp. 1–8, 2003.
61. E. P. Maris, W. C. Ketchie, V. Oleshko, and R. J. Davis, "Metal particle growth during glucose hydrogenation over Ru/SiO_2 evaluated by X-ray absorption spectroscopy and electron microscopy,"Journal of Physical Chemistry B, vol. 110, no. 15, pp. 7869–7876, 2006.
62. E. P. Maris, W. C. Ketchie, M. Murayama, and R. J. Davis, "Glycerol hydrogenolysis on carbon-supported PtRu and AuRu bimetallic catalysts," Journal of Catalysis, vol. 251, no. 2, pp. 281–294, 2007.
63. W. C. Ketchie, E. P. Maris, and R. J. Davis, "In-situ X-ray absorption spectroscopy of supported Ru catalysts in the aqueous phase," Chemistry of Materials, vol. 19, no. 14, pp. 3406–3411, 2007.
64. M. Haruta, T. Kobayashi, H. Sano, and N. Yamada, "Novel gold catalysts for the oxidation of carbon monoxide at a temperature far below 0.DEG.C," Chemistry Letters, vol. 16, no. 2, pp. 405–408, 1987.
65. M. Haruta, N. Yamada, T. Kobayashi, and S. Iijima, "Gold catalysts prepared by coprecipitation for low-temperature oxidation of hydrogen and of carbon monoxide," Journal of Catalysis, vol. 115, no. 2, pp. 301–309, 1989.
66. G. C. Bond, C. Louis, and D. T. Thompson, Catalysis by Gold, Imperial College Press, London, UK, 2006.
67. R. Meyer, C. Lemire, S. K. Shaikhutdinov, and H. Freund, "Surface chemistry of catalysis by gold," Gold Bulletin, vol. 37, no. 1-2, pp. 72–124, 2004.
68. M. Valden, X. Lai, and D. W. Goodman, "Onset of catalytic activity of gold clusters on titania with the appearance of nonmetallic properties," Science, vol. 281, no. 5383, pp. 1647–1650, 1998.
69. N. Lopez, T. V. W. Janssens, B. S. Clausen et al., "On the origin of the catalytic activity of gold nanoparticles for low-temperature CO oxidation," Journal of Catalysis, vol. 223, no. 1, pp. 232–235, 2004.
70. Q. Fu, H. Saltsburg, and M. Flytzani-Stephanopoulos, "Active nonmetallic Au and Pt species on ceria-based water-gas shift catalysts," Science, vol. 301, no. 5635, pp. 935–938, 2003.
71. J. Guzman and B. C. Gates, "Simultaneous presence of cationic and reduced gold in functioning MgO-supported CO oxidation catalysts: evidence from X-ray absorption spectroscopy," Journal of Physical Chemistry B, vol. 106, no. 31, pp. 7659–7665, 2002.
72. L. M. Molina and B. Hammer, "Active role of oxide support during CO oxidation at Au/MgO," Physical Review Letters, vol. 90, no. 20, 4 pages, 2003.
73. C. K. Costello, M. C. Kung, H. S. Oh, Y. Wang, and H. H. Kung, "Nature of the active site for CO oxidation on highly active Au/γ-Al_2O_3," Applied Catalysis A, vol. 232, no. 1-2, pp. 159–168, 2002.
74. H. H. Kung, M. C. Kung, and C. K. Costello, "Supported Au catalysts for low temperature CO oxidation,"Journal of Catalysis, vol. 216, no. 1-2, pp. 425–432, 2003.
75. W. B. Kim, T. Voitl, G. J. Rodriguez-Rivera, and J. A. Dumesic, "Powering fuel cells with CO via aqueous polyoxometalates and gold catalysts," Science, vol. 305, no. 5688, pp. 1280–1283, 2004.
76. M. A. Sanchez-Castillo, C. Couto, W. B. Kim, and J. A. Dumesic, "Gold-nanotube membranes for the oxidation of CO at gas-water interfaces," Angewandte Chemie, vol. 43, no. 9, pp. 1140–1142, 2004.
77. W. C. Ketchie, Y. L. Fang, M. S. Wong, M. Murayama, and R. J. Davis, "Influence of gold particle size on the aqueous-phase oxidation of carbon monoxide and

glycerol," Journal of Catalysis, vol. 250, no. 1, pp. 94–101, 2007.

78. W. C. Ketchie, M. Murayama, and R. J. Davis, "Promotional effect of hydroxyl on the aqueous phase oxidation of carbon monoxide and glycerol over supported Au catalysts," Topics in Catalysis, vol. 44, no. 1-2, pp. 307–317, 2007.

79. N. M. Markoví and P. N. Ross, "Surface science studies of model fuel cell electrocatalysts," Surface Science Reports, vol. 45, no. 4–6, pp. 117–229, 2002.

80. F. B. de Mongeot, M. Scherer, B. Gleich, E. Kopatzki, and R. J. Behm, "CO adsorption and oxidation on bimetallic Pt/Ru(0001) surfaces—a combined STM and TPD/TPR study," Surface Science, vol. 411, no. 3, pp. 249–262, 1998.

81. H. A. Gasteiger, N. Markovic, P. N. Ross, and E. J. Cairns, "Electro-oxidation of small organic molecules on well-characterized PtRu alloys," Electrochimica Acta, vol. 39, no. 11-12, pp. 1825–1832, 1994.

82. N. M. Markovic, H. A. Gasteiger, P. N. Ross, X. Jiang, I. Villegas, and M. J. Weaver, "Electro-oxidation mechanisms of methanol and formic acid on Pt-Ru alloy surfaces," Electrochimica Acta, vol. 40, no. 1, pp. 91–98, 1995.

83. S. Desai and M. Neurock, "A first principles analysis of CO oxidation over Pt and Pt 66.7%Ru33.3% (111) surfaces," Electrochimica Acta, vol. 48, no. 25-26, pp. 3759–3773, 2003.

84. S. K. Desai and M. Neurock, "First-principles study of the role of solvent in the dissociation of water over a Pt-Ru alloy," Physical Review B, vol. 68, no. 7, Article ID 075420, pp. 754201–754207, 2003.

85. B. E. Hayden, D. Pletcher, M. E. Rendall, and J. P. Suchsland, "CO oxidation on gold in acidic environments: particle size and substrate effects," Journal of Physical Chemistry C, vol. 111, no. 45, pp. 17044–17051, 2007.

86. B. E. Hayden, D. Pletcher, and J. P. Suchsland, "Enhanced activity for electrocatalytic oxidation of carbon monoxide on titania-supported gold nanoparticles," Angewandte Chemie, vol. 46, no. 19, pp. 3530–3532, 2007.

87. J. L. Roberts and D. T. Sawyer, "Electrochemical oxidation of carbon monoxide at gold electrodes,"Electrochimica Acta, vol. 10, no. 10, pp. 989–1000, 1965.

88. P. Rodriguez, N. Garcia-Araez, and M. T. M. Koper, "Self-promotion mechanism for CO electrooxidation on gold," Physical Chemistry Chemical Physics, vol. 12, no. 32, pp. 9373–9380, 2010.

89. P. Rodriguez, N. Garcia-Araez, A. Koverga, S. Frank, and M. T. M. Koper, "CO electroxidation on gold in alkaline media: a combined electrochemical, spectroscopic, and DFT study," Langmuir, vol. 26, no. 14, pp. 12425–12432, 2010.

90. I. X. Green, W. Tang, M. Neurock, and J. T. Yates Jr., "Spectroscopic observation of dual catalytic sites during oxidation of CO on a Au/TiO_2 catalyst," Science, vol. 333, no. 6043, pp. 736–739, 2011.

91. M. Neurock, D. Hibbitts, and J. A. Dumesic, "Mechanistic insights into the role of water on the oxidation of CO over Au," to be submitted, 2011.

92. N. Dimitratos, J. A. Lopez-Sanchez, J. M. Anthonykutty et al., "Oxidation of glycerol using gold-palladium alloy-supported nanocrystals," Physical Chemistry Chemical Physics, vol. 11, no. 25, pp. 4952–4961, 2009.

93. N. Dimitratos, J. A. Lopez-Sanchez, S. Meenakshisundaram et al., "Selective formation of lactate by oxidation of 1,2-propanediol using gold palladium alloy supported nanocrystals," Green Chemistry, vol. 11, no. 8, pp. 1209–1216, 2009.

94. M. D. Hughes, Y. J. Xu, P. Jenkins et al., "Tunable gold catalysts for selective hydrocarbon oxidation under mild conditions," Nature, vol. 437, no. 7062, pp. 1132–1135, 2005.

95. G. J. Hutchings, S. Carrettin, P. Landon et al., "New approaches to designing selective oxidation catalysts: Au/C a versatile catalyst," Topics in Catalysis, vol. 38, no. 4, pp. 223–230, 2006.

96. M. Sankar, N. Dimitratos, D. W. Knight et al., "Oxidation of glycerol to glycolate by using supported gold and palladium nanoparticles," ChemSusChem, vol. 2, no. 12, pp. 1145–1151, 2009.

97. B. N. Zope and R. J. Davis, "Influence of reactor configuration on the selective oxidation of glycerol over Au/TiO_2," Topics in Catalysis, vol. 52, no. 3, pp. 269–277, 2009.

98. B. N. Zope, D. D. Hibbitts, M. Neurock, and R. J. Davis, "Reactivity of the gold/water interface during selective oxidation catalysis," Science, vol. 330, no. 6000, pp. 74–78, 2010.

99. J. S. Spendelow, P. K. Babu, and A. Wieckowski, "Electrocatalytic oxidation of carbon monoxide and methanol on platinum surfaces decorated with ruthenium," Current Opinion in Solid State and Materials Science, vol. 9, no. 1-2, pp. 37–48, 2005.

100. J. S. Spendelow and A. Wieckowski, "Electrocatalysis of oxygen reduction and small alcohol oxidation in alkaline media," Physical Chemistry Chemical Physics, vol. 9, no. 21, pp. 2654–2675, 2007.

101. Z. Borkowska, A. Tymosiak-Zielinska, and G. Shul, "Electrooxidation of methanol on polycrystalline and single crystal gold electrodes," Electrochimica Acta, vol. 49, no. 8, pp. 1209–1220, 2004.

102. P. Parpot, A. P. Bettencourt, A. M. Carvalho, and E. M. Belgsir, "Biomass conversion: attempted electrooxidation of lignin for vanillin production," Journal of Applied Electrochemistry, vol. 30, no. 6, pp. 727–731, 2000.

103. P. Parpot, A. P. Bettencourt, G. Chamoulaud, K. B. Kokoh, and E. M. Belgsir, "Electrochemical investigations of the oxidation-reduction of furfural in aqueous medium—application to electrosynthesis," Electrochimica Acta, vol. 49, no. 3, pp. 397–403, 2004.

104. P. Parpot, N. Nunes, and A. P. Bettencourt, "Electrocatalytic oxidation of monosaccharides on gold electrode in alkaline medium: structure-reactivity relationship," Journal of Electroanalytical Chemistry, vol. 596, no. 1, pp. 65–73, 2006.

105. P. Parpot, S. G. Pires, and A. P. Bettencourt,

"Electrocatalytic oxidation of D-galactose in alkaline medium," Journal of Electroanalytical Chemistry, vol. 566, no. 2, pp. 401–408, 2004.

106. P. Parpot, P. R. B. Santos, and A. P. Bettencourt, "Electro-oxidation of d-mannose on platinum, gold and nickel electrodes in aqueous medium," Journal of Electroanalytical Chemistry, vol. 610, no. 2, pp. 154–162, 2007.
107. S. H. Yan, S. C. Zhang, Y. Lin, and G. R. Liu, "Electrocatalytic performance of gold nanoparticles supported on activated carbon for methanol oxidation in alkaline solution," Journal of Physical Chemistry C, vol. 115, no. 14, pp. 6986–6993, 2011.
108. P. A. Christensen, A. Hamnett, and D. Linares-Moya, "Oxygen reduction and fuel oxidation in alkaline solution," Physical Chemistry Chemical Physics, vol. 13, no. 12, pp. 5206–5214, 2011.
109. D. Kardash, C. Korzeniewski, and N. Markovic, "Effects of thermal activation on the oxidation pathways of methanol at bulk Pt-Ru alloy electrodes," Journal of Electroanalytical Chemistry, vol. 500, no. 1-2, pp. 518–523, 2001.
110. A. V. Tripkovi, K. D. Popovi, B. N. Grgur, B. Blizanac, P. N. Ross, and N. M. Markovi, "Methanol electrooxidation on supported Pt and PtRu catalysts in acid and alkaline solutions," Electrochimica Acta, vol. 47, no. 22-23, pp. 3707–3714, 2002.
111. F. Colmati, G. Tremiliosi-Filho, E. R. Gonzalez, A. Berna, E. Herrero, and J. M. Feliu, "The role of the steps in the cleavage of the C-C bond during ethanol oxidation on platinum electrodes," Physical Chemistry Chemical Physics, vol. 11, no. 40, pp. 9114–9123, 2009.
112. F. Colmati, G. Tremiliosi-Filho, E. R. Gonzalez, A. Berna, E. Herrero, and J. M. Feliu, "Surface structure effects on the electrochemical oxidation of ethanol on platinum single crystal electrodes," Faraday Discussions, vol. 140, pp. 379–397, 2008.
113. V. Del Colle, A. Berna, G. Tremiliosi-Filho, E. Herrero, and J. M. Feliu, "Ethanol electrooxidation onto stepped surfaces modified by Ru deposition: electrochemical and spectroscopic studies," Physical Chemistry Chemical Physics, vol. 10, no. 25, pp. 3766–3773, 2008.
114. J. Souza-Garcia, E. Herrero, and J. M. Feliu, "Breaking the C-C bond in the ethanol oxidation reaction on platinum electrodes: effect of steps and ruthenium adatoms," ChemPhysChem, vol. 11, no. 7, pp. 1391–1394, 2010.
115. S. C. S. Lai, S. E. F. Kleijn, F. T. Z. Ozturk et al., "Effects of electrolyte pH and composition on the ethanol electro-oxidation reaction," Catalysis Today, vol. 154, no. 1-2, pp. 92–104, 2010.
116. S. C. S. Lai, S. E. F. Kleyn, V. Rosca, and M. T. M. Koper, "Mechanism of the dissociation and electrooxidation of ethanol and acetaldehyde on platinum as studied by SERS," Journal of Physical Chemistry C, vol. 112, no. 48, pp. 19080–19087, 2008.
117. S. C. S. Lai and M. T. M. Koper, "Electro-oxidation of ethanol and acetaldehyde on platinum single-crystal electrodes," Faraday Discussions, vol. 140, pp. 399–416, 2008.
118. S. C. S. Lai and M. T. M. Koper, "Ethanol electro-oxidation on platinum in alkaline media," Physical Chemistry Chemical Physics, vol. 11, no. 44, pp. 10446–10456, 2009.
119. S. C. S. Lai and M. T. M. Koper, "The influence of surface structure on selectivity in the ethanol electro-oxidation reaction on platinum," Journal of Physical Chemistry Letters, vol. 1, no. 7, pp. 1122–1125, 2010.
120. M. Neurock, M. Janik, and A. Wieckowski, "A first principles comparison of the mechanism and site requirements for the electrocatalytic oxidation of methanol and formic acid over Pt," Faraday Discussions, vol. 140, no. 1, pp. 363–378, 2008.
121. S. R. Brankovic, J. X. Wang, Y. Zhu, R. Sabatini, J. McBreen, and R. R. Adzic, "Electrosorption and catalytic properties of bare and Pt modified single crystal and nanostructured Ru surfaces," Journal of Electroanalytical Chemistry, vol. 524-525, pp. 231–241, 2002.
122. T. Iwasita, "Electrocatalysis of methanol oxidation," Electrochimica Acta, vol. 47, no. 22-23, pp. 3663–3674, 2002.
123. W. C. Ketchie, M. Murayama, and R. J. Davis, "Selective oxidation of glycerol over carbon-supported AuPd catalysts," Journal of Catalysis, vol. 250, no. 2, pp. 264–273, 2007.
124. A. T. Governo, L. Proenca, P. Parpot, M. I. S. Lopes, and I. T. E. Fonseca, "Electro-oxidation of D-xylose on platinum and gold electrodes in alkaline medium," Electrochimica Acta, vol. 49, no. 9-10, pp. 1535–1545, 2004.
125. Y. Kwon and M. T. M. Koper, "Combining voltammetry with HPLC: application to electro-oxidation of glycerol," Analytical Chemistry, vol. 82, no. 13, pp. 5420–5424, 2010.
126. Z. W. Liu, F. Peng, H. J. Wang, H. Yu, W. X. Zheng, and J. A. Yang, "Phosphorus-doped graphite layers with high electrocatalytic activity for the O_2 reduction in an alkaline medium," Angewandte Chemie, vol. 50, no. 14, pp. 3257–3261, 2011.
127. A. C. Chen and J. Lipkowski, "Electrochemical and spectroscopic studies of hydroxide adsorption at the Au(111) electrode," Journal of Physical Chemistry B, vol. 103, no. 4, pp. 682–691, 1999.
128. S. Desai, "Theoretical investigation of solution effects on metal catalyzed hydrogenation and oxidation processes," in Department of Chemical Engineering, University of Virginia, 2002.
129. F. T. Wagner and T. E. Moylan, "Generation of surface hydronium from water and hydrogen coadsorbed on Pt(111)," Surface Science, vol. 206, no. 1-2, pp. 187–202, 1988.
130. N. Kizhakevariam and E. M. Stuve, "Coadsorption of water and hydrogen on Pt(100): formation of adsorbed hydronium ions," Surface Science, vol. 275, no. 3, pp. 223–236, 1992.

CHAPTER 5

RECENT ADVANCES ON ELECTRO-OXIDATION OF ETHANOL ON PT- AND PD-BASED CATALYSTS: FROM REACTION MECHANISMS TO CATALYTIC MATERIALS

Ye Wang, Shouzhong Zou, and Wen-Bin Cai

Keywords: Ethanol, Electrocatalysts, Electro-oxidation, Mechanism, Rational design

INTRODUCTION

Rising demands for energy coupled with concerns over ecosystem damage and growing consumption of non-regenerative fossil energy pose a great need for clean and efficient power sources [1,2,3,4]. Fuel cells are widely considered as sustainable energy conversion devices. Low-temperature fuel cells are undergoing rapid development for mobile applications and in particular for the transport sector. Among different fuels that have been used for fuel cells, hydrogen, methanol, and ethanol have been the most explored and each has its advantages and disadvantages. The choice of the fuel depends on the applications. Proton exchange membrane fuel cells (PEMFCs) using hydrogen as the fuel have the advantages of low operating temperature, sustained operation at high current densities, low weight, compactness, and suitability for discontinuous operation, but face challenges in the production, storage and transport of hydrogen. As an alternative fuel, ethanol which can be produced on a massive scale from biomass feed stocks originating from agriculture (first-generation bioethanol), forestry, and urban residues (second-generation bioethanol), is attracting increasing interest [5,6,7,8]. Compared to another common fuel, methanol, ethanol complements the shortcomings of methanol owing to its non-toxicity, higher boiling point, and most importantly, renewability. In addition, ethanol has a high specific energy of 8.01 $kWh \cdot kg^{-1}$, which is comparable to that of gasoline [9,10]. Nevertheless, the relatively sluggish kinetics for the ethanol oxidation reaction (EOR) presents a major roadblock for the development of direct ethanol fuel cells (DEFCs) [3,7]. Higher performance catalysts are needed to overcome this bottleneck. A detailed understanding of the reaction mechanism and in particular of the rate-limiting step(s) in EOR under continuous reaction conditions is of critical importance for the design of highly active catalysts [11,12]. Although numerous experimental studies using Fourier transform infrared spectroscopy (FTIR) [13,14,15,16,17,18,19,20,21,22,23,24,25,26] or differential electrochemical mass spectrometry (DEMS) [27,28,29,30,31,32,33,34,35,36,37], as well as theoretical studies [38,39,40,41,42,43,44,45] have been conducted to understand the EOR process, a detailed mechanism of EOR remains unclear or even contradictory. Nevertheless, a so-called dual-pathway (C1 and C2) mechanism has been largely agreed upon: the C1 pathway proceeds via adsorbed carbon monoxide (CO_{ads}) intermediate to form CO_2 (or carbonate in alkaline solutions) by delivering 12 electrons, and the C2 pathway mainly leads to the formation of acetic acid (or acetate in alkaline solutions) by delivering four electrons and/or acetaldehyde by delivering two electrons. Though a higher electro-efficiency can be achieved by the C1 pathway, the C2 pathway is generally dominant in the overall EOR [46,47,48,49]. Therefore selectively enhancing the C1 pathway by rational design of high performance catalysts is an effective way to increase the DEFC efficiency.

Pt is the most commonly used catalytic metal in the anode of DEFCs because of its excellent properties in the adsorption and dissociation of ethanol. However, the cost of Pt is a major impediment in the commercialization of fuel cell technology,

Citation: Ye Wang, Shouzhong Zou, and Wen-Bin Cai, Recent Advances on Electro-Oxidation of Ethanol on Pt- and Pd-Based Catalysts: From Reaction Mechanisms to Catalytic Materials, doi:10.3390/catal5031507.

because it alone accounts for approximately 54% of the total fuel cell stack cost [3]. On the other hand, Pd has similar catalytic properties to Pt (in the same group of the periodic table, having the same face centered cubic (fcc) crystal structure and a similar atomic size) [12], but is much lower in material cost. Moreover, the abundance of Pd on the Earth's crust is 200 times higher than that of Pt (0.6 part per billion (ppb) *vs.* 0.003 ppb), making it very attractive for long-term industrial applications [3]. Though Pt and Pd show relatively good activity, a complete oxidation of ethanol in both acidic and/or basic media remains virtually impossible. A large number of studies have reported the enhancement of the electrocatalytic performance of Pt-M and Pd-M binary or ternary catalysts by adding additional elements such as metallic elements Ru, Sn, Ir, Bi, Rh, Mo, Fe, Co, Cu, Ni, Au, Ag and nonmetallic elements, oxides *etc.* Yet, there are many unanswered questions regarding the role of these foreign materials in improving the electrocatalytic activities. To rationally design Pt- or Pd-based materials as anode catalysts and to develop DEFC technology, a better understanding of the structure-electrocatalytic activity relationships in the EOR is a pre-requisite.

In this review, by discussing selected publications on mechanism studies and the development of advanced catalysts, we present an overview of how the achievements in mechanism studies have been used to guide the rational design of catalysts. Recent advancements in fundamental studies as well as in developing promising new anode EOR catalysts are briefly surveyed. Finally, we summarize the key problems in the investigation on catalysts for EOR and provide outlooks for their future development. Because EOR is under active research, it is impossible to cover every aspect of the new developments. We therefore focus our discussion on Pt- and Pd-based catalysts. Selected examples are only used to facilitate the discussion and inevitably we may have omitted other significant contributions in the field.

REACTION MECHANISM OF EOR

Activity, selectivity, and stability are critical issues that need to be addressed for any catalysts. Comprehensive fundamental studies of EOR form the basis of design rules for high efficiency catalysts [2,21,26,47,50]. A great deal of work on the mechanisms of Pt- or Pd-based catalysts have been devoted to solve the long-standing puzzle concerning the intermediates and the products from EOR. The pioneering work on the mechanism of EOR can be traced back to the 1950s [51] and now has been evolved into a commonly accepted dual-pathway mechanism on Pt- or Pd-based catalysts in either acidic or alkaline media as shown in Figure 1. [44,47,52,53,54,55].

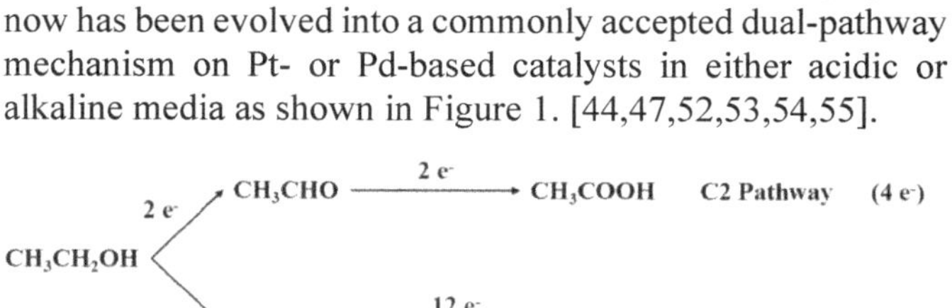

Figure 1. Schematic representation of the parallel pathways for ethanol oxidation on Pt electrodes in acidic media.

The C1 pathway is the complete oxidation of ethanol to CO_2 or carbonates via CO_{ads} intermediate by delivering 12 electrons and the C2 pathway is the partial oxidation of ethanol to acetate by delivering four electrons or to acetaldehyde by delivering two electrons without the breaking of the C–C bond as shown in the following equations:

- **C1 pathway**:

$$CH_3\text{–}CH_2OH + 3H_2O \rightarrow 2CO_2 + 12H^+ + 12e^- \quad (1)$$

$$CH_3\text{–}CH_2OH + 5H_2O \rightarrow 2HCO_3^- + 14H^+ + 12e^- \quad (2)$$

$$CH_3\text{–}CH_2OH + 5H_2O \rightarrow 2CO_3^{2-} + 16H^+ + 12e^- \quad (3)$$

- **C2 pathway**:

$$CH_3\text{–}CH_2OH + H_2O \rightarrow CH_3\text{–}COOH + 4H^+ + 4e^- \quad (4)$$

$$CH_3\text{–}CH_2OH \rightarrow CH_3\text{–}CHO + 2H^+ + 2e^- \quad (5)$$

Adsorbed CO, C1 and C2 hydrocarbon residues have been identified as the major adsorbed intermediates on Pt- or Pd-based catalysts, while acetaldehyde and acetic acid have been detected as the main by-products using techniques such as infrared spectroscopy [13,14,15,16,17,18,19,20,21,22,23,24,25,26], online DEMS [27,28,29,30,31,32,33,34,35,36,37], ion chromatography [56,57], and liquid chromatography [14]. However, EOR has been shown to occur via a series of complex reactions involving a number of sequential and parallel reaction steps, thus resulting in more than 40 possible volatile and adsorbed intermediates or oxidative derivatives [43]. Previous studies agree that CO is a dominant adsorbed species formed during EOR, however, they disagree on details such as the adsorbed state of other intermediates and on the question of the rate limiting steps: the adsorption of intermediate or the cleavage of C–C bond or the formation of OH or oxides [20,23,26,31,38,45,47,54]. In the following we summarize EOR mechanisms developed in the last two decades from experimental as well as DFT calculation studies and discuss the unsolved issues in understanding the EOR mechanism.

Experimental Detection and Quantification of Reaction Intermediates and Products

As mentioned above, the reaction mechanism is complex involving several adsorbed intermediates and numerous products and by-products. Determining product distribution and identifying reactive intermediates are the keys for solving the EOR mechanism puzzle and therefore are always hot topics under active debate. To address these two issues, many efforts have been made to combine traditional electrochemical methods (cyclic voltammetry, chronoamperometry, rotating disc electrodes, *etc.*) with other physicochemical methods, such as *in situ* FTIR [9,13,14,15,16,17,18,19,20,21,22,23,24], broadband sum-frequency generation (BB-SFG) spectroscopy [58], DEMS [16,27,28,29,30,31,32,33,34,35,36], HPLC [56,57], GC [14], electrochemical quartz crystal microbalance (EQCM) [59] and *in situ* NMR [60] to probe the adsorbed intermediates and/or quantify the reaction products and by-products [56]. We devote the following section to summarize and discuss how these techniques were used for product quantification as well as the clarification of intermediates, starting with FTIR studies and followed by mass spectrometric results. In the early days Weaver and co-workers [25,61,62] adopted real-time FTIR spectroscopy to study EOR on Pt surfaces in acidic media, and found that the final reaction products included acetic acid, acetaldehyde along with a smaller amount of CO_2. Their work was the first quantification of specific oxidation products, and provides values of the effective absorption coefficient, ε_{eff} of CO_2, acetic acid, and acetaldehyde which are 3.5×10^4, 5.8×10^3, and 2.2×10^3 $M^{-1}\cdot cm^{-2}$, respectively. The yields of oxidation products were calculated using respective integrated band intensities (A_i), and the amount of a given species Q ($mol\cdot cm^{-2}$) trapped inside the thin layer between the electrode surface and the optical window followed the relationship:

$$Q = \frac{A_i}{\varepsilon_{eff}} \tag{6}$$

To better compare the selectivity and activity of EOR on catalysts, Adzic's [9] group applied *in situ* infrared reflection-absorption spectroscopy (*in situ* IRRAS) to quantify the ratio of C1 pathway to C2 pathway on the ternary Pd–Rh–SnO_2/C electrocatalysts using the following Equation (7):

$$\frac{C_{CO_2}}{C_{CH_3COOH} + C_{CH_3CHO}} = \frac{6 \times Q_{CO_2}}{4 \times Q_{CH_3COOH} + 2 \times Q_{CH_3CHO}} \tag{7}$$

where C_{CO2} and C_{CH3}COOH + C_{CH3}CHO represent the charges associated with the total oxidation pathway (C1) and the partial oxidation pathway (C2), respectively.

As can be seen from reactions 1–5, water, or its adsorption residue (adsorbed OH) is involved in EOR and therefore the product distribution strongly depends on the nature of the electrolyte, such as the concentration of ethanol, pH, or the anion [18,63,64]. Camara and Iwasita [55] systematically investigated the effects of ethanol concentration on the product distribution on polycrystalline Pt by FTIR. They found that the C1 pathway is more pronounced at low ethanol concentration (below 0.1 M) with negligible acetaldehyde. When the ethanol concentration was higher than 0.2 M, the formation of CO_2 and acetic acid was inhibited and acetaldehyde was the main product. Recently the fast development of anion exchange membranes has renewed the interest in the development of alkaline polymer electrolyte fuel cells. In alkaline media, a facile EOR can be achieved less costly with relatively abundant non-Pt metal catalysts. However, it is not clear whether or to what extent the mechanism proposed for EOR on Pt in acidic media can be extended to basic media. By combining electrochemical and spectroscopic techniques (SERS and *in situ* FTIR), Koper's group [18,54,65] showed that the activity of the reaction on Pt electrodes increases significantly when the pH of the electrolyte was higher than 10. Detailed mechanisms were proposed for EOR at low electrolyte pH (<6) and at high electrolyte pH (>11) as shown in Figure 2:

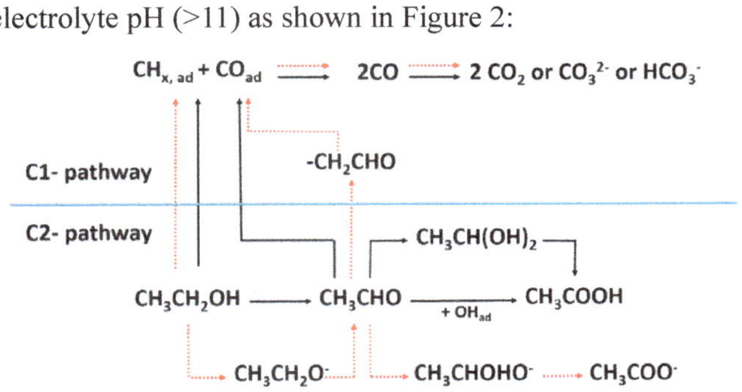

Figure 2. Proposed reaction mechanism for electro-oxidation of ethanol on Pt electrodes. Solid arrows denote the mechanism at low electrolyte pH, while dashed arrows denote the mechanism at high electrolyte pH. Adapted from Reference [18].

On the other hand, based on their *in situ* FTIR spectroscopic results in studying EOR on Pt in alkaline media, Christensen and his co-workers [20] proposed that in contrast to the acidic solutions, under alkaline conditions, the intermediates interact with the surface through O rather than C as shown in Figure 3. Pt_s–CH_2–C(=O)–O–Pt_s was speculated as a new intermediate, and the solution acetate species was the predominant product.

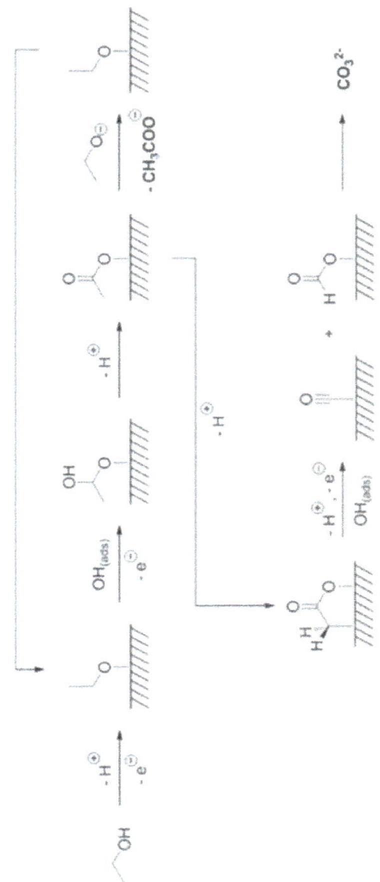

Figure 3. The mechanism of ethanol oxidation at polycrystalline Pt in alkaline solutions proposed by Christensen. Reproduced with permission from Reference [20] Copyright 2012, American Chemical Society.

Pd exhibits a much higher electrocatalytic activity in alkaline media compared to Pt due to its higher oxophilicity and relatively inert nature [17,47,66]. Ethanol oxidation on Pd is dramatically affected by the pH of the solution: virtually no reaction occurs in acidic solutions, while the reaction is fast in alkaline solutions. Through IRRAS studies [19], it was demonstrated that the oxidation of ethanol is incomplete on Pd electrodes and the main product is acetate. In addition, a quantitative FTIR study showed the selectivity for ethanol oxidation to CO_2 is less than 2.5% on Pd in the potential region of −0.60 to 0 V [67], but it is still slightly higher than that of Pt in alkaline media.

Though the external infrared reflection absorption spectroscopy with a thin-layer configuration enables the evaluation of the selectivity of the reaction products (*i.e.*, CO_2, acetate, and acetaldehyde) as a function of the applied potential, it is not sufficiently sensitive to probe low-coverage or weakly adsorbed intermediates and therefore may not provide complete information for understanding the reaction mechanism [68]. In addition the limited mass transport to and from the thin layer can skew or even alter the product distribution [20]. In contrast, attenuated total reflection surface enhanced infrared absorption spectroscopy (ATR-SEIRAS) provides high surface sensitivity and unobstructed mass transport and thus is promising for the complete disclosure of the EOR mechanism [26,69]. Yang *et al.* [23] recently investigated the surface reaction of ethanol on Pd in alkaline media using ATR-SEIRAS and H–D isotope replacement on α-C to shed new light on the self-dissociation and oxidation processes. As illustrated in Figure 4, ethanol may undergo dehydrogenation at α-C to form adsorbed acetyl rather than acetaldehyde, followed by successive decomposition to form C1 species, including CO_{ad} and CH_x at open circuit potential or lower potentials. Moreover, ATR-FTIR was also adopted to clarify the rate limiting steps on Pd thin film electrodes. It was found that at higher potential the subsequent dissociation (C–C bond breaking) of the adsorbed acetyl species is the rate limiting step rather than the formation of adsorbed acetyl [70]. Thus, despite controversy over the details, the reaction pathways for EOR with Pd electrodes in alkaline media are more or less similar to that with Pt electrodes in basic and acidic media.

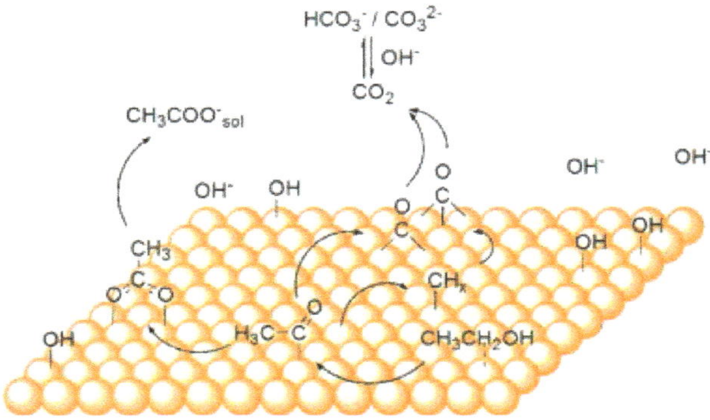

Figure 4. Reaction pathways for interfacial CH_3CH_2OH at Pd electrodes in alkaline media. Reproduced with permission from Reference [23]. Copyright 2014, American Chemical Society.

Despite the fact that external IRRAS has been frequently used in identifying reaction intermediates and products, the thin-layer configuration limits its capability for quantification. First, the thin layer structure is not reproducible from experiment to experiment. Second, product accumulation and diffusion out of the thin layer take place at the same time with species-dependent rates, *i.e.*, the diffusion rate varies with different species in the thin layer. CO_2 is the most volatile among the three major products and diffuses out faster than CH_3COOH and CH_3CHO, resulting in its lower estimation. Complimentary to IR spectroscopy, on-line DEMS can provide accurate quantitative information on ethanol oxidation products. In fact, mass spectrometry has often been used to study EOR in conjunction with IR spectroscopy. Studies using DEMS [28,29,30,31,32,33,34,35,36,37], electrochemical thermal desorption mass spectroscopy (ECTDMS) [71], and multipurpose electrochemical mass spectrometry studies (MPEMS) [29] have drawn the conclusion that all mono- or bi-aliphatic alcohols, except tertiary ones, yield minor amounts of CO_2 with the corresponding aldehydes or keto-compounds as major products during their electro-oxidation. Typically, on-line DEMS, especially under well-defined transport and diffusion conditions, has the ability to quantitatively determine kinetic parameters (reaction orders, activation energies, steady-state rates) for the overall EOR, and also for the partial reactions leading to the individual reaction products such as CO_2, acetic acid, and acetaldehyde [16,30,31,32,33,35,72].

By combining cyclic voltammetry and potential step measurements of the reaction transients with DEMS, Behm and co-workers performed a thorough investigation on the EOR products on a carbon-supported Pt nanoparticle catalyst at reaction temperature (23–60 °C) [30], and on PtRu and Pt_3Sn catalysts [33]. Absolute rates for CO_2 and acetaldehyde formation were determined via the doubly ionized carbon dioxide at $m/z = 22$ and the CHO^+ fragment at $m/z = 29$ from the calibrated mass spectrometric currents, whereas acetic acid yields were determined indirectly by calculating the difference between the measured Faradaic current and the partial currents of ethanol oxidation to CO_2 and acetaldehyde. More importantly, they convincingly showed that Pt-based catalysts exhibit selectivity towards CO_2 ranging from 0.5%–7.5%, which is far below the selectivity needed for economic implementation of the DEFC technology. They further explored the reaction in a wider temperature range (up to 100 °C) and found that the current efficiency for CO_2 formation increased significantly with the temperature while it decreased with increasing potential [37]. The latter observation suggests that the rate limiting step was changed from CO_{ad} oxidation at lower potential to C–C bond breaking at higher potentials. This transition is reasonable because the cleavage of the C–C bond becomes vital when the surface has the ability to easily remove poison species at higher potentials.

Theoretical Studies

Over the past decade the theoretical description of surface reactions has undergone a radical development [38,39]. Advances in density functional theory make it now possible to describe catalytic reactions on surfaces with the detail and accuracy required, so that computational results compare favorably with experiments. Simulations and theoretical studies have helped to advance our understanding of the EOR including predictions of vital intermediates, the preferable pathway, or the underlying electron transport process [40,41,42]. Some of these theoretical works corroborate with works that have been proposed by experimentalists, including the confirmation of some adsorbed intermediates in the step-wise mechanism, which are not detectable, probably due to the limited time resolution of the current experimental techniques. Therefore, theoretical studies may help to reconcile controversies and to understand variations in catalytic activity from one catalyst to another [73].

A higher percentage of C1 pathway is desirable for high efficiency DEFCs, but the production of CO_2 can be as low as 0.5% as shown by DEMS [31]. DFT was used to elucidate the reasons for the low efficiency of EOR on Pt [38] and the study found that a higher percentage of C1-pathway requires a careful control of oxidant surface coverage to allow facile C–C bond cleavage. Ethanol oxidation shows significant structure sensitivity in that the defect sites activate both the O–H and the C–C bonds [74]. The presence of OH or O species will considerably increase the energy barrier of the C–C bond cleavage as shown by DFT calculations, therefore the C1 pathway will be largely reduced compared to the corresponding clean surface. On a clean Pt surface with defects at low applied potentials, and thus low oxidant coverages, the formation of acetic acid and CO/CO_2 are energetically favorable and, interestingly, comparable. This finding is consistent with the experimental observations [18,20,23,66] that CO formation is indeed reasonably facile at low applied potentials on clean Pt surfaces. Increasing the applied potential, results in increasing oxidant surface coverage and leads to a large reduction in the rate of C–C bond cleavage. However, surface oxidants are required for conversion of CO to CO_2. These two competing processes explain the inability of pure platinum catalysts to act as efficient DEFC catalysts.

There has been some debate over whether C–C bond breaking is the rate limiting step in EOR. Very recently Anderson's group calculated the reversible potentials for the reaction intermediates of EOR on Pt (111) [45] using DFT. They found that surface potentials for the path to CO_2 were low and close to the calculated 0.004 V reversible potential for the 12 electron oxidation of ethanol. The main activation energy in the total oxidation of ethanol to CO_2 comes from the formation of OH_{ads} from H_2O with a reversible potential of 0.49 V, the highest potential as shown in Figure 5. The favorable path to CO_2 takes the right hand branch to $OCCH_3$ and then to $OCCH_{ads}$ [45]. OH_{ads} is essential for the oxidation of CO_{ads} and CH_{ads}, which leaves OH_{ads} formation to be the rate limiting step. Accordingly, an ideal catalyst would have the ability to adsorb most intermediates weakly but OH more strongly. This conclusion agrees with the observation that the maximum rate of adsorbed acetyl decomposition into CO_{ads} and $CH_{x,ads}$ appeared at 0.3–0.4 V *vs.* RHE [70], which is significantly lower than the onset potential of OH_{ads} formation (~0.5 V) [75].

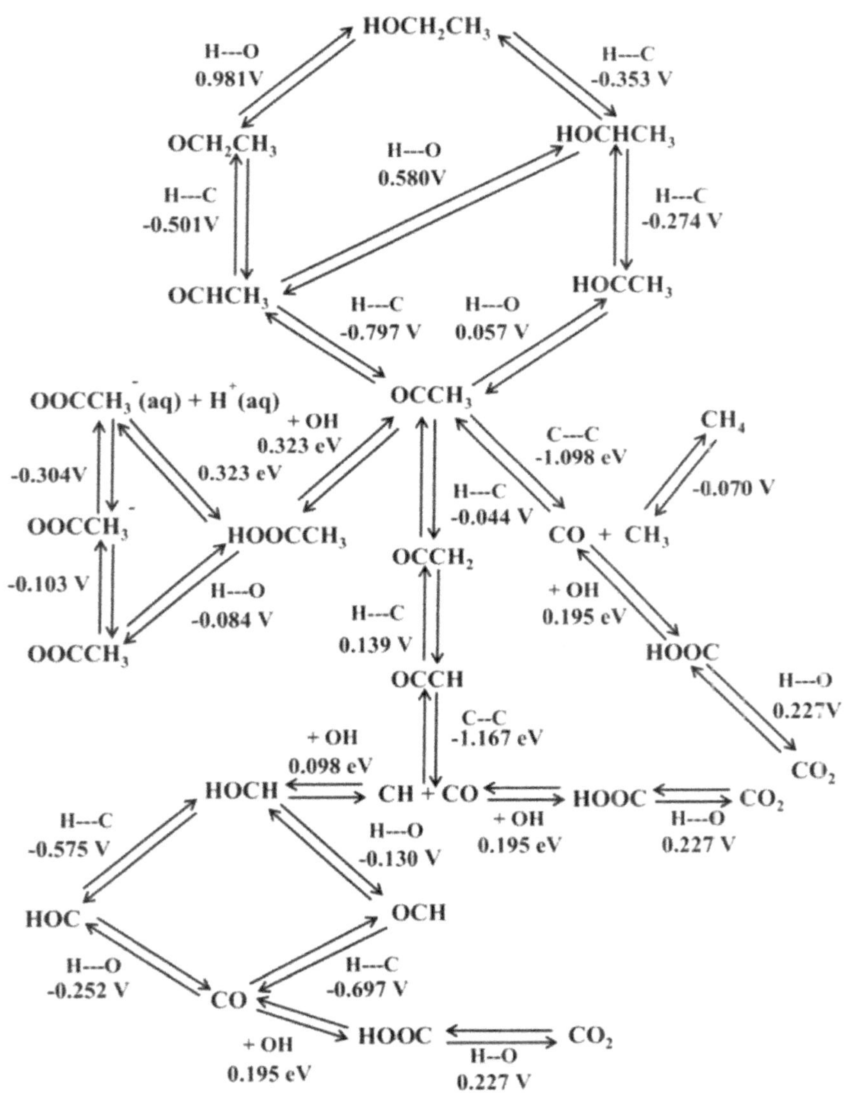

Figure 5. Reaction network calculated for ethanol electro-oxidation to CO_2, methane, and acetate. Reproduced with permission from Reference [45]. Copyright 2015, the Electrochemical Society.

The role of water and hydroxyls during EOR on Pd electrodes in alkaline media was further investigated by Lin's group [39] by acquiring first principle calculations. The possible pathways for the formation of acetate from acetaldehyde were evaluated by comparing the reaction barriers (E_a) as well as thermodynamic (ΔE) and structural parameters. Their results suggested that acetaldehyde is first hydrated in water to form germinal diol, and then the dehydrogenation of germinal diol produces acetate. According to this study, the OH^- anion acts as the center in the concerted-like dehydrogenation path as shown in Figure 6 confirming what has been found in experiments.

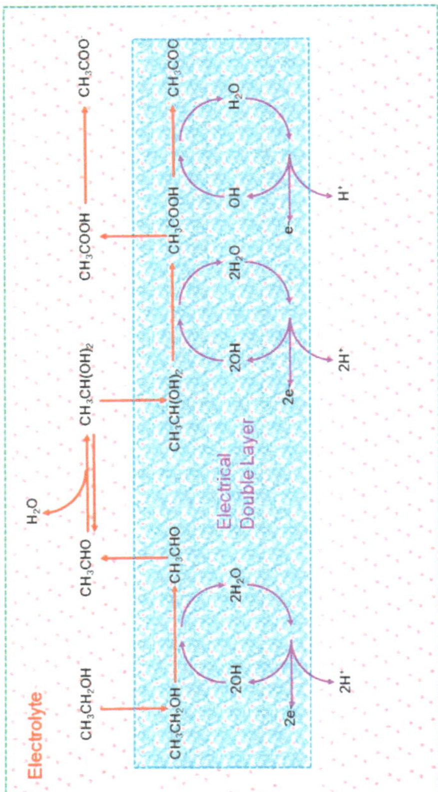

Figure 6. General scheme describing the ethanol oxidation reaction on Pd electrodes in the presence of an electrical double layer proposed by first principle calculations. Reproduced with permission from Reference [39]. Copyright 2014, American Chemical Society.

The total oxidation current at lower potentials was found to be rather structure sensitive where the presence of steps enhances the rupture of the C–C bond and the complete oxidation to CO_2 [74]. Theoretical studies have identified the platinum monoatomic steps as the most likely sites for full ethanol oxidation and concluded that the close-packed surfaces are unsuitable [76]. Liu's group [43,44] clarified the location of the transition state and saddle points for most surface reactions during EOR on different Pt surfaces based on gradient-corrected DFT as shown in Figure 7. Their results suggest the EOR is a structure-sensitive reaction that is influenced by two key reaction steps: (i) the initial dehydrogenation of ethanol and (ii) the oxidation of acetyl (CH_3CO). By simulating three typical Pt surfaces, namely close-packed Pt (111), monatomic stepped Pt (211), and open Pt (100), these authors demonstrated for the first time that the selectivity of ethanol oxidation on Pt is highly structure sensitive among which Pt (100) is the best surface to fully oxidize ethanol to CO_2 at low coverages. It shows that CO_2 and acetic acid originate from the same surface intermediate *i.e.*, CH_3CO as experimentally evidenced by our group [23], but acetaldehyde is from ethanol directly. The cleavage of the C–C bond occurs through the strongly chemisorbed precursor CH_2CO or CHCO only at low-coordinated surface sites, not from CH_3CO as proposed by FTIR study [13,14]. Acetaldehyde is produced via the one-step concerted dehydrogenation of ethanol, which occurs mainly on close-packed (111), and is enhanced by increased CH_x coverage. Acetic acid is the dominant oxidation product on Pt(111) at oxidative conditions, but its formation is significantly inhibited by the monoatomic steps.

In summary, EOR is a very complex reaction that can proceed via different pathways. The product distribution and the dominant reaction pathway depend on many factors including temperature, catalyst material and structure, applied potential, ethanol concentration, reaction media, *etc.* [20,30,31,32,37,45,46,47]. The interplay of these factors affects EOR greatly. Typically, the adsorption of hydroxyl is vital in both acidic and alkaline media [18,39,40,45,46,47]. In acidic media, at low potentials C–C bond cleavage occurs readily to form CO [30,43]. Owing to the unavailability of oxidants to remove CO_{ads}, the surface is poisoned and CO_2 production is limited. At higher potentials, there are abundant oxidants, but C–C bond cleavage is inhibited by the high coverage of oxidants, thus leading again to small CO/CO_2 production [45]. Similarly in alkaline media, the dissociative adsorption of ethanol proceeds rather quickly and the rate-determining step is the removal of the adsorbed species, which varies among literature, by the adsorbed hydroxyl. At higher potentials the kinetics is not only affected by the electro-adsorption of OH^- ions, but also by the formation of the inactive surface oxide layer [20,23,39]. Therefore EOR on Pt- or Pd-based catalysts proceeds predominantly with the C2 pathway and acetic acid or acetate is the main product. Carbon dioxide or carbonate is relatively low in the product distribution [3,30,37,57]. In general EOR proceeds through similar reaction pathways but some differences have been discussed. In acidic media, the initial bond breaking step is dehydrogenation at α-carbon, while in strong alkaline media it is the O–H bond cleavage [18,20,43,45].

Figure 7. Calculated reaction network and reaction barriers (units,

eV) for ethanol oxidation (**A**) on Pt (111) and acetyl oxidation (**B**) on Pt (211) (data on the left) and Pt (100) (data on the right). Reprinted and adapted with permission from Ref. [43]. Copyright 2008, American Chemical Society.

Despite the diverse reaction mechanisms for EOR presented in the literature, some characteristics of a highly efficient catalyst for complete oxidation of ethanol to CO_2 have emerged. The rationally designed catalysts would have: (i) suitable surface sites for C–C bond breaking; (ii) a suitable surface composition to increase selectivity for CO_2 formation; (iii) a bifunctional effect to facilitate the adsorption and activation of water to form OH_{ads} for the removal of CO and -CH_x species [45,76]. Utilizing these design guidelines, researchers have developed many new catalysts with higher selectivity and activity, and longer durability as demonstrated in the following studies.

CATALYTIC ROLE OF THE ELECTRODE MATERIALS

As discussed above, both experimental and theoretical studies suggest that the lower electro-efficiency C2 pathway is the dominant for EOR on Pt and Pd surfaces. Therefore a good tactic to obtain a better electrocatalytic performance is to increase the ratio of C1 to C2 pathway to achieve a more complete ethanol oxidation, which as shown above requires active surface sites for C–C bond breaking, CO and –CH_x species removal together with a suitable surface structure to increase selectivity for CO_2 formation [77]. Moreover, C1 pathway always involves the participation of water or its adsorption residue, a good electrocatalyst must be able to activate both ethanol and water adsorption, which can be achieved by varying the composition and structure of the rationally designed catalysts. In the following we first summarize some design principles and then use selected examples to elucidate the applications of these principles to obtain high performance catalysts.

Principles in Rational Design

Electronic effect: when Pt or Pd is alloyed or modified with another metal, the electronic interaction between Pt or Pd and the other metal results in the changes in their valence electronic structure through ligand effect and strain effect which can be described as the shift in the d-band center (ε_d) as proposed by Norskov and co-workers [50,73,78,79] and reviewed by Demirci [10]. The d-band center directly relates to the binding energy of surface poison or reactive intermediates. A higher-lying ε_d suggests a more reactive surface that tends to bind adsorbates more strongly while a surface with a lower-lying ε_d tends to bind adsorbates more weakly and facilitates the formation of bonds among them. In the EOR mechanism mentioned above, a suitable ε_d with a moderate binding energy of CO, CH_x, and acetyl or acetaldehyde, but higher energy for OH binding is needed by adjusting the electronic effect.

A bifunctional effect originates from Pt-Ru alloys and is extended to Pt/Sn and Pt/SnO_x [1]. In this mechanism, the presence of Ru, Sn, or SnO_x aids in the activation of water dissociation to form surface hydroxides, which can more readily oxidize CO and CH_x intermediates and therefore exhibit relatively higher EOR performance [80]. These alloys also tend to promote the partial oxidation of acetaldehyde to acetic acid. However, catalysts with a bifunctional effect do not particularly enhance C–C bond cleavage during EOR [81].

Surface-structure effect is another important parameter that significantly changes the activity of catalysts. EOR is a surface sensitive reaction and its efficiency largely depends on the crystal orientation of the catalyst surface [48]. In alkaline media Pt(111) electrodes display the highest current and lowest onset oxidation potential, however with little CO_2 production. On the other hand, Pt(100) electrodes are considered to be more active in the breaking of C–C bond and the formation of CO [43,48] no matter whether acidic or alkaline media. Construction of Pt or Pd based catalysts with well-defined morphology and a tunable surface is therefore another way to gain higher electrocatalytic performance catalysts.

Pt and Pd Based Electrocatalysts

Pt Based Catalysts

Previous studies have shown a monometallic catalyst such as Pt exhibits a selectivity of oxidation of ethanol to CO_2 of 0.5%–7.5% in acidic media [30], which falls short for the commercialization of DEFCs. Pure Pt can easily be poisoned by intermediates (including CO) generated during EOR. Regardless of the media, Pt-based materials represent the benchmark catalysts for ethanol oxidation. Elements such as Ru [32,33,35,60,82,83], Sn [15,32,33,35,80,84,85,86], Pb [21,87,88,89,90,91], Bi [36,89,90,92,93], Re [86], Sb [93], Ir [94], Au [95], Ce [96], Rh [81,97], Pd [81,98,99], Fe [100], Ni [95,101], P [102], Mo [83,103] have been widely employed to enhance the activity and selectivity. In addition, metal oxides, such as MgO, CeO_2, ZrO_2, SnO_x [76,94,104] RuO_2, PbO_x [88] have also been investigated for facilitating the removal of CO. The enhancement has been attributed to the steric hindrance of the surface, or electronic and/or bifunctional effects when the adatoms are directly involved in the catalytic process.

Up to now, PtRu [32,33,35,60,82,83] and PtSn [15,32,33,35,80,84,85,86] have been regarded as some of the most efficient catalysts for EOR according to the bifunctional effects. The distribution of products is well-studied by FTIR [35] and DEMS [33,35,55]. The addition of Sn or Ru, though beneficial for the overall activity of EOR, and the partial oxidation of acetaldehyde to acetic acid, does not enhance the activity for C–C breaking [35]. The higher current is mainly contributed from the higher yields of C2 products. These alloys aid in the oxidation of CO but actually lower the total conversion to CO_2 because they slow down the breaking of the C–C bond. Because Pt is the active metal for C–C bond activation, alloying decreases the amount of Pt and its ability to activate the C–C bond. To further promote selectivity and activity of EOR, Adzic's group developed Pt monolayer (Pt_{ML}) electrocatalysts comprising a one atom thick layer of Pt placed on selected extended or nanoparticle surfaces. They observed a correlation between substrate-induced lateral strains in the Pt monolayer and its activity/selectivity towards EOR. In agreement with previous theories [10,73], a positive- or tensile-surface strain in the metal overlayer tends to upshift ε_d and therefore facilitates OH_{ads} formation resulting in an enhanced EOR as shown in Figure 8. The IRRAS spectra showed that acetic acid is the predominant product [105]. This work demonstrates nicely the importance of electronic effect in tuning electrocatalytic activity.

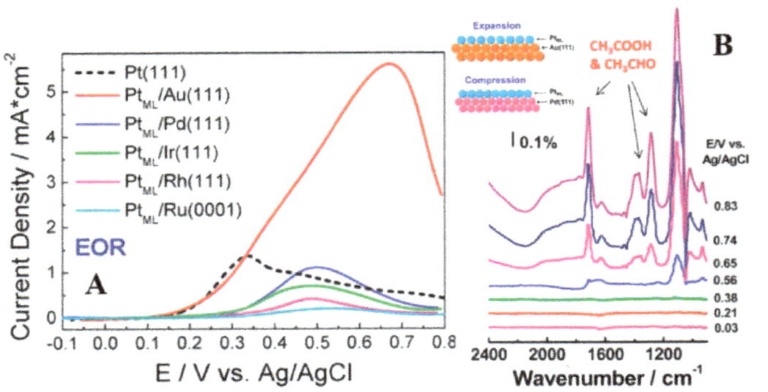

Figure 8. (**A**) Positive voltammetric scans for Pt(111) and Pt$_{ML}$ supported on five different substrates in 0.1 M $HClO_4$ containing 0.5 M ethanol; (**B**) *In situ* infrared reflection-absorption spectroscopy (IRRAS) spectra recorded during EOR on the Pt$_{ML}$/Au(111) electrode in 0.1 M $HClO_4$ containing 0.5 M ethanol. Inserted are models of pseudomorphic monolayers of Pt on two different substrates of Au(111) and Pd(111). Reprinted and adapted with permission from Reference [105]. Copyright 2012, American Chemical Society.

Along this line, various Pt-Au alloys have been proposed and have shown relatively high EOR performance [41,106]. Recently Pt-Au hetero-nanostructures were synthesized by varying the reduction kinetics of a gold precursor to obtain dimer Pt-Au or core-satellite (Pt@Au) structures. These catalysts show high selectivity and improved efficiency in alkaline media compared to their monometallic counterparts [107].

Ternary nanoalloys incorporating suitable metal and metalloid components are expected to exhibit more flexibility in tuning the geometric and electronic properties of Pt surfaces, thus are promising to achieve a higher electrocatalytic performance. Adzic's group found the ternary-electrocatalysts, Pt-Rh-SnO_2/C [9,76] can effectively split the C–C bond in ethanol at room temperature in acidic solutions and the highest activity was obtained with a composition of Pt:Rh:Sn = 3:1:4 [9]. As shown in Figure 9, the integrated band intensities of CO_2 (2343 cm^{-1}), CH_3CHO (933 cm^{-1}), and CH_3COOH (1280 cm^{-1}) for both Pt–Rh–SnO_2/C and Pt–SnO_2/C samples proved the enhanced cleavage of the C–C bond in ethanol and all three constituents Pt, Rh, and SnO_2 are needed to gain the synergistic effect in facilitating the total oxidation of ethanol.

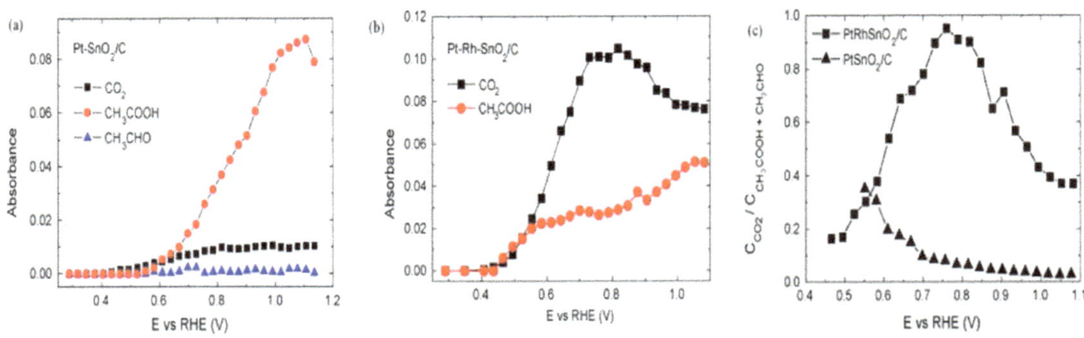

Figure 9. Integrated band intensities of CO_2, CH_3CHO and CH_3COOH in IRRAS spectra from (**a**). Pt–SnO_2/C with the atomic ratio Pt:Sn = 3:4; (**b**). Pt–Rh–SnO_2/C with the atomic ratio Pt:Rh:Sn = 3:1:4; (**c**) The charge ratio of the total oxidation pathway (C_{CO2}) over the partial oxidation pathway (C_{CH3}COOH + C_{CH3}CHO) as a function of electrode potential for both electrocatalysts in 0.1 M $HClO_4$ and 0.1 M ethanol. Reprinted and adapted with permission from Reference [9]. Copyright 2010, Elsevier.

Electrode arrays of 91 combinations of Pt–Sn–M (M = Fe, Ni, Pd, and Ru) were prepared and screened by a fluorescence assay (Figure 10) to optimize the catalysts with the highest electrocatalytic activity by Abruña and co-workers [101]. They found that Fe-containing catalysts exhibited the highest activity followed by Ni- and Pd-containing materials with similar results. This work shows that the variation and combination of different components can exert better electro-activity performance with different electronic effects.

EOR strongly depends on the electronic and surface structures of the catalysts. The different EOR activities observed on Pt surfaces with different crystallographic orientations offer the possibility of optimizing activities of nanoscale practical catalysts by controlling the particle shape [48,49,74]. By using cubic Pt nanoparticles, on which (100) surface sites are predominant, the performance of DEFCs can be increased from 14–24 mW per mg of Pt when compared with cuboctahedral nanoparticles. Moreover, the open circuit potential shifts about 50 mV toward more positive potentials [108]. Pt nanoparticles with 24 high-index facets such as (730), (210), and/or (520) surfaces were synthesized and showed enhancement on EOR compared with commercial Pt/C as well as a higher selectivity for the cleavage of the C–C bond [109]. Bimetallic or ternary nanocrystals have also been studied to explore the optimal combination of higher activity surface facets and electronic effect. The composition-varied (100)-terminated Pt–Pd–Rh nanocubes (NCs) and (111)-terminated Pt–Pd–Rh nanotruncated-oxtahedrons (NTOs) were synthesized with the help of halides [81]. Owing to the synergistic effects resulted from appropriate surface composition and exposed facets, the Pt–Pd–Rh NTOs exhibited the highest selectivity to CO_2 and PtPdRh NCs possessed the best durability. With the help of *in situ* FTIR and DFT calculations, the influence of the exposed facet and surface composition on the capability of C–C bond cleavage was examined. It was found that (100)-bounded surface is favorable to the cleavage of C–C bond while (111)-bounded surface tends to oxidize CO more easily.

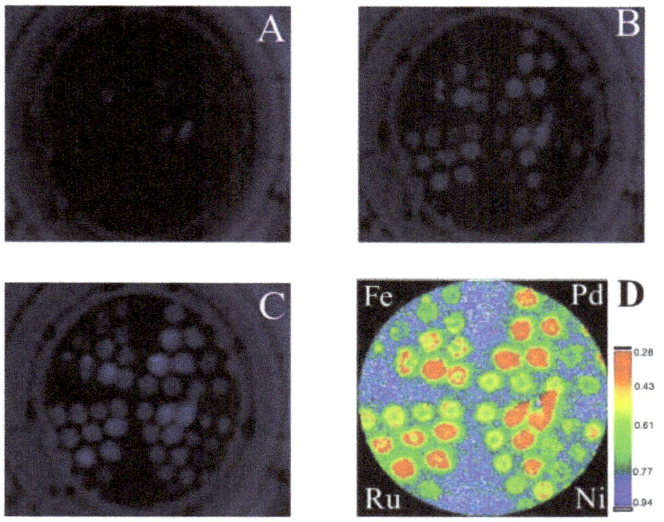

Figure 10. Combinatorial array and screening results by fluorescence imaging of PtSnM electrode arrays in 6.0 M ethanol and quinine fluorescent indicator. Active compositions for ethanol electro-oxidation are shown as bright spots: fluorescence image at (**A**) lower overpotential (~0.27V *vs.* RHE); (**B**) intermediate overpotential (~0.46 V *vs.* RHE) and (**C**) high overpotential (~0.93 V *vs.* RHE); (**D**) Fluorescence onset potential for ethanol electro-oxidation on PtM and PtSnM library. Reproduced with permission from Reference [101]. Copyright 2015, Elsevier.

Pd Based Catalysts

There has been a surge of interest in developing Pd-based catalysts mainly because facile EOR kinetics are expected in alkaline media on the less costly and more abundant Pd [17,47,66]. However, Pd itself cannot meet the practical demand because of shortage of activity and durability. In particular, a great deal of interest has been focused on the use of Pd-based alloy catalysts as an alternative for EOR where the catalytic activity may be further increased by the addition of a second metal or metal oxide promoters through the effects mentioned above. Various Pd-based catalysts have been synthesized with the addition of one or more elements including Ni [110,111], Ag [112,113], Au [114,115,116], P [117], Co [118], Sn [80,110,119,120], Ru [53], Zn [121] as well as metal oxides like SnO_2, CeO_x, Co_3O_4, Mn_3O_4, NiO [102,122,123,124,125] or on various substrates like graphene [126], carbon microspheres [127], nanowire [128], carbon fiber [118], *etc.*

Sn can promote the catalytic activity of Pd toward EOR by providing oxygen-containing species at more negative potentials due to the bifunctional effect and electronic effects, and has therefore been widely studied [80,119,120]. Teng's group synthesized carbon supported Pd–Sn electrocatalysts with different amounts of Sn, and determined that the optimum Sn content in Pd–Sn for EOR was 14%. The promotional effect of Sn on EOR activity was confirmed by kinetic study and DFT calculations [80]. The reaction energies of the initial steps of EOR (H removal) were studied with several catalysts using DFT (a Pd surface, a Pd–Sn_1, and a Pd–Sn_5 surface) in this work. As shown in Figure 11, Pd–Sn alloy structures resulted in lower reaction energies for the dehydrogenation of ethanol compared to pure Pd. Despite that the DFT calculations in reference [80] only involve the initial steps of EOR, their experimental results support the premise that Pd-Sn may be a better catalyst than Pd for EOR.

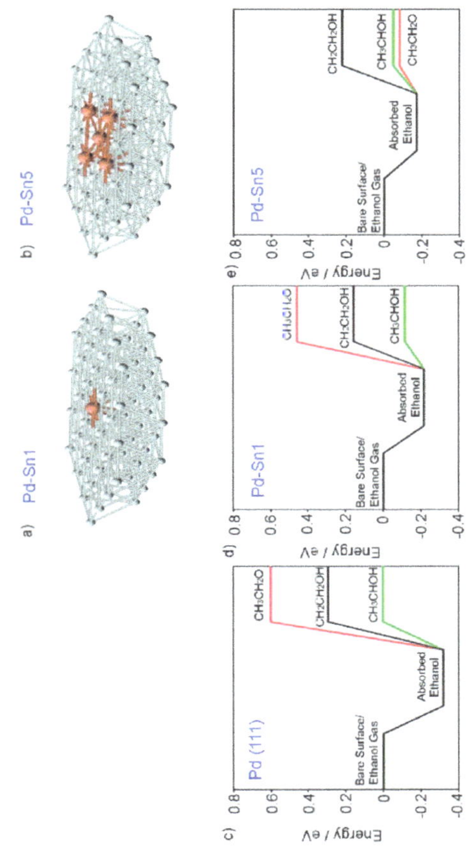

Figure 11. Surface cluster models of (**a**) Pd–Sn_1 and (**b**) Pd–Sn_5 used for modeling ethanol dehydrogenation, and reaction energies for H removal from ethanol over (**c**) Pd (111); (**d**) Pd–Sn_1, (**e**) Pd–Sn_5 surfaces. Reproduced with permission from Reference [80]. Copyright 2012, American Chemical Society.

Au is another promising candidate for promoting EOR on Pd catalysts. Zhao's group [114] synthesized a monolayer or a sub-monolayer of Pd adatoms decorated on Au/C with different Pd:Au atomic ratios in the precursors via a chemical epitaxial growth method and found Pd_1Au_4 shows the highest specific activity due to the electronic effect between the Au support and the Pd decoration, and the enhanced poison resistance. To further tune the electronic and geometric effect of the catalysts, ternary catalysts PdNiAu [116] were synthesized which possessed a peak power density about three times that of the monometallic Pd catalyst, and twice that of the bimetallic PdNi catalyst. A relatively preferable C2 pathway on the Pd–Au–Ni catalyst compared to its single and binary counterparts in alkaline media was estimated by ion chromatography.

Doping of various oxides to Pd catalysts is another strategy to enhance the efficiency of EOR. Xu *et al.* [124] demonstrated that the addition of oxides such as CeO_2, NiO, Co_3O_4, and Mn_3O_4 significantly promoted catalytic activity and stability. Most importantly these oxides have the ability to lower the onset potential for EOR. A recent study showed a maximum energy efficiency of about 7% from room temperature air breathing DEFCs with Pd on a mixed CeO_2–C support as the anode catalysts [125]. The use of a mixed carbon-CeO_2 support extends the stability of the Pd catalyst under working conditions by promoting ethanol electro-oxidation at lower anode potentials.

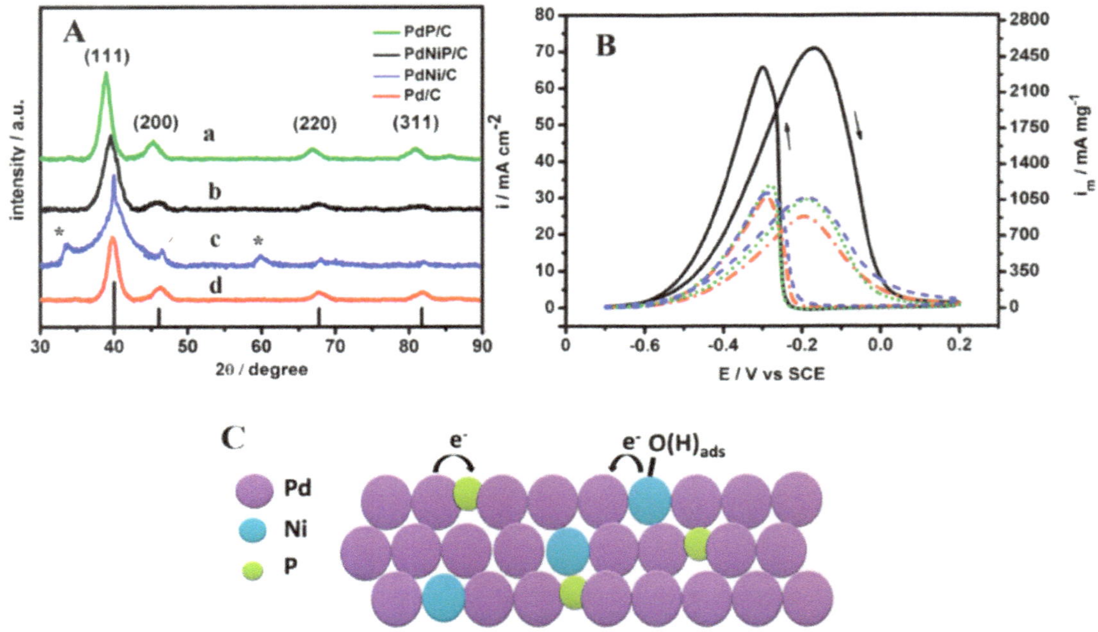

Figure 12. (**A**) XRD patterns of the Pd–P/C (curve a), Pd–Ni–P/C (curve b), Pd–Ni/C (curve c) and Pd/C (curve d). (**B**) Cyclic voltammograms for Pd-based catalysts in 0.5 M NaOH and 1 M C_2H_5OH at 50 mV·s^{-1}; (**C**) Scheme of Pd–Ni–P atomic arrangement. The asterisks at 33.5° and 59.2° mark the peaks from $Ni(OH)_2$ (100) and (110) facets. Reprinted and adapted with permission from Reference [117]. Copyright 2013, Elsevier.

Nonmetal component can also play a role in the rational design of catalysts [129,130,131,132]. Recently, our group discovered that Pd–Ni–P/C ternary nanocatalysts showed a remarkable enhancement towards EOR compared to Pd–Ni/C, Pd–P/C and Pd/C [117]. After a careful analysis of the structure and oxidation state of the catalysts, we found apparently opposite effects of alloying the elements P and Ni. P expands the Pd lattice with the incorporation of P atoms in the interstice of Pd lattice, while alloying with Ni partially replaced Pd sites with smaller Ni atoms, resulting in a contraction of the Pd lattice. In addition, P accepts electrons from the surrounding Pd atoms while Ni donates, resulting in a slight downshift of the Pd d-band center. Alloying Pd with Ni and P creates appropriate electronic and geometric modifications to Pd, leading to a modestly weakened adsorption of intermediates on Pd sites. In addition, the oxophilic nature of Ni provides OH_{ads} at lower potentials which facilitates the removal of surface poisons. (Figure 12).

Furthermore, by de-alloying some of the Ni and P in an electrodeposited Pd–Ni–P film, we investigated the de-alloying effect on EOR performance in alkaline media. We found that the enhancement of electrocatalytic activity does not simply originate from the increase of active surface area, but is due to the variation of relative contributions of the two pathways as evidenced by *in situ* infrared spectroscopic results shown in Figure 13 [24]. CO and acetate band intensities (Figure 13B,D) can be used to approximate the relative contributions of the C1 and C2 pathways. Notably, the more pronounced CO_{ad} and acetate bands observed on the de-alloyed Pd–Ni–P film provide molecular spectral evidence supporting the assumption that suitable de-alloying can enhance both C–C cleavage in the C1 pathway with CO formation and production of acetate in the C2 pathway, correlating well with the observed higher EOR current on the de-alloyed film. Furthermore, the relative intensities of the CO bands *versus* the acetate bands appeared to be much higher on the de-alloyed film as compared to those on the as-deposited film, indicating that the C1 pathway is relatively more favorable after the de-alloying treatment.

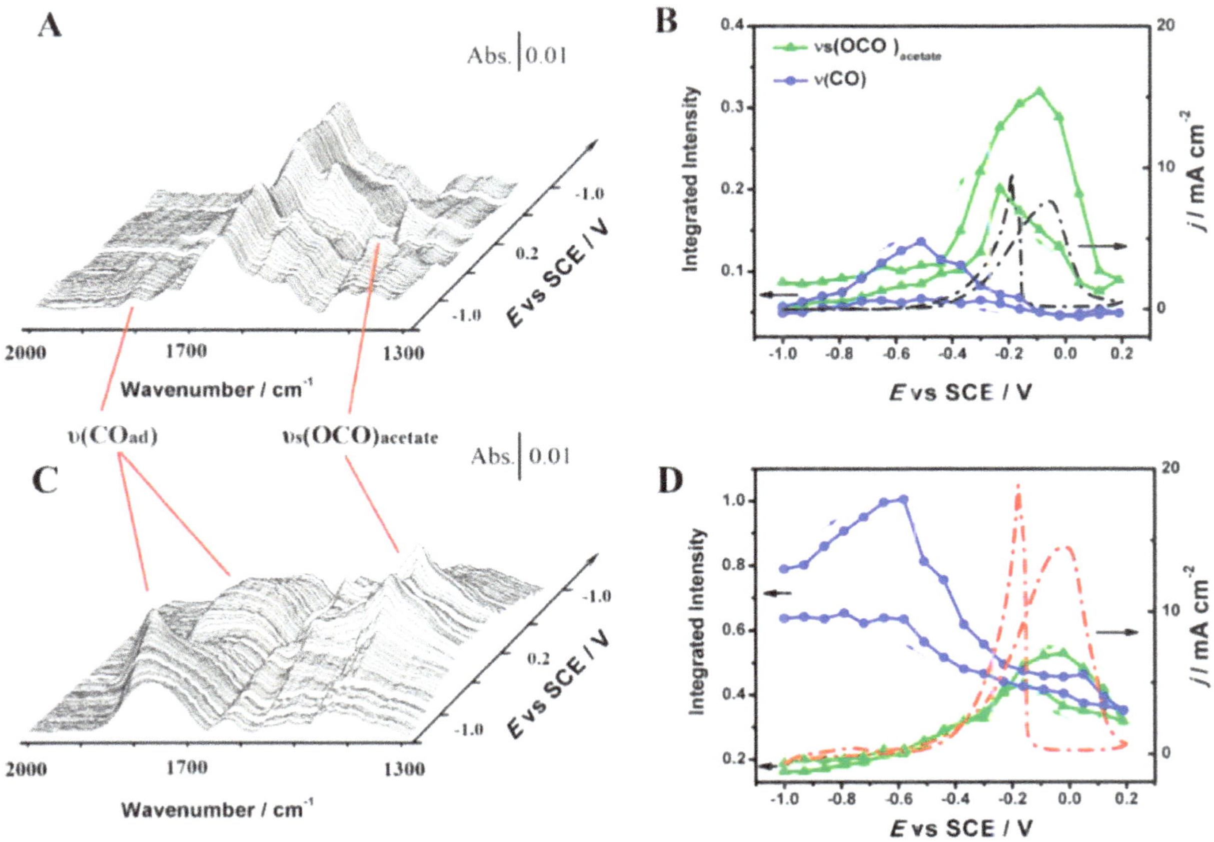

Figure 13. Potentiodynamic ATR-SEIRAS spectra on the as-deposited Pd–Ni–P film (**A**) and the de-alloyed film (**C**) in 0.1 M NaOH + 0.5 M ethanol; Potential-dependent band intensities for $v(CO_{ad})$ (blue) and $v_{s(OCO)}$ of adsorbed acetate (green) with corresponding CVs recorded at 5 mV·s^{-1} on the as-deposited Pd–Ni–P film (**B**) and the de-alloyed film (**D**) in 0.1 M NaOH + 0.5 M ethanol. Reproduced with permission from Reference [24]. Copyright 2014, Elsevier.

CONCLUSIONS AND OUTLOOK

From the survey of experimental and theoretical studies, some basic principles for rational design of high performance EOR catalysts can be obtained. The catalysts should have (i) active surface sites for C–C bond breaking; (ii) a suitable surface composition to increase selectivity for CO_2 formation; (iii) the ability to facilitate the adsorption and activation of water for the removal of CO and $-CH_x$ species. Utilizing these design guidelines, researchers have developed many new catalysts with higher selectivity, activity, and longer durability. These advances have significantly propelled the development of DEFCs.

Despite this progress, several fundamental and practical issues of catalysts remain to be addressed. From a fundamental point of view, the EOR mechanism is far from solidified. Some key aspects need to be clarified. These include finding the key factors in determining whether EOR goes through the C1 pathway or the C2 pathway; identifying intermediates during the C–C cleavage step in the C1 pathway; and understanding the nature of the intermediates, especially their adsorption mode on Pt and Pd surfaces. Until these issues are satisfactorily addressed, the rational design of high performance catalysts for DEFCs will remain in its infancy. From a practical point of view, the reported methods for synthesizing high performance catalysts are not suitable for large scale commercial production. We believe efforts in these directions are essential for the further development and deployment of commercially viable DEFCs.

Acknowledgments

This work is supported by the 973 Program (No. 2015CB932303) of MOST and NSFC (Nos. 21473039 and 21273046) to W.-B.C and the US-NSF (CHE 1156425) to SZ.

Author Contributions

Wen-Bin Cai conceived the project. Ye Wang wrote the first draft of the manuscript which was then revised by Shouzhong Zou and Wen-Bin Cai. All the authors researched the literature.

Conflicts of Interest

authors declare no conflict of interest.

REFERENCES

1. Antolini, E.; Gonzalez, E.R. Alkaline direct alcohol fuel cells. *J. Power Sources* **2010**, *195*, 3431–3450.
2. Kamarudin, M.Z.F.; Kamarudin, S.K.; Masdar, M.S.; Daud, W.R.W. Review: Direct ethanol fuel cells. *Int. J. Hydrogen Energy* **2013**, *38*, 9438–9453.
3. Teng, X. Anodic Catalyst Design for the Ethanol Oxidation Fuel Cell. Available online: http://www.formatex.info/energymaterialsbook/book/473-484.pdf (accessed on 17 August 2015).
4. Yao, L.X.; Chang, Y.H. Shaping china's energy security: The impact of domestic reforms. *Energy Policy* **2015**, *77*,

131–139.

5. Xuan, J.; Leung, M.K.; Leung, D.Y.; Ni, M. A review of biomass-derived fuel processors for fuel cell systems. *Renew. Sustain. Energy Rev.* **2009**, *13*, 1301–1313.

6. Wee, J.-H. Applications of proton exchange membrane fuel cell systems. *Renew. Sustain. Energy Rev.* **2007**, *11*, 1720–1738.

7. Antolini, E. Catalysts for direct ethanol fuel cells. *J. Power Sources* **2007**, *170*, 1–12.

8. Rao, L.; Jiang, Y.; Zhang, B.; You, L.; Li, Z.; Sun, S. Electrocatalytic oxidation of ethanol. *Prog. Chem.* **2014**, *26*, 727–736.

9. Li, M.; Kowal, A.; Sasaki, K.; Marinkovic, N.; Su, D.; Korach, E.; Liu, P.; Adzic, R.R. Ethanol oxidation on the ternary Pt–Rh–SnO_2/C electrocatalysts with varied Pt:Rh:Sn ratios. *Electrochim. Acta* **2010**, *55*, 4331–4338.

10. Demirci, U.B. Theoretical means for searching bimetallic alloys as anode electrocatalysts for direct liquid-feed fuel cells. *J. Power Sources* **2007**, *173*, 11–18.

11. Bianchini, C.; Shen, P.K. Palladium-based electrocatalysts for alcohol oxidation in half cells and in direct alcohol fuel cells. *Chem. Rev.* **2009**, *109*, 4183–4206.

12. Antolini, E. Palladium in fuel cell catalysis. *Energy Environ. Sci.* **2009**, *2*, 915–931.

13. Chang, S.C.; Leung, L.W.H.; Weaver, M.J. Metal crystallinity effects in electrocatalysis as probed by real-time FTIR spectroscopy: Electrooxidation of formic acid, methanol, and ethanol on ordered low-index platinum surfaces. *J. Phys. Chem.* **1990**, *94*, 6013–6021.

14. Hitmi, H.; Belgsir, E.; Léger, J.-M.; Lamy, C.; Lezna, R. A kinetic analysis of the electro-oxidation of ethanol at a platinum electrode in acid medium. *Electrochim. Acta* **1994**, *39*, 407–415.

15. Vigier, F.; Coutanceau, C.; Hahn, F.; Belgsir, E.; Lamy, C. On the mechanism of ethanol electro-oxidation on Pt and PtSn catalysts: Electrochemical and *in situ* IR reflectance spectroscopy studies. *J. Electroanal. Chem.* **2004**, *563*, 81–89.

16. Raskó, J.; Dömök, M.; Baán, K.; Erdőhelyi, A. FTIR and mass spectrometric study of the interaction of ethanol and ethanol-water with oxide-supported platinum catalysts. *Appl. Catal. A* **2006**, *299*, 202–211.

17. Fang, X.; Wang, L.; Shen, P.K.; Cui, G.; Bianchini, C. An *in situ* fourier transform infrared spectroelectrochemical study on ethanol electrooxidation on Pd in alkaline solution. *J. Power Sources* **2010**, *195*, 1375–1378.

18. Lai, S.C.S.; Kleijn, S.E.F.; Ozturk, F.T.Z.; Vellinga, V.C.V.; Koning, J.; Rodriguez, P.; Koper, M.T.M. Effects of electrolyte pH and composition on the ethanol electro-oxidation reaction. *Catal. Today* **2010**, *154*, 92–104.

19. Zhou, Z.Y.; Wang, Q.A.; Lin, J.L.; Tian, N.; Sun, S.G. *In situ* FTIR spectroscopic studies of electrooxidation of ethanol on Pd electrode in alkaline media. *Electrochim. Acta* **2010**, *55*, 7995–7999.

20. Christensen, P.A.; Jones, S.W.M.; Hamnett, A. *In situ* FTIR studies of ethanol oxidation at polycrystalline Pt in alkaline solution. *J. Phys. Chem. C* **2012**, *116*, 26109–26109.

21. Christensen, P.A.; Jones, S.W.; Hamnett, A. An *in situ* FTIR spectroscopic study of the electrochemical oxidation of ethanol at a Pb-modified polycrystalline Pt electrode immersed in aqueous KOH. *Phys. Chem. Chem. Phys.* **2013**, *15*, 17268–17276.

22. Anjos, D.M.; Hahn, F.; Leger, J.M.; Kokoh, K.B.; Tremiliosi, G. *In situ* FTIRS studies of the electrocatalytic oxidation of ethanol on Pt alloy electrodes. *J. Solid State Electrochem.* **2007**, *11*, 1567–1573.

23. Yang, Y.-Y.; Ren, J.; Li, Q.-X.; Zhou, Z.-Y.; Sun, S.-G.; Cai, W.-B. Electrocatalysis of ethanol on a Pd electrode in alkaline media: An *in situ* attenuated total reflection surface-enhanced infrared absorption spectroscopy study. *ACS Catal.* **2014**, *4*, 798–803.

24. Wang, Y.; Jiang, K.; Cai, W.-B. Enhanced electrocatalysis of ethanol on dealloyed Pd–Ni–P film in alkaline media: An infrared spectroelectrochemical investigation. *Electrochim. Acta* **2015**, *162*, 100–107.

25. Leung, L.W.H.; Chang, S.C.; Weaver, M.J. Real-time FTIR spectroscopy as an electrochemical mechanistic probe—Electrooxidation of ethanol and related species on well-defined Pt(111) surfaces. *J. Electroanal. Chem.* **1989**, *266*, 317–336.

26. Shao, M.H.; Adzic, R.R. Electrooxidation of ethanol on a Pt electrode in acid solutions: *In situ* ATR-SEIRAS study. *Electrochim. Acta* **2005**, *50*, 2415–2422.

27. Willsau, J.; Heitbaum, J. Elementary steps of ethanol oxidation on Pt in sulfuric acid as evidenced by isotope labelling. *J. Electroanal. Chem. Interface* **1985**, *194*, 27–35.

28. Iwasita, T.; Pastor, E. A DEMS and FTIR spectroscopic investigation of adsorbed ethanol on polycrystalline platinum. *Electrochim. Acta* **1994**, *39*, 531–537.

29. Wang, J.; Wasmus, S.; Savinell, R. Evaluation of ethanol, 1-propanol, and 2-propanol in a direct oxidation polymer-electrolyte fuel cell a real-time mass spectrometry study. *J. Electrochem. Soc.* **1995**, *142*, 4218–4224.

30. Wang, H.; Jusys, Z.; Behm, R.J. Ethanol electrooxidation on a carbon-supported Pt catalyst: Reaction kinetics and product yields. *J. Phys. Chem. B* **2004**, *108*, 19413–19424.

31. Wang, H.; Jusys, Z.; Behm, R.J. Ethanol and acetaldehyde adsorption on a carbon-supported Pt catalyst: A comparative DEMS study. *Fuel Cells* **2004**, *4*, 113–125.

32. Colmenares, L.; Wang, H.; Jusys, Z.; Jiang, L.; Yan, S.; Sun, G.Q.; Behm, R.J. Ethanol oxidation on novel, carbon supported Pt alloy catalysts-model studies under defined diffusion conditions. *Electrochim. Acta* **2006**, *52*, 221–233.

33. Wang, H.; Jusys, Z.; Behm, R. Ethanol electro-oxidation on carbon-supported Pt, PtRu and Pt_3Sn catalysts: A quantitative DEMS study. *J. Power Sources* **2006**, *154*, 351–359.

34. Rao, V.; Cremers, C.; Stimming, U. Investigation of the ethanol electro-oxidation in alkaline membrane electrode assembly by differential electrochemical mass spectrometry. *Fuel Cells* **2007**, *7*, 417–423.

35. Wang, Q.; Sun, G.Q.; Jiang, L.H.; Xin, Q.; Sun, S.G.; Jiang, Y.X.; Chen, S.P.; Jusys, Z.; Behm, R.J. Adsorption and oxidation of ethanol on colloid-based Pt/C, PtRu/C and Pt_3Sn/C catalysts: *In situ* FTIR spectroscopy and on-line DEMS studies. *Phys. Chem. Chem. Phys.* **2007**, *9*, 2686–2696.

36. Figueiredo, M.C.; Aran-Ais, R.M.; Feliu, J.M.; Kontturi, K.; Kallio, T. Pt catalysts modified with Bi: Enhancement of the catalytic activity for alcohol oxidation in alkaline media. *J. Catal.* **2014**, *312*, 78–86.

37. Sun, S.; Halseid, M.C.; Heinen, M.; Jusys, Z.; Behm, R.J. Ethanol electrooxidation on a carbon-supported Pt catalyst at elevated temperature and pressure: A high-temperature/high-pressure DEMS study. *J. Power Sources* **2009**, *190*, 2–13.

38. Kavanagh, R.; Cao, X.M.; Lin, W.F.; Hardacre, C.; Hu, P. Origin of low CO_2 selectivity on platinum in the direct ethanol fuel cell. *Angew. Chem. Int. Ed. Engl.* **2012**, *51*, 1572–1575.

39. Sheng, T.; Lin, W.F.; Hardacre, C.; Hu, P. Role of water and adsorbed hydroxyls on ethanol electrochemistry on Pd: New mechanism, active centers, and energetics for direct ethanol fuel cell running in alkaline medium. *J. Phys. Chem. C* **2014**, *118*, 5762–5772.

40. Sheng, T.; Lin, W.-F.; Hardacre, C.; Hu, P. Significance of β-dehydrogenation in ethanol electro-oxidation on platinum doped with Ru, Rh, Pd, Os and Ir. *Phys. Chem. Chem. Phys.* **2014**, *16*, 13248–13254.

41. Neurock, M. First-Principles Modeling for the Electro-Oxidation of Small Molecules. In *Handbook of Fuel Cells*; Vielstich, W.G., Lamm, A., Yokokawa, H., Eds.; John Wiley & Sons, Ltd.: Hoboken, NJ, USA, 2010.

42. Cui, G.F.; Song, S.Q.; Shen, P.K.; Kowal, A.; Bianchini, C. First-principles considerations on catalytic activity of Pd toward ethanol oxidation. *J. Phys. Chem. C* **2009**, *113*, 15639–15642.

43. Wang, H.F.; Liu, Z.P. Comprehensive mechanism and structure-sensitivity of ethanol oxidation on platinum: New transition-state searching method for resolving the complex reaction network. *J. Am. Chem. Soc.* **2008**, *130*, 10996–11004.

44. Wang, H.F.; Liu, Z.P. Selectivity of direct ethanol fuel cell dictated by a unique partial oxidation channel. *J. Phys. Chem. C* **2007**, *111*, 12157–12160.

45. Asiri, H.A.; Anderson, A.B. Mechanisms for ethanol electrooxidation on Pt(111) and adsorption bond strengths defining an ideal catalyst. *J. Electrochem. Soc.* **2015**, *162*, F115–F122.

46. Zhiani, M.; Majidi, S.; Rostami, H.; Taghiabadi, M.M. Comparative study of aliphatic alcohols electrooxidation on zero-valent palladium complex for direct alcohol fuel cells. *Int. J. Hydrogen Energy* **2015**, *40*, 568–576.

47. Liang, Z.X.; Zhao, T.S.; Xu, J.B.; Zhu, L.D. Mechanism study of the ethanol oxidation reaction on palladium in alkaline media. *Electrochim. Acta* **2009**, *54*, 2203–2208.

48. Buso-Rogero, C.; Herrero, E.; Feliu, J.M. Ethanol oxidation on Pt single-crystal electrodes: Surface-structure effects in alkaline medium. *Chemphyschem* **2014**, *15*, 2019–2028.

49. Zhou, W.J.; Li, M.; Zhang, L.; Chan, S.H. Supported PtAu catalysts with different nano-structures for ethanol electrooxidation. *Electrochim. Acta* **2014**, *123*, 233–239.

50. Hammer, B.; Nørskov, J.K. Theoretical Surface Science and Catalysis-Calculations and Concepts. *Adv. Catal.* **2000**, *45*, 71–129.

51. Srinivasan, S.; Dave, B.B.; Murugesamoorthi, K.A.; Parthasarathy, A.; Appleby, A.J. Overview of Fuel Cell Technology. In *Fuel cell Systems*; Plenum Press: New York, NY, USA, 1993.

52. Lamy, C.; Belgsir, E.; Leger, J. Electrocatalytic oxidation of aliphatic alcohols: Application to the direct alcohol fuel cell (DAFC). *J. Appl. Electrochem.* **2001**, *31*, 799–809.

53. Zhou, W.; Zhou, Z.; Song, S.; Li, W.; Sun, G.; Tsiakaras, P.; Xin, Q. Pt based anode catalysts for direct ethanol fuel cells. *Appl. Catal. B* **2003**, *46*, 273–285.

54. Lai, S.C.S.; Koper, M.T.M. Ethanol electro-oxidation on platinum in alkaline media. *Phys. Chem. Chem. Phys.* **2009**, *11*, 10446–10456.

55. Camara, G.A.; Iwasita, T. Parallel pathways of ethanol oxidation: The effect of ethanol concentration. *J. Electroanal. Chem.* **2005**, *578*, 315–321.

56. Belgsir, E.M.; Bouhier, E.; Yei, H.E.; Kokoh, K.B.; Beden, B.; Huser, H.; Leger, J.M.; Lamy, C. Electrosynthesis in aqueous medium: A kinetic study of the electrocatalytic oxidation of oxygenated organic molecules. *Electrochim. Acta* **1991**, *36*, 1157–1164.

57. Tarnowski, D.J.; Korzeniewski, C. Effects of surface step density on the electrochemical oxidation of ethanol to acetic acid. *J. Phys. Chem. B* **1997**, *101*, 253–258.

58. Kutz, R.B.; Braunschweig, B.; Mukherjee, P.; Dlott, D.D.; Wieckowski, A. Study of ethanol electrooxidation in alkaline electrolytes with isotope labels and sum-frequency generation. *J. Phys. Chem. Lett.* **2011**, *2*, 2236–2240.

59. Ke, X.; Deng, L.L.; Shen, P.K.; Cu, G.F. Electrochemical quartz crystal microbalance (EQCM) characterization of electrodeposition and catalytic activity of Pd-based electrocatalysts for ethanol oxidation. *Chem. Res. Chin. Univ.* **2010**, *26*, 443–448.

60. Huang, L.; Sorte, E.; Sun, S.-G.; Tong, Y.Y.J. A straightforward implementation of *in situ* solution electrochemical ^{13}C NMR spectroscopy for studying reactions on commercial electrocatalysts: Ethanol oxidation. *Chem. Commun.* **2015**, *51*, 8086–8088.

61. Leung, L.W.H.; Weaver, M.J. Real-time FTIR spectroscopy as a quantitative kinetic probe of competing electrooxidation pathways of small organic molecules. *J. Phys. Chem.* **1988**, *92*, 4019–4022.

62. Gao, P.; Chang, S.C.; Zhou, Z.H.; Weaver, M.J. Electrooxidation pathways of simple alcohols at platinum in pure nonaqueous and concentrated aqueous environments as studied by real-time FTIR spectroscopy. *J. Electroanal. Chem.* **1989**, *272*, 161–178.

63. Buso-Rogero, C.; Grozovski, V.; Vidal-Iglesias, F.J.; Solla-Gullon, J.; Herrero, E.; Feliu, J.M. Surface

structure and anion effects in the oxidation of ethanol on platinum nanoparticles. *J. Mater. Chem. A* **2013**, *1*, 7068–7076.

64. Paulino, M.E.; Nunes, L.M.; Gonzalez, E.R.; Tremiliosi-Filho, G. *In situ* FTIR spectroscopic study of ethanol oxidation on Pt (111)/Rh/Sn surface: The anion effect. *Electrochem. Commun.* **2015**, *52*, 85–88.
65. Lai, S.C.; Koper, M.T. Electro-oxidation of ethanol and acetaldehyde on platinum single-crystal electrodes. *Faraday Discuss.* **2009**, *140*, 399–416.
66. Ma, L.; Chu, D.; Chen, R. Comparison of ethanol electro-oxidation on Pt/C and Pd/C catalysts in alkaline media. *Int. J. Hydrogen Energy* **2012**, *37*, 11185–11194.
67. Zhou, Z.-Y.; Chen, D.-J.; Li, H.; Wang, Q.; Sun, S.-G. Electrooxidation of dimethoxymethane on a platinum electrode in acidic solutions studied by *in situ* FTIR spectroscopy. *J. Phys. Chem. C* **2008**, *112*, 19012–19017.
68. Osawa, M. Surface-Enhanced Infrared Absorption Spectroscopy. In *Handbook of Vibrational Spectroscopy*; Osawa, M.C., Griffiths, P.R., Eds.; Wiley: New York, NY, USA; Chichester, UK, 2002; Volume 1, pp. 785–799.
69. Yang, Y.Y.; Zhang, H.X.; Cai, W.B. Recent experimental progresses on electrochemical ATR-SEIRAS. *J. Electrochem.* **2013**, *19*, 6–16.
70. Heinen, M.; Jusys, Z.; Behm, R.J. Ethanol, acetaldehyde and acetic acid adsorption/electrooxidation on a Pt thin film electrode under continuous electrolyte flow: An *in situ* ATR-FTIRs flow cell study. *J. Phys. Chem. C* **2010**, *114*, 9850–9864.
71. Bittins-Cattaneo, B.; Wilhelm, S.; Cattaneo, E.; Buschmann, H.W.; Vielstich, W. Intermediates and products of ethanol oxidation on platinum in acid solution. *Ber. Bunsenges. Phys. Chem.* **1988**, *92*, 1210–1218.
72. Lai, S.C.; Koper, M.T. The influence of surface structure on selectivity in the ethanol electro-oxidation reaction on platinum. *J. Phys. Chem. Lett.* **2010**, *1*, 1122–1125.
73. Norskov, J.K.; Bligaard, T.; Rossmeisl, J.; Christensen, C.H. Towards the computational design of solid catalysts. *Nat. Chem.* **2009**, *1*, 37–46.
74. Da Silva, S.G.; Silva, J.C.M.; Buzzo, G.S.; de Souza, R.F.B.; Spinace, E.V.; Neto, A.O.; Assumpcao, M.H.M.T. Electrochemical and fuel cell evaluation of PtAu/C electrocatalysts for ethanol electro-oxidation in alkaline media. *Int. J. Hydrogen Energy* **2014**, *39*, 10121–10127.
75. Climent, V.; Gómez, R.; Orts, J.M.; Feliu, J.M. Thermodynamic analysis of the temperature dependence of oh adsorption on Pt(111) and Pt(100) electrodes in acidic media in the absence of specific anion adsorption. *J. Phys. Chem. B* **2006**, *110*, 11344–11351.
76. Kowal, A.; Li, M.; Shao, M.; Sasaki, K.; Vukmirovic, M.; Zhang, J.; Marinkovic, N.; Liu, P.; Frenkel, A.; Adzic, R. Ternary Pt/Rh/SnO_2 electrocatalysts for oxidizing ethanol to CO_2. *Nat. Mater.* **2009**, *8*, 325–330.
77. Assumpção, M.; Nandenha, J.; Buzzo, G.; Silva, J.; Spinacé, E.; Neto, A.; de Souza, R. The effect of ethanol concentration on the direct ethanol fuel cell performance and products distribution: A study using a single fuel cell/attenuated total reflectance-fourier transform infrared spectroscopy. *J. Power Sources* **2014**, *253*, 392–396.
78. Greeley, J.; Norskov, J.K.; Mavrikakis, M. Electronic structure and catalysis on metal surfaces. *Annu. Rev. Phys. Chem.* **2002**, *53*, 319–348.
79. Rabis, A.; Rodriguez, P.; Schmidt, T.J. Electrocatalysis for polymer electrolyte fuel cells: Recent achievements and future challenges. *ACS Catal.* **2012**, *2*, 864–890.
80. Du, W.X.; Mackenzie, K.E.; Milano, D.F.; Deskins, N.A.; Su, D.; Teng, X.W. Palladium-Tin alloyed catalysts for the ethanol oxidation reaction in an alkaline medium. *ACS Catal.* **2012**, *2*, 287–297.
81. Zhu, W.; Ke, J.; Wang, S.-B.; Ren, J.; Wang, H.-H.; Zhou, Z.-Y.; Si, R.; Zhang, Y.-W.; Yan, C.-H. Shaping single-crystalline trimetallic Pt–Pd–Rh nanocrystals toward high-efficiency C–C splitting of ethanol in conversion to CO_2. *ACS Catal.* **2015**, *5*, 1995–2008.
82. Camara, G.A.; de Lima, R.B.; Iwasita, T. Catalysis of ethanol electrooxidation by PtRu: The influence of catalyst composition. *Electrochem. Commun.* **2004**, *6*, 812–815.
83. Neto, A.O.; Giz, M.J.; Perez, J.; Ticianelli, E.A.; Gonzalez, E.R. The electro-oxidation of ethanol on Pt–Ru and Pt–Mo particles supported on high-surface-area carbon. *J. Electrochem. Soc.* **2002**, *149*, A272–A279.
84. Lamy, C.; Rousseau, S.; Belgsir, E.M.; Coutanceau, C.; Leger, J.M. Recent progress in the direct ethanol fuel cell: Development of new platinum-tin electrocatalysts. *Electrochim. Acta* **2004**, *49*, 3901–3908.
85. Zhou, W.J.; Song, S.Q.; Li, W.Z.; Zhou, Z.H.; Sun, G.Q.; Xin, Q.; Douvartzides, S.; Tsiakaras, P. Direct ethanol fuel cells based on PtSn anodes: The effect of Sn content on the fuel cell performance. *J. Power Sources* **2005**, *140*, 50–58.
86. Tayal, J.; Rawat, B.; Basu, S. Effect of addition of rhenium to Pt-based anode catalysts in electro-oxidation of ethanol in direct ethanol PEM fuel cell. *Int. J. Hydrogen Energy* **2012**, *37*, 4597–4605.
87. He, Q.G.; Shyam, B.; Macounova, K.; Krtil, P.; Ramaker, D.; Mukerjee, S. Dramatically enhanced cleavage of the C–C bond using an electrocatalytically coupled reaction. *J. Am. Chem. Soc.* **2012**, *134*, 8655–8661.
88. Suffredini, H.B.; Salazar-Banda, G.R.; Avaca, L.A. Enhanced ethanol oxidation on PbO_x-containing electrode materials for fuel cell applications. *J. Power Sources* **2007**, *171*, 355–362.
89. Huang, Y.; Cai, J.; Liu, M.; Guo, Y. Fabrication of a novel PtPbBi/C catalyst for ethanol electro-oxidation in alkaline medium. *Electrochim. Acta* **2012**, *83*, 1–6.
90. Matsumoto, F. Ethanol and methanol oxidation activity of PtPb, PtBi, and $PtBi_2$ intermetallic compounds in alkaline media. *Electrochemistry* **2012**, *80*, 132–138.
91. Gunji, T.; Tanabe, T.; Jeevagan, A.J.; Usui, S.; Tsuda, T.; Kaneko, S.; Saravanan, G.; Abe, H.; Matsumoto, F. Facile route for the preparation of ordered intermetallic Pt_3Pb–PtPb core-shell nanoparticles and its enhanced activity for alkaline methanol and ethanol oxidation. *J. Power Sources* **2015**, *273*, 990–998.

92. Tusi, M.M.; Polanco, N.S.; da Silva, S.G.; Spinacé, E.V.; Neto, A.O. The high activity of PtBi/C electrocatalysts for ethanol electro-oxidation in alkaline medium. *Electrochem. Commun.* **2011**, *13*, 143–146.

93. Figueiredo, M.C.; Santasalo-Aarnio, A.; Vidal-Iglesias, F.J.; Solla-Gullon, J.; Feliu, J.M.; Kontturi, K.; Kallio, T. Tailoring properties of platinum supported catalysts by irreversible adsorbed adatoms toward ethanol oxidation for direct ethanol fuel cells. *Appl Catal. B* **2013**, *140*, 378–385.

94. Li, M.; Cullen, D.A.; Sasaki, K.; Marinkovic, N.S.; More, K.; Adzic, R.R. Ternary electrocatalysts for oxidizing ethanol to carbon dioxide: Making Ir capable of splitting C–C bond. *J. Am. Chem. Soc.* **2013**, *135*, 132–141.

95. Dutta, A.; Ouyang, J.Y. Ternary niaupt nanoparticles on reduced graphene oxide as catalysts toward the electrochemical oxidation reaction of ethanol. *ACS Catal.* **2015**, *5*, 1371–1380.

96. Jacob, J.M.; Corradini, P.G.; Antolini, E.; Santos, N.A.; Perez, J. Electro-oxidation of ethanol on ternary Pt–Sn–Ce/C catalysts. *Appl. Catal. B* **2015**, *165*, 176–184.

97. Zhao, Y.H.; Wang, R.Y.; Han, Z.X.; Li, C.Y.; Wang, Y.S.; Chi, B.; Li, J.Q.; Wang, X.J. Electrooxidation of methanol and ethanol in acidic medium using a platinum electrode modified with lanthanum-doped tantalum oxide film.*Electrochim. Acta* **2015**, *151*, 544–551.

98. Datta, J.; Dutta, A.; Mukherjee, S. The beneficial role of the cometals Pd and Au in the carbon-supported PtPdAu catalyst toward promoting ethanol oxidation kinetics in alkaline fuel cells: Temperature effect and reaction mechanism. *J. Phys. Chem. C* **2011**, *115*, 15324–15334.

99. Yang, X.; Yang, Q.; Xu, J.; Lee, C.-S. Bimetallic PtPd nanoparticles on nafion-graphene film as catalyst for ethanol electro-oxidation. *J. Mater. Chem.* **2012**, *22*, 8057–8062.

100. Liao, F.L.; Lo, T.W.B.; Sexton, D.; Qu, J.; Wu, C.T.; Tsang, S.C.E. PdFe nanoparticles as selective catalysts for C–C cleavage in hydrogenolysis of vicinal diol units in biomass-derived chemicals. *Catal. Sci. Technol.* **2015**, *5*, 887–896.

101. Almeida, T.S.; van Wassen, A.R.; van Dover, R.B.; de Andrade, A.R.; Abruña, H.D. Combinatorial PtSnM (M = Fe, Ni, Ru and Pd) nanoparticle catalyst library toward ethanol electrooxidation. *J. Power Sources* **2015**, *284*, 623–630.

102. Carmo, M.; Sekol, R.C.; Ding, S.Y.; Kumar, G.; Schroers, J.; Taylor, A.D. Bulk metallic glass nanowire architecture for electrochemical applications. *ACS Nano* **2011**, *5*, 2979–2983.

103. Li, L.; Yuan, X.X.; Xia, X.Y.; Du, J.; Ma, Z.; Ma, Z.F. Effects of Mo doping on properties of Pt/C as catalyst towards electro-oxidation of ethanol. *J. Inorg. Mater.* **2014**, *29*, 1044–1048.

104. Jiang, L.; Colmenares, L.; Jusys, Z.; Sun, G.Q.; Behm, R.J. Ethanol electrooxidation on novel carbon supported Pt/SnO_x/C catalysts with varied Pt:Sn ratio. *Electrochim. Acta* **2007**, *53*, 377–389.

105. Li, M.; Liu, P.; Adzic, R.R. Platinum monolayer electrocatalysts for anodic oxidation of alcohols. *J. Phys. Chem. Lett.***2012**, *3*, 3480–3485.

106. Cheng, F.L.; Dai, X.C.; Wang, H.; Jiang, S.P.; Zhang, M.; Xu, C.W. Synergistic effect of Pd-Au bimetallic surfaces in Au-covered Pd nanowires studied for ethanol oxidation. *Electrochim. Acta* **2010**, *55*, 2295–2298.

107. Mourdikoudis, S.; Chirea, M.; Zanaga, D.; Altantzis, T.; Mitrakas, M.; Bals, S.; Liz-Marzán, L.M.; Perez-Juste, J.; Pastoriza-Santos, I. Governing the morphology of Pt–Au heteronanocrystals with improved electrocatalytic performance. *Nanoscale* **2015**, *7*, 8739–8747.

108. Figueiredo, M.C.; Solla-Gullón, J.; Vidal-Iglesias, F.J.; Nisula, M.; Feliu, J.M.; Kallio, T. Carbon-supported shape-controlled Pt nanoparticle electrocatalysts for direct alcohol fuel cells. *Electrochem. Commun.* **2015**, *55*, 47–50.

109. Tian, N.; Zhou, Z.-Y.; Sun, S.-G.; Ding, Y.; Wang, Z.L. Synthesis of tetrahexahedral platinum nanocrystals with high-index facets and high electro-oxidation activity. *Science* **2007**, *316*, 732–735.

110. Sheikh, A.; Silva, E.; Moares, L.; Antonini, L.; Abellah, M.Y.; Malfatti, C. Pd-based catalysts for ethanol oxidation in alkaline electrolyte. *Am. J. Min. Metal.* **2014**, *2*, 64–69.

111. Kumar, K.S.; Haridoss, P.; Seshadri, S.K. Synthesis and characterization of electrodeposited Ni–Pd alloy electrodes for methanol oxidation. *Surf. Coat. Technol.* **2008**, *202*, 1764–1770.

112. Hosseini-Sarvari, M.; Khanivar, A.; Moeini, F. Magnetically recoverable nano Pd/Fe_3O_4/ZnO catalyst: Preparation, characterization, and application for the synthesis of 2-oxazolines and benzoxazoles. *J. Mater. Sci.* **2015**, *50*, 3065–3074.

113. Li, G.; Jiang, L.; Jiang, Q.; Wang, S.; Sun, G. Preparation and characterization of Pd_xAg_y/C electrocatalysts for ethanol electrooxidation reaction in alkaline media. *Electrochim. Acta* **2011**, *56*, 7703–7711.

114. Zhu, L.D.; Zhao, T.S.; Xu, J.B.; Liang, Z.X. Preparation and characterization of carbon-supported sub-monolayer palladium decorated gold nanoparticles for the electro-oxidation of ethanol in alkaline media. *J. Power Sources* **2009**,*187*, 80–84.

115. Wang, H.; Xu, C.W.; Cheng, F.L.; Jiang, S.P. Pd nanowire arrays as electrocatalysts for ethanol electrooxidation. *Electrochem. Commun.* **2007** *9*, 1212–1216.

116. Dutta, A.; Datta, J. Outstanding catalyst performance of pdauni nanoparticles for the anodic reaction in an alkaline direct ethanol (with anion-exchange membrane) fuel cell. *J. Phys. Chem. C* **2012**, *116*, 25677–25688.

117. Wang, Y.; Shi, F.F.; Yang, Y.Y.; Cai, W.B. Carbon supported Pd–Ni–P nanoalloy as an efficient catalyst for ethanol electro-oxidation in alkaline media. *J. Power Sources* **2013**, *243*, 369–373.

118. Bahemmat, S.; Ghassemzadeh, M.; Afsharpour, M.; Harms, K. Synthesis, characterization and crystal structure of a Pd(ii) complex containing a new bis-1,2,4-triazole ligand: A new precursor for the preparation of Pd(0) nanoparticles.*Polyhedron* **2015**, *89*, 196–202.

119. Da Silva, S.G.; Assumpcao, M.H.M.T.; Silva, J.C.M.; de Souza, R.F.B.; Spinace, E.V.; Neto, A.O.; Buzzo, G.S. PdSn/C electrocatalysts with different atomic ratios for ethanol electro-oxidation in alkaline media. *Int. J. Electrochem. Sci.* **2014**, *9*, 5416–5424.

120. Mao, H.M.; Wang, L.L.; Zhu, P.P.; Xu, Q.J.; Li, Q.X. Carbon-supported PdSn SnO_2 catalyst for ethanol electro-oxidation in alkaline media. *Int. J. Hydrogen Energy* **2014**, *39*, 17583–17588.

121. Bambagioni, V.; Bianchini, C.; Filippi, J.; Oberhauser, W.; Marchionni, A.; Vizza, F.; Psaro, R.; Sordelli, L.; Foresti, M.L.; Innocenti, M. Ethanol oxidation on electrocatalysts obtained by spontaneous deposition of palladium onto nickel-zinc materials. *ChemSusChem* **2009**, *2*, 99–112.

122. Takeguchi, T.; Anzai, Y.; Kikuchi, R.; Eguchi, K.; Ueda, W. Preparation and characterization of CO-tolerant Pt and Pd anodes modified with SnO_2 nanoparticles for PEFC. *J. Electrochem. Soc.* **2007**, *154*, B1132–B1137.

123. Mao, H.; Huang, T.; Yu, A.-S. Facile synthesis of trimetallic $Cu_1Au_{0.15}Pd_{1.5}$/C catalyst for ethanol oxidation with superior activity and stability. *J. Mater. Chem. A* **2014**, *2*, 16378–16380.

124. Shen, P.K.; Xu, C. Alcohol oxidation on nanocrystalline oxide Pd/C promoted electrocatalysts. *Electrochem. Commun.* **2006**, *8*, 184–188.

125. Wang, L.; Lavacchi, A.; Bevilacqua, M.; Bellini, M.; Fornasiero, P.; Filippi, J.; Innocenti, M.; Marchionni, A.; Miller, H.A.; Vizza, F. Energy efficiency of alkaline direct ethanol fuel cells employing nanostructured palladium electrocatalysts. *Chemcatchem* **2015**, *7*, 2214–2221.

126. Wang, Y.; Wang, X.M.; Wang, Y.Z.; Li, J.P. Acid-treatment-assisted synthesis of Pt–Sn/graphene catalysts and their enhanced ethanol electro-catalytic activity. *Int. J. Hydrogen Energy* **2015**, *40*, 990–997.

127. Xu, C.; Cheng, L.; Shen, P.; Liu, Y. Methanol and ethanol electrooxidation on Pt and Pd supported on carbon microspheres in alkaline media. *Electrochem. Commun.* **2007**, *9*, 997–1001.

128. Wang, A.L.; He, X.J.; Lu, X.F.; Xu, H.; Tong, Y.X.; Li, G.R. Palladium-Cobalt nanotube arrays supported on carbon fiber cloth as high-performance flexible electrocatalysts for ethanol oxidation. *Angew. Chem. Int. Ed.* **2015**, *54*, 3669–3673.

129. Zhao, M.; Abe, K.; Yamaura, S.-I.; Yamamoto, Y.; Asao, N. Fabrication of Pd–Ni–P metallic glass nanoparticles and their application as highly durable catalysts in methanol electro-oxidation. *Chem. Mater.* **2014**, *26*, 1056–1061.

130. Jiang, R.Z.; Tran, D.T.; McClure, J.P.; Chu, D. A class of (Pd–Ni–P) electrocatalysts for the ethanol oxidation reaction in alkaline media. *ACS Catal.* **2014**, *4*, 2577–2586.

131. Shao, A.F.; Wang, Z.B.; Chu, Y.Y.; Jiang, Z.Z.; Yin, G.P.; Liu, Y. Evaluation of the performance of carbon supported Pt–Ru–Ni–P as anode catalyst for methanol electrooxidation. *Fuel Cells* **2010**, *10*, 472–477.

132. Lo, Y.L.; Hwang, B.J. Kinetics of ethanol oxidation on electroless Ni–P/SnO_2/Ti electrodes in KOH solutions. *J. Electrochem. Soc.* **1995**, *142*, 445–450.

CHAPTER 6

RECENT DEVELOPMENT OF PD-BASED ELECTROCATALYSTS FOR PROTON EXCHANGE MEMBRANE FUEL CELLS

Hui Meng, Dongrong Zeng and Fangyan Xie

Keywords: Fuel Cells, Catalyst, Palladium

INTRODUCTION

Fuel cells convert chemical energy directly into electrical current without combustion. The first article illustrating such a device was published at the end of the 1830s [1], and the interest in this field has been growing since the 1950s [2]. Among various types of the low-temperature fuel cells, proton exchange membrane fuel cells (PEMFCs) are attractive power sources for portable, automotive and stationary applications due to their high energy density, high efficiency and low operating temperature. Comparing with H_2 as fuel, the liquid fuels such as formic acid and ethanol have special advantage in storage and transport, which can find better applications in portable devices and make use of current gasoline system. The direct formic acid fuel cells (DFAFCs) have an open circuit potential of 1.190 V and energy density of 2086 Wh L^{-1}. The formic acid is non-toxic and has small crossover flux. The shortcoming of DFAFCs is their relatively low energy density. Compared with formic acid, ethanol has much higher energy density of 8030 Wh kg^{-1}. However, ethanol suffers from the sluggish reaction kinetics.

The electrochemical oxidation of fuels requires the use of a catalyst to achieve the high current densities for practical applications. Platinum (Pt) is the mostly used catalyst in the PEMFCs. However, the vast commercialization of fuel cell is hindered by the high cost and low reserve of Pt. The kinetics of the oxygen reduction reaction (ORR), which is the cathode reaction of a fuel cell is slow on Pt Moreover, Pt is easy to be poisoned without recovery by the intermediates of the reaction of the impurities from the fuel or oxidant. Particularly in some types of PEMFCs, Pt is not the best choice. For example, in direct formic acid fuel cells (DFAFCs), the oxidation of formic acid is quite low due to the poisoning of the Pt by the CO-like intermediates during the reaction. In direct ethanol fuel cells (DEFCs), the poor utilization and the poisoning of Pt catalyst particularly in alkaline solution also limite its applications. The electrocatalytic activities of the ethanol oxidation reaction could be significantly improved in alkaline media on Pd-based catalyst which have comparable or even better electrocatalytic activity than that of Pt-based catalyst. The ORR is one of the key reactions in fuel cells with the higher overpotential compared with the anode reactions. Pt group metal based catalysts are currently used for PEMFCs to reduce the large ORR overpotential. Unfortunately, even on the active Pt surface, the overpotential is over 200 mV at open circuit voltage (OCV). Pd is another active metal for the ORR. Binary Pd-base metal systems have been identified as promising PEMFC cathode catalyst with the enhanced activity for ORR and stability compared with Pd alone [3].

Pd has an electronic configuration identical to Pt and forms a not very strong bond to most absorbates. The key differences are that the *d* bands of Pd are closer to the cores than that of Pt. There are less *d* electron densities available for bonding. This leads to weaker interactions with *d* bonds, which allows unique chemistry to occur. Pd has higher oxidation potential than Pt and the Pd oxides are more stable. Weak inter-atomic bonds between Pd atoms compared with Pt lead to easier formation of the subsurface species. Also, Pd has a very similar lattice

Citation: Hui Meng , Dongrong Zeng and Fangyan Xie, Recent Development of Pd-Based Electrocatalysts for Proton Exchange Membrane Fuel Cells, doi:10.3390/catal5031221.

constant to that of Pt. The electrocatalysis of formic acid on a Pd single crystal surface could be significantly enhanced as the *d*-band centre of Pd shifted down with an appropriate value due to the modest lattice compressive strain. All these electronic properties make Pd a promising alternative to Pt or even better than Pt in many situations.

From the recent 5-year price change of the Pd, its price has changed from one forth to two fifth of the Pt [4]. However, Pd still has some advantages considering the reservation and price. As one of the most studied materials, Pd has attracted considerable interest for its applications in many fields. Similar to Pt, most of the Pd is used in the automotive industry for catalytic converters to reduce the toxicity of emissions from a combustion engine. Pd also has vast applications in electronic, dental and jewelry. Only a small percentage of Pd is used in chemistry. In electrochemistry, Pd nanoparticles are very important catalyst, especially for the oxidation of formic acid, ehanol oxidation in alkaline solution, hydrogen oxidation and the ORR. Pd-based catalyst for alcohol oxidation have been reviewed in 2009 [5]. Another review paper in 2009 reviewed the application of Pd in fuel cell anode and cathode [6]. However, the Pd-based fuel cells are still very hot in recent years. We have surveyed the 270 published papers on the Pd-based catalysts in recent four years from 2009 to 2014 and the distribution of the papers was as shown in Figure 1. Seven review papers were published after 2009. Two review papers published in 2010 partly concerned the application of Pd in fuel cells [7,8]. Morozan *et al.*, rediewed the application of Pd in fuel cells [9]. Recently Shao reviewed Pd as catalyst for hydrogen oxidation and oxygen reduction reaction [3]. Zhao *et al.*, reviewed the catalysts for direct methanol fuel cells including the application of Pd [10]. Adams and Chen's review paper focused on the application of Pd in hydrogen storage [11]. The nano-structure of Pd for catalysis and hydrogen storage was also reviewed recently by Zhou in *Chem. Soc. Rev.* [12]. Analyzing recent review papers on Pd-based catalysts, it was found that there was no comprehensive review paper concerning the application of Pd in fuel cells. This review analyzed the latest four years' publications on Pd catalyst for proton exchange membrane fuel cells and provides a comprehensive review on the recent development of the Pd-based catalysts for formic acid oxidation, alcohol oxidation and oxygen reduction reaction.

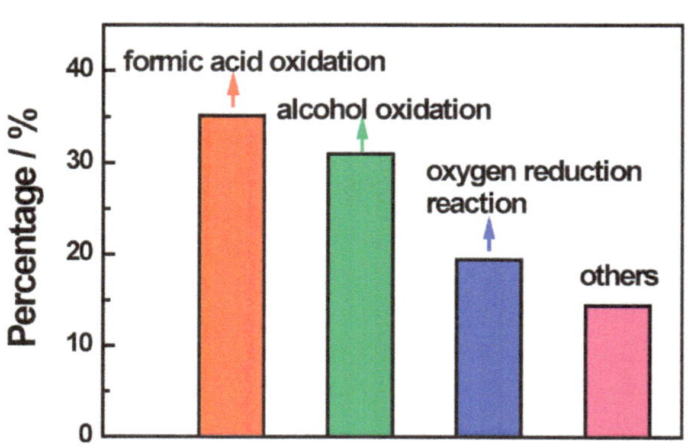

Figure 1. The distribution of Pd in recent four years in fuel cell technologies.

Based on our survey, in all the published papers on Pd-based catalysts, 35.12% is for the the formic acid oxidation, 30.99% for alcohol oxidation and 19.42% for ORR as shown in Figure 1. Therefore, this review article mainly focuses on the recent development of the Pd-based catalysts, particularly, on these three most important reactions. This survey result also shows the possible areas where Pd can compete with or replace Pt in PEMFCs. Besides above aspects, the control of the morphology and crystallography of Pd and the corresponding effect on catalysis are also reviewed.

PD NANOSTRUCTURES

The development of nanotechnology makes it possible to control the morphology and crystallography of Pd nanosturctures, which has been proven to affect the catalytic activity. Crystalline surfaces with a high density of low-coordinated atoms are generally superior in catalytic activity and stability to flat planes that are composed of closely packed surface atoms. This makes nanoparticle possible to show high activity and stability at very small Pd loadings if the nanoparticles are controllably synthesized as an open-structured surfaces with high density of low-coordinated atoms. The availability of the open-structured surfaces comes from the morphology of the material, so the shape control of the nanocrystals has become one of the crucial challenges.

The 0-D Pd Strucutres

Zero-dimensional (0-D) Pd structures include nano-particles, quantum dots, atomic cluster and nanoclusters and so on. The nano-particles are the most often used structure for fuel cell applications, Pd can be prepared into nano-particles which are usually called "Pd black" or Pd nano-particles loaded on support materials. Both Pd black and the supported Pd nano-particles are widely used as fuel cell catalysts. Other than the common ball-like particles, the particles with multifacets attracted much attention since their unique crystallographic and morphologic structures, which greatly improved the activity and stability when used as fuel cell catalysts. With a square-wave potential method Tian *et al.* [13] electrodeposited tetrahexahedral Pd nanocrystals with {730} high-index facets. With similar technique, Zhou *et al.* [14] prepared the Pd NCs not limited to the tetrahexahedral crystals. Shen *et al.* [15] used a sonoelectrochemical method to prepare Pd spherical nanoparticles, multitwinned particles, and spherical spongelike particles. Ding *et al.* [16] prepared single crystalline Pd nanocubes with the polyol method. Porous Pd nanoflowers were prepared by a liquid phase approach [17]. The above Pd particles all show improved mass activity or specific activity in fuel cell reactions, for example the tetrahexahedral Pd shows 4-6 times enhancement of specific activity and 1.5–3 times enhancement of mass activity in ethanol oxidation compared with commercial Pd particles on carbon [13].

The morphology of the crystal is determined by the internal features of a crystal, but the relationship between the crystalline structure and crystalline shape is still an unresolved problem. It still needs a correlation between crystallographic structure, morphological evolution and resultant shape of the nanocrystals. Zhou *et al.* [14] proposed the correlation between crystalline planes and nanocrystalline shape. There is a triangle relationship between the unit stereographic of fcc single-crystal and the surface atomic arrangement. There is also an intrinsic triangle that coordinates the crystalline surface index and the shape of the metal, which is shown in Figure 2.

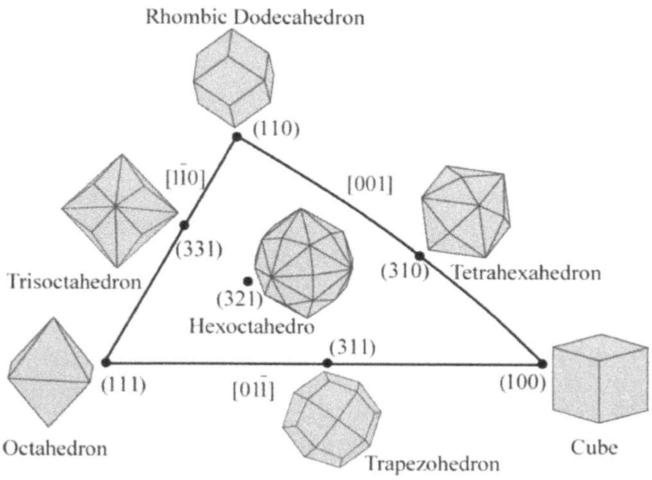

Figure 2. Unit stereographic triangle of polyhedral nanocrystals bounded by different crystal planes. Reproduced with permission from reference [14].

The 1-D Pd Structures

Compared with nanoparticess, one-dimensional (1-D) materials such as the nanowires, nanothorns and nanotubes offer unique benefits including (1) anisotropic morphology; (2) thin metal catalyst layer which leads to higher mass transport of the reactants; (3) high aspect ratio which is immune to surface energy driven coalescence via crystal migration; (4) less vulnerable to dissolution, Ostwald ripening and aggregation during the electrocatalytic process due to their micrometer-sized length; (5) high electrochemical active areas. Therefore, the 1-D structures are also intensively studied. Meng *et al.*[18,19] synthesized Pd single-crystal nanothorns along the <220> direction with a square wave electrochemical reduction method as shown in Figure 3. The nanothorn was made by a succession of epitaxic dodecahedrons of decreasing sizes aligned in the direction of the (111) plane and the growth of the thorn occurs along the (220) plane. Tian *et al.* [20] obtained five fold twinned Pd nanorods with high-index facets of {hkk} or {hk0}. Patra *et al.* [21] prepared Pd dendrite branches growing along the <110> directions. According to the mechanism proposed by different authors, the Pd thorns obtained from different methods might share similar mechanism in the crystal growth [21]. It has been proposed that the breaking of symmetry leads to the formation of one-dimensional nanostructure, which facilitates the formation of one specific facet as the bounding side facet in the nanostructure. The formation of a tapered structure implies that the specific facet has a high surface step density to accommodate the change in diameter.

Besides the preparation of nanothorns without templates or sufactants, other 1-D structures such as the nanorods, nanotube and nanowires can be prepared based on the templates such as anodic aluminum oxide (AAO) [22]. Du *et al.* [23] prepared porous Pd-based alloy nanowires with AAO. Lee *et al.* [24] prepared Pd nanotubes with ZnO nanowires as sacrificial templates. The nanotube structures have large surface areas and it is expected to show enhanced catalytic efficiencies. Wen *et al.* [25] used Te NWs as sacrificial template to prepare ultrathin Pd nanowires. The Pd nanowire or nanotube forms nanoporous metallic structure with high surface area, unique chemical properties and interconnected structures that do not require any support to avoid the corrosion and detachment problems common for carbon supported catalysts.

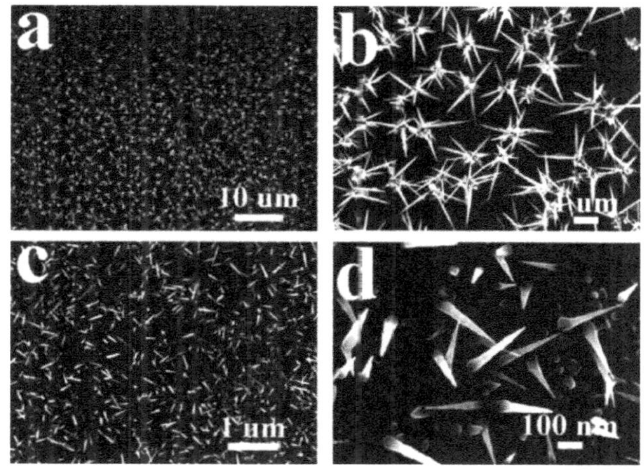

Figure 3. SEM micrographs of pure Pd thorn clusters (**a**,**b**) and the mixture of Pd thorns and Pd particles (**c**,**d**). Reproduced with permission from reference [19].

The 3-D Pd Structures

Compared with two-dimensional (2-D) structure, the three-dimensional (3-D) structure often has high porosity which will lead to higher surface area, especially higher electrochemical active surface area. Usually, the 2-D structure is very few used in the fuel cell application. Jena *et al.* [26] explored the synthesis of 3-D porous Pd nanostructures with a various shapes and morphologies. These structures have high surface roughness and surface steps which can contribute to the increased accessibility of reactant species and are more attractive for enhancing catalytic application. Zhou *et al.* [27] and Yu *et al.* [28] produced dendritic structures by electrochemical deposition. The authors concluded that the morphology of the electrochemically deposited nanostructured Pd could be solely controlled by tuning the depositing potentials. Fang *et al.*[29] reported an electrochemical route to synthesize Pd nanourchins. Li *et al.* [30] synthesized Pd/Au hollow cone-like microstructures by electrodeposition. Ye *et al.* [31,32] fabricated a three-dimensional mesoporous Pd networks by a simple reduction method in solution using a face centered cubic silica super crystal as template. From above analysis, it is concluded that the electrochemical synthesis is a useful tool in the preparation of Pd nanostructures without templates or surfactants. Many structures such as the high index facets enclosed nanoparticles, nanothorns and dendritic structures were prepared by the electrodeposition.

Hollow or Core-Shell Structures

The hollow nanomaterials have big potentials for further reducing the cost. Moreover, they have distinguished chemical and physical properties resulting from their special shape and composition [33,34]. Liu *et al.* [35] and Bai *et al.* [36] prepared raspberry hollow Pd nanospheres by a galvanic replacement reaction involving Co nanoparticles as sacrificial template. The raspberry surface might be helpful in increasing the surface area for catalysis. The core-shell structure is effective to reduce cost by reducing the amount of Pd used in the catalyst. Fang *et al.* [37] prepared the Au@Pd@Pt structure with a gold core, a Pd shell and Pt clusters on the shell as shown in Figure 4. The optimized structure had only two atomic layers of Pd and a half-monolayer equivalent of Pt. The activity was critically dependent upon the Pd-shell thickness and the Pt-cluster coverage. The high activity originated from the synergistic effect existing between the three different nanostructure components (sphere, shell and islands). Ksar *et al.* [38,39,40] synthesized bimetallic Pd-Au nanostructures with a core rich in gold and a Pd porous shell. This structure greatly reduced the loading of Pd. The addition of Au to Pd catalysts was not only improved the catalytic activity and selectivity but also enhanced the resistance to poisoning.

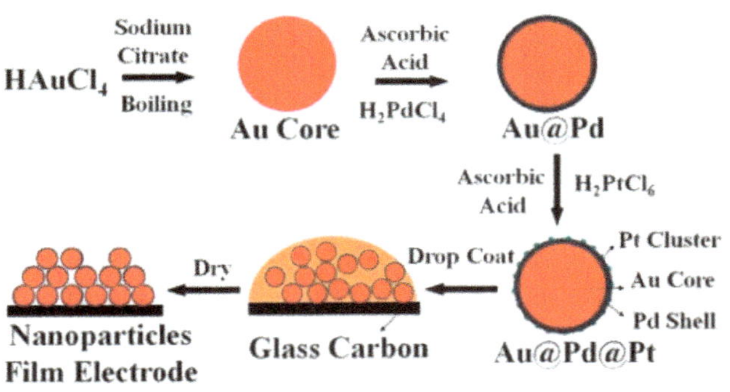

Figure 4. The procedure used to prepare an Au@Pd@Pt NP film on a glass carbon (GC) electrode. Reproduced with permission from reference [37].

Conclusions and Perspective Discussions

With the development of nanotechnology, researchers are able to control the morphology of Pd in nanoscale. Different morphologies such as particles with multi-facets, cubes, twinned particles, assembled particles, nanorods, nanothorns, nanowires, nanotubes, dendritic structure, networks and hollow/core-shell structures and so on have been successfully prepared. Most of the Pd nanomaterials have high performance in fuel cell half cell characterization, which is caused by the unique crystallography and morphology of the material. Both the activity and stability of the electrochemical reaction could be improved when catalyzed by the Pd nanomaterials to significantly reduce the usage of Pt in fuel cells. However, the research on Pd nanostructures is still in the fundermental stage and great effortsl need to be done to apply these nanomaterials into the real fuel cell applications. With the development of nano-synthesis technology there will be more mophologies with unique advantages prepared, emphasis should be put on the controlled synthesis and application in fuel cell membrane electrode assembly application.

PD-BASED CATALYSTS FOR THE DIRECT FORMIC ACID FUEL CELLS (DFAFCS)

The direct formic acid fuel cells (DFAFCs) have a high theoretical open circuit potential of 1.450 V compared with 1.229 V for H_2 proton exchange membrane fuel cells (PEMFCs) and 1.190 V for direct methanol fuel cells (DMFCs) at room temperature. The formic acid is a non-toxic liquid fuel and lower crossover fiux than methanol and ethanol. Although the net energy density of formic acid (2086 Wh L^{-1}) is lower than that of methanol (4690 Wh L^{-1}), high concentrated formic acid can be used as fuel, e.g., 20 M (70 wt. %), compared with lower methanol concentration, e.g., 1–2 M. Therefore, formic acid carries more energy per volume than methanol. The formic acid itself is an electrolyte and can facilitate proton transport within anode compartment [41].

It was reported that Pd/C exhibited much better activity compared with Pt/C, however, the activity was not satisfactory and more importantly the durability of Pd/C catalyst was in urgent needed for further improvement. Generally, formic acid oxidation on Pd or Pt surface follows a dual pathway mechanism, namely, a dehydration pathway [42,43,44]:

$$HCOOH \rightarrow CO_{ads} + H_2O \quad (1)$$

$$CO_{ads} + H_2O \rightarrow CO_2 + 2H^+ + 2e^- \quad (2)$$

and a dehydrogenation one:

$$HCOOH \rightarrow CO_2 + 2H^+ + 2e^- \quad (3)$$

On Pt catalyst, formic acid oxidation also took place via the third pathway with bridge bonded adsorbed formate intermediate being in equilibrium with the solution formats [36,37,38].

$$HCOOH \rightarrow HCOO_{ads} + H^+ + e^- \quad (4)$$

$$HCOO_{ads} \rightarrow CO_2 + H^+ + e^- \quad (5)$$

Pd has a propensity to break only the O–H bonds of the HCOOH molecule in the entire potential region, while, Pt has a propensity to break the C–O and/or C–H bond (at low overpotential) and the O–H bond (at high overpotential). Consequently, formic acid oxidation on Pd surfaces proceeds exclusively through the dehydrogenation reaction step, whereas, on Pt surfaces the dehydration pathway is predominant at low potentials. It should be noted that one of the major problems in formic acid oxidation is that the intermediary species formed during the oxidation of HCOOH cause catalyst poisoning. These intermediate species could be mostly CO, which strongly interacts with the active sites on the electrode surface and requires a higher overpotential for oxidation to CO_2. On a Pt catalyst, formic acid decomposition proceeds mainly via a dehydration pathway where the strongly adsorbed CO greatly hinders the catalytic activity. On pure Pt surfaces, CO poisoning due to the dehydration of formic acid on at least two or more contiguous Pt atoms hinders the direct dehydrogenation oxidation of formic acid at lower overpotentials. In contrast, formic acid decomposition on a Pd catalyst mainly proceeds in a dehydrogenation pathway. The generation of minor CO on the surfaces of Pd nanoparticles leads to the rapid decay of the catalytic activity. Hence, it is of great importance to design a Pd catalyst with excellent CO tolerance. For HCOOH oxidation, Pd shows superior initial performance compared with Pt. However, the high performance cannot be sustained, as Pd dissolves in acidic solutions and is vulnerable towards intermediate species. Modification of Pd with foreign metal has been considered as an effective method to enhance the activity and durability towards HCOOH oxidation. Studies have shown that the Pd oxidation activity for formic acid is also strongly influenced by the morphology and size of the Pd nanoparticles.

Table 1 listed the available data of the formic acid oxidation on Pd-based catalyst described in the literature. Three factors were compared: the electrochemical active surface area (EASA), the peak potential of formic acid oxidation and the peak current density of formic acid oxidation. According to the results provided by different authors, the EASA varied from the minimum of 15.7 $m^2\ g^{-1}$ to the maximum of 208.2 $m^2\ g^{-1}$. The peak current density varied from 3900 to 56.5 mA mg^{-1}_{Pd}, or from 159 to 12.4 mA cm^{-2}. The peak potential analysis showed that the alloying of the Pd with other metals could reduce the overpotential for formic acid oxidation as evidenced by the negative shift of the peak potential on the alloys. Alloying with other metals can modify the electronic structure and induce tensile strain of the Pd clusters, and finally influence their catalytic activities. The highest peak current density observed on $Pd_{0.9}Pt_{0.1}/C$ is nearly twice as high as that on Pd/C and six times as that on the commercial Pt/C [45]. The anodic peak current density obtained by using PdCo/MWCNTs as catalyst is 3 and 4.3 times higher than Pd/MWCNTs and Pd/XC-72 [46]. $PdNi_2$ alloy had almost three times the activity of Pd, even if the molar Pd content in $PdNi_2$ alloy was only one third of pure Pd [47].

Table 1. Details of the half-cell performance of the Pd-based catalysts.

Catalysts	EASA/$m^2\ g^{-1}$	Peak Potential/V RHE	Peak Current Density/mA mg^{-1}_{Pd}	Conditions H_2SO_4/Formic Acid/Scan Rate	References
Pt@Pd/C	156.5	0.277	3900	0.5 M/2 M/10 mV s^{-1}	[48]
Pt/C	198	0.687	/	0.5 M/2 M/10 mV s^{-1}	[48]
Pd/C	208.2	0.417	1200	0.5 M/2 M/10 mV s^{-1}	[48]
butylphenyl-stabilized Pd	122	0.38	3390	0.1 M/0.1 M/100 mV s^{-1}	[49]
Pd black	33.6	0.46	750	0.1 M/0.1 M/100 mV s^{-1}	[49]
butylphenyl-stabilized Pd	122	0.38	3390	0.1 M/0.1 M/100 mV s^{-1}	[49]
PdPt/C	49	/	2500	0.5 M/0.5 M/50 mV s^{-1}	[50]
Pd/C	/	/	2400	0.1 M/0.5 M/20 mV s^{-1}	[51]
Pd/C	107.2	/	1426.5	0.5 M/0.5 M/50 mV s^{-1}	[52]
PdSn/C	/	/	1420	0.5 M/0.5 M/20 mV s^{-1}	[53]
Pd/C	/	/	610	0.5 M/0.5 M/20 mV s^{-1}	[53]
Pd/CMRT	67.27	/	1140 at 0.44 V *	0.5 M/0.5 M/50 mV s^{-1}	[54]
Pd/RT	41.47	/	670 at 0.44 V *	0.5 M/0.5 M/50 mV s^{-1}	[54]
Pd/C	63.72	/	440 at 0.44 V *	0.5 M/0.5 M/50 mV s^{-1}	[54]
Pd/HPMo-PDDA-MWCNT	/	/	945	0.5 M/0.5 M/20 mV s^{-1}	[55]
Pd/AO-MWCNTs	/	/	554	0.5 M/0.5 M/20 mV s^{-1}	[55]
Pd/C	/	/	373	0.5 M/0.5 M/20 mV s^{-1}	[55]
$Pd_{57}Ni_{43}$	/	/	830	0.5 M/0.5 M/50 mV s^{-1}	[23]
Pd/C	/	/	700	0.5 M/0.5 M/50 mV s^{-1}	[23]
Pd-PANI	/	/	822	0.5 M/0.2 M/100 mV s^{-1}	[56]
PdAu	90	/	800	0.5 M/0.5 M/20 mV s^{-1}	[57]
Pd/C	43	/	250	0.5 M/0.5 M/20 mV s^{-1}	[57]
Pd/graphene	72.72	/	446.3	0.5 M/0.5 M/50 mV s^{-1}	[58]
Pd/C	31.85	/	191.9	0.5 M/0.5 M/50 mV s^{-1}	[58]
Pd/graphene	/	0.39	300	0.5 M/0.5 M/50 mV s^{-1}	[59]
Pd/C	/	/	193	0.5 M/0.5 M/50 mV s^{-1}	[59]
Pd networks	/	/	275.4	0.5 M/0.5 M/50 mV s^{-1}	[31]
Nanoporous Pd	23	/	262	0.5 M/0.5 M/10 mV s^{-1}	[60]
Pd/CNT	/	0.44	200 at 0.27 V *	0.5 M/0.5 M/50 mV s^{-1}	[61]
Pt/CNT	/	0.92	30 at 0.27 V *	0.5 M/0.5 M/50 mV s^{-1}	[61]
Pt Pd/CNT	/	0.64	50 at 0.27 V *	0.5 M/0.5 M/50 mV s^{-1}	[61]
Pt_1Pd_3/CNT	/	0.64	125 at 0.27 V *	0.5 M/0.5 M/50 mV s^{-1}	[61]
PdSn/C	64.7	/	170.2	0.5 M/1 M/10 mV s^{-1}	[62]
Pd/C	39.8	/	102.3	0.5 M/1 M/10 mV s^{-1}	[62]
Pd black	/	/	56.5	0.5 M/0.5 M/50 mV s^{-1}	[31]
Pd/untreated-MWCNT	46.2	0.517	159 mA cm^{-2}	0.5 M/0.5 M/50 mV s^{-1} with glutamate	[63]
Pd/acid-oxidized MWCNT	35.7	/	108 mA cm^{-2}	0.5 M/0.5 M/50 mV s^{-1}	[63]
Pd/untreated-MWCNT	21.4	/	72 mA cm^{-2}	0.5 M/0.5 M/50 mV s^{-1} without glutamate	[63]
PdNi/C	/	0.247	105.1 mA cm^{-2}	0.5 M/1 M/10 mV s^{-1}	[64]

Pd/C	/	0.287	72.9 mA cm^{-2}	0.5 M/1 M/10 mV s^{-1}	[64]
PdCo/MWCNTs	/	0.28	107 mA cm^{-2}	0.5 M/0.1 M/20 mV s^{-1}	[46]
Pd/MWCNTs	/	0.32	35.6 mA cm^{-2}	0.5 M/0.1 M/20 mV s^{-1}	[46]
Pd/XC-72	/	0.38	24.8 mA cm^{-2}	0.5 M/0.1 M/20 mV s^{-1}	[46]
$Pd_{0.9}Pt_{0.1}/C$	83	0.35	87.5 mA cm^{-2}	0.5 M/0.5 M/50 mV s^{-1}	[45]
Pt/C	85.6	0.52	15.1 mA cm^{-2}	0.5 M/0.5 M/50 mV s^{-1}	[45]
Pd/C	89.7	0.48	42.5 mA cm^{-2}	0.5 M/0.5 M/50 mV s^{-1}	[45]
Pd-B/C	87.6	/	65.4 mA cm^{-2}	0.5 M/0.5 M/50 mV s^{-1}	[65]
Pd/C	90	/	36.0 mA cm^{-2}	0.5 M/0.5 M/50 mV s^{-1}	[65]
Pd–Au/C	/	0.37	18.6 mA cm^{-2}	0.5 M/0.5 M	[66]
Pd/C	/	0.48	12.4 mA cm^{-2}	0.5 M/0.5 M	[66]
Pd/phen-MWCNTs	37.6	/	/	0.5 M/1 M/50 mV s^{-1}	[67]
Pd/AO-MWCNTs	15.7	/	/	0.5 M/1 M/50 mV s^{-1}	[67]
Pd/graphene	44	0.367	/	1 M/1 M/10 mV s^{-1}	[68]
Pd/Vulcan C	35	0.377	/	1 M/1 M/10 mV s^{-1}	[68]

Pd Supported on Carbon Materials

Pd on Carbon Powders

An active catalyst should be dispersed on a convenient support to stabilize the catalytic nanoparticles, to obtain optimum catalyst utilization and to reduce the amount of precious metal used, reducing the catalyst cost. In DFAFCs, highly conductive carbon materials such as Vulcan XC-72 carbon (Cabot, Boston, MA, USA) provide a high dispersion of metal nanoparticles to facilitate electron transfer, resulting in better catalytic activity. The supported electrocatalyst is a practical means to achieve high utilization of expensive noble metals and to maintain good life-time. Other carbon materials such as the carbon fiber are also studied as support material for palladium [69,70,71].

There are still aspects to improve in the Pd/C as catalyst in fuel cell applications. The works usually focus on the Pd particle size control via novel synthesis techniques to achieve high electrochemical active surface area and improve the utilization efficiency of the Pd catalyst. Cheng *et al.* [72] prepared highly dispersed Pd/C catalyst through an ambient aqueous way instead of the traditional high temperature polyol process in ethylene glycol. The Pd/C catalyst without stabilizer had a higher oxidation activity toward formic acid compared with that of a traditionally prepared Pd/C catalyst. Liang *et al.* [52] synthesized a highly dispersed and ultrafine carbon-supported Pd nanoparticle catalyst which exhibited significantly high electrochemical active surface area and high electrocatalytic performance for formic acid oxidation with four times larger formic acid oxidation current compared with that prepared by general $NaBH_4$ reduction method. The large electrochemical specific surface may be due to the high dispersion and small particle size of Pd/C catalyst. Suo *et al.* [73] used a simple and stabilizer-free ethylene glycol reduction method to prepare Pd/C catalyst. Size-dependent electrochemical property was observed and electrochemical evaluation showed that Pd/C with a particle size of 6.1 nm performed the highest activity for formic acid oxidation. The performance of the Pd/C catalyst for the oxidation of formic acid could be greatly promoted with 3.19 times enhancement in catalytic stability and 1.57 times improvement in the catalytic activity by simply introducing vanadium ions in very low concentration to the electrolyte [51]. The improvement in the catalytic performance may be attributed to the facilitating of formic acid oxidation due to the existence of VO^{2+}/V^{3+} redox pair and the ensemble effect induced by the adsorption of vanadium ions onto the surface of Pd.

Pd alloys on Carbon Powders

Pd displays an initial high activity for the oxidation of HCOOH. However, its long term performance is poor. Deactivation of Pd activity has been assigned to catalyst poisoning by CO, although, other poisoning species such as anions from the electrolyte can also effective. Alloying Pd with the second metal to change the surface electronic state, an ensemble effect could occur, which could possibly reduce the catalyst poisoning and increase the activity and lifetime of the catalyst.

(1) PdPt alloys

It was found $Pd_{0.9}Pt_{0.1}/C$ was the optimum catalyst for the desired formic acid oxidation reaction [45]. The highest peak current density observed on $Pd_{0.9}Pt_{0.1}/C$ was nearly twice as high as that on Pd/C and six times as that on the commercial Pt/C, which is superior to any reported carbon black-supported Pt/C, Pd/C, and Pt_xPd_{1-x}/C catalysts. There is also a large negative shift (up to 0.21 V) in peak potential for a Pd_xPt_{1-x}/C *versus* that for Pd/C. The high performance of the $Pd_{0.9}Pt_{0.1}$nanoalloy can be ascribed to the effectively inhibited CO poisoning at largely separated Pt sites and appropriately lowered*d*-band centre of Pd sites. The addition of Pt to Pd considerably improved the steady-state activity of Pd [74,75,76]. Chronoamperometric measurements showed that the most active catalyst was $Pd_{0.5}Pt_{0.5}$ with the particle size of 4 nm. Wu *et al.* [48] prepared a Pd decorated Pt/C catalyst, Pt@Pd/C, with a small amount of Pt as core. It was found that the catalyst showed excellent activity toward anodic oxidation of formic acid at room temperature and its activity was 60% higher than that of Pd/C. It is speculated that the high performance of Pt@Pd/C may result from the unique core-shell structure and synergistic effect of Pt and Pd at the interface. Wang *et al.* [50] decorated Pd/C with Pt nanoparticles where the amount of three neighbouring Pt or Pd atoms markedly decreased. As a result, discontinuous Pd and Pt atoms suppressed CO formation and exhibited unprecedented catalytic activity and stability toward formic acid oxidation.

(2) PdSn alloys

Liu *et al.* [77] prepared Pd and PdSn nanoparticles supported on Vulcan XC-72 carbon by a microwave-assisted polyol process. It was found that the addition of Sn to Pd could increase the lattice parameter of the Pd (fcc) crystal. The PdSn/C catalysts have higher electrocatalytic activity for formic acid oxidation than a comparative Pd/C catalyst. The Pd_2Sn_1/C catalyst exhibited higher current density and enhanced electrocatalytic stability compared with Pd/C. There was also a negative shift of the peak potential on Pd_2Sn_1/C than that of Pd/C. Zhang *et al.* [53] synthesised PdSn/C catalysts with different atomic ratios of Pd to Sn. The alloy catalysts exhibited significantly higher catalytic activity and stability for formic acid oxidation than that of Pd/C catalyst. Pd was modified by Sn through an electronic effect which could decrease the adsorption strength of the poisonous intermediates on Pd and thus promote the formic acid oxidation. Tu *et al.* [62] prepared a carbon-supported PdSn (PdSn/C) catalyst with greatly improved performance for formic acid oxidation compared with that of Pd/C. Adding Sn as a small ratio into the carbon-supported Pd catalyst could largely increase the current density of the formic acid oxidation and shift the onset potential toward the negative compared with that of Pd/C. The reason for the improvement of the catalyst was likely attributed to the high dispersion of the Pd and due to the change in the electronic properties of the Pd.

(3) Other alloys

Zhang *et al.* [66] prepared carbon-supported PdAu catalysts with different alloying degree. The electrocatalytic activity of PdAu/C catalyst for the formic acid oxidation was strongly dependent on the alloying degree of Pd-Au nanoparticles. The PdAu/C catalyst with higher alloying degree showed a higher electrocatalytic activity and stability for the formic acid oxidation compared with the PdAu/C catalyst at lower alloying degree, which can be ascribed to the enhancement of CO tolerance and possible suppression of the dehydration pathway in the course of formic acid oxidation. The catalytic activity of the PdAu/C catalyst was also found to be affected by the nature of the supporting materials [78,79,80,81]. Gao *et al.*[64,82,83] synthesized carbon-supported PdNi catalyst by sodium borohydride reduction reaction. The performance of the PdNi/C catalyst for formic acid oxidation was significantly improved compared with that of Pd/C. The potential of the main anodic peak of formic acid at PdNi/C catalyst electrode was about 40 mV more negative than that at Pd/C catalyst electrode.

The onset potential of formic acid oxidation at PdNi/C catalyst electrode was 30 mV more negative than that at Pd/C catalyst electrode. The reason for the promotion effect may be due to that Ni can contribute to the adsorption of oxygen-containing species, which is conducive to the oxidation of formic acid and a change in electronic properties of Pd. Yu *et al.* [84] prepared carbon supported bimetallic PdPb catalysts which were found to be more resistant to deactivation in the DFAFC than Pd/C and to consistently show better long-term performance. The addition of Pb to Pd stabilized it significantly to deactivation during formic acid oxidation. Wang *et al.* [65] synthesized highly dispersed boron-doped Pd nanoparticles supported on carbon black with high Pd loadings (*ca.* 40 wt. % Pd) by using NaBH4 as the reductant. The as-prepared Pd-B/C catalyst showed extraordinary high activity toward formic acid oxidation compared with that of a commercially available Pd/C catalyst. Thermal treatment further enhanced the durability of the oxidation current on Pd-B/C. The superior performance of the Pd-B/C catalyst may arise from uniformly dispersed nanoparticles within optimal size ranges, the increase in surface-active sites, and the electronic modification effect of boron species. The Pd–Co, $PdCeO_x/C$ [85] and PtRu/C [86,87,88] catalyst also exhibited excellent catalytic activity and stability in the oxidation of formic acid.

Pd Supported on Carbon Nanotubes

It remains a chanllege to load Pd nanoparticels on the surface of carbon nanotubes because of the graphitization of carbon nanotubes. However, due to the advantages of carbon nanotubes especially in the contribution to stability of the catalyst, many works are devoted to the deposition of Pd on carbon nanotubes [89,90]. Hu *et al.* [63] synthesized Pd nanoparticles supported on untreated multiwalled carbon nanotubes (MWCNTs). The Pd/MWCNT catalyst displayed superior electrocatalytic activity and stability in formic acid oxidation. Chakraborty *et al.* [91] synthesized nanosized Pd particles supported on MWCNTs. Bai *et al.* [67,92] functionalized MWCNTs with 1,10-phenanthroline (phen-MWCNTs) as a catalyst support for Pd nanoparticles. It was found that Pd nanoparticles were evenly deposited without obvious agglomeration, and the average particle size of the Pd nanoparticles was only as small as 2.3 nm. The as-prepared Pd/phen-MWCNTs catalyst had a better electrocatalytic activity and stability for the oxidation of formic acid than Pd catalyst on acid treated MWCNTs. Phen made a strong impact on the electrocatalytic activity of the catalyst through the functionalization of the MWCNTs and the formation of the active Pd-N sites. Therefore, the dispersivity and the ESA of the Pd nanoparticles were obviously enhanced in the presence of phen, resulting in better electrocatalytic activity and utilization efficiency of the catalyst. Cui *et al.* [55] loaded Pd nanopartilces on phosphomolybdic (HPMo) acid functionalized multiwalled carbon nanotubes supports. The catalysts exhibited a much higher electrocatalytic activity and stability for formic acid oxidation reaction as compared with that on traditional Pd/C. The high electrocatalytic activities were most likely related to highly dispersed and fine Pd nanoparticles as well as synergistic effect between Pd and HPMo immobilized on functionalized MWCNTs.

Pd Alloys Supported on Carbon Nanotubes

(1) PdPt alloys

Selvaraj *et al.* [93,94] prepared PtPd nanoparticles supported on purified singlewalled carbon nanotubes. The modified electrode exhibited significantly high electrocatalytic activity toward formic acid oxidation due to the uniform dispersion of nanoparticles on SWCNTs and the efficacy of Pd species in Pt-Pd system. Winjobi *et al.* [61] prepared Pt, Pd and Pt_xPd_y alloy nanoparticles supported on carbon nanotubes with high and uniform dispersion. With increasing Pd amount of the catalysts, the mass activity of formic acid oxidation reaction on the CNT supported catalysts increased. A direct oxidation pathway of formic acid oxidation occurred on the Pd surface, while, the formic acid oxidation was through CO_{ads} intermediate pathway on the Pt surface. The Pd/CNT demonstrated 7 times better mass activity than that of Pt/CNT at an applied potential of 0.27 V (vs. RHE) in the chronoamperometry test.

(2) PdAu alloys

Chen *et al.* [95] prepared PdAu/multiwalled carbon nanotubes (Pd–Au/MWCNTs) to increase the stability and performance of the Pd-based catalysts in DFAFCs. The catalyst was highly active in formic acid oxidation, due to the hydrogen treated catalysts have smaller metal particles and better contact with MWCNTs support. When Pd was alloyed with Au the leaching of Pd was considerably slower, which may be caused by much slower Pd leaching from Pd-Au alloy than from Pd. The Pd-Au/MWCNTs had higher current and lower onset potential for formic acid oxidation than Pd/MWCNTs. Chen *et al.* [96] prepared multiwalled carbon nanotubes (MWCNTs) supported Pd–Au catalyst for oxidation of formic acid and compared with a similarly prepared Pd/MWCNTs and a commercial Pt-Ru/C catalyst. Both the Pd-Au/MWCNTs and the Pd/MWCNTs catalysts used were more active than that of a commercial Pt-Ru/C catalyst. The specific activity of Pd in the novel Au-Pd/MWCNTs catalyst was over two times higher than that on the Pd/MWCNTs catalyst. Mikolajczuk *et al.* [97] found that the Pd-Au/MWCNTs catalyst exhibited higher activity and more stable in oxidation reaction of formic acid. The higher initial catalytic activity of Pd-Au/MWCNTs catalyst than Pd/MWCNTs catalyst in formic acid oxidation reaction was attributed to the electronic effect of gold in Pd-Au alloy [98,99].

(3) Other alloys

Morales-Acosta *et al.* [46,100] compared the Pd-Co and Pd catalysts prepared by the impregnation synthesis method on MWCNTs. The current density achieved with the PdCo/MWCNTs catalyst was 3 times higher than that of the Pd/MWCNTs catalyst. The onset potential for formic acid oxidation on PdCo/MWCNTs catalyst showed a negative shift *ca.* 50 mV compared with Pd/MWCNTs. The anodic peak current density obtained by using PdCo/MWCNTs was 3 and 4.3 times higher than Pd/MWCNTs and Pd/C, respectively. The PdCo/MWCNTs exhibited good stability in acidic media, higher current density and more negative anodic potential associated to this reaction than that of Pd/MWCNTs and Pd/C catalysts. The difference could be attributed to a better dispersion of the metallic nanoparticles with a lower particle size achieved with the MWCNTs-supported materials. Specific surface area obtained from PdCo and Pd supported on MWCNTs was higher than that obtained on Pd supported on Vulcan XC-72 carbon.

Pd Supported on Graphene

Graphene is a rapidly rising star on the horizon of the materials science and technology which has attracted tremendous attention and holds great promise for advanced fields. Especially in fuel cells, the emergence of graphene has opened a new avenue for utilizing two-dimensional planer carbon material as a catalytic support owing to its high conductivity, unique graphitized basal plane structure. This is propitious to not only maximize the availability of nanosized electrocatalyst surface area for electron transfer but also provide better mass transport of reactants to the electro-catalyst [101]. Fu *et al.* [59] developed an effective surfactant-free strategy for the small-sized Pd particles deposited on graphen. The Pd/GN catalyst exhibited excellent catalytic activity and stability toward formic acid oxidation compared with the Pd/C catalysts [102]. The enhanced performance of the Pd/GN catalyst was contributed to the small size and high dispersion of Pd NPs and the stabilizing effect of the graphene support. While the group has also fabricated the Pd/graphene hybrid via sacrifice template route [58]. Bong *et al.* [68] synthesized high loadings of 80 wt% Pd on graphene catalysts which showed significantly enhanced electrocatalytic activity and stability for formic acid oxidation compared with Pd/C catalysts.

Pd Supported on Oxides

As alternatives to improve the catalytic activity and stability of the Pd-based catalysts, the carbon support materials were modified with semiconducting oxides, such as TiO_2, MoO_x and SnO_2 [103,104,105]. The influence of TiO_2 support on the electronic effect and the bifunctional mechanism can be summarized as follows. (1) TiO_2 support imposes an electronic effect in which the hypo-*d*-electronic titanium ions promote the electrocatalytic properties of hyper-*d*-electronic noble metal surface atoms, thus lowering the adsorption energy of CO intermediates and increasing the mobility of the CO group on Pd nanostructures; (2) Adsorption of OH species (OH_{ad}) on TiO_2 can facilitate the conversion of the catalytically poisonous CO intermediates into CO_2, thereby improving the durability of Pd catalysts. Additionally, TiO_2 can facilitate the dispersion of noble metal nanoparticles to anchor them. Based on the theory that TiO_2 nanoparticles in the catalyst can adsorb OH_{ads} species and promote the oxidation reaction on the electrode, Xu *et al.* [44] prepared highly dispersed carbon supported Pd-TiO_2 catalyst with intermittent microwave irradiation. The activity of Pd-TiO_2/C catalyst for the oxidation of formic acid was higher than that of the Pd/C catalyst [106]. Wang *et al.* [54] investigated Pd nanoparticles supported on carbon-modified rutile TiO_2 (CMRT) as catalyst for formic acid oxidation. The Pd/CMRT showed three times the catalytic activity of Pd/C, as well as better catalytic stability toward formic acid oxidation. The enhanced catalytic property of Pd/CMRT mainly arised from the improved electronic conductivity of the carbon-modified rutile TiO_2, the dilated lattice constant of Pd nanoparticles, an increasing of surface steps and kinks in the microstructure of Pd nanoparticles and slightly better tolerance to the adsorption of poisonous intermediates [107]. Wang *et al.* [108] investigated the influence of the crystal structure of TiO_2 supporting material on formic acid oxidation. TiO_2 with the rutile structure improved the catalytic activity of Pd nanoparticles toward formic acid oxidation. The enhancement of Pd/TiO_2 (rutile) catalytic activity arose from uniform dispersion of Pd nanoparticles, an increase in surface-active sites, and good tolerance to the adsorption of poisonous intermediates (such as CO_{ad}, $COOH_{ad}$ and so on). This study proved that rutile TiO_2 was a better supporting material than anatase TiO_2 or composites.

Pd Supported on other Supporting Materials

At present, carbon nanotubes (CNTs) and Vulcan XC-72 carbon are the most often used supporting materials to load Pd nanoparticles. However, the rate of formic acid oxidation is lower partly due to lower Pd utilization on such conventional carbon supports, which is related to the lower electrochemically accessible surface area for the deposition of Pd particles. Tremendous efforts have been devoted to search for new catalyst supports to achieve good dispersion, utilization, activity and stability [109]. Bai *et al.* [110] supported Pd nanoparticles on polypyrrole-modified fullerene and the Pd/ppy-C_{60} catalyst showed a good electrocatalytic activity and stability for the oxidation of formic acid. Pd on polyaniline (Pd-PANI) nanofiber film [56], on polypyrrole (PPy) film [111] and PMo_{12} [112] showed excellent catalytic activity in the oxidation reaction of formic acid in acidic media. Qin *et al.* [113] synthesized highly dispersed and active Pd/carbon nanofiber (Pd/CNF) catalyst which exhibited good catalytic activity and stability for the oxidation of formic acid. Cheng *et al.* [114] loaded Pd nanoparticles on the carbon nanoparticle-chitosan host, the chitosan matrix was shown to be beneficial in making nanosized catalyst convenient, effective, and reproducible. The Pd on carbon nanoparticle-chitosan host catalysts was highly active for the oxidation of formic acid.

Unsupported Pd

Based on the pratical application of Pd based catalysts in fuel cells, Pd is usually supported on other materials. However, since the nanotechnology provides the possibility to control the morphology of Pd material, and the intrinsic properties of the nanostructured material can be dramatically enhanced by shape and structural variations, there are also some research on the preparation of unsupported Pd materials for fuel cell application [115,116,117]. The one-dimensional Pd nanostructures like nanowires, nanobelts and nanotubes have attracted significant interest due to their exotic technological applications. Wang *et al.* [60] and Ye *et al.* [31] fabricated unsupported nanoporous Pd networks which exhibited high electrochemical active specific surface area, and high catalytic activity for oxidation of formic acid. Pd hollow nanostructures is an effeicient way to reduce the cost of catalsyt, for example Pd/Au hollow cone-like microstructures [30] and Pd clusters on highly dispersed Au nanoparticles [57] are found to show superior performance at much lower cost. Alloyed nanowires such as $PdNi_2$ [47] and $Pd_{57}Ni_{43}$ [23] alloy nanowires also shows improved performance. The size and nature of surface structures, such as crystalline planes and surface ligands is one key issue to improve the mass activity. Zhou *et al.* [49] synthesized monodispersed butylphenyl-functionalized Pd (Pd-BP) nanoparticles with unique surface functionalization and a high specific electrochemical surface area (122 m^2 g^{-1}), the Pd–BP nanoparticles exhibited a mass activity of 4.5 times as that of commercial Pd black for HCOOH oxidation as shown in Figure 5. The Pd-BP catalyst shows obvious improvement in both activity and stability compared with commercial Pd-black.

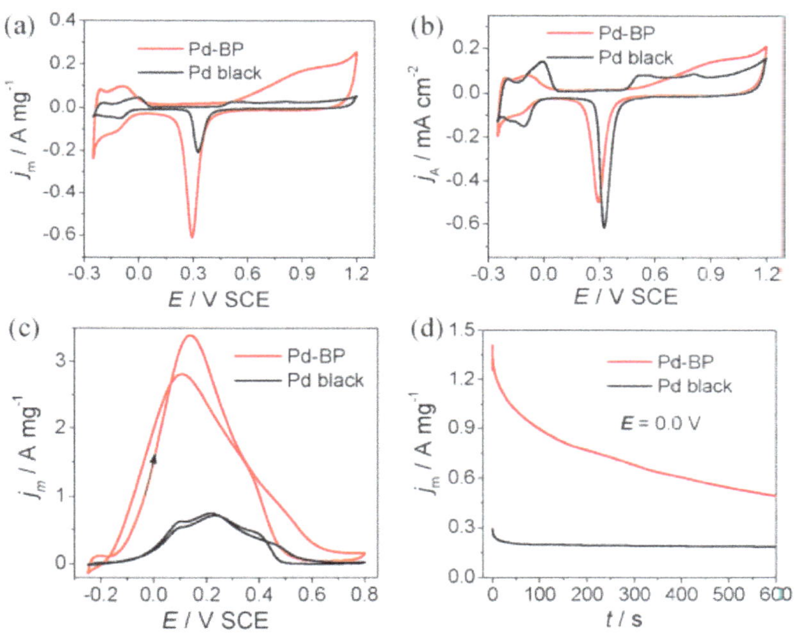

Figure 5. Cyclic voltammograms of butylphenyl-functionalized Pd (Pd-BP) nanoparticles and commer-cial Pd black in 0.1 M H_2SO_4, with the currents normalized (**a**) by the mass loadings of Pd and (**b**) by the effective electrochemical surface areas at a potential scan rate of 100 mV s^{-1}; Panels (**c**) and (**d**) depict the cyclic voltammograms and current-time curves acquired at 0.0 V for HCOOH oxidation, respectively, at the Pd-BP nanoparticles and Pd black-modified electrode in 0.1 M HCOOH + 0.1 M H_2SO_4at room temperature. Reproduced with permission from reference [49].

Single Fuel Cell Characterization

Single Fuel Cell Performance

The aim of the study on the Pd based catalyst is to realize its pratical application in fuel cells. However, because of the difference in the preparation technique of the membrane electrode assembly (MEA) it is relatively difficult to compare the performance of a catalyst. The single fuel cell performance of the different catalysts was selectively summarized in Figure 6. Meng *et al.* [118] designed a novel MEA structure for DFAFCs where Pd nanothorns were directly electrodeposited onto the carbon paper to form the anode catalyst layer. The novel MEA provided 2.4 times higher peak power density than that of the conventional MEA. The increase in the performance was due to the improved mass transport of the formic acid in the catalyst and diffusion layers, better Pd utilization and higher electroactivity of the Pd single crystal nanothorns. Cheng *et al* [119] electrodeposited Pd on graphite felt (GF) and the resulting catalyst was compared with Pt-Ru/

GF for the oxidation of formic acid. The Pd/GF anode reached 852 W m^{-2} compared to 392 W m^{-2} with a commercial Pd catalyst-coated membrane (CCM). Mikołajczuk *et al.* [42] prepared a new carbon black supported Pd catalyst for DFAFC applications. The maximum power density of the novel 10 wt. % Pd catalyst was only 23% lower than that of the commercial 20 wt. % Pd/C. Pd-Au on multiwalled carbon nanotubes (MWCNTs) exhibited higher power density and better stability in DFAFC than that of the similar Pd/MWCNTs catalyst [79]. The 17.8 wt. % Pd/MWCNTs catalyst reached three times of the peak power density compared with that of a 20 wt. % Pt-Ru/C catalyst [120]. A MEA of the Pd/MWCNTs catalyst showed a power density of 3.3 mW cm^{-2} with 50% less Pd loadings than that of commercial Pd/C [121]. The addition of Pb and Sb into the Pd black catalyst caused a strong promotion for the formic acid oxidation in a cell [43]. Yu *et al.* [122] systematically evaluated and compared a number of carbon-supported Pt-based and Pd-based catalysts with commercial Pd/C, PtRu/C, and Pt/C catalysts in a multi-anode DFAFC. It is found that the PdBi/C provided higher stability than that of the commercial Pd/C catalyst, while both of the PdMo/C and PdV/C catalysts provided poor cell performances. The results provided strong evidence that both Mo and V poison Pd through electronic and/or chemical effects. Based on above analysis and Figure 6 it can be concluded that alloying with another metal is an efficient way to improve the performance of Pd based catalyst in direct formic acid fuel cell full cell performance, as is evidenced by the PdSn or Pd Sb alloy [98,123]. Next to the alloy is the Pd black with lower performance than the alloy. Pd nanothorns prepared by a novel electrodeposition technique shows higher performance than most Pd black, showing the effect of morphology control on catalytic activity. However the unexpected conclusion is Pd or Pd alloy supported on novel carbon supports such as graphene and carbon nanotube does not show much improvement in the performance. The conclusion points out futher aspects to be explored in this field: try to realize the morphology or crystallography control of Pd nanostructure aiming at high activity and stability, try to load the novel nanostructure on carbon support material aiming at reduce the loading amount of Pd, finally reducing the cost.

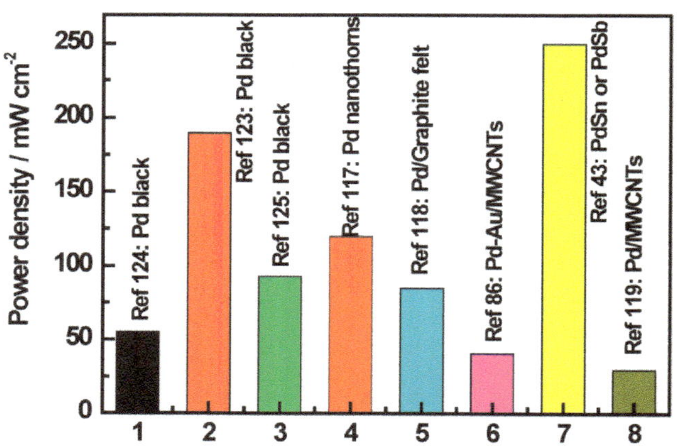

Figure 6. Comparison of the power density of the single fuel cell performance with selected catalysts.

Deactivation Mechanism and Reactivation

The limited durability of Pd-based anode for DFAFCs has seriously restricted their practical application. The mechanistic clarification on how to remove the poisoning species and to reactivate the Pd-based anode still remains a big challenge. The major factors causing performance degradation in the DFAFCs are an increment in the anode charge-transfer resistance and a growth in the particle size of the Pd anode catalyst [124]. The anode charge-transfer resistance, confirmed by EIS, increases with the operation time due to poisoning of the catalyst surface. The performance loss caused by surface poisoning could be completely recovered by the reactivation process. However, the increase in the catalyst size induces a reduction in active surface area and the performance loss caused by the growth in catalyst size cannot be recovered by the reactivation process. The deactivation of Pd/C increases sharply with increasing the formic acid concentration but only depends on the potential at high cell voltages. Reactivation can be achieved by driving the cell voltage to a reverse polarity of −0.2 V or higher. Although the reason for the activity loss is still unclear, it has been found that almost full activity can be recovered by applying an anodic potential of *ca.* 1.0 V *vs.* RHE or more. Ren *et al.* [125] found that the Pd oxides/hydrous oxides (POHOs) play a crucial role in promoting better performance and minimizing performance degradation of the Pd-based DFAFCs. The intrinsic presence or introduction of Pd oxides/hydrous oxides during catalysis of formic acid oxidation was found to promote elimination of poisoning species, thereby leading to a better performance of DFAFCs. Zhou *et al.* [126] found that Pd catalyst poisoned at the anode of a DFAFC under constant current discharging could be fully regenerated by a non-electrochemical method, *i.e.*, just switching pure water to DFAFC for 1 h. The voltage variation during the regeneration showed that one platform of 0.35 V was formed by the intermediate species of formic acid oxidation, which is proven to be critical for cell performance regeneration. The results indicated that the absorption of poisoning species on Pd was the main reason for the decaying of cell performance. Yu *et al.* [127] systematically studied the deactivation and electrochemical reactivation of a carbon supported Pd catalyst. The reactivation can be accomplished within a matter of seconds at ≥1.0 V *vs.* DHE (cell voltage ≤−0.3 V). However, reactivation at a cell voltage of 0 V or higher is required from a practical perspective. Analyzing above results it is concluded that the decaying of the cell performance is mainly caused by two factors: one is the growth of Pd particle size and the other is the absorption of poisoning species on Pd surface. The performance loss caused by surface poisoning could be recovered by the reactivation process, while the loss caused by particle growth can not be recovered. So how to limit the growth of particle size becomes one crucial problem in the stability study of the Pd based catalyst.

PD-BASED CATALYSTS FOR THE DIRECT ETHANOL FUEL CELLS (DEFCS)

Among the alcohols, methanol and ethanol are the two most commonly used fuels for direct alcohol fuel cells (DAFCs). Direct ethanol fuel cell (DEFC) has attracted significant attention since ethanol is non-toxic, naturally available, renewable, and higher power density comparing to methanol (8030 Wh kg^{-1} for ethanol and 6100 Wh kg^{-1} for methanol). However, the sluggish reaction kinetics of ethanol oxidation is still a challenge to the commercialization of DEFC. There are many difficulties associated with ethanol oxidation and consequently in its use in fuel cell. Complete oxidation of ethanol to CO_2 involves 12 electrons and the process involves the scission of a C–C bond thus demanding high activation energies to be overcome. Many of the intermediates (mainly CO and –CHO) produced during the oxidation reaction poison the anode catalyst and in turn reduce the catalytic efficiency. Pt-based catalysts are recognized as the best catalysts for low temperature fuel cells. However, the high cost and limited resource of Pt limited its use as catalysts. At the same time, the poor utilization and the poisoning of Pt catalyst particularly in alkaline solution also limited its applications. A great deal of interest has recently been focused on the cheaper materials than platinum and the use of non-noble transition metals in alkaline media, in particular, the performance of alloys of Pd with non-noble metals for the oxidation of ethanol. In acidic environment, the complete oxidation of ethanol is difficult and the catalytic activities of the catalysts for ethanol oxidation reaction (EOR) could be significantly improved in alkaline media, where Pd-based catalyst has comparable or even better catalytic activity compared with Pt-based catalysts for ethanol oxidation [128,129]. This is why that the Pd-based materials are intensively studied for the oxidation of ethanol.

Table 2 summerized the key factors of the EOR including the specific activity, the mass activity, the onset potential and the peak potential. It was found that the mass activity of Pd for ethanol oxidation could reach as high as 2200 A g^{-1}_{Pd} [130]. Such high activity originated from the higher utilization of Pd and at the same time the reduced metal content. The mass activity could be even higher for the alloys if the current was only account by the mass of Pd. Alloying with other metals, especially, Ni could improve the kinetics of the ethanol oxidation as revealed by the negatively shifted onset potential and peak potential [131,132,133]. The construction of the core-shell structure was found to be an effective way to reduce the overpoential [38]. It is found that the Ru@PtPd/C [134] and Pd–Ni–Zn/C [135] show highest mass activity in the EOR, implying the advantages of multi-alloy and core-shell structure. Two components alloying of Pd with other metals have been intensively studied, while the multi-alloying still needs to be put more attention, and the muli-alloying also provides vast variations in the choice and combination of different metals. Novel nanostructure of Pd is still a popular topic in this field [136,137,138,139,140,141].

Table 2. Details of the ethanol oxidation on Pd catalysts. (SA: Specific Activity; MA: Mass Activity; OP: Onset Potential and PP: Peak Potential).

Catalysts	SA[1]/mA cm^{-2}	MA[2]/A g^{-1}_{Pd}	OP[3]/V RHE	PP[4]/V RHE	Conditions Ethanol/KOH/Scan Rate	References
Dendritic Pd	17	/	−0.353	−0.003	1 M/1 M/20 mV s^{-1}	[28]
Pd/CNTs	/	800	−0.353	0.007	0.5 M/0.5 M/50 mV s^{-1}	[35]
Pd shell/Au core	0.890	/	−0.582	−0.15	1 M/1 M/50 mV s^{-1}	[38]
Pd/HCHs	42	2200	−0.352	−0.072	2 M/1 M/50 mV s^{-1}	[130]
PdNPs/CFCNT	16	/	−0.24	0.14	1 M/1 M/60 mV s^{-1}	[136]
Pd–In_2O_3/CNTs	61	/	−0.347	−0.022	1 M/0.5 M/50 mV s^{-1}	[137]
Pd/CNFs	74.5	/	−0.541	−0.021	1 M/1 M/50 mV s^{-1}	[138]
Pd/CNFs	66.1	1187	−0.511	−0.057	1 M/1 M/50 mV s^{-1}	[139]
Pd/CNFs	80	1400	−0.46	−0.04	1 M/1 M/50 mV s^{-1}	[140]
Pd-Ni/CNFs	200	/	−0.602	−0.072	1 M/1 M/10 mV s^{-1}	[141]
Pd/SnO_2-GNS	46.1	/	−0.403	0.227	0.25 M/0.25 M/50 mV s^{-1}	[142]
Pd-PANI	/	1300	−0.46	0.07	1 M/0.5 M/100 mV s^{-1}	[56]
Pd/polyamide 6	70	/	/	−0.072	0.5 M/1.5 M/50 mV s^{-1}	[143]
Pd/Nickel foam	107.7	/	−0.442	0.078	1 M/1 M/50 mV s^{-1}	[144]
Pd-TiN	2.87	59.2	−0.474	0.022	0.5 M/1 M/20 mV s^{-1}	[145]
Pd-NiO/MgO@C	69.3	/	−0.602	−0.052	1 M/1 M/10 mV s^{-1}	[146]
Ru@PtPd/C	/	3600	−0.402	−0.052	1 M/1 M/30 mV s^{-1}	[134]
Pd particles	151	/	−0.548	/	1 M/1 M/50 mV s^{-1}	[147]
Nanoporous Pd	90.64	227.7	−0.403	0	0.5 M/1 M/10 mV s^{-1}	[148]
Pd NWs	/	7.96 *	/	−0.01	0.5 M/1 M/50 mV s^{-1}	[149]
Pd NWs	/	2.16 *	−0.566	−0.068	1 M/1 M/50 mV s^{-1}	[150]
Pd/Au NWA	199	/	−0.402	0.138	1 M/1 M/50 mV s^{-1}	[151]
PdPt	34	478	−0.523	0.077	1 M/1 M/50 mV s^{-1}	[152]
Pd-Sn/C	121.59		−0.46	0.135	3 M/0.5 M/50 mV s^{-1}	[153]
Pd-Ru-Sn/C	65	/	−0.536	0.097	3 M/0.5 M/50 mV s^{-1}	[153]
$Pd_{91}Sn_9$/C	130	/	−0.205	0.347	1 M/1 M/50 mV s^{-1}	[154]
Pd_7 Ir/C	103	/	−0.584	0.008	1 M/1 M/50 mV s^{-1}	[155]
Pd-Ni	6	/	/	−0.03	0.5 M/1 M/50mV s^{-1}	[156]
Pd_{40} Ni_{60}	180	/	−0.712	0.098	1 M/1 M/50mV s^{-1}	[157]
Pd-Ni/CNF	199.8	/	−0.602	−0.0.072	1 M/1 M/10mV s^{-1}	[131]
Pd-Ni-Zn/C	78.5	3600	−0.423	0.057	10 wt%/2 M/50mV s^{-1}	[135]
Pd-Ni-Zn-P/C	108.7	3030	−0.373	0.097	10wt%/2 M/50 mV s^{-1}	[135]
Pd- Ag/C	3.7**	/	−0.602	0.018	1 M/1 M/50mV s^{-1}	[158]
Pd-Ag film	5 **	/		0.02	1 M/1 M/20 mV s^{-1}	[159]
Pd–Pb/C	4.25 **	/	−0.452	−0.0.012	1 M/1 M/20 mV s^{-1}	[160]
PdAu/C	165	/	−0.402	0.0273	1 M/1 M/50 mV s^{-1}	[132]
Pd_4Au/C	/	/	0.5	0.88	1 M/0.25 M/10 mV s^{-1}	[161]
$Pd_{2.5}$Sn/C	/	/	0.55	0.86	1 M/0.25 M/10 mV s^{-1}	[161]
PdAu nanowire	83.7	/	−0.472	−0.001	1 M/1 M/50 mV s^{-1}	[133]
Pd@Au/C	/	800	−0.41	0.04	1 M/1 M/50 mV s^{-1}	[162]

* Unit: A·cm^{-2}·mg^{-1}; ** This value was divided by the EASA.

Pd Supported on Carbon Materials

To improve the utilization efficiency of Pd, the commonly used catalysts are Pd nanoparticles loaded on supports. Various carbon materials such as carbon nanotubes, carbon nanospheres, carbon nanowires, carbon nanofibers, porous carbon, fullerene and graphene have been used as the support. The supporting materials are required to have high surface area, low density, high chemical stability and excellent electrical conductivity.

Pd Supported on Carbon Spheres

In a series of work, Yan *et al.* [163,164,165] synthesized hollow carbon spheres/hemispheres and used them as catalyst support. The hollow carbon hemispheres (HCHs) provided high surface area (up to 1095.59 m^2 g^{-1}) at reduced volume to improve the dispersion of the nanoparticles of the noble metal. At the same time, the hemispherical structure with hollow shell resulted in the improvement in the mass transfer, which leads to greatly improved stability. The peak current density of the ethanol oxidation on the Pd/carbon spheres catalyst reached almost four times higher than that of Pd/C catalyst.Figure 7 shows the typical results of the materials for the alcohol oxidation in alkaline solution. The catalyst showed the best performance for ethanol oxidation. It was revealed that on the same carbon support, the morphology of Pd greatly influenced the activity of ethanol oxidation, the Pd nanobars on carbon have much negative oxidation peak of ethanol than Pd particle on carbon [166].

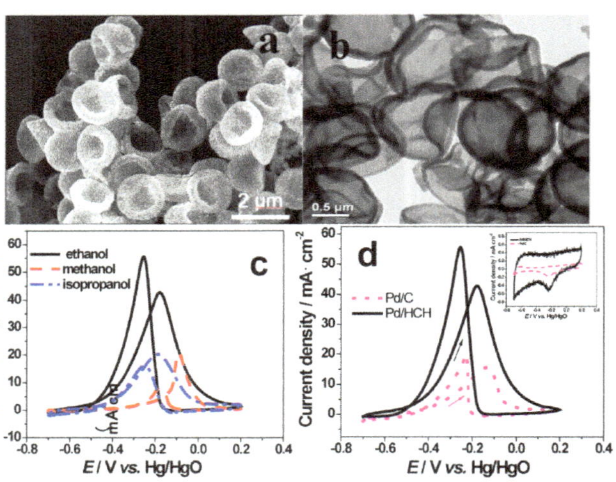

Figure 7. SEM microgram (**a**) and TEM image (**b**) of the HCHs with the mass ratio of PSs to glucose of 1:2 and the cyclic voltammograms of (**c**) different alcohol oxidation on Pd/HCH electrodes in 1.0 mol dm^{-3} KOH/1.0 mol dm^{-3} alcohol solution at 303 K, scan rate: 50 mV s^{-1} and (**d**) ethanol oxidation on Pd/HCH and Pd/C electrodes in 1.0 mol dm^{-3} KOH/1.0 mol dm^{-3} ethanol solution at 303 K, scan rate: 50 mV s^{-1}. The inset in Figure 7d is the cyclic voltammograms of Pd/HCH and Pd/C in 1.0 mol dm^{-3} KOH solution at 303 K, scan rate: 50 mV s^{-1}. Reproduced with permission from reference [130].

Pd Supported on Carbon Nanotubes

Carbon nanotubes (CNTs) can work as ideal substrate to modify the electrode surface used in electrochemistry. Immobilizing metal nanoparticles on CNTs has turned into an interesting field mainly due to the key role of CNTs and metal nanoparticles in the field of electrocatalysis. There are three typical methods for generating Pd nanoparticles on a CNTs surface: chemical reduction reaction, thermal decomposition and electrochemical reduction reaction. Chen *et al.* [136] synthesized Pd nanoparticles-carboxylic functional carbon nanotubes without surfactant. The material revealed high electrochemical activity and excellent catalytic characteristic for alcohol oxidation. Ding *et al.* [167,168] prepared Pd nanoparticles supported on multiwalled carbon nanotubes. Chu *et al.* [169] prepared Pd-In_2O_3/CNTs composite catalysts. The results showed that the addition of nanoparticles of In_2O_3 into Pd catalysts could significantly promote the catalytic activity for ethanol oxidation. Carbon nanotubes have higher graphitization degree than amorphous carbon which leads to higher stability in electrical environment, but higher graphitization also brings a drawback: it is not easy to load Pd nanoparticles on the surface of carbon nanotube evenly. The resolve of the dilemma will greatly accelerate the application of carbon naotube as catalyst support material in fuel cell.

Pd Supported on Carbon Nanofibers

Carbon nanofibers (CNFs) as a novel carbon material offered an ideal opportunity as catalyst support due to their superior electronic conductivity, anti-corrosion ability, and high surface area. Qin *et al.* [170,171,172] prepared Pd catalyst supported on carbon nanofibers (CNFs). The structure of the CNFs significantly affected the catalytic activity of the catalyst because the CNFs had high ratio of edge atoms to basal atoms and correspondingly faster electrode kinetics and stronger Pd-CNFs interaction. Maiyalagan *et al.* [131] prepared carbon nanofibers (CNF) supported Pd–Ni nanoparticles. The onset potential was 200 mV lower and the peak current density four times higher for ethanol oxidation for Pd–Ni/CNF compared to that for Pd/C. Cabon fibers prepared with novel nanotechnology can take the advantage of carbon nanotubes such as high graphitization and high surface area, at the same time it is also possible to construct functional groups on the surface of carbon fibers which is beneficial of the deposition of Pd nanoparticles, so carbon fiber will be a potential candidate as support material of Pd.

Pd Supported on Graphene

The combination of the high specific surface area (theoretical value of 2600 m^2 g^{-1}), excellent electronic conductivity, high chemical stability, unique graphitized basal plane structure and potentially low manufacturing cost, graphene nanosheets (GNS) can thus be exploited as an alternative material for catalyst support in fuel cells. Recently, graphene has received great attention as the catalyst support for fuel cell application. Wen *et al.* [173] prepared Tin oxide (SnO_2)/GNS composite as the catalyst support for direct ethanol fuel cells. Compared with Pd/GNS, the Pd/SnO_2-GNS catalyst showed superior electrocatalytic activity for ethanol oxidation. Chen *et al.* [174] prepared ultrafine Pd nanoparticles on graphene oxide (GO) surfaces. The as-made catalyst expressed high electrocatalytic ability in ethanol oxidation relative to a commercial Pd/C catalyst. Singh *et al.* [142] used graphene nanosheets (GNS) as a catalyst support of palladium nanoparticles for the electrooxidation of ethanol. The Pd nanoparticles dispersed on GNS were more active compared to those dispersed on nanocarbon particles (NC) or multiwall carbon nanotubes

(MWCNTs) for electrooxidation under similar experimental conditions. The enhanced electrochemical activity of Pd/GNS toward alcohol oxidation can be ascribed to the greatly enhanced electrochemical active surface area of Pd nanoparticles on the GNS support.

Pd Supported on Non-Carbon Supports

Conventional carbon supports are prone to undergo corrosion in aggressive electrolytes that are very often encountered in fuel cells. The corroded carbon support cannot hold the catalyst on its surface, leading to aggregation or sintering of noble metal particles (reduces electrochemical active surface area) and often resulting in oxidation and subsequent leaching of the catalyst. Corrosion of the support/catalyst happens mainly because they are exposed to aggressive electrolytes, high temperature and pressure, and high humidity. Carbon is known to undergo corrosion even at open circuit voltages of the fuel cell. So there are some attempts on the application of other materials other than carbon to use as support material of Pd.

Pd Supported on Conducting Polymers

Polyaniline (PANI) is one of the most important conducting polymers remarkable for its high stability, solution processability and tunable electrical conductivity, which can be controlled by simple doping of the polymer. Pandey *et al.*[56] deposited a porous Pd-polyaniline (Pd-PANI) nanofiber film on conducting surfaces and the Pd-PANI showed excellent electrocatalytic activity towards the oxidation of ethanol. Su *et al.* [143] prepared Pd/polyamide 6 (Pd/PA6) nanofibers with high surface area using a simple electroless plating method. The Pd/PA6 showed excellent mechanical property, good conductivity, and high porosity. The large surface area and reduced diffusion resistance of the free-standing Pd/PA6 nanofibers led to a superior catalytic property.

Pd Supported on Zeolite

Zeolite has a large specific area with strongly organized microporous channel systems in which both a regular and high dispersion of metal nanoparticles can be obtained. El-Shafei *et al.* [175] prepared Pd-zeolite graphite (Pd-ZG) electrodes for ethanol oxidation. The Pd-ZG electrodes showed a better activity as well as poisoning tolerance during ethanol oxidation in alkaline medium in comparison with Pd electrode.

Pd Supported on Metal Supports

Nickel foam has the advantages of extinguished electronic conductivity, low weight, and 3-D cross-linked grid structure which provids high porosity and surface area. It can be used as an ideal support of catalyst. The nickel foam would not only reduce the diffusion resistance of the electrolyte but also enhance the facility of ion transportation and maintain the very smooth electron pathways in the rapid electrochemical reactions. Wang *et al.* [144] fabricated a three-dimensional, hierarchically structured Pd electrode by direct electrodepositing. The improved electrocatalytic activity and excellent stability of the Pd/Nickel foam electrode make it a favorable platform for direct ethanol fuel cell applications. Metal nanowire or nanotube array architecture can be a potential substrate to improve noble metal utilization efficiency [176]. Besides it can act as a template, which means that no stabilizer is required, it also serves as an excellent current collector. Cherevko *et al.* [151] prepared highly ordered Pd decorated Au nanowire arrays (Pd/Au NWA). The maximum current densities were several times higher on the modified electrodes than on the unmodified Pd NWA. The highly active electrode showed almost 4-fold increase in the peak current for ethanol oxidation. The synergistic effect between substrate and deposited materials was a most important factor infecting such unusually high activity.

Pd Supported on Nitrides

Titanium nitride (TiN) is a very hard, conducting ceramic material used as an abrasive coating for engineering components. It possesses metal-like electronic conductivity with a very reproducible surface for electron transfer. Thotiyl *et al.* [145] studied the excellent metal-support interaction between Pd and TiN and found the efficient ethanol oxidation coupled with excellent stability of the Pd-TiN catalyst.

Performance of Pd Novel Nano Structures

Core-Shell Structure

Hybrid nanomaterials, particularly, the core-shell structured hybrid nanomaterials are promising due to their multi-functional and designable properties. The core/shell structured nanomaterials has the ability to improve the stability and surface chemistry of the core materials. It is possible to obtain unique structures and properties for applications via a combination of the different characteristics of the components that are not available with their single-component counterparts. The carbon-coated nanomaterials are of great interest due to their stability toward oxidation and degradation [177]. The creation of the core/shell nanostructures containing bi-metal oxides greatly enhanced the catalytic efficiency of these structures over pure single metal oxide particles [178]. Mahendiran *et al.* [146] synthesized carbon coated NiO/MgO in a core/shell nanostructure. The results indicated that the Pd-NiO/MgO@C catalyst has excellent electrocatalytic activity and stability. Gao *et al.* [134] synthesized a core-shell structured Ru@PtPd/C catalyst, with PtPd on the surface and a Ru as core. The ethanol oxidation activities of the Ru@PtPd/C catalysts were 1.3, 3, 1.4 and 2.0 times as high as that of PtPd/C, PtRu/C, Pd/C and Pt/C with same PtPd loadings, respectively. The stability of the Ru@PtPd/C was higher than that of Pt/C and PtPd/C. Ksar *et al.* [38] synthesized bimetallic Pd-Au nanostructures with a core rich in gold and a Pd porous shell. The Pd shell-Au core nanostructures synthesized in mesophases were promising for application in direct ethanol fuel cells as they exhibited a very good electrocatalytic activity and a high stability. The core-shell structure has some essential advantages such as the synergistic effect between the core material and the shell which can improve the kinetics of the reaction, the porous structure which greatly improves the mass transfer in the electrode and the reduced cost because of the less use of the noble metal. So vast efforts have been devoted to this field, but the core-shell structure is limited by the preparation technique and its application in the membrane electrode assembly of fuel cell is still a chanllenge.

Other Structures

Yu *et al.* [28] prepared dendritic Pd nanostructures, which exhibited high catalytic activity toward ethanol oxidation in alkaline media. Yin *et al.* [147] prepared Pd nanoparticles and the ethanol oxidation on the Pd catalysts took place at a more negative anodic potential, implying a reduced overpotential. Wang *et al.* [148] fabricated nanoporous Pd composites through chemical dealloying of the $Al_{70}Pd_{30}$ alloy. The nanoporous Pd composites had high electrochemical active surface areas and exhibited remarkable catalytic activities toward ethanol oxidation in alkaline media. Liu *et al.* [149] synthesized Pd nanowires with high catalytic activity and long-term stability toward the oxidation of alcohols. Ksar *et al.* [150] synthesized Pd nanowires with length of a few tens of nanometers. The Pd nanowires exhibited both a very important catalytic activity for ethanol oxidation and a very high stability. Liu *et al.* [31] prepared raspberry hollow Pd nanospheres (HPNs)-decorated carbon nanotube (CNT) for the oxidation of ethanol in alkaline media. The catalyst was fabricated by attaching HPNs onto the surface of the functionalized CNT. The hybrid nanostructure exhibited higher mass activity toward ethanol oxidation which increased the utilization of Pd. Pd dentric structure, Pd nanowires, hollow raspberry spheres, hollow spheres [179], nanoparticles and nanomembrane [180] are prepared and all showed superior catalytic activity in ehthanol oxidation, but they share the same problem with the core-shell structure: how to a material with high half-cell performance into a material with high full cell performance, which still needs great efforts in the preparation technique.

Pd Alloys

Pd is well known to be very active for ethanol oxidation in alkaline. Alloying Pd with another metal M (M = Au, Sn, Ru, Ag, Ni, Pb and Cu) is expected to increase the activity and at the same time the stability of the catalyst for the EOR in alkaline media [181].

PdNi Alloys

Qiu *et al.* [156] prepared bimetallic Pd-Ni thin film on glass carbon electrodes (GCEs). The high catalytic activity and the low cost of the Pd-Ni films enable them to be promising catalyst for the oxidation of methanol and ethanol in alkaline media. Qi *et al.* [157] fabricated $Pd_{40}Ni_{60}$ alloy catalyst with an enhanced catalytic performance toward ethanol oxidation in alkaline media comoared with nanoporous Pd. Maiyalagan *et al.* [131] prepared carbon nanofibers (CNF) supported Pd–Ni nanoparticles. The onset potential was 200 mV lower and the peak current density was four times higher for ethanol oxidation on Pd–Ni/CNF compared with that of Pd/C. Shen *et al.* [182] synthesized carbon-supported PdNi catalysts for the ethanol oxidation reaction in alkaline direct ethanol fuel cells. The Pd_2Ni_3/C catalyst exhibited higher activity and stability for the EOR in an alkaline medium than that on Pd/C catalyst. Bambagioni *et al.* [135] prepared Pd–(Ni–Zn)/C and Pd–(Ni–Zn–P)/C catalysts which provided excellent results in terms of the specific current and onset potential at room temperature.

PdAu Alloys

Xu *et al.* [132] prepare Pd-Au alloy catalyst for the EOR in an alkaline medium. The catalyst samples were in sequence of Pd/C > Pd_3Au/C > Pd_7Au/C > PdAu/C in terms of the peak current density. However, the stability tests demonstrated that the catalyst samples were in sequence of PdAu/C > Pd_3Au/C > Pd_7Au/C > Pd/C. Cheng *et al.* [133] prepared highly ordered PdAu nanowire arrays (NWAs). The onset potential of ethanol oxidation on the PdAuNWAs electrode was 123 mV more negative compared with that on the Pd NWAs due to the synergistic effect of Pd-Au bimetallic alloy. He *et al.* [161] prepared carbon-supported Pd_4Au and $Pd_{2.5}Sn$-alloyed nanoparticles, the results suggested that the Pd-based alloy catalysts represented promising candidates for the oxidation of ethanol. The Pd_4Au/C displayed the best catalytic activity among the series for the ethanol oxidation in alkaline media. Zhu *et al.* [162] decorated carbon-supported gold nanoparticles with mono- or sub-monolayer Pd atoms, the Pd@Au/C had higher specific activities than that of Pd/C for the oxidation of ethanol in alkaline media. This suggested that the Pd utilization was improved with such a surface-alloyed nanostructure. Several other alloys such as the PdPt alloy [152,183,184,185],

PdSn alloy [153,186], PdIr alloy [155,187], PdAg alloy [158,159,188], PdTi alloy [189] and PdPb alloy [160] were also studied as the catalysts for the EOR.

Single Fuel Cell Characterizations

Although there were quite a lot reports about the Pd-based catalysts for ethanol oxidation, especially, in alkaline media, the performance of a direct ethanol fuel cell operating in alkaline membrane was rarely reported. The reason is that the anion-exchange membrane is not commercially available. The Tokuyama company in Japan is one of the pioneers in the anion-exchange membrane and most reported work were using their membranes. In recent years, Zhao's group [132,182,190,191] has done most of the work on the anion-exchange membrane direct ethanol fuel cells (AEMDEFCs). Other researchers such as Antolini [192] and Bambagioni [193] also tested the performance of the AEMDEFC, the results of these groups were compared in Figure 8.

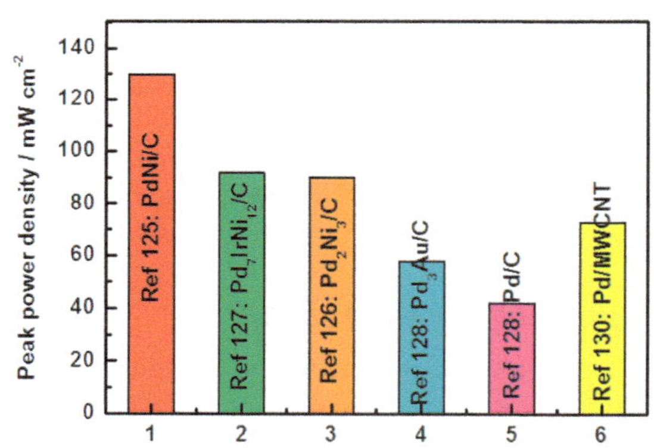

Figure 8. Peak powder density of single fuel cell performance of the Pd catalysts in anion-exchange membrane direct ethanol fuel cells (AEMDEFCs).

It is found that the PdNi alloy had the highest activity which was in accordance with the half cell testing. Among all catalysts, the Pd/C had the lowest activity. The change of carbon powders to carbon nanotubes improved the performance. A promising way was to alloy Pd with other metals, especially, with Ni. The highest activity of the PdNi alloy catalyst reached more than 3 times higher than that of Pd/C catalyst. The conclusion is similar with the half-cell results: alloying with other metals will improve the performance, attention should be put to multi-alloys. Another conclusion reached is the advantage of carbon nanotube *versus* carbon powder with enhanced performance.

PD-BASED CATALYSTS FOR THE OXYGEN REDUCTION REACTION (ORR)

Oxygen reduction reaction (ORR) is one of the important catalytic reactions due to its role in metal corrosion and electrochemical energy converters and, in particularly, fuel cells. Pt group metal based catalysts were currently used for PEMFCs to reduce the large ORR overpotential. Unfortunately, even on the most active Pt surface, the overpotential was over 250 mV at open circuit voltage (OCV). The thermodynamic efficiency droped from 83% (1.229 V) to 66% at an OCV value of 0.98 V under standard conditions. Since the exchange current density of the ORR is several orders of magnitude slower than that of the hydrogen oxidation reaction (HOR) (10^{-9} A cm^{-2} *vs.* 10^{-3} A cm^{-2}), the operating voltage must be largely reduced to about 0.65 V at a reasonable current density, making the electronic efficiency of the PEMFCs only at 44% [194]. Pt is widely used as the catalyst for ORR due to its high activity and excellent chemical stability. However, Pt is expensive and the limited supply of Pt poses serious problems to widespread commercialization of the fuel cell technology. Thus, research efforts in the development of cathode catalyst have been focused on decreasing the Pt content or replacing it with less expensive materials while maintaining high ORR activity. An alternative approach is to replace Pt with less expensive, catalytically active, and relatively stable noble metals and to alloy the noble metals with base metals to enhance their stability and activity via electronic modification. Pd is the second most active metal for the ORR. Pd possesses a similar valence electronic configuration and lattice constant to Pt and highly methanol-tolerant ability. However its mass activity for the ORR is approximately five times lower than that of Pt. In an acidic electrolyte, the exchange current density of Pd for ORR is 10^{-10} A cm^{-2}, which is one magnitude lower than that of Pt (10^{-9} A cm^{-2}). Binary Pd-base metal (BM) (where BM = Co, Ni, Fe, Cu, W and Mo) systems have been identified as promising PEFC cathode catalyst with enhanced activity for ORR and stability compared to Pd alone. The origin of the enhanced activity has been linked to the modification of the electronic structure of Pd upon bonding with the alloying metal. In addition to enhancing activity, the dissolution potentials of the noble metals may be shifted to higher potentials, stabilizing the catalyst against dissolution in acidic medium.

Table 3. Comparison of the onset potential and mass activity of Pd for oxygen reduction reaction (ORR).

Catalysts	Mass Activity/mA mg^{-1}_{Pd}	Onset Potential/V *vs.* RHE	Conditions Solution **/Scan Rate/Rotating Speed	References
Pd/C	/	0.87	A/5 mV s^{-1}/1600 rpm	[195]
Pt/C (JM)	/	1.05	A/5 mV s^{-1}/1600 rpm	[195]
Pd–PPy/C	/	0.82 (NHE)	A/5 mV s^{-1}/1600 rpm	[196]
40% E-TEK Pt/C	/	0.98 (NHE)	A/5 mV s^{-1}/1600 rpm	[196]
Pd/Vulcan XC-72R	47 at 0.7 V *	0.82	A/5 mV s^{-1}/1600 rpm	[197]
20% E-TEK Pt/C		0.92	A/5 mV s^{-1}/1600 rpm	[197]
E-Tek 20% Pd/C	23 at 0.7 V *	0.77	A/5 mV s^{-1}/1500 rpm	[197]
PdFe-WC/C	/	0.91	A/10 mV s^{-1}/1600 rpm	[198]
$Pt_{50}Au_{50}/Ce_xC$	4.1 at 0.7 V *	/	A/5 mV s^{-1}/1500 rpm	[180]
PdCoMo/CDX975	4.1 at 0.7 V *	0.915	A/5 mV s^{-1}/1600 rpm	[199]
Pd/CDX975	1.6 at 0.7 V *	0.84	A/5 mV s^{-1}/1600 rpm	[199]
$Pd_{100-x}W_x$	/	0.85	A/5 mV s^{-1}/1600 rpm	[200]
Pd	/	0.7	A/5 mV s^{-1}/1600 rpm	[200]
PdPt/C	/	0.98	A/5 mV s^{-1}/1600 rpm	[201]
Pt rich-core Pd rich-shell	9 at 0.85 V * (catalyst)	/	A/5 mV s^{-1}/1600 rpm	[202]
JM20 Pt/C	5.8 at 0.85 V * (catalyst)	/	A/5 mV s^{-1}/1600 rpm	[202]
Pd/Au	0.34 at 0.8 V *	/	A/10 mV s^{-1}	[203]
	1.1 at 0.8 V *	/	B/10 mV s^{-1}	[203]
Pd/MWCNTs	/	0.76 (SHE)	B/10 mV s^{-1}	[91]
PdCu/C	23 at 0.9 V *	0.96	B/10 mV s^{-1}/1600 rpm	[204]
Pd–Cu film	/	0.86	B/5 mV s^{-1}/1000 rpm	[205]
Pd–Co/C	3.6 at 0.7 V *	/	B/5 mV s^{-1}/1600 rpm	[206]
Pt–Pd/C	/	0.92	B/5 mV s^{-1}/1600 rpm	[207]

Catalysts	Mass Activity/mA $mg^{-1}{}_{Pd}$	Onset Potential/V *vs.* RHE	Conditions Solution **/Scan Rate/Rotating Speed	References
Pt–Pd/C	114.87 at 0.9 V * (Pt)	1.05	B/5 mV s^{-1}/1600 rpm	[208]
Pt/C	73.87 at 0.9 V * (Pt)	1.0	B/5 mV s^{-1}/1600 rpm	[208]
PdFe@PdPt/C	1.92 at 0.8 V * (Pt)	/	B/20 mV s^{-1}/1600 rpm	[209]
	1.2 at 0.8 V * (Pt)	/	B/20 mV s^{-1}/1600 rpm	[209]
PdCo@PdPt/C	65 at 0.9 V * (Pt)	/	B/20 mV s^{-1}/1600 rpm	[210]
Pt/C	14 at 0.9 V * (Pt)	/	B/20 mV s^{-1}/1600 rpm	[210]
PdFe Nanorods	284 at 0.85 V *	/	B/10 mV s^{-1}/1600 rpm	[211]
Pt/C	265 at 0.85 V * (Pt)	/	B/10 mV s^{-1}/1600 rpm	[211]
Pt/Pd/Pd3Fe	1.8 at 0.9 V *	/	B/20 mV s^{-1}/1600 rpm	[212]
Pt(111)	0.8 at 0.9 V *	/	B/20 mV s^{-1}/1600 rpm	[212]
Pd/SnO2–KB	0.75 at 0.8 V *	0.88	B/5 mV s^{-1}/900 rpm	[64]
Pd/KB	0.28 at 0.8 V *	0.86	B/5 mV s^{-1}/900 rpm	[213]
Tanaka Pt/C	/	0.95	B/5 mV s^{-1}/900 rpm	[213]
Pd/GNS	280 at 0.9 V *	1.06	C/10 mV s^{-1}/1600 rpm	[188]
Pt/GNS	110 at 0.9 V *	1.0	C/10 mV s^{-1}/1600 rpm	[188]
Pd@MnO_2/C	450 at 0.9 V *	1.02	D/10 mV s^{-1}/2500 rpm	[214]
Pd black	180 at 0.9 V *	1.07	D/10 mV s^{-1}/2500 rpm	[214]

* The potentials were *vs.* RHE; **solution A: 0.5 M H_2SO_4, solution B: 0.1 M $HClO_4$, solution C: 0.1 M NaOH and solution D: 0.1 M KOH.

With the rapid development of alkaline anion exchange membranes for substituting the conventional aqueous alkaline electrolyte, attention has been drawn to the study of ORR catalysts in alkaline media. More ORR catalysts are available for operation in alkaline than in acidic media due to the excessive corrosion in acidic solutions. In alkaline solution, Pd-based alloys are suitable alternative of Pt not only due to its lower costs and more abundance but also to the lower activity for the adsorption and oxidation of methanol in direct methanol fuel cells which tends to crossover to the cathode compartment and inhibits ORR. Recent reports have shown that ORR activity on the Pd alloys is comparable or slightly better than that on Pt/C. Table 3 summarized the performance of Pd and Pd alloys as catalysts for ORR. In acidic solution, the most positive onset potential on Pd catalyst was only 0.87 V *vs.* RHE which was almost 0.2 V negative compared with Pt catalyst. In alkaline solution, the onset potential on Pd catalyst was comparable to that of Pt catalyst. The onset potential could be improved to 1.05 V when alloyed with Pt, the highest record for alloying with non-noble metal was 0.96 V in half cell testing. Fuel cell testing gave similar conclusions. The performance of Pd catalyst could not surpass that of Pt catalyst in both PEMFC and direct alcohol fuel cells, but when Pd was alloyed the performance could be slightly better than or comparable to the commercial Pt catalyst.

Pd Supported on Carbon Materials

Pd on Carbon Powder

Although Pd is the second most active metal for the ORR, its mass activity is approximately five times lower than that of Pt since the ORR exchange current density is one magnitude lower than that of Pt. Tang *et al.* [195] synthesized a carbon supported Pd/C catalyst which showed high activity for the ORR. However, the performance of the Pd/C was still much poorer than that of the commercial Pt/C. There was a big difference in onset potential for the ORR in acidic solution. Single cell with Pd/C as the cathode displayed a maximum power density of 508 mW cm^{-2}, which was almost half of that of the commercial Pt/C catalyst. Different supporting materials such as highly ordered pyrolytic graphite and modified carbon was studied [215]. Jeyabharathi *et al.* [196] synthesized carbon-supported Pd-polypyrrole (Pd-PPy/C) nanocomposite. The introduction of Pd in the conducting PPy/C matrix gave higher catalytic activity toward ORR with resistance to methanol oxidation. The performance of the Pd-PPy/C catalyst was still inferior to the Pt/C catalyst. Kumar *et al.* [197] studied the influence of the chemical pretreatment of carbon support for ORR on Pd nanoparticles in acidic electrolyte. They found that the chemical treatment significantly changed the surface chemical properties and surface area of the carbon support. The kinetics of ORR on these catalysts predominantly involved a four-electron step. The performance of the Pd on pretreated carbon support was found to be much higher than the commercial E-Tek 20% Pd/C catalyst with the mass activity of 47 mA $mg^{-1}{}_{Pd}$ at 0.7 V RHE compared to 23 mA $mg^{-1}{}_{Pd}$ of E-Tek catalyst. It is concluded that limited by the intrinsic kinetic of ORR on different metals, the performance of the most popular Pd/C catalyst is much less than the Pt/C catalyst. However, considering the difference in the cost, it is still meaningful to study the Pd/C catalyst, but new material or structure should be put more attention.

Pd Supported on Carbon Nanotubes

Chakraborty *et al.* [91] synthesized nanosized Pd particles supported on multiwalled carbon nanotubes (MWCNTs). The authors concluded that the catalytic reduction of the oxygen followed a four electron pathway on Pd-based catalysts. Jukk *et al.* [216] found enhanced electrocatalytic activity of PdNP/MWCNT modified GC electrodes and the oxygen electroreduction kinetics were higher compared with those of bulk palladium electrodes. The number of electrons transferred per oxygen molecule was calculated to be 4. Kim *et al.* [217] studied the influence of counter ions on the oxygen reduction of Pd catalyst on functionalized carbon nanotube and found that the electrocatalytic activity is affected by the nature of anion of imidazolium salt. As is known the activity of Pd on carbon powder is less than Pt/C, not too much work is done on the attempts of loading Pd on carbon nanotubes for ORR application. However if carbon nanotubes can be helpful in the stability there will be new interest in this field.

Pd Supported on Graphene

Graphene is found to have vast applications in many fields where carbon powder or carbon nanotubes are used, for example as catalyst support in electrocatalysis. There was no report to prove that the activity of Pd catalyst for the ORR could surpass that of Pt catalyst in acidic solution. However, it is possible for the Pd catalyst to perform better than that of the Pt catalyst for the ORR in alkaline solution when supported on graphene. Seo *et al.* [218] studied graphene supported Pd catalyst in alkaline media. The graphene-supported Pd catalyst (Pd/GNS) showed significantly high catalytic activity for the ORR with higher mass activity and surface area for ORR in an alkaline solution. The catalyst was more favorable for ORR than that of the Pt/GNS catalyst at high metal loadings. The mass activity at 0.85 V *vs.* RHE of the Pd/GNS reached 0.84 mA μg^{-1}_{Pd}, while, the value was 0.35 mA μg^{-1}_{Pd} for the Pt/GNS. Pd metal decorated graphene oxide was found to have high ORR activity via the direct four electron pathway [219]. Gotoh *et al.* [220,221] also proved that Pd supported on graphene oxide showed better performance in ORR than Pt.

Pd Supported on Oxides

Oxides such as manganese dioxide was used as support to form Pd@MnO_2 catalyst [214]. The ORR onset potential on the Pd@MnO_2 catalyst positively shifted for more than 250 mV compared with the MnO_2 catalyst without Pd. Both the ORR onset potential and the limit current density obtained by the rotating disk electrode (RDE) measurements on the Pd@ MnO_2catalyst were close to those on the Pd black catalyst. The mass activity of the Pd@MnO_2 catalysts (normalized by Pd mass) was 2.5 times higher than that of the Pd black catalyst. The PtPd/TiO_2 electrocatalyst with a proper ratio of Pt/Pd showed activity comparable to that of a commercial Pt/C catalyst [222]. The interaction between Pd and highly dispersed TiO_2 is proven to improve the catalytic activity of Pd supported on TiO_2-modified carbons [223].

Pd Alloys

The intrinsic catalytic activity of Pd for the ORR is lower than that of Pt and the long-term stability at high potentials is also not as good as that of Pt. It has been proven that the ORR takes place on Pd in the same manner as that on Pt. As like Pt, the oxygen intermediate species will be covered on Pd surface at technically relevant potentials regioning from 0.7 to 0.9 V*vs.* NHE and hinders oxygen reduction. So the development of the Pd alloy catalysts that inhibit the adsorption of oxy/ hydroxy species and enhance the ORR activity has been of interest. Bimetallic catalysts often exhibit notably different catalytic and chemical properties than their corresponding monometallic component. Bimetallic systems often provide enhanced selectivity, stability, and/or activity [224,225]. Several mechanisms to explain the catalytic properties of bimetallic catalysts have been proposed including geometric or ensemble effects, formation of bifunctional surfaces and electronic modification of the surface sites [226].

PdFe Alloys

Yin *et al.* [198] synthesized a PdFe-WC/C cathodic catalyst for ORR. The ORR activity of the PdFe–WC/C catalyst in acidic solution was found to be comparable to that of Pt/C catalyst. It was believed that the high catalytic activity as a Pt-free catalyst originated from the synergistic effect between PdFe and WC. The alcohol-tolerance and selectivity of the PdFe-WC/C catalyst are favorable for the ORR. Tungsten-based materials as novel supports have been intensively studied due to their chemical and/or electrochemical activities for various reactions as they exhibit a synergistic effect in many reactions. The use of WC as the catalyst support for PEMFCs and DMFCs has also been reported [227,228,229,230,231] PdFe/C catalyst was also found to have better performance in ORR than Pd/C catalyst [232,233].

PdCu Alloys

Gobal *et al.* [234,235] prepared CuPd alloys with different compositions on nickel. The number of transferred electrons involved in the ORR on Pd-Cu alloys is four, which is the same as Pt. A 60 mV/dec Tafel slope for the ORR was found for all the PdCu alloys. The enhancement of the activity of the alloy toward ORR was attributed to the change in geometric and electronic structures of Pd caused by the insertion of Cu. Kariuki *et al.* [204] prepared monodispersed PdCu alloy nanoparticles, which showed high ORR activity in acidic electrolyte. Fouda-Onana *et al.* [205] found $Pd_{50}Cu_{50}$ exhibited the high activity in ORR. The enhancement was attributed to an optimal *d*-band property that made the OOH dissociative adsorption easier, which was considered as chemical rate determining step (RDS) for the ORR [236,237].

PdAg Alloys

Ag catalysts have lower activity in ORR than those of Pt catalysts because of their weak interaction for binding O_2. However, the inexpensive Ag nanoparticles have been shown to have higher stability than pure Pt cathodes during long-term operation. Lee *et al.* [238] supported AgPd alloy on multiwalled carbon nanotubes and found that the ORR proceeded through a two-electron pathway, while according to other authors the ORR on AgPd alloy was a four-electron process [239,240]. In alkaline medium the electrode reaction kinetics is higher than that in the acidic medium, enabling the use of Pt-free catalysts. Oliveira *et al.* [159] evaluated PdAg alloys toward the ORR in alkaline medium and found that alloying Pd with Ag leaded to an increase in the ORR kinetics relative to Pd.

PdAu Alloys

Gold has been used as support material for studying the catalytic behaviour of Pd. Sarapuu *et al.* [203] evaluated the influence of the Pd film thickness and Au substrate to the ORR activity of Pd. The ORR proceeded through 4-electron pathway on all PdAu electrodes. The specific activity of ORR was lower in H_2SO_4 solution and decreased slightly with decreasing the Pd film thickness. In $HClO_4$, the specific activity was higher and was not significantly dependent on the film thickness. Xu *et al.* [241] found Au-modified Pd catalyst exhibited increased catalytic activity for ORR in alkaline media, which was 1.4 times higher than that with the mono-Pd catalyst.

PdCo Alloys

There was an in-depth understanding of the various factors that ifluence the catalytic activity of the PdCo nanoalloys [242,243,244]. It was found that a mild annealing of the alloys at moderate temperatures (350 °C) was desirable to clean the surface and maximize the catalytic activity and durability [245,246,247]. Tominaka *et al.* [248] synthesized a mesoporous PdCo sponge-like nanostructure with a most desirable lattice contraction into a Pd catalyst for the ORR. The mesoporous PdCo catalyst had a higher specific activity than that of the Pt catalyst. Wei *et al.* [206] obtained PdCo/C alloy catalysts with an atomic ratio of 3:1. They found the well-formed PdCo alloy showed excellent ORR activity. Serov *et al.* [249] investigated PdCo catalysts for ORR in a direct methanol fuel cell. Such a non-Pt catalyst showed comparable power density with a commercial MEA prepared using Pt cathode. Rao *et al.* [199] prepared PdCoMo alloy nanoparticles with better catalytic activity compared with Pd.

PdPt Alloys

The nanosized Pd supported on carbon black will gradually grow larger during long-term operation, thus reducing the electrochemical active surface area and resulting in irreversible performance loss. Many works had been tried to improve the performance and durability of the catalyst, alloying with other metal is one important method. Lots of works had been done on the alloys of Pd with other noble metal, for example Pt [250,251,252]. Pd and Pt had a face-centered cubic (fcc) phase with a unit length of 3.92 Å for Pt and 3.89 Å for Pd. The small lattice mismatch meant that the epitaxial growth should be favored. Figure 9 shows a typical result of PdPt alloy for the ORR. The PdPt nanodendrites were two and a half times more active than the state-of-the-art Pt/C catalyst [253,254,255].

The PtPd/C showed a comparable performance and better durability than that of the Pt/C [201]. Thanasilp *et al.* [256] demonstrated that the different Pt:Pd atomic ratios had a significant effect on the catalyst activity. Decreasing the Pt:Pd atomic ratio led to an increase in the particle size and decrease in the electrochemical activity. Fıçıcılar *et al.* [257] found that when the particle size of Pd increased with the content and a lower Pd content exhibited a considerable activity and increased stability Chang *et al.* [202] found that the Pt_3Pd_1/C nanocatalyst has a 50% enhancement in ORR due to the synergistic effect. Ohashi *et al.* [207] prepared various PtPd/C bimetallic catalysts with a higher tolerance to ripening induced by potential cycling. Peng *et al.* [208] designed a Pt particle-on-Pd structure to address both the activity and stability issues. Other alloys of Pd–W, [200] Pd–V, [258] Pd–Ni [259,260], Pd–Sn [261,262] and Pt–Ir–Re [263] were also studied as catalysts for the ORR.

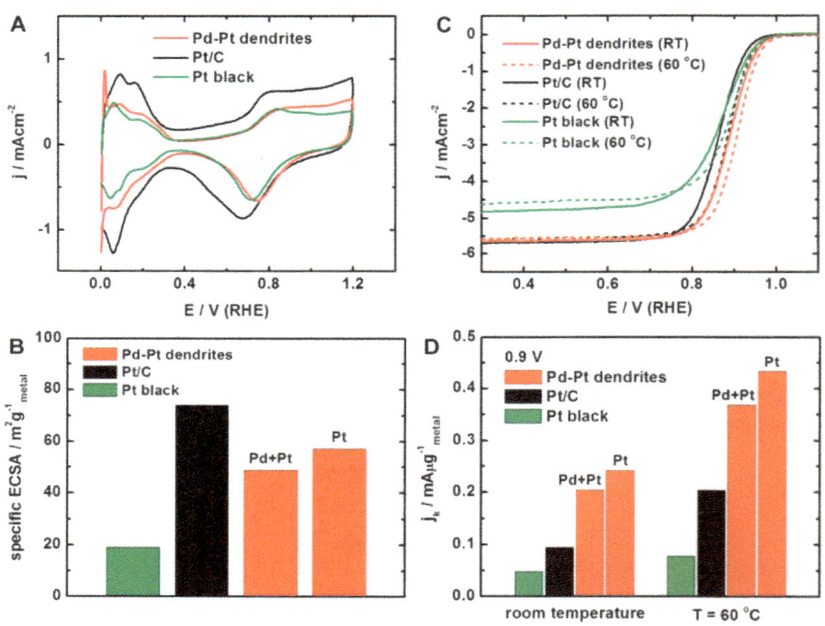

Figure 9. Comparison of the catalytic properties of the Pd-Pt nanodendrites, Pt/C catalyst (E-Tek), and Pt black (Aldrich). (**A**) The CV curves recorded at room temperature in an Ar-purged 0.1 M $HClO_4$ solution with a sweep rate of 50 mV s^{-1}; (**B**) specific ECSAs (electrochemical active surface area) for the Pd-Pt nanodendrites, Pt/C catalyst, and Pt black; (**C**) ORR polarization curves for the Pd-Pt nanodendrites, Pt/C catalyst and Pt black recorded at room temperature and 60 °C in an O_2-saturated 0.1 M $HClO_4$ solution with a sweep rate of 10 mV s^{-1} and a rotation rate of 1600 rpm and (**D**) mass activity at 0.9 V *versus* RHE (reversible hydrogen electrode) for these three catalysts. Reproduced with permission from reference [253].

Novel Nanostructures

Recent attention has been drawn to the strong dependence of the catalytic properties of Pd on their surface morphologies. For example, the specific activity of the Pd nanorods prepared by electrodeposition was found to be close to that of Pt in acidic solution and was 10 times higher than that of electrodeposited Pd nanoparticles [264]. The higher activity was attributed to the exposure of Pd (110) surface facet. The activity of the Pd with low index planes increased in the following order: Pd(110) < Pd(111) < Pd(100) [265,266,267]. The Pd(100) single crystal plane was even more active than Pt(110) in 0.1 M $HClO_4$. The low indexed Pd became favorable in the electrochemical reaction, controlling the morphology is one of the commonly used methods to get low indexed metal.

Core-Shell Structures

The heterogeneous core shell structure has many advantages. First of all, the core-shell structure can greatly reduce the cost of the catalyst. Second, the strain caused by the lattice mismatch between the surface and core components may be used to modify the electronic properties of the surface metal atoms, most notably their *d*-band centres, which affect the rates of one or more elementary steps in the overall catalytic reaction [268,269,270].

Lim *et al.* [253] synthesized PdPt bimetallic nanodendrites consisting of Pt branches as shell and Pd as core, the nanodendrites were two and half times more active for the ORR than the state-of-the-art Pt/C catalyst. Peng *et al.* [208] also found improved stability for ORR with a Pt-on-Pd core-shell nanostructure. It was calculated that the optimal coverage of Pt on Pd (111) surface was on the order of two monolayers [271]. Yang *et al.* [209] prepared a core-shell structure with a PdFe core and a PdPt shell, which showed four times ORR activity compared with a commercial Pt/C catalyst. Yang *et al.* [210] constructed a PdPt shell on a PdCo core with six fold increases in the activity and with much higher stability. Sasaki *et al.*[272] illustrated a core/shell catalyst with Pd and Pd_9Au_1 alloy as core and Pt monolayer as shell with high activity and very high stability. The origin of the improved activity and stability of the core-shell catalyst was studied. The Pd core not only assured the long-term stability of the monolayer Pt shell, but increased the activity of Pt by causing it to contract slightly, lowering its *d*-band centre energy and reducing the bond strength of the adsorbed oxygen intermediates. These effects decreased the bonding of OH and O to Pt that inhibited the ORR kinetics and also stabilize Pt against oxidation and dissolution.

Unsupported Pd

Besides the most frequently used supported Pd catalysts, the application of nano-structured Pd as catalyst for the ORR has also been studied. For example, PdFe nanorods [211], $PdCo_xCN_y$ [273], Pt monolayer supported on Pd/Pd_3Fe [212] and Pd/SnO_2 [213] demonstrated significantly increased ORR activity. The catalytic activity of Pd could become comparable to that of Pt upon appropriate modification of its electronic structure. The surface specific activity of Pd nanorods (Pd-NRs) toward the ORR was found to be not only 10-fold higher than that of Pd nanoparticles (Pd-NPs), but also comparable to that of Pt at operating potentials of fuel cell cathodes [264]. Zhang *et al.* [194] prepared PdFe-nanoleaves with Pd-rich nanowires surrounded by Fe-rich sheets. The structure demonstrated three times increased specific activity and 2.7 times increased mass activity compared witha commercial Pt/C catalyst.5.5. Singal Fuel Cell Characterizations

Two kinds of fuel cells, the PEMFC feeding with H_2/O_2 and direct alcohol (ethanol of methanol) fuel cells, were studied by using above mentioned catalysts. In PEMFC applications, Tang *et al.* [195] proved the performance of Pd/C was much less than that of the commercial Pt/C. Single cell with Pd/C as the cathode displayed a maximum power density of 508 mW cm^{-2}which was almost half of that of commercial Pt/C catalyst. This result was the same with the half cell test, that is, the performance of Pd/C catalyst was much inferior to that of the Pt/C catalyst for the ORR. Thanasilp *et al.* [256] studied the influence of Pt:Pd atomic ratios on a carbon supported upon its suitability as a cathode for a PEMFC. Although the different Pt:Pd atomic ratios had a significant effect on the performance in a H_2/O_2 fuel cell, the performance of the PdPt alloy was still much lower than that of Pt/C catalyst. Rosa *et al.* [274] directly sprayed Pd ink on carbon paper to form a novel oxygen diffusion electrode for the PEMFC, but the utilization efficiency of Pd was not satisfactory. With the help of nanotechnology, the performance of Pd catalyst with novel nano-structures as the PEMFC cathode was greatly improved. Li *et al.* [211] synthesized PdFe nanorods with tunable length which showed a better PEMFC performance than that of the commercial Pt/C due to their high intrinsic activity to ORR at reduced cell inner resistance and improved mass transport.

In direct alcohol fuel cells, Xu *et al.* [241] studied Au-modified Pd catalysts on carbon nanotubes which yielded a peak power density of 1.4 times higher than that with the mono-Pd cathode but was still less than Pt cathode. Pd alloys such as PdCoMo alloy [199] and PdCo alloy [199] showed comparable performance with that of commercial Pt/C, the PdNi alloy showed much higher performance than of Pd/C but still inferior to that of Pt/C catalyst [260]. It turns out that alloying Pd with other metals like Fe, Co and Ni is possible to improve the performance, reduce the cost and improve the stability of fuel cell.

CONCLUSIONS AND FUTURE PERSPECTIVE

This article reviewed the latest advances in Pd-based catalysts for fuel cells. The review focused on Pd nanostructure, Pd catalysts for formic acid oxidation, alcohol oxidation and oxygen reduction reaction.

Different Pd morphologies were prepared. Due to the intrinsic advantages in crystallography and morphology, most of the Pd nanomaterials have high performance in fuel cell half cell characterization. Both the activity and stability of the catalysts could be improved, which would significantly reduce the usage of Pd in fuel cells. But the nano-structured Pd could not be vast applied in fuel cell, because of the limited yield and the difficulty in the MEA preparation. Futhure studies should be conducted to realize the mass production and find ways for efficient MEA preparing techniques.

As fuel cell catalysts for formic acid oxidation, methanol oxidation and oxyren reduction, Pd was loaded on carbon powders or other novel supports such as graphene and carbon nanotubes to achieve high electrochemical active surface area and improve the utilization efficiency of the Pd catalyst. The nature of the support materials also had great influence on the activity of Pd catalyst. Supported Pd nano-strucutures often had better performance than particles. Alloying with other metals could modify the electronic structure and induce tensile strain of the Pd clusters and finally infiuence their catalytic activities. There was great potential in the development of Pd alloy catalysts especially with non-noble metals to perform improved performance and stability and at the same time the reduced cost. Considering the cost and comparable activity with Pt, the Pd-based catalysts are potential candidates as main catalysts for fuel cells to reduce the use of Pt and the cost for commercialization.

Acknowledgments

This work was supported by National Natural Science Foundation of China (21106190, 21476096, 51303217); Pearl River S&T Nova Program of Guangzhou (2013J2200040); Blue Ocean Talent Project of Nanhai, Foshan; 44th Scientific Research Foundation for the Returned Overseas Chinese Scholars, State Education Ministry; Key Laboratory of Functional Inorganic Material Chemistry (Heilongjiang University), Ministry of Education; the Fundamental Research Funds for the Central Universities.

Author Contributions

Hui Meng worte the review and designed the structure. Fangyan Xie and Dongrong Zeng collected and classified the references.

Conflicts of Interest

The authors declare no conflict of interest.

REFERENCES

1. Grove, W.R. XXIV. On voltaic series and the combination of gases by platinum. *Philos. Mag. Ser.* **1839**, *14*, 127–130.
2. Soszko, M.; Lukaszewski, M.; Mianowska, Z.; Czerwinski, A. Electrochemical characterization of the surface and methanol electrooxidation on Pt–Rh–Pd ternary alloys. *J. Power Sources* **2011**, *196*, 3513–3522.
3. Shao, M. Palladium-based electrocatalysts for hydrogen oxidation and oxygen reduction reactions. *J. Power Sources***2011**, *196*, 2433–2444.
4. Palladium Chart—Last 5 years. Available online: http://www.kitco.com/charts/popup/pd1825nyb.html (accessed on 1 May 2015).
5. Bianchini, C.; Shen, P.K. Palladium-Based Electrocatalysts for Alcohol Oxidation in Half Cells and in Direct Alcohol Fuel Cells. *Chem. Rev.* **2009**, *109*, 4183–4206.
6. Antolini, E. Palladium in fuel cell catalysis. *Energy Environ. Sci.* **2009**, *2*, 915–931.
7. Mazumder, V.; Lee, Y.; Sun, S.H. Recent Development of Active Nanoparticle Catalysts for Fuel Cell Reactions. *Adv. Funct. Mater.* **2010**, *20*, 1224–1231.
8. Antolini, E.; Gonzalez, E.R. Alkaline direct alcohol fuel cells. *J. Power Sources* **2010**, *195*, 3431–3450.
9. Morozan, A.; Jousselme, B.; Palacin, S. Low-platinum and platinum-free catalysts for the oxygen reduction reaction at fuel cell cathodes. *Energy Environ. Sci.* **2011**, *4*, 1238–1254.
10. Zhao, X.; Yin, M.; Ma, L.; Liang, L.; Liu, C.P.; Liao, J.H.; Lu, T.H.; Xing, W. Recent advances in catalysts for direct methanol fuel cells. *Energy Environ. Sci.* **2011**, *4*, 2736–2753.
11. Adams, B.D.; Chen, A.C. The role of palladium in a hydrogen economy. *Mater. Today* **2011**, *14*, 282–289.
12. Zhou, Z.Y.; Tian, N.; Li, J.T.; Broadwell, I.; Sun, S.G. Nanomaterials of high surface energy with exceptional properties in catalysis and energy storage. *Chem. Soc. Rev.* **2011**, *40*, 4167–4185.
13. Tian, N.; Zhou, Z.Y.; Yu, N.F.; Wang, L.Y.; Sun, S.G. Direct Electrodeposition of Tetrahexahedral Pd Nanocrystals with High-Index Facets and High Catalytic Activity for Ethanol Electrooxidation. *J. Am. Chem. Soc.* **2010**, *132*, 7580–7581.
14. Zhou, Z.Y.; Tian, N.; Huang, Z.Z.; Chen, D.J.; Sun, S.G. Nanoparticle catalysts with high energy surfaces and enhanced activity synthesized by electrochemical method. *Faraday Discuss.* **2008**, *140*, 81–92.
15. Shen, Q.M.; Min, Q.H.; Shi, J.J.; Jiang, L.P.; Zhang, J.R.; Hou, W.H.; Zhu, J.J. Morphology-Controlled Synthesis of Palladium Nanostructures by Sonoelectrochemical Method and Their Application in Direct Alcohol Oxidation. *J. Phys. Chem. C* **2009**, *113*, 1267–1273.
16. Ding, H.; Shi, X.Z.; Shen, C.M.; Hui, C.; Xu, Z.C.; Li, C.; Tian, Y.A.; Wang, D.K.; Gao, H.J. Synthesis of monodisperse palladium nanocubes and their catalytic activity for methanol electrooxidation. *Chin. Phys. B* **2010**, *19*.
17. Yin, Z.; Zheng, H.J.; Ma, D.; Bao, X.H. Porous Palladium Nanoflowers that Have Enhanced Methanol Electro-Oxidation Activity. *J. Phys. Chem. C* **2009**, *113*, 1001–1005.
18. Meng, H.; Sun, S.; Masse, J.P.; Dodelet, J.P. Electrosynthesis of Pd Single-Crystal Nanothorns and Their Application in the Oxidation of Formic Acid. *Chem. Mater.* **2008**, *20*, 6998–7002.
19. Meng, H.; Wang, C.X.; Shen, P.K.; Wu, G. Palladium thorn clusters as catalysts for electrooxidation of formic acid.*Energy Environ. Sci.* **2011**, *4*, 1522–1526.
20. Tian, N.; Zhou, Z.Y.; Sun, S.G. Electrochemical preparation of Pd nanorods with high-index facets. *Chem. Commun.***2009**, 1502–1504.
21. Patra, S.; Viswanath, B.; Barai, K.; Ravishankar, N.; Munichandraiah, N. High-Surface Step Density on Dendritic Pd Leads to Exceptional Catalytic Activity for Formic Acid Oxidation. *ACS Appl. Mater. Interfaces* **2010**, *2*, 2965–2969.
22. Zhang, J.; Cheng, Y.; Lu, S.F.; Jia, L.C.; Shen, P.K.; Jiang, S.P. Significant promotion effect of carbon nanotubes on the electrocatalytic activity of supported Pd NPs for ethanol oxidation reaction of fuel cells: The role of inner tubes. *Chem. Commun.* **2014**, *50*, 13732–13734.
23. Du, C.Y.; Chen, M.; Wang, W.G.; Yin, G.P. Nanoporous PdNi Alloy Nanowires As Highly Active Catalysts for the Electro-Oxidation of Formic Acid. *ACS Appl. Mater. Interfaces* **2011**, *3*, 105–109.
24. Lee, Y.W.; Lim, M.A.; Kang, S.W.; Park, I.; Han, S.W. Facile synthesis of noble metal nanotubes by using ZnO nanowires as sacrificial scaffolds and their electrocatalytic properties. *Chem. Commun.* **2011**, *47*, 6299–6301.
25. Wen, D.; Guo, S.J.; Dong, S.J.; Wang, E.K. Ultrathin Pd nanowire as a highly active electrode material for sensitive and selective detection of ascorbic acid. *Biosens. Bioelectron.* **2010**, *26*, 1056–1061.
26. Jena, B.K.; Sahu, S.C.; Satpati, B.; Sahu, R.K.; Behera,

D.; Mohanty, S. A facile approach for morphosynthesis of Pd nanoelectrocatalysts. *Chem. Commun.* **2011**, *47*, 3796–3798.

27. Zhou, R.; Zhou, W.Q.; Zhang, H.M.; Du, Y.K.; Yang, P.; Wang, C.Y.; Xu, J.K. Facile template-free synthesis of pine needle-like Pd micro/nano-leaves and their associated electro-catalytic activities toward oxidation of formic acid.*Nanoscale Res. Lett.* **2011**, *6*.

28. Yu, J.S.; Fujita, T.; Inoue, A.; Sakurai, T.; Chen, M.W. Electrochemical synthesis of palladium nanostructures with controllable morphology. *Nanotechnology* **2010**, *21*.

29. Fang, Y.X.; Guo, S.J.; Zhu, C.Z.; Dong, S.J.; Wang, E.K. Twenty Second Synthesis of Pd Nanourchins with High Electrochemical Activity through an Electrochemical Route. *Langmuir* **2010**, *26*, 17816–17820.

30. Li, L.Q.; E, Y.; Yuan, J.M.; Luo, X.Y.; Yang, Y.; Fan, L.Z. Electrosynthesis of Pd/Au hollow cone-like microstructures for electrocatalytic formic acid oxidation. *Electrochim. Acta* **2011**, *56*, 6237–6244.

31. Ye, L.; Wang, Y.; Chen, X.Y.; Yue, B.; Tsang, S.C.; He, H.Y. Three-dimensionally ordered mesoporous Pd networks templated by a silica super crystal and their application in formic acid electrooxidation. *Chem. Commun.* **2011**, *47*, 7389–7391.

32. Carrera-Cerritos, R.; Fuentes-Ramirez, R.; Cuevas-Muniz, F.M.; Ledesma-Garcia, J.; Arriaga, L.G. Performance and stability of Pd nanostructures in an alkaline direct ethanol fuel cell. *J. Power Sources* **2014**, *269*, 370–378.

33. Choi, S.I.; Shao, M.H.; Lu, N.; Ruditskiy, A.; Peng, H.C.; Park, J.; Guerrero, S.; Wang, J.G.; Kim, M.J.; Xia, Y.N. Synthesis and Characterization of Pd@Pt–Ni Core-Shell Octahedra with High Activity toward Oxygen Reduction.*ACS Nano* **2014**, *8*, 10363–10371.

34. Lim, Y.; Kim, S.K.; Lee, S.C.; Choi, J.; Nahm, K.S.; Yoo, S.J.; Kim, P. One-step synthesis of carbon-supported Pd@Pt/C core-shell nanoparticles as oxygen reduction electrocatalysts and their enhanced activity and stability. *Nanoscale* **2014**,*6*, 4038–4042.

35. Liu, Z.L.; Zhao, B.; Guo, C.L.; Sun, Y.J.; Shi, Y.; Yang, H.B.; Li, Z.A. Carbon nanotube/raspberry hollow Pd nanosphere hybrids for methanol, ethanol, and formic acid electro-oxidation in alkaline media. *J. Colloid Interface Sci.* **2010**, *351*, 233–238.

36. Bai, Z.Y.; Yang, L.; Li, L.; Lv, J.; Wang, K.; Zhang, J. A Facile Preparation of Hollow Palladium Nanosphere Catalysts for Direct Formic Acid Fuel Cell. *J. Phys. Chem. C* **2009**, *113*, 10568–10573.

37. Fang, P.P.; Duan, S.; Lin, X.D.; Anema, J.R.; Li, J.F.; Buriez, O.; Ding, Y.; Fan, F.R.; Wu, D.Y.; Ren, B.; *et al.* Tailoring Au-core Pd-shell Pt-cluster nanoparticles for enhanced electrocatalytic activity. *Chem. Sci.* **2011**, *2*, 531–539.

38. Ksar, F.; Ramos, L.; Keita, B.; Nadjo, L.; Beaunier, P.; Remita, H. Bimetallic Palladium-Gold Nanostructures: Application in Ethanol Oxidation. *Chem. Mater.* **2009**, *21*, 3677–3683.

39. Hsu, C.J.; Huang, C.W.; Hao, Y.W.; Liu, F.Q. Au/Pd core-shell nanoparticles with varied hollow Au cores for enhanced formic acid oxidation. *Nanoscale Res. Lett.* **2013**, *8*.

40. Kim, D.Y.; Kang, S.W.; Choi, K.W.; Choi, S.W.; Han, S.W.; Im, S.H.; Park, O.O. Au@Pd nanostructures with tunable morphologies and sizes and their enhanced electrocatalytic activity. *Crystengcomm* **2013**, *15*, 7113–7120.

41. Miao, F.J.; Tao, B.R.; Sun, L.; Liu, T.; You, J.C.; Wang, L.W.; Chu, P.K. Preparation and characterization of novel nickel-palladium electrodes supported by silicon microchannel plates for direct methanol fuel cells. *J. Power Sources* **2010**,*195*, 145–150.

42. Mikolajczuk, A.; Borodzinski, A.; Kedzierzawski, P.; Stobinski, L.; Mierzwa, B.; Dziura, R. Deactivation of carbon supported palladium catalyst in direct formic acid fuel cell. *Appl. Surface Sci.* **2011**, *257*, 8211–8214.

43. Haan, J.L.; Stafford, K.M.; Masel, R.I. Effects of the Addition of Antimony, Tin, and Lead to Palladium Catalyst Formulations for the Direct Formic Acid Fuel Cell. *J. Phys. Chem. C* **2010**, *114*, 11665–11672.

44. Xu, W.F.; Gao, Y.; Lu, T.H.; Tang, Y.W.; Wu, B. Kinetic Study of Formic Acid Oxidation on Highly Dispersed Carbon Supported Pd-TiO_2 Electrocatalyst. *Catal. Lett.* **2009**, *130*, 312–317.

45. Zhang, H.X.; Wang, C.; Wang, J.Y.; Zhai, J.J.; Cai, W.B. Carbon-Supported Pd-Pt Nanoalloy with Low Pt Content and Superior Catalysis for Formic Acid Electro-oxidation. *J. Phys. Chem. C* **2010**, *114*, 6446–6451.

46. Morales-Acosta, D.; Ledesma-Garcia, J.; Godinez, L.A.; Rodriguez, H.G.; Alvarez-Contreras, L.; Arriaga, L.G. Development of Pd and Pd–Co catalysts supported on multi-walled carbon nanotubes for formic acid oxidation. *J. Power Sources* **2010**, *195*, 461–465.

47. Du, C.Y.; Chen, M.; Wang, W.G.; Yin, G.P.; Shi, P.F. Electrodeposited $PdNi_2$ alloy with novelly enhanced catalytic activity for electrooxidation of formic acid. *Electrochem. Commun.* **2010**, *12*, 843–846.

48. Wu, Y.N.; Liao, S.J.; Su, Y.L.; Zeng, J.F.; Dang, D. Enhancement of anodic oxidation of formic acid on palladium decorated Pt/C catalyst. *J. Power Sources* **2010**, *195*, 6459–6462.

49. Zhou, Z.Y.; Kang, X.W.; Song, Y.; Chen, S.W. Butylphenyl-functionalized palladium nanoparticles as effective catalysts for the electrooxidation of formic acid. *Chem. Commun.* **2011**, *47*, 6075–6077.

50. Wang, X.M.; Wang, M.E.; Zhou, D.D.; Xia, Y.Y. Structural design and facile synthesis of a highly efficient catalyst for formic acid electrooxidation. *Phys. Chem. Chem. Phys.* **2011**, *13*, 13594–13597.

51. Ge, J.J.; Chen, X.M.; Liu, C.P.; Lu, T.H.; Liao, J.H.; Liang, L.A.; Xing, W. Promoting effect of vanadium ions on the anodic Pd/C catalyst for direct formic acid fuel cell application. *Electrochim. Acta* **2010**, *55*, 9132–9136.

52. Liang, Y.; Zhu, M.N.; Ma, J.; Tang, Y.W.; Chen, Y.; Lu, T.H. Highly dispersed carbon-supported Pd nanoparticles catalyst synthesized by novel precipitation-reduction method for formic acid electrooxidation. *Electrochim.*

Acta **2011**,*56*, 4696–4702.

53. Zhang, Z.H.; Ge, J.J.; Ma, L.A.; Liao, J.H.; Lu, T.H.; Xing, W. Highly Active Carbon-supported PdSn Catalysts for Formic Acid Electrooxidation. *Fuel Cells* **2009**, *9*, 114–120.

54. Wang, X.M.; Wang, J.; Zou, Q.Q.; Xia, Y.Y. Pd nanoparticles supported on carbon-modified rutile TiO_2 as a highly efficient catalyst for formic acid electrooxidation. *Electrochim. Acta* **2011**, *56*, 1646–1651.

55. Cui, Z.M.; Kulesza, P.J.; Li, C.M.; Xing, W.; Jiang, S.P. Pd nanoparticles supported on HPMo-PDDA-MWCNT and their activity for formic acid oxidation reaction of fuel cells. *Int. J. Hydrogen Energy* **2011**, *36*, 8508–8517.

56. Pandey, R.K.; Lakshminarayanan, V. Electro-Oxidation of Formic Acid, Methanol, and Ethanol on Electrodeposited Pd-Polyaniline Nanofiber Films in Acidic and Alkaline Medium. *J. Phys. Chem. C* **2009**, *113*, 21596–21603.

57. Park, I.S.; Lee, K.S.; Yoo, S.J.; Cho, Y.H.; Sung, Y.E. Electrocatalytic properties of Pd clusters on Au nanoparticles in formic acid electro-oxidation. *Electrochim. Acta* **2010**, *55*, 4339–4345.

58. Zhao, H.; Yang, J.; Wang, L.; Tian, C.G.; Jiang, B.J.; Fu, H.G. Fabrication of a palladium nanoparticle/graphene nanosheet hybrid via sacrifice of a copper template and its application in catalytic oxidation of formic acid. *Chem. Commun.* **2011**, *47*, 2014–2016.

59. Yang, J.; Tian, C.G.; Wang, L.; Fu, H.G. An effective strategy for small-sized and highly-dispersed palladium nanoparticles supported on graphene with excellent performance for formic acid oxidation. *J. Mater. Chem.* **2011**, *21*, 3384–3390.

60. Wang, X.G.; Wang, W.M.; Qi, Z.; Zhao, C.C.; Ji, H.; Zhang, Z.H. High catalytic activity of ultrafine nanoporous palladium for electro-oxidation of methanol, ethanol, and formic acid. *Electrochem. Commun.* **2009**, *11*, 1896–1899.

61. Winjobi, O.; Zhang, Z.Y.; Liang, C.H.; Li, W.Z. Carbon nanotube supported platinum-palladium nanoparticles for formic acid oxidation. *Electrochim. Acta* **2010**, *55*, 4217–4221.

62. Tu, D.D.; Wu, B.; Wang, B.X.; Deng, C.; Gao, Y. A highly active carbon-supported PdSn catalyst for formic acid electrooxidation. *Appl. Catal. B* **2011**, *103*, 163–168.

63. Hu, C.G.; Bai, Z.Y.; Yang, L.; Lv, J.; Wang, K.; Guo, Y.M.; Cao, Y.X.; Zhou, J.G. Preparation of high performance Pd catalysts supported on untreated multi-walled carbon nanotubes for formic acid oxidation. *Electrochim. Acta* **2010**, *55*, 6036–6041.

64. Gao, Y.W.; Wang, G.; Wu, B.; Deng, C.; Gao, Y. Highly active carbon-supported PdNi catalyst for formic acid electrooxidation. *J. Appl. Electrochem.* **2011**, *41*, 1–6.

65. Wang, J.Y.; Kang, Y.Y.; Yang, H.; Cai, W.B. Boron-Doped Palladium Nanoparticles on Carbon Black as a Superior Catalyst for Formic Acid Electro-oxidation. *J. Phys. Chem. C* **2009**, *113*, 8366–8372.

66. Zhang, G.J.; Wang, Y.E.; Wang, X.; Chen, Y.; Zhou, Y.M.; Tang, Y.W.; Lu, L.D.; Bao, J.C.; Lu, T.H. Preparation of Pd-Au/C catalysts with different alloying degree and their electrocatalytic performance for formic acid oxidation. *Appl. Catal. B* **2011**, *102*, 614–619.

67. Bai, Z.Y.; Guo, Y.M.; Yang, L.; Li, L.; Li, W.J.; Xu, P.L.; Hu, C.G.; Wang, K. Highly dispersed Pd nanoparticles supported on 1,10-phenanthroline-functionalized multi-walled carbon nanotubes for electrooxidation of formic acid.*J. Power Sources* **2011**, *196*, 6232–6237.

68. Bong, S.; Uhm, S.; Kim, Y.R.; Lee, J.; Kim, H. Graphene Supported Pd Electrocatalysts for Formic Acid Oxidation.*Electrocatalysis* **2010**, *1*, 139–143.

69. Hu, G.Z.; Nitze, F.; Barzegar, H.R.; Sharifi, T.; Mikolajczuk, A.; Tai, C.W.; Borodzinski, A.; Wagberg, T. Palladium nanocrystals supported on helical carbon nanofibers for highly efficient electro-oxidation of formic acid, methanol and ethanol in alkaline electrolytes. *J. Power Sources* **2012**, *209*, 236–242.

70. Nitze, F.; Mazurkiewicz, M.; Malolepszy, A.; Mikolajczuk, A.; Kedzierzawski, P.; Tai, C.W.; Hu, G.Z.; Kurzydlowski, K.J.; Stobinski, L.; Borodzinski, A.; *et al.* Synthesis of palladium nanoparticles decorated helical carbon nanofiber as highly active anodic catalyst for direct formic acid fuel cells. *Electrochim. Acta* **2012**, *63*, 323–328.

71. Feng, Y.Y.; Yin, Q.Y.; Lu, G.P.; Yang, H.F.; Zhu, X.; Kong, D.S.; You, J.M. Enhanced catalytic performance of Pd catalyst for formic acid electrooxidation in ionic liquid aqueous solution. *J. Power Sources* **2014**, *272*, 606–613.

72. Cheng, N.C.; Lv, H.F.; Wang, W.; Mu, S.C.; Pan, M.; Marken, F. An ambient aqueous synthesis for highly dispersed and active Pd/C catalyst for formic acid electro-oxidation. *J. Power Sources* **2010**, *195*, 7246–7249.

73. Suo, Y.; Hsing, I.M. Size-controlled synthesis and impedance-based mechanistic understanding of Pd/C nanoparticles for formic acid oxidation. *Electrochim. Acta* **2009**, *55*, 210–217.

74. Lee, J.Y.; Kwak, D.H.; Lee, Y.W.; Lee, S.; Park, K.W. Synthesis of cubic PtPd alloy nanoparticles as anode electrocatalysts for mehtanol and formic acid oxidation reactions. *Phys. Chem. Chem. Phys.* **2015**, *17*, 8642–8648.

75. Baranova, E.A.; Miles, N.; Mercier, P.H.J.; le Page, Y.; Patarachao, B. Formic acid electro-oxidation on carbon supported Pd_xPt_{1-x} ($0 \leq x \leq 1$) nanoparticles synthesized via modified polyol method. *Electrochim. Acta* **2010**, *55*, 8182–8188.

76. Matin, M.A.; Jang, J.H.; Lee, E.; Kwon, Y.U. Sonochemical synthesis of Pt-doped Pd nanoparticles with enhanced electrocatalytic activity for formic acid oxidation reaction. *J. Appl. Electrochem.* **2012**, *42*, 827–832.

77. Liu, Z.L.; Zhang, X.H. Carbon-supported PdSn nanoparticles as catalysts for formic acid oxidation. *Electrochem. Commun.* **2009**, *11*, 1667–1670.

78. Celorrio, V.; de Oca, M.G.M.; Plana, D.; Moliner, R.; Lazaro, M.J.; Fermin, D.J. Effect of Carbon Supports on Electrocatalytic Reactivity of Au-Pd Core-Shell

Nanoparticles. *J. Phys. Chem. C* **2012**, *116*, 6275–6282.

79. Celorrio, V.; de Oca, M.G.M.; Plana, D.; Moliner, R.; Fermin, D.J.; Lazaro, M.J. Electrochemical performance of Pd and Au-Pd core-shell nanoparticles on surface tailored carbon black as catalyst support. *Int. J. Hydrogen Energy* **2012**, *37*, 7152–7160.

80. Wang, H.; Ge, X.B. Facile Fabrication of Porous Pd-Au Bimetallic Nanostructures for Electrocatalysis. *Electroanalysis***2012**, *24*, 911–916.

81. Shi, R.R.; Wang, J.S.; Cheng, N.C.; Sun, X.L.; Zhang, L.; Zhang, J.J.; Wang, L.C. Electrocatalytic activity and stability of carbon nanotubes-supported Pt-on-Au, Pd-on-Au, Pt-on-Pd-on-Au, Pt-on-Pd, and Pd-on-Pt catalysts for methanol oxidation reaction. *Electrochim. Acta* **2014**, *148*, 1–7.

82. Li, G.Q.; Feng, L.G.; Chang, J.F.; Wickman, B.; Gronbeck, H.; Liu, C.P.; Xing, W. Activity of Platinum/Carbon and Palladium/Carbon Catalysts Promoted by Ni_2P in Direct Ethanol Fuel Cells. *Chemsuschem* **2014**, *7*, 3374–3381.

83. Liu, H.Q.; Koenigsmann, C.; Adzic, R.R.; Wong, S.S. Probing Ultrathin One-Dimensional Pd–Ni Nanostructures As Oxygen Reduction Reaction Catalysts. *ACS Catal.* **2014**, *4*, 2544–2555.

84. Yu, X.W.; Pickup, P.G. Novel Pd-Pb/C bimetallic catalysts for direct formic acid fuel cells. *J. Power Sources* **2009**, *192*, 279–284.

85. Feng, L.G.; Yang, J.; Hu, Y.; Zhu, J.B.; Liu, C.P.; Xing, W. Electrocatalytic properties of $PdCeO_x$/C anodic catalyst for formic acid electrooxidation. *Int. J. Hydrogen Energy* **2012**, *37*, 4812–4818.

86. Liu, Z.L.; Zhang, X.H.; Tay, S.W. Nanostructured PdRu/C catalysts for formic acid oxidation. *J. Solid State Electrochem.***2012**, *16*, 545–550.

87. Arikan, T.; Kannan, A.M.; Kadirgan, F. Binary Pt–Pd and ternary Pt–Pd–Ru nanoelectrocatalysts for direct methanol fuel cells. *Int. J. Hydrogen Energy* **2013**, *38*, 2900–2907.

88. Awasthi, R.; Singh, R.N. Graphene-supported Pd–Ru nanoparticles with superior methanol electrooxidation activity.*Carbon* **2013**, *51*, 282–289.

89. Qin, Y.H.; Jia, Y.B.; Jiang, Y.; Niu, D.F.; Zhang, X.S.; Zhou, X.G.; Niu, L.; Yuan, W.K. Controllable synthesis of carbon nanofiber supported Pd catalyst for formic acid electrooxidation. *Int. J. Hydrogen Energy* **2012**, *37*, 7373–7377.

90. Li, Y.H.; Xu, Q.Z.; Li, Q.Y.; Wang, H.Q.; Huang, Y.G.; Xu, C.W. Pd deposited on MWCNTs modified carbon fiber paper as high-efficient electrocatalyst for ethanol electrooxidation. *Electrochim. Acta* **2014**, *147*, 151–156.

91. Chakraborty, S.; Raj, C.R. Electrocatalytic performance of carbon nanotube-supported palladium particles in the oxidation of formic acid and the reduction of oxygen. *Carbon* **2010**, *48*, 3242–3249.

92. Liu, J.; Liu, R.; Yuan, C.L.; Wei, X.P.; Yin, J.L.; Wang, G.L.; Cao, D.X. Pd-Co/MWCNTs Catalyst for Electrooxidation of Hydrazine in Alkaline Solution. *Fuel Cells* **2013**, *13*, 903–909.

93. Selvaraj, V.; Grace, A.N.; Alagar, M. Electrocatalytic oxidation of formic acid and formaldehyde on nanoparticle decorated single walled carbon nanotubes. *J. Colloid Interface Sci.* **2009**, *333*, 254–262.

94. Limpattayanate, S.; Hunsom, M. Electrocatalytic activity of Pt-Pd electrocatalysts for the oxygen reduction reaction in proton exchange membrane fuel cells: Effect of supports. *Renew. Energy* **2014**, *63*, 205–211.

95. Chen, C.H.; Liou, W.J.; Lin, H.M.; Wu, S.H.; Mikolajczuk, A.; Stobinski, L.; Borodzinski, A.; Kedzierzawski, P.; Kurzydlowski, K. Carbon nanotube-supported bimetallic palladium-gold electrocatalysts for electro-oxidation of formic acid. *Phys. Status Solidi Appl. Mater. Sci.* **2010**, *207*, 1160–1165.

96. Chen, C.H.; Liou, W.J.; Lin, H.M.; Wu, S.H.; Borodzinski, A.; Stobinski, L.; Kedzierzawski, P. Palladium and Palladium Gold Catalysts Supported on MWCNTs for Electrooxidation of Formic Acid. *Fuel Cells* **2010**, *10*, 227–233.

97. Mikolajczuk, A.; Borodzinski, A.; Stobinski, L.; Kedzierzawski, P.; Lesiak, B.; Laszlo, K.; Jozsef, T.; Lin, H.M. Study of Pd-Au/MWCNTs formic acid electrooxidation catalysts. *Phys. Status Solidi B* **2010**, *247*, 2717–2721.

98. Qin, Y.H.; Jiang, Y.; Niu, D.F.; Zhang, X.S.; Zhou, X.G.; Niu, L.; Yuan, W.K. Carbon nanofiber supported bimetallic PdAu nanoparticles for formic acid electrooxidation. *J. Power Sources* **2012**, *215*, 130–134.

99. Wang, X.G.; Tang, B.; Huang, X.B.; Ma, Y.; Zhang, Z.H. High activity of novel nanoporous Pd-Au catalyst for methanol electro-oxidation in alkaline media. *J. Alloys Compd.* **2013**, *565*, 120–126

100. Takenaka, S.; Tsukamoto, T.; Matsune, H.; Kishida, M. Carbon nanotube-supported Pd–Co catalysts covered with silica layers as active and stable cathode catalysts for polymer electrolyte fuel cells. *Catal. Sci. Technol.* **2013**, *3*, 2723–2731.

101. Yang, S.D.; Shen, C.M.; Lu, X.J.; Tong, H.; Zhu, J.J.; Zhang, X.G.; Gao, H.J. Preparation and electrochemistry of graphene nanosheets-multiwalled carbon nanotubes hybrid nanomaterials as Pd electrocatalyst support for formic acid oxidation. *Electrochim. Acta* **2012**, *62*, 242–249.

102. Qu, K.G.; Wu, L.; Ren, J.S.; Qu, X.G. Natural DNA-Modified Graphene/Pd Nanoparticles as Highly Active Catalyst for Formic Acid Electro-Oxidation and for the Suzuki Reaction. *ACS Appl. Mater. Interfaces* **2012**, *4*, 5001–5009.

103. Li, R.; Hao, H.; Huang, T.; Yu, A.S. Electrodeposited Pd–MoO_x catalysts with enhanced catalytic activity for formic acid electrooxidation. *Electrochim. Acta* **2012**, *76*, 292–299.

104. Lu, H.T.; Fan, Y.; Huang, P.; Xu, D.L. SnO_2 nanospheres supported Pd catalyst with enhanced performance for formic acid oxidation. *J. Power Sources* **2012**, *215*, 48–52.

105. Kim, I.T.; Choi, M.; Lee, H.K.; Shim, J. Characterization of methanol-tolerant Pd–WO_3 and Pd–SnO_2 electrocatalysts for the oxygen reduction reaction in direct methanol fuel

cells. *J. Ind. Eng. Chem.* **2013**, *19*, 813–818.

106. Maheswari, S.; Sridhar, P.; Pitchumani, S. Pd–TiO_2/C as a methanol tolerant catalyst for oxygen reduction reaction in alkaline medium. *Electrochem. Commun.* **2013**, *26*, 97–100.

107. Matos, J.; Borodzinski, A.; Zychora, A.M.; Kedzierzawski, P.; Mierzwa, B.; Juchniewicz, K.; Mazurkiewicz, M.; Hernandez-Garrido, J.C. Direct formic acid fuel cells on Pd catalysts supported on hybrid TiO_2–C materials. *Appl. Catal. B* **2015**, *163*, 167–178.

108. Wang, X.M.; Xia, Y.Y. The influence of the crystal structure of TiO_2 support material on Pd catalysts for formic acid electrooxidation. *Electrochim. Acta* **2010**, *55*, 851–856.

109. Liao, M.Y.; Hu, Q.; Zheng, J.B.; Li, Y.H.; Zhou, H.; Zhong, C.J.; Chen, B.H. Pd decorated Fe/C nanocatalyst for formic acid electrooxidation. *Electrochim. Acta* **2013**, *111*, 504–509.

110. Bai, Z.Y.; Yang, L.; Guo, Y.M.; Zheng, Z.; Hu, C.G.; Xu, P.L. High-efficiency palladium catalysts supported on ppy-modified C-60 for formic acid oxidation. *Chem. Commun.* **2011**, *47*, 1752–1754.

111. Ding, K.Q.; Jia, H.T.; Wei, S.Y.; Guo, Z.H. Electrocatalysis of Sandwich-Structured Pd/Polypyrrole/Pd Composites toward Formic Acid Oxidation. *Ind. Eng. Chem. Res.* **2011**, *50*, 7077–7082.

112. Zhao, X.; Zhu, J.B.; Liang, L.; Liu, C.P.; Liao, J.H.; Xing, W. Enhanced electroactivity of Pd nanocrystals supported on $H_3PMo_{12}O_{40}$/carbon for formic acid electrooxidation. *J. Power Sources* **2012**, *210*, 392–396.

113. Qin, Y.H.; Yue, J.; Yang, H.H.; Zhang, X.S.; Zhou, X.G.; Niu, L.; Yuan, W.K. Synthesis of highly dispersed and active palladium/carbon nanofiber catalyst for formic acid electrooxidation. *J. Power Sources* **2011**, *196*, 4609–4612.

114. Cheng, N.C.; Webster, R.A.; Pan, M.; Mu, S.C.; Rassaei, L.; Tsang, S.C.; Marken, F. One-step growth of 3–5 nm diameter palladium electrocatalyst in a carbon nanoparticle-chitosan host and characterization for formic acid oxidation. *Electrochim. Acta* **2010**, *55*, 6601–6610.

115. Zhang, B.A.; Ye, D.D.; Li, J.; Zhu, X.; Liao, Q. Electrodeposition of Pd catalyst layer on graphite rod electrodes for direct formic acid oxidation. *J. Power Sources* **2012**, *214*, 277–284.

116. Liu, S.L.; Han, M.; Shi, Y.; Zhang, C.Z.; Chen, Y.; Bao, J.C.; Dai, Z.H. Gram-Scale Synthesis of Multipod Pd Nanocrystals by a Simple Solid-Liquid Phase Reaction and Their Remarkable Electrocatalytic Properties. *Eur. J. Inorg. Chem.* **2012**, *2012*, 3740–3746.

117. Xu, C.X.; Liu, Y.Q.; Wang, J.P.; Geng, H.R.; Qiu, H.J. Nanoporous PdCu alloy for formic acid electro-oxidation. *J. Power Sources* **2012**, *199*, 124–131.

118. Meng, H.; Xie, F.Y.; Chen, J.; Shen, P.K. Electrodeposited palladium nanostructure as novel anode for direct formic acid fuel cell. *J. Mater. Chem.* **2011**, *21*, 11352–11358.

119. Cheng, T.T.; Gyenge, E.L. Novel catalyst-support interaction for direct formic acid fuel cell anodes: Pd electrodeposition on surface-modified graphite felt. *J. Appl. Electrochem.* **2009**, *39*, 1925–1938.

120. Mikolajczuk, A.; Borodzinski, A.; Stobinski, L.; Kedzierzawski, P.; Lesiak, B.; Kover, L.; Toth, J.; Lin, H.M. Physicochemical characterization of the Pd/MWCNTs catalysts for fuel cell applications. *Phys. Status Solidi B* **2010**,*247*, 3063–3067.

121. Morales-Acosta, D.; Rodriguez, H.; Godinez, L.A.; Arriaga, L.G. Performance increase of microfluidic formic acid fuel cell using Pd/MWCNTs as catalyst. *J. Power Sources* **2010**, *195*, 1862–1865.

122. Yu, X.W.; Pickup, P.G. Screening of PdM and PtM catalysts in a multi-anode direct formic acid fuel cell. *J. Appl. Electrochem.* **2011**, *41*, 589–597.

123. Yan, Z.X.; Xie, J.M.; Shen, P.K.; Zhang, M.M.; Zhang, Y.; Chen, M. Pd supported on 2–4 nm MoC particles with reduced particle size, synergistic effect and high stability for ethanol oxidation. *Electrochim. Acta* **2013**, *108*, 644–650.

124. Jung, W.S.; Han, J.; Yoon, S.P.; Nam, S.W.; Lim, T.H.; Hong, S.A. Performance degradation of direct formic acid fuel cell incorporating a Pd anode catalyst. *J. Power Sources* **2011**, *196*, 4573–4578.

125. Ren, M.J.; Kang, Y.Y.; He, W.; Zou, Z.Q.; Xue, X.Z.; Akins, D.L.; Yang, H.; Feng, S.L. Origin of performance degradation of palladium-based direct formic acid fuel cells. *Appl. Catal. B* **2011**, *104*, 49–53.

126. Zhou, Y.; Liu, J.G.; Ye, J.L.; Zou, Z.G.; Ye, J.H.; Gu, J.; Yu, T.; Yang, A.D. Poisoning and regeneration of Pd catalyst in direct formic acid fuel cell. *Electrochim. Acta* **2010**, *55*, 5024–5027.

127. Yu, X.W.; Pickup, P.G. Deactivation/reactivation of a Pd/C catalyst in a direct formic acid fuel cell (DFAFC): Use of array membrane electrode assemblies. *J. Power Sources* **2009**, *187*, 493–499.

128. Chen, Y.X.; Lavacchi, A.; Chen, S.P.; di Benedetto, F.; Bevilacqua, M.; Bianchini, C.; Fornasiero, P.; Innocenti, M.; Marelli, M.; Oberhauser, W.; *et al.* Electrochemical Milling and Faceting: Size Reduction and Catalytic Activation of Palladium Nanoparticles. *Angew. Chem. Int. Ed.* **2012**, *51*, 8500–8504.

129. Wu, Q.M.; Rao, Z.X.; Yuan, L.Z.; Jiang, L.H.; Sun, G.Q.; Ruan, J.M.; Zhou, Z.C.; Sang, S.B. Carbon supported PdO with improved activity and stability for oxygen reduction reaction in alkaline solution. *Electrochim. Acta* **2014**, *150*, 157–166.

130. Yan, Z.X.; Hu, Z.F.; Chen, C.; Meng, H.; Shen, P.K.; Ji, H.B.; Meng, Y.Z. Hollow carbon hemispheres supported palladium electrocatalyst at improved performance for alcohol oxidation. *J. Power Sources* **2010**, *195*, 7146–7151.

131. Maiyalagan, T.; Scott, K. Performance of carbon nanofiber supported Pd–Ni catalysts for electro-oxidation of ethanol in alkaline medium. *J. Power Sources* **2010**, *195*, 5246–5251.

132. Xu, J.B.; Zhao, T.S.; Shen, S.Y.; Li, Y.S. Stabilization of the palladium electrocatalyst with alloyed gold for ethanol oxidation. *Int. J. Hydrogen Energy* **2010**, *35*,

6490–6500.

133. Cheng, F.L.; Dai, X.C.; Wang, H.; Jiang, S.P.; Zhang, M.; Xu, C.W. Synergistic effect of Pd–Au bimetallic surfaces in Au-covered Pd nanowires studied for ethanol oxidation. *Electrochim. Acta* **2010**, *55*, 2295–2298.

134. Gao, H.L.; Liao, S.J.; Liang, Z.X.; Liang, H.G.; Luoa, F. Anodic oxidation of ethanol on core-shell structured Ru@PtPd/C catalyst in alkaline media. *J. Power Sources* **2011**, *196*, 6138–6143.

135. Bambagioni, V.; Bianchini, C.; Filippi, J.; Oberhauserlal, W.; Marchionni, A.; Vizza, F.; Psaro, R.; Sordelli, L.; Foresti, M.L.; Innocenti, M. Ethanol Oxidation on Electrocatalysts Obtained by Spontaneous Deposition of Palladium onto Nickel-Zinc Materials. *Chemsuschem* **2009**, *2*, 99–112.

136. Chen, X.M.; Lin, Z.J.; Jia, T.T.; Cai, Z.M.; Huang, X.L.; Jiang, Y.Q.; Chen, X.; Chen, G.N. A facile synthesis of palladium nanoparticles supported on functional carbon nanotubes and its novel catalysis for ethanol electrooxidation. *Anal. Chim. Acta* **2009**, *650*, 54–58.

137. Fu, S.F.; Zhu, C.Z.; Du, D.; Lin, Y.H. Facile One-Step Synthesis of Three-Dimensional Pd–Ag Bimetallic Alloy Networks and Their Electrocatalytic Activity toward Ethanol Oxidation. *ACS Appl. Mater. Interfaces* **2015**, *7*, 13842–13848.

138. Lv, J.J.; Wisitruangsakul, N.; Feng, J.J.; Luo, J.; Fang, K.M.; Wang, A.J. Biomolecule-assisted synthesis of porous PtPd alloyed nanoflowers supported on reduced graphene oxide with highly electrocatalytic performance for ethanol oxidation and oxygen reduction. *Electrochim. Acta* **2015**, *160*, 100–107.

139. Wang, A.L.; He, X.J.; Lu, X.F.; Xu, H.; Tong, Y.X.; Li, G.R. Palladium-Cobalt Nanotube Arrays Supported on Carbon Fiber Cloth as High-Performance Flexible Electrocatalysts for Ethanol Oxidation. *Angew. Chem. Int. Ed.* **2015**, *54*, 3669–3673.

140. Ma, J.W.; Wang, J.; Zhang, G.H.; Fan, X.B.; Zhang, G.L.; Zhang, F.B.; Li, Y. Deoxyribonucleic acid-directed growth of well dispersed nickel-palladium-platinum nanoclusters on graphene as an efficient catalyst for ethanol electrooxidation. *J. Power Sources* **2015**, *278*, 43–49.

141. Peng, C.; Hu, Y.L.; Liu, M.R.; Zheng, Y.X. Hollow raspberry-like PdAg alloy nanospheres: High electrocatalytic activity for ethanol oxidation in alkaline media. *J. Power Sources* **2015**, *278*, 69–75.

142. Singh, R.N.; Awasthi, R. Graphene support for enhanced electrocatalytic activity of Pd for alcohol oxidation. *Catal. Sci. Technol.* **2011**, *1*, 778–783.

143. Su, L.; Jia, W.Z.; Schempf, A.; Ding, Y.; Lei, Y. Free-Standing Palladium/Polyamide 6 Nanofibers for Electrooxidation of Alcohols in Alkaline Medium. *J. Phys. Chem. C* **2009**, *113*, 16174–16180.

144. Wang, Y.L.; Zhao, Y.Q.; Xu, C.L.; Zhao, D.D.; Xu, M.W.; Su, Z.X.; Li, H.L. Improved performance of Pd electrocatalyst supported on three-dimensional nickel foam for direct ethanol fuel cells. *J. Power Sources* **2010**, *195*, 6496–6499.

145. Thotiyl, M.M.O.; Kumar, T.R.; Sampath, S. Pd Supported on Titanium Nitride for Efficient Ethanol Oxidation. *J. Phys. Chem. C* **2010**, *114*, 17934–17941.

146. Mahendiran, C.; Maiyalagan, T.; Scott, K.; Gedanken, A. Synthesis of a carbon-coated NiO/MgO core/shell nanocomposite as a Pd electro-catalyst support for ethanol oxidation. *Mater. Chem. Phys.* **2011**, *128*, 341–347.

147. Yi, Q.F.; Niu, F.J.; Sun, L.Z. Fabrication of novel porous Pd particles and their electroactivity towards ethanol oxidation in alkaline media. *Fuel* **2011**, *90*, 2617–2623.

148. Wang, X.G.; Wang, W.M.; Qi, Z.; Zhao, C.C.; Ji, H.; Zhang, Z.H. Fabrication, microstructure and electrocatalytic property of novel nanoporous palladium composites. *J. Alloys Compd.* **2010**, *508*, 463–470.

149. Liu, R.; Liu, J.F.; Jiang, G.B. Use of Triton X-114 as a weak capping agent for one-pot aqueous phase synthesis of ultrathin noble metal nanowires and a primary study of their electrocatalytic activity. *Chem. Commun.* **2010**, *46*, 7010–7012.

150. Ksar, F.; Surendran, G.; Ramos, L.; Keita, B.; Nadjo, L.; Prouzet, E.; Beaunier, P.; Hagege, A.; Audonnet, F.; Remita, H. Palladium Nanowires Synthesized in Hexagonal Mesophases: Application in Ethanol Electrooxidation. *Chem. Mater.* **2009**, *21*, 1612–1617.

151. Cherevko, S.; Xing, X.L.; Chung, C.H. Pt and Pd decorated Au nanowires: Extremely high activity of ethanol oxidation in alkaline media. *Electrochim. Acta* **2011**, *56*, 5771–5775.

152. Lin, S.C.; Chen, J.Y.; Hsieh, Y.F.; Wu, P.W. A facile route to prepare PdPt alloys for ethanol electro-oxidation in alkaline electrolyte. *Mater. Lett.* **2011**, *65*, 215–218.

153. Modibedi, R.M.; Masombuka, T.; Mathe, M.K. Carbon supported Pd-Sn and Pd–Ru–Sn nanocatalysts for ethanol electro-oxidation in alkaline medium. *Int. J. Hydrogen Energy* **2011**, *36*, 4664–4672.

154. Jou, L.H.; Chang, J.K.; Whang, T.J.; Sun, I.W. Electrodeposition of Palladium-Tin Alloys from 1-Ethyl-3-methylimidazolium Chloride-Tetrafluoroborate Ionic Liquid for Ethanol Electro-Oxidation. *J. Electrochem. Soc.* **2010**,*157*, D443–D449.

155. Shen, S.Y.; Zhao, T.S.; Xu, J.B. Carbon-supported bimetallic PdIr catalysts for ethanol oxidation in alkaline media.*Electrochim. Acta* **2010**, *55*, 9179–9184.

156. Qiu, C.C.; Shang, R.; Xie Y.F.; Bu, Y.R.; Li, C.Y.; Ma, H.Y. Electrocatalytic activity of bimetallic Pd-Ni thin films towards the oxidation of methanol and ethanol. *Mater. Chem. Phys.* **2010**, *120*, 323–330.

157. Qi, Z.; Geng, H.R.; Wang, X.G.; Zhao, C.C.; Ji, H.; Zhang, C.; Xu, J.L.; Zhang, Z.H. Novel nanocrystalline PdNi alloy catalyst for methanol and ethanol electro-oxidation in alkaline media. *J. Power Sources* **2011**, *196*, 5823–5828.

158. Nguyen, S.T.; Law, H.M.; Nguyen, H.T.; Kristian, N.; Wang, S.Y.; Chan, S.H.; Wang, X. Enhancement effect of Ag for Pd/C towards the ethanol electro-oxidation in alkaline media. *Appl. Catal. B* **2009**, *91*, 507–515.

159. Oliveira, M.C.; Rego, R.; Fernandes, L.S.; Tavares, P.B.

Evaluation of the catalytic activity of Pd–Ag alloys on ethanol oxidation and oxygen reduction reactions in alkaline medium. *J. Power Sources* **2011**, *196*, 6092–6098.

160. Wang, Y.; Nguyen, T.S.; Liu, X.W.; Wang, X. Novel palladium-lead (Pd–Pb/C) bimetallic catalysts for electrooxidation of ethanol in alkaline media. *J. Power Sources* **2010**, *195*, 2619–2622.

161. He, Q.G.; Chen, W.; Mukerjee, S.; Chen, S.W.; Laufek, F. Carbon-supported PdM (M = Au and Sn) nanocatalysts for the electrooxidation of ethanol in high pH media. *J. Power Sources* **2009**, *187*, 298–304.

162. Zhu, L.D.; Zhao, T.S.; Xu, J.B.; Liang, Z.X. Preparation and characterization of carbon-supported sub-monolayer palladium decorated gold nanoparticles for the electro-oxidation of ethanol in alkaline media. *J. Power Sources* **2009**,*187*, 80–84.

163. Hu, Z.F.; Yan, Z.X.; Shen, P.K.; Zhong, C.J. Nano-architectures of ordered hollow carbon spheres filled with carbon webs by template-free controllable synthesis. *Nanotechnology* **2012**, *23*, 48–53.

164. Yan, Z.X.; Meng, H.; Shen, P.K.; Meng, Y.Z.; Ji, H.B. Effect of the templates on the synthesis of hollow carbon materials as electrocatalyst supports for direct alcohol fuel cells. *Int. J. Hydrogen Energy* **2012**, *37*, 4728–4736.

165. Yan, Z.X.; Meng, H.; Shi, L.; Li, Z.H.; Shen, P.K. Synthesis of mesoporous hollow carbon hemispheres as highly efficient Pd electrocatalyst support for ethanol oxidation. *Electrochem. Commun.* **2010**, *12*, 689–692.

166. Cerritos, R.C.; Guerra-Balcazar, M.; Ramirez, R.F.; Ledesma-Garcia, J.; Arriaga, L.G. Morphological Effect of Pd Catalyst on Ethanol Electro-Oxidation Reaction. *Materials* **2012**, *5*, 1686–1697.

167. Ding, K.Q.; Yang, G.K. HCl-assisted pyrolysis of $PdCl_2$ to immobilize palladium nanoparticles on multi-walled carbon nanotubes. *Mater. Chem. Phys.* **2010**, *123*, 498–501.

168. Ding, K.Q.; Yang, G.K. Using RTILs of EMIBF4 as "water" to prepare palladium nanoparticles onto MWCNTs by pyrolysis of $PdCl_2$. *Electrochim. Acta* **2010**, *55*, 2319–2324.

169. Chu, D.B.; Wang, J.; Wang, S.X.; Zha, L.W.; He, J.G.; Hou, Y.Y.; Yan, Y.X.; Lin, H.S.; Tian, Z.W. High activity of Pd–In_2O_3/CNTs electrocatalyst for electro-oxidation of ethanol. *Catal. Commun.* **2009**, *10*, 955–958.

170. Qin, Y.H.; Yang, H.H.; Zhang, X.S.; Li, P.; Ma, C.A. Effect of carbon nanofibers microstructure on electrocatalytic activities of Pd electrocatalysts for ethanol oxidation in alkaline medium. *Int. J. Hydrogen Energy* **2010**, *35*, 7667–7674.

171. Qin, Y.H.; Li, H.C.; Yang, H.H.; Zhang, X.S.; Zhou, X.G.; Niu, L.; Yuan, W.K. Effect of electrode fabrication methods on the electrode performance for ethanol oxidation. *J. Power Sources* **2011**, *196*, 159–163.

172. Qin, Y.H.; Yang, H.H.; Zhang, X.S.; Li, P.; Zhou, X.G.; Niu, L.; Yuan, W.K. Electrophoretic deposition of network-like carbon nanofibers as a palladium catalyst support for ethanol oxidation in alkaline media. *Carbon* **2010**, *48*, 3323–3329.

173. Wen, Z.L.; Yang, S.D.; Liang, Y.Y.; He, W.; Tong, H.; Hao, L.A.; Zhang, X.G.; Song, Q.J. The improved electrocatalytic activity of palladium/graphene nanosheets towards ethanol oxidation by tin oxide. *Electrochim. Acta* **2010**, *56*, 139–144.

174. Chen, X.M.; Wu, G.H.; Chen, J.M.; Chen, X.; Xie, Z.X.; Wang, X.R. Synthesis of "Clean" and Well-Dispersive Pd Nanoparticles with Excellent Electrocatalytic Property on Graphene Oxide. *J. Am. Chem. Soc.* **2011**, *133*, 3693–3695.

175. El-Shafei, A.A.; Elhafeez, A.M.A.; Mostafa, H.A. Ethanol oxidation at metal-zeolite-modified electrodes in alkaline medium. Part 2: Palladium-zeolite-modified graphite electrode. *J. Solid State Electrochem.* **2010**, *14*, 185–190.

176. Hasan, M.; Newcomb, S.B.; Rohan, J.F.; Razeeb, K.M. Ni nanowire supported 3D flower-like Pd nanostructures as an efficient electrocatalyst for electrooxidation of ethanol in alkaline media. *J. Power Sources* **2012**, *218*, 148–156.

177. Wang, W.J.; Zhang, J.; Yang, S.C.; Ding, B.J.; Song, X.P. Au@Pd Core-Shell Nanobricks with Concave Structures and Their Catalysis of Ethanol Oxidation. *Chemsuschem* **2013**, *6*, 1945–1951.

178. Li, S.J.; Cheng, D.J.; Qiu, X.G.; Cao, D.P. Synthesis of Cu@Pd core-shell nanowires with enhanced activity and stability for formic acid oxidation. *Electrochim. Acta* **2014**, *143*, 44–48.

179. Li, C.L.; Su, Y.; Lv, X.Y.; Shi, H.J.; Yang, X.G.; Wang, Y.J. Enhanced ethanol electrooxidation of hollow Pd nanospheres prepared by galvanic exchange reactions. *Mater. Lett.* **2012**, *69*, 92–95.

180. Wu, H.X.; Li, H.J.; Zhai, Y.J.; Xu, X.L.; Jin, Y.D. Facile Synthesis of Free-Standing Pd-Based Nanomembranes with Enhanced Catalytic Performance for Methanol/Ethanol Oxidation. *Adv. Mater.* **2012**, *24*, 1594–1597.

181. Ozturk, Z.; Sen, F.; Sen, S.; Gokagac, G. The preparation and characterization of nano-sized Pt-Pd/C catalysts and comparison of their superior catalytic activities for methanol and ethanol oxidation. *J. Mater. Sci.* **2012**, *47*, 8134–8144.

182. Shen, S.Y.; Zhao, T.S.; Xu, J.B.; Li, Y.S. Synthesis of PdNi catalysts for the oxidation of ethanol in alkaline direct ethanol fuel cells. *J. Power Sources* **2010**, *195*, 1001–1006.

183. Zhu, C.Z.; Guo, S.J.; Dong, S.J. PdM (M = Pt, Au) Bimetallic Alloy Nanowires with Enhanced Electrocatalytic Activity for Electro-oxidation of Small Molecules. *Adv. Mater.* **2012**, *24*, 2326–2331.

184. Seweryn, J.; Lewera, A. Electrooxidation of ethanol on carbon-supported Pt–Pd nanoparticles. *J. Power Sources* **2012**,*205*, 264–271.

185. Lv, J.J.; Zheng, J.N.; Zhang, H.B.; Lin, M.; Wang, A.J.; Chen, J.R.; Feng, J.J. Simple synthesis of platinum-palladium nanoflowers on reduced graphene oxide and their enhanced catalytic activity for oxygen reduction reaction. *J. Power Sources* **2014**, *269*, 136–143.

186. Ding, L.X.; Wang, A.L.; Ou, Y.N.; Li, Q.; Guo, R.; Zhao, W.X.; Tong, Y.X.; Li, G.R. Hierarchical Pd–Sn Alloy Nanosheet Dendrites: An Economical and Highly Active Catalyst for Ethanol Electrooxidation. *Sci. Rep.* **2013**, *3*.

187. Assumpcao, M.; da Silva, S.G.; de Souza, R.F.B.; Buzzo, G.S.; Spinace, E.V.; Santos, M.C.; Neto, A.O.; Silva, J.C.M. Investigation of PdIr/C electrocatalysts as anode on the performance of direct ammonia fuel cell. *J. Power Sources* **2014**,*268*, 129–136.

188. Nguyen, S.T.; Yang, Y.H.; Wang, X. Ethanol electro-oxidation activity of Nb-doped-TiO_2 supported PdAg catalysts in alkaline media. *Appl. Catal. B* **2012**, *113*, 261–270.

189. Liu, Y.Q.; Xu, C.X. Nanoporous PdTi Alloys as Non-Platinum Oxygen-Reduction Reaction Electrocatalysts with Enhanced Activity and Durability. *Chemsuschem* **2013**, *6*, 78–84.

190. Li, Y.S.; Zhao, T.S. A high-performance integrated electrode for anion-exchange membrane direct ethanol fuel cells.*Int. J. Hydrogen Energy* **2011**, *36*, 7707–7713.

191. Shen, S.Y.; Zhao, T.S.; Xu, J.B.; Li, Y.S. High performance of a carbon supported ternary PdIrNi catalyst for ethanol electro-oxidation in anion-exchange membrane direct ethanol fuel cells. *Energy Environ. Sci.* **2011**, *4*, 1428–1433.

192. Antolini, E.; Colmati, F.; Gonzalez, E.R. Ethanol oxidation on carbon supported (PtSn)(alloy)/SnO_2 and (PtSnPd)(alloy)/SnO_2 catalysts with a fixed Pt/SnO_2 atomic ratio: Effect of the alloy phase characteristics. *J. Power Sources* **2009**,*193*, 555–561.

193. Bambagioni, V.; Bianchini, C.; Marchionni, A.; Filippi, J.; Vizza, F.; Teddy, J.; Serp, P.; Zhiani, M. Pd and Pt-Ru anode electrocatalysts supported on multi-walled carbon nanotubes and their use in passive and active direct alcohol fuel cells with an anion-exchange membrane (alcohol = methanol, ethanol, glycerol). *J. Power Sources* **2009**, *190*, 241–251.

194. Zhang, Z.Y.; More, K.L.; Sun, K.; Wu, Z.L.; Li, W.Z. Preparation and Characterization of PdFe Nanoleaves as Electrocatalysts for Oxygen Reduction Reaction. *Chem. Mater.* **2011**, *23*, 1570–1577.

195. Tang, Y.F.; Zhang, H.M.; Zhong, H.X.; Ma, Y.W. A facile synthesis of Pd/C cathode electrocatalyst for proton exchange membrane fuel cells. *Int. J. Hydrogen Energy* **2011**, *36*, 725–731.

196. Jeyabharathi, C.; Venkateshkumar, P.; Mathiyarasu, J.; Phani, K.L.N. Carbon-Supported Palladium-Polypyrrole Nanocomposite for Oxygen Reduction and Its Tolerance to Methanol. *J. Electrochem. Soc.* **2010**, *157*, B1740–B1745.

197. Kumar, S.M.S.; Herrero, J.S.; Irusta, S.; Scott, K. The effect of pretreatment of Vulcan XC-72R carbon on morphology and electrochemical oxygen reduction kinetics of supported Pd nano-particle in acidic electrolyte. *J. Electroanal. Chem.***2010**, *647*, 211–221.

198. Yin, S.B.; Cai, M.; Wang, C.X.; Shen, P.K. Tungsten carbide promoted Pd–Fe as alcohol-tolerant electrocatalysts for oxygen reduction reactions. *Energy Environ. Sci.* **2011**, *4*, 558–563.

199. Rao, C.V.; Viswanathan, B. Carbon supported Pd–Co–Mo alloy as an alternative to Pt for oxygen reduction in direct ethanol fuel cells. *Electrochim. Acta* **2010**, *55*, 3002–3007.

200. Sarkar, A.; Murugan, A.V. Manthiram, A. Low cost Pd-W nanoalloy electrocatalysts for oxygen reduction reaction in fuel cells. *J. Mater. Chem.* **2009**, *19*, 159–165.

201. Zhou, Z.M.; Shao, Z.G.; Qin, X.P.; Chen, X.G.; Wei, Z.D.; Yi, B.L. Durability study of Pt–Pd/C as PEMFC cathode catalyst. *Int. J. Hydrogen Energy* **2010**, *35*, 1719–1726.

202. Chang, S.H.; Su, W.N.; Yeh, M.H.; Pan, C.J.; Yu, K.L.; Liu, D.G.; Lee, J.F.; Hwang, B.J. Structural and Electronic Effects of Carbon-Supported Pt_xPd_{1-x} Nanoparticles on the Electrocatalytic Activity of the Oxygen-Reduction Reaction and on Methanol Tolerance. *Chem. Eur. J.* **2010**, *16*, 11064–11071.

203. Sarapuu, A.; Kasikov, A.; Wong, N.; Lucas, C.A.; Sedghi, G.; Nichols, R.J.; Tammeveski, K. Electroreduction of oxygen on gold-supported nanostructured palladium films in acid solutions. *Electrochim. Acta* **2010**, *55*, 6768–6774.

204. Kariuki, N.N.; Wang, X.P.; Mawdsley, J.R.; Ferrandon, M.S.; Niyogi, S.G.; Vaughey, J.T.; Myers, D.J. Colloidal Synthesis and Characterization of Carbon-Supported Pd–Cu Nanoparticle Oxygen Reduction Electrocatalysts. *Chem. Mater.* **2010**, *22*, 4144–4152.

205. Fouda-Onana, F.; Bah, S.; Savadogo, O. Palladium-copper alloys as catalysts for the oxygen reduction reaction in an acidic media I: Correlation between the ORR kinetic parameters and intrinsic physical properties of the alloys. *J. Electroanal. Chem.* **2009**, *636*, 1–9.

206. Wei, Y.C.; Liu, C.W.; Chang, Y.W.; Lai, C.M.; Lim, P.Y.; Tsai, L.D.; Wang, K.W. The structure-activity relationship of Pd–Co/C electrocatalysts for oxygen reduction reaction. *Int. J. Hydrogen Energy* **2010**, *35*, 1864–1871.

207. Ohashi, M.; Beard, K.D.; Ma, S.G.; Blom, D.A.; St-Pierre, J.; van Zee, J.W.; Monnier, J.R. Electrochemical and structural characterization of carbon-supported Pt–Pd bimetallic electrocatalysts prepared by electroless deposition. *Electrochim. Acta* **2010**, *55*, 7376–7384.

208. Peng, Z.M.; Yang, H. Synthesis and Oxygen Reduction Electrocatalytic Property of Pt-on-Pd Bimetallic Heteronanostructures. *J. Am. Chem. Soc.* **2009**, *131*, 7542–7543.

209. Yang, J.H.; Zhou, W.J.; Cheng, C.H.; Lee, J.Y.; Liu, Z.L. Pt-Decorated PdFe Nanoparticles as Methanol-Tolerant Oxygen Reduction Electrocatalyst. *ACS Appl. Mater. Interfaces* **2010**, *2*, 119–126.

210. Yang, J.; Cheng, C.H.; Zhou, W.; Lee, J.Y.; Liu, Z. Methanol-Tolerant Heterogeneous PdCo@PdPt/C Electrocatalyst for the Oxygen Reduction Reaction. *Fuel Cells* **2010**, *10*, 907–913.

211. Li, W.Z.; Haldar, P. Supportless PdFe nanorods as highly active electrocatalyst for proton exchange membrane fuel cell. *Electrochem. Commun.* **2009**, *11*, 1195–1198.

212. Zhou, W.P.; Yang, X.F.; Vukmirovic, M.B.; Koel, B.E.;

Jiao, J.; Peng, G.W.; Mavrikakis, M.; Adzic, R.R. Improving Electrocatalysts for O_2 Reduction by Fine-Tuning the Pt-Support Interaction: Pt Monolayer on the Surfaces of a $Pd_3Fe(111)$ Single-Crystal Alloy. *J. Am. Chem. Soc.* **2009**, *131*, 12755–12762.

213. Jin, S.A.; Kwon, K.; Pak, C.; Chang, H. The oxygen reduction electrocatalytic activity of intermetallic compound of palladium-tin supported on tin oxide-carbon composite. *Catal. Today* **2011**, *164*, 176–180.

214. Sun, W.; Hsu, A.; Chen, R.R. Palladium-coated manganese dioxide catalysts for oxygen reduction reaction in alkaline media. *J. Power Sources* **2011**, *196*, 4491–4498.

215. Arroyo-Ramirez, L.; Rodriguez, D.; Otano, W.; Cabrera, C.R. Palladium Nanoshell Catalysts Synthesis on Highly Ordered Pyrolytic Graphite for Oxygen Reduction Reaction. *Acs Appl. Mater. Interfaces* **2012**, *4*, 2018–2024.

216. Jukk, K.; Alexeyeva, N.; Johans, C.; Kontturi, K.; Tammeveski, K. Oxygen reduction on Pd nanoparticle/multi-walled carbon nanotube composites. *J. Electroanal. Chem.* **2012**, *666*, 67–75.

217. Kim, Y.S.; Shin, J.Y.; Chun, Y.S.; Lee, C.; Lee, S.G. Anion-Dependent Electrocatalytic Activity of Supported Palladium Catalysts onto Imidazolium Salt-Functionalized Carbon Nanotubes in Oxygen Reduction Reaction. *Bull. Korean Chem. Soc.* **2011**, *32*, 3209–3210.

218. Seo, M.H.; Choi, S.M.; Kim, H.J.; Kim, W.B. The graphene-supported Pd and Pt catalysts for highly active oxygen reduction reaction in an alkaline condition. *Electrochem. Commun.* **2011**, *13*, 182–185.

219. Kim, D.; Ahmed, M.S.; Jeon, S. Different length linkages of graphene modified with metal nanoparticles for oxygen reduction in acidic media. *J. Mater. Chem.* **2012**, *22*, 16353–16360.

220. Gotoh, K.; Kawabata, K.; Fujii, E.; Morishige, K.; Kinumoto, T.; Miyazaki, Y.; Ishida, H. The use of graphite oxide to produce mesoporous carbon supporting Pt, Ru, or Pd nanoparticles. *Carbon* **2009**, *47*, 2120–2124.

221. Du, S.F.; Lu, Y.X.; Steinberger-Wilckens, R. PtPd nanowire arrays supported on reduced graphene oxide as advanced electrocatalysts for methanol oxidation. *Carbon* **2014**, *79*, 346–353.

222. Huang, S.Y.; Ganesan, P.; Popov, B.N. Electrocatalytic Activity and Stability of Titania-Supported Platinum-Palladium Electrocatalysts for Polymer Electrolyte Membrane Fuel Cell. *ACS Catal.* **2012**, *2*, 825–831.

223. Bae, S.J.; Nahm, K.S.; Kim, P. Electroreduction of oxygen on Pd catalysts supported on Ti-modified carbon. *Curr. Appl. Phys.* **2012**, *12*, 1476–1480.

224. Ramanathan, M.; Ramani, V.; Prakash, J. Kinetics of the oxygen reduction reaction on Pd_3M (M = Cu, Ni, Fe) electrocatalysts synthesized at elevated annealing temperatures. *Electrochim. Acta* **2012**, *75*, 254–261.

225. Wen, M.; Zhou, B.; Fang, H.; Wu, Q.S.; Chen, S.P. Novel-Phase Structural High-Efficiency Anode Catalyst for Methanol Fuel Cells: α-$(NiCu)_3Pd$ Nanoalloy. *J. Phys. Chem. C* **2014**, *118*, 26713–26720.

226. Lee, K.R.; Jung, Y.; Woo, S.I. Combinatorial Screening of Highly Active Pd Binary Catalysts for Electrochemical Oxygen Reduction. *ACS Comb. Sci.* **2012**, *14*, 10–16.

227. Meng, H.; Shen, P.K. The beneficial effect of the addition of tungsten carbides to Pt catalysts on the oxygen electroreduction. *Chem. Commun.* **2005**.

228. Meng, H.; Shen, P.K. Tungsten carbide nanocrystal promoted Pt/C electrocatalysts for oxygen reduction. *J. Phys. Chem. B* **2005**, *109*, 22705–22709.

229. Nie, M.; Shen, P.K.; Wu, M.; Wei, Z.D.; Meng, H. A study of oxygen reduction on improved Pt–WC/C electrocatalysts.*J. Power Sources* **2006**, *162*, 173–176.

230. Cui, G.F.; Shen, P.K.; Meng, H.; Zhao, J.; Wu, G. Tungsten carbide as supports for Pt electrocatalysts with improved CO tolerance in methanol oxidation. *J. Power Sources* **2011**, *196*, 6125–6130.

231. He, G.Q.; Yan, Z.X.; Ma, X.M.; Meng, H.; Shen, P.K.; Wang, C.X. A universal method to synthesize nanoscale carbides as electrocatalyst supports towards oxygen reduction reaction. *Nanoscale* **2011**, *3*, 3578–3582.

232. Pan, Y.; Zhang, F.; Wu, K.; Lu, Z.Y.; Chen, Y.; Zhou, Y.M.; Tang, Y.W.; Lu, T.H. Carbon supported Palladium-Iron nanoparticles with uniform alloy structure as methanol-tolerant electrocatalyst for oxygen reduction reaction. *Int. J. Hydrogen Energy* **2012**, *37*, 2993–3000.

233. Han, B.H.; Xu, C.X. Nanoporous PdFe alloy as highly active and durable electrocatalyst for oxygen reduction reaction.*Int. J. Hydrogen Energy* **2014**, *39*, 18247–18255.

234. Gobal, F.; Arab, R. A preliminary study of the electro-catalytic reduction of oxygen on Cu–Pd alloys in alkaline solution. *J. Electroanal. Chem.* **2010**, *647*, 66–73.

235. Yang, R.Z.; Bian, W.Y.; Strasser, P.; Toney, M.F. Dealloyed $PdCu_3$ thin film electrocatalysts for oxygen reduction reaction. *J. Power Sources* **2013**, *222*, 169–176.

236. You, D.J.; Jin, S.A.; Lee, K.H.; Pak, C.; Choi, K.H.; Chang, H. Improvement of activity for oxygen reduction reaction by decoration of Ir on PdCu/C catalyst. *Catal. Today* **2012**, *185*, 138–142.

237. Xiong, L.; Huang, Y.X.; Liu, X.W.; Sheng, G.P.; Li, W.W.; Yu, H.Q. Three-dimensional bimetallic Pd-Cu nanodendrites with superior electrochemical performance for oxygen reduction reaction. *Electrochim. Acta* **2013**, *89*, 24–28.

238. Lee, C.L.; Chiou, H.P.; Chang, K.C.; Huang, C.H. Carbon nanotubes-supported colloidal Ag–Pd nanoparticles as electrocatalysts toward oxygen reduction reaction in alkaline electrolyte. *Int. J. Hydrogen Energy* **2011**, *36*, 2759–2764.

239. Godinez-Garcia, A.; Perez-Robles, J.F.; Martinez-Tejada, H.V.; Solorza-Feria, O. Characterization and electrocatalytic properties of sonochemical synthesized PdAg nanoparticles. *Mater. Chem. Phys.* **2012**, *134*, 1013–1019.

240. Xu, L.; Luo, Z.M.; Fan, Z.X.; Zhang, X.; Tan, C.L.; Li, H.; Zhang, H.; Xue, C. Triangular Ag–Pd alloy nanoprisms: Rational synthesis with high-efficiency for

electrocatalytic oxygen reduction. *Nanoscale* **2014**, *6*, 11738–11743.

241. Xu, J.B.; Zhao, T.S.; Li, Y.S.; Yang, W.W. Synthesis and characterization of the Au-modified Pd cathode catalyst for alkaline direct ethanol fuel cells. *Int. J. Hydrogen Energy* **2010**, *35*, 9693–9700.

242. Kim, D.S.; Kim, J.H.; Jeong, I.K.; Choi, J.K.; Kim, Y.T. Phase change of bimetallic PdCo electrocatalysts caused by different heat-treatment temperatures: Effect on oxygen reduction reaction activity. *J. Catal.* **2012**, *290*, 65–78.

243. Son, D.N.; Takahashi, K. Selectivity of Palladium-Cobalt Surface Alloy toward Oxygen Reduction Reaction. *J. Phys. Chem. C* **2012**, *116*, 6200–6207.

244. Ren, Y.B.; Zhang, S.C.; Fang, H.; Wei, X.; Yang, P.H. Investigation of Co_3O_4 nanorods supported Pd anode catalyst for methanol oxidation in alkaline solution. *J. Energy Chem.* **2014**, *23*, 801–808.

245. Liu, H.; Li, W.; Manthiram, A. Factors influencing the electrocatalytic activity of $Pd_{100-x}Co_x$ ($0 \leq x \leq 50$) nanoalloys for oxygen reduction reaction in fuel cells. *Appl. Catal. B-Environ.* **2009**, *90*, 184–194.

246. Oishi, K.; Savadogo, O. Electrochemical investigation of Pd-Co thin films binary alloy for the oxygen reduction reaction in acid medium. *J. Electroanal. Chem.* **2013**, *703*, 108–116.

247. Barakat, N.A.M.; Abdelkareem, M.A.; Shin, G.; Kim, H.Y. Pd-doped Co nanofibers immobilized on a chemically stable metallic bipolar plate as novel strategy for direct formic acid fuel cells. *Int. J. Hydrogen Energy* **2013**, *38*, 7438–7447.

248. Tominaka, S.; Hayashi, T.; Nakamura, Y.; Osaka, T. Mesoporous PdCo sponge-like nanostructure synthesized by electrodeposition and dealloying for oxygen reduction reaction. *J. Mater. Chem.* **2010**, *20*, 7175–7182.

249. Serov, A.; Nedoseykina, T.; Shvachko, O.; Kwak, C. Effect of precursor nature on the performance of palladium-cobalt electrocatalysts for direct methanol fuel cells. *J. Power Sources* **2010**, *195*, 175–180.

250. Alia, S.M.; Jensen, K.O.; Pivovar, B.S.; Yan, Y.S. Platinum-Coated Palladium Nanotubes as Oxygen Reduction Reaction Electrocatalysts. *ACS Catal.* **2012**, *2*, 858–863.

251. Slanac, D.A.; Li, L.; Mayoral, A.; Yacaman, M.J.; Manthiram, A.; Stevenson, K.J.; Johnston, K.P. Atomic resolution structural insights into PdPt nanoparticle-carbon interactions for the design of highly active and stable electrocatalysts. *Electrochim. Acta* **2012**, *64*, 35–45.

252. Chen, X.T.; Jiang, Y.Y.; Sun, J.Z.; Jin, C.H.; Zhang, Z.H. Highly active nanoporous Pt-based alloy as anode and cathode catalyst for direct methanol fuel cells. *J. Power Sources* **2014**, *267*, 212–218.

253. Lim, B.; Jiang, M.J.; Camargo, P.H.C.; Cho, E.C.; Tao, J.; Lu, X.M.; Zhu, Y.M.; Xia, Y.N. Pd–Pt Bimetallic Nanodendrites with High Activity for Oxygen Reduction. *Science* **2009**, *324*, 1302–1305.

254. Tang, Y.F.; Gao, F.M.; Yu, S.X.; Li, Z.P.; Zhao, Y.F. Surfactant-free synthesis of highly methanol-tolerant, polyhedral Pd–Pt nanocrystallines for oxygen reduction reaction. *J. Power Sources* **2013**, *239*, 374–381.

255. Limpattayanate, S.; Hunsom, M. Effect of supports on activity and stability of Pt–Pd catalysts for oxygen reduction reaction in proton exchange membrane fuel cells. *J. Solid State Electrochem.* **2013**, *17*, 1221–1231.

256. Thanasilp, S.; Hunsom, M. Effect of Pt: Pd atomic ratio in Pt–Pd/C electrocatalyst-coated membrane on the electrocatalytic activity of ORR in PEM fuel cells. *Renew. Energy* **2011**, *36*, 1795–1801.

257. Ficicilar, B.; Bayrakceken, A.; Eroglu, I. Effect of Pd loading in Pd-Pt bimetallic catalysts doped into hollow core mesoporous shell carbon on performance of proton exchange membrane fuel cells. *J. Power Sources* **2009**, *193*, 17–23.

258. Ang, S.Y.; Walsh, D.A. Palladium-vanadium alloy electrocatalysts for oxygen reduction: Effect of heat treatment on electrocatalytic activity and stability. *Appl. Catal. B* **2010**, *98*, 49–56.

259. Li, B.; Prakash, J. Oxygen reduction reaction on carbon supported Palladium-Nickel alloys in alkaline media. *Electrochem. Commun.* **2009**, *11*, 1162–1165.

260. Zhao, J.; Sarkar, A.; Manthiram, A. Synthesis and characterization of Pd–Ni nanoalloy electrocatalysts for oxygen reduction reaction in fuel cells. *Electrochim. Acta* **2010**, *55*, 1756–1765.

261. Miah, M.R.; Masud, J.; Ohsaka, T. Kinetics of oxygen reduction reaction at electrochemically fabricated tin-palladium bimetallic electrocatalyst in acidic media. *Electrochim. Acta* **2010**, *56*, 285–290.

262. Kim, J.; Park, J.E.; Momma, T.; Osaka, T. Synthesis of Pd-Sn nanoparticles by ultrasonic irradiation and their electrocatalytic activity for oxygen reduction. *Electrochim. Acta* **2009**, *54*, 3412–3418.

263. Karan, H.I.; Sasaki, K.; Kuttiyiel, K.; Farberow, C.A.; Mavrikakis, M.; Adzic, R.R. Catalytic Activity of Platinum Mono layer on Iridium and Rhenium Alloy Nanoparticles for the Oxygen Reduction Reaction. *ACS Catal.* **2012**, *2*, 817–824.

264. Xiao, L.; Zhuang, L.; Liu, Y.; Lu, J.T.; Abruna, H.D. Activating Pd by Morphology Tailoring for Oxygen Reduction. *J. Am. Chem. Soc.* **2009**, *131*, 602–608.

265. Kondo, S.; Nakamura, M.; Maki, N.; Hoshi, N. Active Sites for the Oxygen Reduction Reaction on the Low and High Index Planes of Palladium. *J. Phys. Chem. C* **2009**, *113*, 12625–12628.

266. Erikson, H.; Sarapuu, A.; Alexeyeva, N.; Tammeveski, K.; Solla-Gullon, J.; Feliu, J.M. Electrochemical reduction of oxygen on palladium nanocubes in acid and alkaline solutions. *Electrochim. Acta* **2012**, *59*, 329–335.

267. Lee, C.L.; Chiou, H.P. Methanol-tolerant Pd nanocubes for catalyzing oxygen reduction reaction in H_2SO_4 electrolyte. *Appl. Catal. B* **2012**, *117*, 204–211.

268. Wang, D.L.; Xin, H.L.; Wang, H.S.; Yu, Y.C.; Rus, E.; Muller, D.A.; DiSalvo, F.J.; Abruna, H.D. Facile Synthesis of Carbon-Supported Pd–Co Core-Shell Nanoparticles as Oxygen Reduction Electrocatalysts and

Their Enhanced Activity and Stability with Monolayer Pt Decoration. *Chem. Mater.* **2012**, *24*, 2274–2281.

269. Jang, J.H.; Pak, C.; Kwon, Y.U. Ultrasound-assisted polyol synthesis and electrocatalytic characterization of Pd_xCo alloy and core-shell nanoparticles. *J. Power Sources* **2012**, *201*, 179–183.

270. Koenigsmann, C.; Sutter, E.; Adzic, R.R.; Wong, S.S. Size- and Composition-Dependent Enhancement of Electrocatalytic Oxygen Reduction Performance in Ultrathin Palladium-Gold ($Pd_{1-x}Au_x$) Nanowires. *J. Phys. Chem. C* **2012**, *116*, 15297–15306.

271. Wang, J.X.; Inada, H.; Wu, L.J.; Zhu, Y.M.; Choi, Y.M.; Liu, P.; Zhou, W.P.; Adzic, R.R. Oxygen Reduction on Well-Defined Core-Shell Nanocatalysts: Particle Size, Facet, and Pt Shell Thickness Effects. *J. Am. Chem. Soc.* **2009**, *131*, 17298–17302.

272. Sasaki, K.; Naohara, H.; Cai, Y.; Choi, Y.M.; Liu, P.; Vukmirovic, M.B.; Wang, J.X.; Adzic, R.R. Core-Protected Platinum Monolayer Shell High-Stability Electrocatalysts for Fuel-Cell Cathodes. *Angew. Chem.-Int. Ed.* **2010**, *49*, 8602–8607.

273. Di Noto, V.; Negro, E. Synthesis, characterization and electrochemical performance of tri-metal Pt-free carbon nitride electrocatalysts for the oxygen reduction reaction. *Electrochim. Acta* **2010**, *55*, 1407–1418.

274. Rego, R.; Oliveira, M.C.; Alcaide, F.; Alvarez, G. Development of a carbon paper-supported Pd catalyst for PEMFC application. *Int. J. Hydrogen Energy* **2012**, *37*, 7192–7199.

CHAPTER 7

TERNARY PT-RU-NI CATALYTIC LAYERS FOR METHANOL ELECTROOXIDATION PREPARED BY ELECTRODEPOSITION AND GALVANIC REPLACEMENT

Athanasios Papaderakis, Nikolaos Pliatsikas, Chara Prochaska, Kalliopi M. Papazisi, Stella P. Balomenou, Dimitrios Tsiplakides, Panagiotis Patsalas and Sotiris Sotiropoulos

INTRODUCTION

Electrochemistry offers a clean and efficient route for energy production and storage: the electron is a clean reactant/product and electrochemical cells do not suffer from the Carnot cycle limitations of thermal engines (Denno, 1988; Narayan and Viswanathan, 1998; Chouhan and Liu, 2012). Among the main electrochemical energy conversion systems (fuel cells, solar cells, and batteries), fuel cells are characterized by high energy density and speed of recharging-refueling (Srinivasan, 2006; Viswanathan and Scibioh, 2007; Kreuer, 2013). The oxidant in a fuel cell is almost always oxygen (in few systems, hydrogen peroxide can act as such too) while the most common fuel in low temperature applications is hydrogen gas. For microfuel cell applications however (i.e., portable devices such as music players, laptops etc) liquid fluids are being explored, with methanol, borohydride, and formic acid being the best candidates (Yeom et al., 2005; La O' et al., 2007;Kamarudin et al., 2009). Direct methanol fuel cells (DMFCs) have a number of advantages, including a large theoretical energy density, a low operating temperature and the simplicity of storing, distributing, and handling a liquid chemical (Lamy et al., 2000; Hamnett, 2003). The main parameters that limit their performance and widespread establishment are methanol crossover to the cathode (whereby its spontaneous oxidation lowers the cathode potential) and anode poisoning by the adsorption of CO formed as an intermediate (Antolini et al., 2008; Yong et al., 2009;Brouzgou et al., 2012; Ehteshami and Chan, 2013; Tiwari et al., 2013). The first problem can be tackled by the development of improved membrane separators between the anode and cathode departments (Yong et al., 2009; Brouzgou et al., 2012) as well as the search for methanol-tolerant cathodes (Antolini et al., 2008); the second problem has prompted decades of electrochemical research (both theoretical and experimental) into modified Pt-based catalysts that are CO-tolerant (Petry et al., 1965; Beden et al., 1981; Shibata and Motoo, 1986, 1987; Hamnett and Kennedy, 1988;Ishikawa et al., 2004; Demirci, 2007). Both binary (Page et al., 2000; Park et al., 2002; Zhang et al., 2004; Koffi et al., 2005; Martz et al., 2005; Salgado et al., 2005, 2006; Yang et al., 2005;Antolini et al., 2006a; Royochowdhury et al., 2006; Stassi et al., 2006; Yuan et al., 2006; Kristian and Wang, 2008) and ternary (Lima et al., 2001; Lamy et al., 2002; Cooper and McGinn, 2006; Liu et al., 2006; Wang et al., 2006; Antolini, 2007; Zhang et al., 2007; Huang et al., 2009; Kim et al., 2012; Poh et al., 2012; Zhang and McGinn, 2012; Lee et al., 2013; Li et al., 2013; Tusi et al., 2013) Pt-based systems have been tested, with Ru being accepted almost universally as the second metal, mainly due to its ability to form surface oxides and hydroxides that can oxidize and remove the CO intermediate. As far as the third metal is concerned, a variety of other candidates have been tested (usually transition metals) (Lima et al., 2001; Lamy et al., 2002; Cooper and McGinn, 2006;Antolini, 2007; Zhang and McGinn, 2012).

The most common preparation method for fuel cell catalysts is the impregnation/reduction method, whereby the support (high surface area carbons) is dispersed in a solution containing the metal catalyst in ionic form which, following its adsorption on the support, is reduced by gaseous hydrogen at high temperature or by dissolved agents in solution (Antolini,

Citation: Papaderakis A, Pliatsikas N, Prochaska C, Papazisi KM, Balomenou SP, Tsiplakides D, Patsalas P and Sotiropoulos S (2014) Ternary Pt-Ru-Ni catalytic layers for methanol electrooxidation prepared by electrodeposition and galvanic replacement. Front. Chem. 2:29. doi: 10.3389/fchem.2014.00029.

2003, 2007; Rao and Trivedi, 2005). In another family of methods, the metal particles are formed first in a colloidal form by chemical reduction at moderate temperatures [using organic stabilizers such as in the Bonnemann or microemulsion procedures (Boutonnet et al., 1982; Bonnemann et al., 1991)] and then adsorbed on the high surface area carbon supports.

Since 2001 an alternative method for the preparation of poly-metallic catalyst layers (known as galvanic replacement or transmetalation) has been proposed (Brankovic et al., 2001a,b,c;Kokkinidis et al., 2001). It comprises of the spontaneous deposition of a noble metal from its ionic solutions onto a less noble metallic layer or substrate which in turn dissolves as an ionic species. The early variant of the method involved the complete replacement of a Cu underpotentially deposited monolayer on a metallic substrate (Brankovic et al., 2001a,b,c), while later ones expanded to the partial replacement of Pb and Cu electrodeposited polylayers (Kokkinidis et al., 2001; Van Brussel et al., 2002, 2003; Papadimitriou et al., 2007; Tegou et al., 2007), the partial replacement of other transition metals (e.g., Fe, Co, Ni) (Papadimitriou et al., 2008, 2010; Tegou et al., 2008, 2009, 2010, 2011) and the preparation of practical catalysts by the use of high area carbon supports as the substrate (where the less noble precursor metal was chemically deposited and the more noble metal (Pt) was subsequently anchored) (Mintsouli et al., 2013a,b). By its very nature, the method tends to produce structures of a noble-metal (M)-rich shell and of a less-noble-metal (M′)-rich core (denoted as M(M′) hereafter). This structure has the practical advantage of minimizing the use of the expensive noble metal (since it will be restricted to the surface) and the fundamental virtue of allowing the study of electronic only effects of the less noble metal on the noble metal catalytic activity since, as the former resides in the core of the system, it will not be in contact with the reaction medium but only with the noble metal over-layers.

Despite some papers on using transmetalation to deposit binary precious metal systems from mixed solutions of their ions (e.g., Pt-Au, Pt-Pd etc), only a couple of papers using ternary Pt-Ru(Co) (Zhao et al., 2008) and Pt-Ru(Pb) (Ando et al., 2009) catalysts for methanol oxidation have appeared. At the same time, although Pt-Ru-Ni ternary systems have also been tried in methanol oxidation (Liu et al., 2006; Wang et al., 2006; Zhang et al., 2007; Lee et al., 2013; Li et al., 2013;Tusi et al., 2013), these were prepared by co-reduction methods. The feasibility of preparing Ru(Ni) layers has also been proven, but the material was tested in the hydrogen evolution reaction (Bianchi et al., 2005). Thus, there is scope for investigating the possibility of preparing Pt-Ru-Ni ternary systems of the Pt-Ru(Ni) type, by the partial galvanic replacement of pre-formed Ni layers by Pt and Ru, upon the contact of the former with ionic solutions of the latter. In line with the strategy followed in our group, before departing on the development of practical Pt(M) catalysts supported on high surface area carbons (where M is chemically deposited), we investigate the replacement of M layers electrodeposited on flat glassy carbon substrates.

The aim of this work has been to prepare Pt-Ru(Ni) coatings by the surface transmetalation of electrodeposited Ni layers on glassy carbon (GC) supports and correlate their structure with their electrocatalytic activity toward the oxidation of methanol. In that direction, specific objectives of this research have been: (i) The microscopic (SEM, AFM) and spectroscopic (EDS, XPS) characterization of the catalysts to obtain information about the transmetalation mechanism and the interactions of the catalyst components; (ii) The characterization of the catalyst electroactive surface area by electrochemical (voltammetric) experiments in pure acidic solutions; (iii) The evaluation of the catalyst activity in the electrooxidation of methanol (by means of current vs. applied potential curves) and its correlation to its properties.

EXPERIMENTAL

Preparation of Nickel and Platinized Nickel Layers on Glassy Carbon

Ni has been electrodeposited on a glassy carbon, GC, disc electrode at −1.10 V vs. a Saturated Calomel Electrode (SCE), from a 0.01 M Ni sulfamate + 0.227 mM $NiCl_2$ + 0.025 M H_3BO_3deaerated solution. The cathodic charge passed during the constant potential electrolysis was controlled to be 380 mC cm^{-2}, resulting in an even deposit [see SEM pictures in Papadimitriou et al. (2008, 2010)] of a ca 130 nm thickness [based on Faraday's law, assuming a 100% Ni deposition current efficiency and taking into account the density of Ni (Callister et al., 1997)].

The Ni-coated GC electrodes, Ni/GC, thus prepared were subsequently immersed for 30 min in either a 0.1 M HCl + 5 × 10^{-5} M K_2PtCl_6 solution or a 0.1 M HCl + 2.5 × 10^{-5} M $RuCl_3$ + 2.5 × 10^{-5}M K_2PtCl_6 mixed solution. As previously reported (Bianchi et al., 2005), $RuCl_3$ stock solutions were aged for a week before they were used. Spontaneous Pt and/or Ru deposition by partial Ni replacement may then occur according to the following transmetalation reactions:

$$2Ni/GC + PtCl_6^{2-} \rightarrow Pt/GC + 2Ni^{2+} + 6Cl^- \quad (1)$$

$$3Ni/GC + 2RuCl_3 \rightarrow 2Ru/GC + 3Ni^{2+} + 6Cl^- \quad (2)$$

During this treatment the solution was continuously bubbled with purified N_2, since the complete absence of O_2 proved to be crucial toward the co-deposition of Pt and Ru particles onto the Ni surface (Bianchi et al., 2005). This is because oxygen reduction, having a high standard potential, is highly competitive with Pt and Ru deposition for coupling with Ni dissolution. Note that at the higher concentrations of Pt complexes used in cases that a simple bi-metallic Pt(Ni) catalyst was prepared (Mintsouli et al., 2013b), there was no need for oxygen exclusion since, once the latter is consumed, the cathodic reaction supporting Ni dissolution becomes exclusively that of Pt deposition. (See also Discussion at the end of Microscopic and spectroscopic characterization of Pt(Ni) and Pt-Ru(Ni) deposits on GC as to why low Pt and Ru concentrations have been used in this work).

Microscopic and Spectroscopic Characterization of Coatings

Scanning Electron Microscopy (SEM) micrographs were acquired with a JEOL 6300 microscope and the composition of the electrode coatings was determined by the companion Oxford ISIS 2000 X-ray, EDS (EDAX) facility. True three-dimensional morphological features of the deposits as well their thickness were obtained with a NT-MDT Solver Pro Atomic Force Microscope, in the tapping (semi-contact) mode.

X-ray photoelectron spectroscopy (XPS) analysis of the catalyst surface layers was carried out in an Axis Ultra DLD system by Kratos Analytical using a monochromated Al-Ka1 X-ray beam as the excitation source. The analyzed area had an elliptical shape with the two axes being ~400 and 700 μm. The pass energy was 80 eV for survey scans and 20 eV for HR spectra; for the latter case, the pass energy resulted in a broadening (FWHM) of less than 500 meV for the Ag-3d line. The studied surfaces were cleaned of adventitious Carbon and other surface contaminants by using a 4 kV Ar^+ion beam for 2 min. Data interpretation was performed with the Kratos-Vision software.

Electrochemical Characterization of Coatings

Linear and cyclic voltammetry (continuous current monitoring as the potential is varied linearly with time, either to one direction or via a periodic change of direction between two limits) was performed with the Autolab 100 system (EcoChemie, now Metrohm), using a three-compartment cell. The working electrode (Pt, Pt(Ni)/GC or Pt-Ru(Ni)/GC) was placed in the central chamber above a special capillary (luggin capillary) to which the second chamber, containing the Saturated Calomel Electrode (SCE) reference electrode ended. The tip was positioned underneath the center of the disc and at a constant distance of 2 mm (this was to ensure minimal and reproducible uncompensated ohmic losses, as set by established electrochemical practice whose foundations can be found in Newman (1996). The uncompensated resistance, measured by both the current-interrupt and positive feedback methods, translated to ohmic losses always lower than 2 mV. The third chamber contained a Pt coil that served as the counter electrode, placed in a compartment separated from that of the working electrode by a glass frit. To ensure minimum contamination, each different set of experiments was carried out in dedicated cells. Hence, different cells were used for the initial Ni deposition, the electrode activation-cleaning in acid, the study of surface electrochemistry and that of methanol oxidation.

Following Ni deposition and immersion into the platinum or platinum-ruthenium exchange solutions, the as prepared Pt(Ni)/GC or Pt-Ru(Ni)/GC electrodes were scanned repeatedly (typically, more than 60 times) at 1 V s^{-1} in a deaerated 0.1 M $HClO_4$ solution between −0.30 and +0.60 V vs. SCE, ensuring that any unreacted surface Ni was anodically dissolved or passivated during exposure to positive potentials. The electrodes were then scanned again at 500 mV s^{-1} in a fresh deaerated 0.1 M $HClO_4$ solution until a steady state voltammogram, typical of Pt or Pt-Ru surface electrochemistry, was recorded (typically, the voltammetric picture stabilized after less than 10 cycles). As an independent determination of their electroactive surface area and also as an evaluation of their catalytic activity toward CO oxidation, the electrodes were immersed in 0.1 M $HClO_4$ saturated with CO (>99.99%; Air Liquide), held at +0.20 V vs. SCE for 5 min (for adsorption of CO to take place) and then, after dissolved CO had been removed by 10 min of N_2 purging, scanned to more positive potentials at 500 mV s^{-1} so that the CO monolayer was oxidized. For the study of methanol oxidation, the electrodes were transferred to a 0.1 M $HClO_4$ + 0.5 M MeOH deaerated solution where linear potential sweep voltammetry was carried out at 5 mV s^{-1}, between +0.10 V and +0.60 V vs. SCE for three times so that a stabilized picture was obtained.

Electrode Materials and Chemicals

These were similar to those of (Papadimitriou et al., 2010). Glassy carbon from Alfa Aesar (1 mm thick) was cut into 5 mm diameter discs and sealed into glass tubes with epoxy resin glue. Electrodes to be used for SEM/EDS and XPS experiments were prepared by connecting the carbon disc (Alfa Aesar 1 mm, cut into 3 mm diameter specimens) to a glass tube via a bridge made of a shrinkable thermoplastic tube. The latter was filled with mercury to ensure electrical contact between the disc and a commercial wire inserted from the open end of the glass tube.

Ni sulfamate (p.a. 99%) from Fluka, $NiCl_2$ (puriss> 97%) from Merck and H_3BO_3 (puriss 98%)from Aldrich were used in the preparation of Ni deposition solutions. H_2PtCl_6 hexahydrate from Sigma-Aldrich (ACS reagent, ≥37.50% as Pt) was employed for the Pt exchange solution. $RuCl_3$ hydrate was from Merck. HClO4 from Riedel, (puriss p.a., ACS reagent, ≥70%) was used in the working solutions and MeOH was also from Riedel (Chromasolv®, for HPLC, gradient grade, ≥99.9%).

RESULTS AND DISCUSSION

Microscopic and Spectroscopic Characterization of pt(Ni) and Pt-Ru(Ni) Deposits on GC

Figure 1A shows the SEM micrograph of a Pt(Ni) coating on a glassy carbon (GC) substrate, prepared by the immersion of a Ni/GC electrodeposit into a 5×10^{-5} M chloroplatinic acid exchange solution, after it has been electrochemically activated by cycling its potential between the onset of the hydrogen evolution region (−0.30 V) and the onset of surface oxide formation (+0.60 V). The morphology of the coating is similar to that observed in Papadimitriou et al. (2008, 2010), i.e., there is overall a uniform coverage of the substrate, decorated by nanopores and a few uncovered large patches. The pores and the patches result from significant Ni dissolution, either via the transmetalation process (coupled with Pt deposition at nearby locations) or by Ni electrochemical dissolution (induced by the positive potential applied to the electrode during its activation). Figure 1B shows the SEM micrograph of a Pt-Ru(Ni) coating, prepared by the immersion of the Ni/GC electrodeposit into a mixed, equimolar (2.5×10^{-5} M) chloroplatinic acid + ruthenium chloride solution, following its electrochemical activation. A less homogeneous and rougher coating is obtained, with both the deposit and uncovered areas organized in irregular patches. EDS elemental analysis of the deposits gives a Pt ÷ Ru ÷ Ni % bulk atomic composition ratio of 37 ÷ 12 ÷ 51 (and 47 ÷ 53 for the binary Pt-Ni control system).

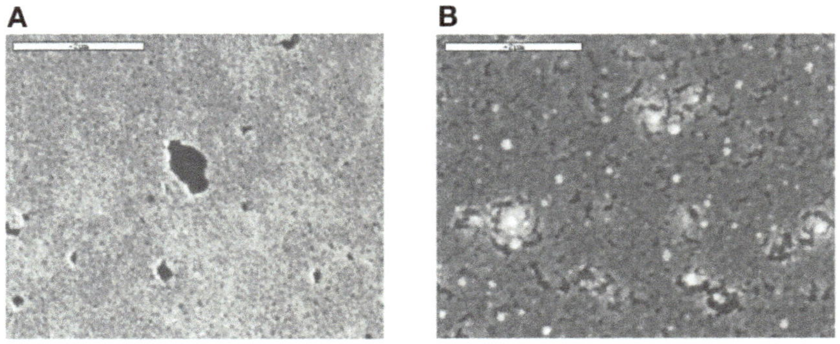

Figure 1. SEM micrographs of: **(A)** A Pt(Ni) coating of a 47% Pt-53%Ni atom composition, prepared after the immersion of Ni electrodeposits on GC (deposition charge density of 380 mC cm^{-2}) in a 0.1 M HCl + 5 × 10^{-5} M K_2PtCl_6 solution, followed by electrochemical conditioning in the −0.3 ± 0.6 V vs. SCE potential range; **(B)** A Pt-Ru(Ni) coating of a 37% Pt-12% Ru-51%Ni atom composition, prepared from Ni electrodeposits [same as those of **(A)**] after their immersion in a 0.1 M HCl + 2.5 × 10^{-5} M $RuCl_3$ + 2.5 × 10^{-5} M K_2PtCl_6 mixed solution and activated in a manner similar to **(A)**. Scale bar: 2 μm.

A direct estimate of the coating thickness is not possible by the SEM pictures in either case. It could be however inferred by the true topographical information supplied by AFM measurements. Figure2 depict three dimensional representations of the low resolution AFM image taken over a 5 × 5 μ m area of the Pt(Ni) and Pt-Ru(Ni) deposits.

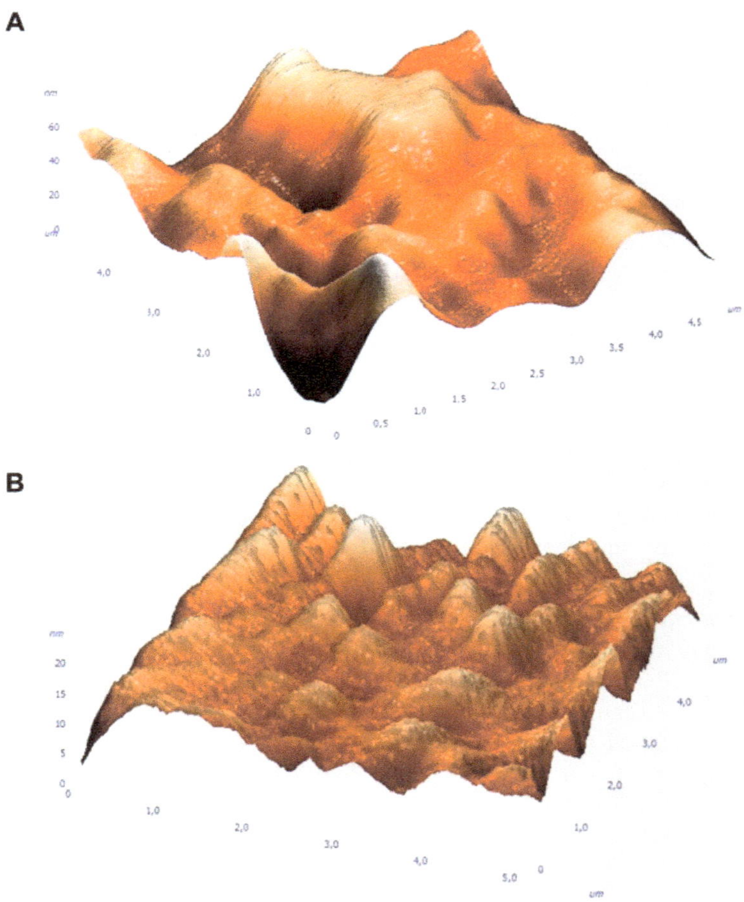

Figure 2. AFM micrographs of 5 × 5 μm sample areas of: **(A)** the Pt(Ni) coating of Figure 1A above; **(B)** the Pt-Ru(Ni) coating of Figure 1B above.

It can be seen that Pt(Ni) is organized in hundreds of nm to few μ m nodules, further aggregated in larger formations and separated by wells. Assuming that the deepest wells correspond to the GC substrate (darkest and largest areas of the corresponding SEM picture of Figure 1A), the maximum deposit height can be estimated to be in the 60–70 nm range. This is to be compared with the thickness of the initially electrodeposited smooth Ni layer, estimated as 130 nm (from the charge passed and Faraday's law, and by the density of bulk Ni-see also Experimental above). Both the perforated morphology and the thickness of the resulting Pt(Ni) layer point to a "thinning" process that occurs during the transmetalation step. At the low concentration levels of the Pt-containing solution of this study (so that Ru is also deposited-see discussion below), the rate of Ni replacement by Pt becomes slow due to mass transfer limitations (despite the relatively high standard potential difference of the two metals-see also below); hence the protection of the dissolving Ni from the outer Pt shell takes time before it is completed, allowing for significant Ni losses. In the case of Pt-Ru(Ni), the AFM picture of Figure 2B shows a deposit consisting of smaller nodules and of a maximum thickness in the range of 20–30 nm, indicating larger Ni losses during the transmetalation step. This is to be expected since in the mixed, equimolar solution, Pt is partially substituted by Ru which has a lower standard potential, hence lower kinetics of deposition; this is because not only the thermodynamics but also the kinetics of the transmetalation reactions (1) and (2) (see Experimental above) depend on the $E^0_{PtCl_6^{-4}/Pt} - E^0_{Ni^{2+}/Ni}$ and $E^0_{Ru(III)/Ru} - E^0_{Ni^{2+}/Ni}$ differences (Papadimitriou et al., 2008). [Note that, according to (Lovrecek et al., 1985), $E^0_{PtCl_6^{-4}/Pt} = + 0.774$ V vs. SHE, $E^0_{Ru(III)Ru} = +0.386$ V vs SHE, and $E^0_{Ni^{2+}/Ni} = -0.257$ V vs. SHE]. At this point we would like to stress that the reason of using solutions of low Pt concentration [in contrast to our previous studies (Papadimitriou et al., 2007, 2008, 2010; Tegou et al., 2007, 2008, 2009,2010, 2011; Mintsouli et al., 2013a,b)] has been the deliberate introduction of mass transfer limitations so that the large difference in Pt and Ru deposition kinetics is attenuated and Ru co-deposition is accomplished. (Attempts to use higher Pt and Ru concentrations failed to produce mixed Pt-Ru deposits and instead, only Pt ones were obtained).

In order to obtain further insight into the chemical state, the interactions and the composition of the outer layers of the deposits, XPS and X-Ray excited Auger Electron Spectroscopy (XAES) (two more surface sensitive techniques) were used. Figure 3 presents the combined XPS/XAES spectra for Ni.

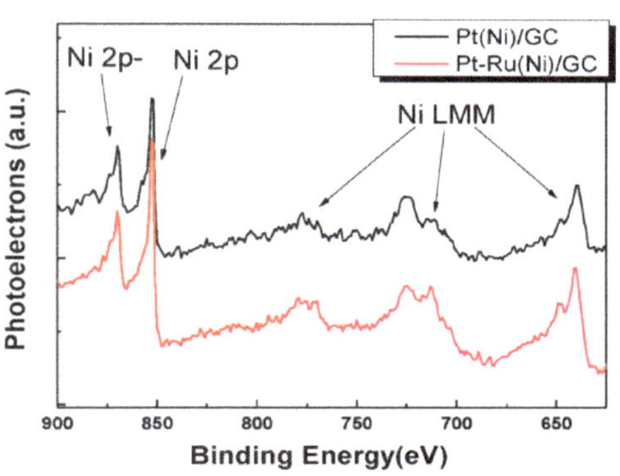

Figure 3. XPS spectra in the Ni2p binding energy range for Pt(Ni) (upper curve) and Pt-Ru(Ni) (lower curve) coatings.

It can be seen that the two main peaks observed in the 850–900 eV range are those corresponding to metallic Ni and no "satellite peaks" (attributed to Ni oxides and hydroxides Park et al., 2002;Mintsouli et al., 2013b) are pronounced in the binding energy region between these two peaks. This indicates that even in the few outer atomic layers that are probed by XPS, Ni is not present as an oxide, further hinting to its absence from the outmost surface layer (where, in contact with the solution, it would have been converted to oxides/hydroxides) and the formation of a complete, protective Pt or Pt-Ru shell.

Figure 4 shows the spectrum segment in the C1s and Ru 3d electron binding energies for the Pt(Ni) and Pt-Ru(Ni) catalysts respectively.

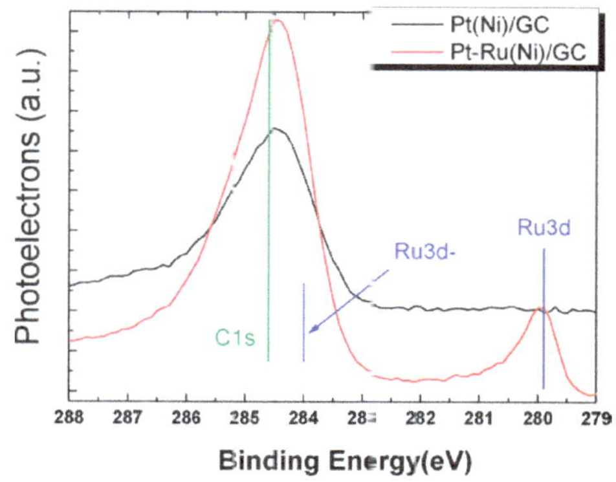

Figure 4. XPS spectra in the C1s and Ru3d binding energy range for Pt(Ni) (upper curve) and Pt-Ru(Ni) coatings (lower curve).

The C1s line corresponds to adventitious carbon, determined separately, set to the value of 284.8 eV reported in the literature (Jackson and Nuzzo, 1995) and used as a reference point for all the spectra recorded. The position of the C1s peak of Figure 4 is at 284.5 eV, as expected for the glassy carbon substrate (Leiro et al., 2003), accessible at uncovered areas. The peak around 280 eV observed in Figure 4 proves unequivocally the presence of Ru in the outer layers of the Pt-Ru(Ni) catalyst.

Figure 5 presents the XPS spectra in the Pt 4f electron energy range for the Pt(Ni) and Pt-Ru(Ni) deposits.

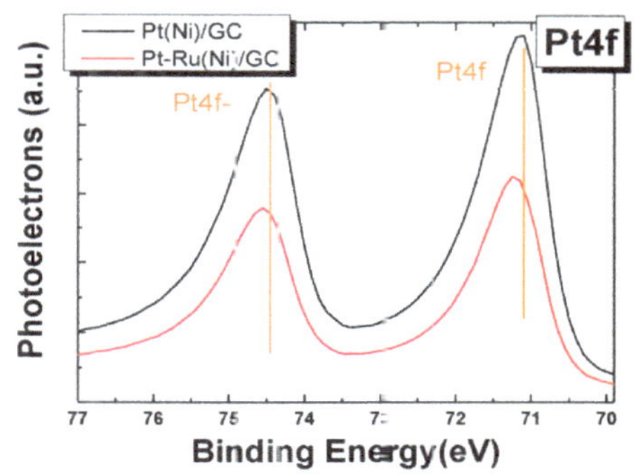

Figure 5. XPS spectra in the Pt4f binding energy range for Pt(Ni) (upper curve) and Pt-Ru(Ni) (lower curve) coatings.

In the former case, there is hardly any significant change in peak position (i.e., it is less than 0.1 eV), in accordance with our previous findings for similar Pt(Ni)/C systems (Mintsouli et al., 2013b), indicating weak interactions (undetected by XPS) of the Pt outer shell with the Ni-rich core. On the contrary, measurable 0.1 and 0.2 eV shifts to higher binding energy values (for the higher and lower binding energy Ptf peaks respectively) are observed in the case of the Pt-Ru(Ni) deposit (in accordance with extensive literature for Pt-Ru systems [see for example (Wakisaka et al., 2006;Zhou et al., 2007; Xu et al., 2010)], indicating strong Pt-Ru interactions. The origin of Pt electronic modification by smaller or less electronegative metal atoms is the strain experienced by Pt atoms residing on top of atoms with a smaller Wagner-Sweitz radius or partial charge transfer respectively. The theoretical work of Nørskov and co-workers (Ruban et al., 1997; Kitchin et al., 2004; Nørskov et al., 2004) specifies this modification as a downshift of the electron d-band center [confirmed by UPS measurements (Stamenkovic et al., 2006, 2007)], whereas Watanabe and co-workers further specify this as a lowering of core electron energy levels (confirmed by XPS measurements) (Wakisaka et al., 2006, 2008). Since Ni atoms are smaller and less electronegative than Ru, one would expect that Pt modification would be pronounced in the case of Pt(Ni) systems too. The fact that it is only observed for Pt-Ru(Ni) indicates that either the Pt and Pt-Ru shells are too thick for any effects of the underlying Ni to be operative at all or that the composition of the films under the shells is such that these effects are too weak to be measured by XPS (that could be due to the fact that the underlying Ni is intermixed with Pt or Pt and Ru and hence its modifying effect attenuated). Arguments against the existence of a very thick Pt shell can be found inPapadimitriou et al. (2010) where Auger electrons originating from Ni can be detected during AES experiments on similar Pt(Ni) samples as well as from the combined XPS/AES results of Figure 3. Based on this finding and the inelastic mean free path of the corresponding Auger electrons through Pt, it has been estimated in Papadimitriou et al. (2010) that the outer Pt shell could be thinner than the 4–5 layers to which, according to (Meier et al., 2004; Kibler, 2008), the strain effect of the under-layers can extend.

Evidence for the enrichment of the outer layers with Pt and Pt-Ru and their depletion from Ni is provided by XPS elemental analysis which gives a Pt ÷ Ni % surface atomic composition ratio of 85 ÷ 15 and a Pt ÷ Ru ÷ Ni ratio of 61 ÷ 12 ÷ 27, i.e., a clear Pt and Pt-Ru enrichment of the outer layers when compared to the bulk composition (obtained by EDS and quoted above as 47 ÷ 53 and 37 ÷ 12 ÷ 51 respectively). On the contrary, the co-existence of Ru in the outer layers of the deposit (a result of the galvanic replacement process itself) allows for its stronger interaction with Pt on the surface, reflected in a shift of its peaks.

Electrochemical Characterization of Pt(Ni)/GC and Pt-Ru(Ni)/GC Electrodes

The electrochemical properties of the catalysts' surface were probed by means of current vs. potential curves recorded during the periodic variation of the potential within a given potential range, i.e., by means of cyclic voltammetry. Figure 6 presents the cyclic voltammograms (obtained at fast potential sweep rates of 500 mV s^{-1}) for the Pt(Ni)/GC and Pt-Ru(Ni)/GC electrodes in a deaerated 0.1 M $HClO_4$ solution, whereas Figure 7 that for a Pt disk electrode.

For the Pt(Ni)/GC electrode features similar to those of pure polycrystalline Pt [(Tegou et al., 2007) and Figure 7] are seen. These comprise of the cathodic (negative current) peaks, attributed to hydrogen electrosorption on Pt as an atomic layer (at ca. 0.00 and −0.20 V) and their conjugate anodic (positive current) peaks, attributed to its oxidative desorption upon potential reversal. At positive potentials the anodic currents that start to increase at ca 0.50 V correspond to the onset of Pt surface oxide formation while the cathodic peak at the same potential corresponds to its reductive removal/stripping.

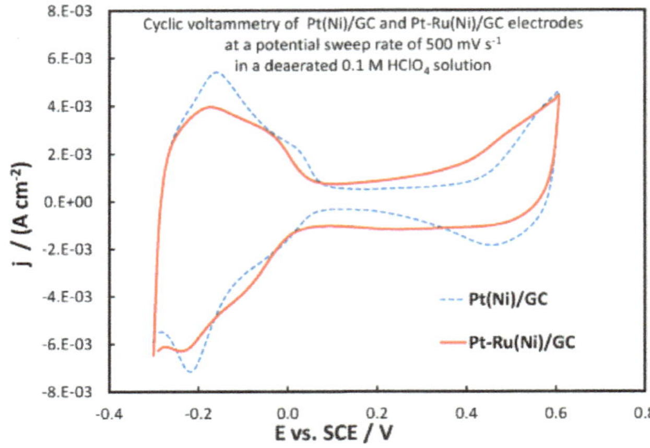

Figure 6. Voltammograms of Pt(Ni)/GC and Pt-Ru(Ni)/GC electrodes (as indicated in the graph), in a deaerated 0.1 M $HClO_4$ **solution at 500 mV** s^{-1}potential scan rate.

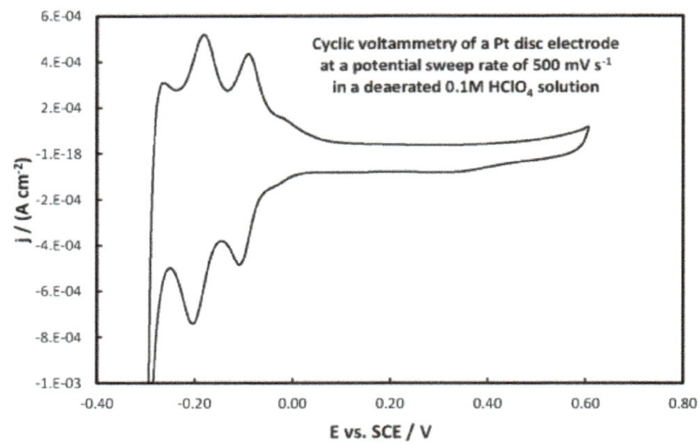

Figure 7. Voltammogram of a Pt disc electrode in a deaerated 0.1 M $HClO_4$solution at 500 mV s^{-1} **potential scan rate.**

The similarity of the voltammetric features with those of pure Pt is further evidence of the absence of any Ni from the catalyst surface (as also inferred from the Ni XPS spectrum discussed in the previous section). The slight difference between Pt(Ni) and Pt with respect to the relative heights of the strongly bound (at ca. 0.00 V) and weakly bound H (at ca. −0.20 V), was also observed in Tegou et al. (2009, 2010, 2011, Papadimitriou et al., 2010) and should be due to the effect of Ni on the adsorption strength of Pt (see also discussion for methanol oxidation below). Using the charge density associated to the formation or stripping of a H monolayer [210 μ C cm^{-2} (Tegou et al., 2007)] and the charge corresponding to H stripping from the catalyst (calculated from the area under the curve segment between the H desorption peaks, after subtraction of the capacitive current recorded in the double layer region, between 0.10 and 0.30 V, assuming that this remains constant during the desorption process), an estimate of the Pt electroactive surface area, A_e, can be made. From that, the ratio $r = A_e/A_g$(known as roughness factor, with A_g being the substrate geometric area) can be estimated as 12.8 cm^2 of Pt per 1 cm^2 of substrate geometric area, in accordance with the micro-particulate/rough texture of the deposit, as shown in the SEM and AFM pictures above. At first sight, the use of the hydrogen adsorption/desorption charge and the value of 210 μ C cm^{-2}, to find the electroactive area of Pt(Ni) electrodes, contradicts the findings of (Fowler et al., 2008) for PtNi(111) and of (Hoffmannová et al., 2013) for nanoparticulate PtNi alloys, in which the authors reported a H coverage less than one. However, those electrodes were different than the ones prepared for this work: those of (Fowler et al., 2008) had a monocrystalline underlayer with the one in contact with the Pt shell being a pure Ni layer; those of (Hoffmannová et al., 2013) were conditioned in the hydrogen adsorption region and had Ni surfacing to the outer shell too. An indication that the approach used to find the electroactive area is legitimate in the case of our Pt(Ni) electrodes is provided by the fact that a very similar electroactive area could be estimated by using the charge corresponding to the oxidation of a CO monolayer in the experiment depicted in Figure 8.

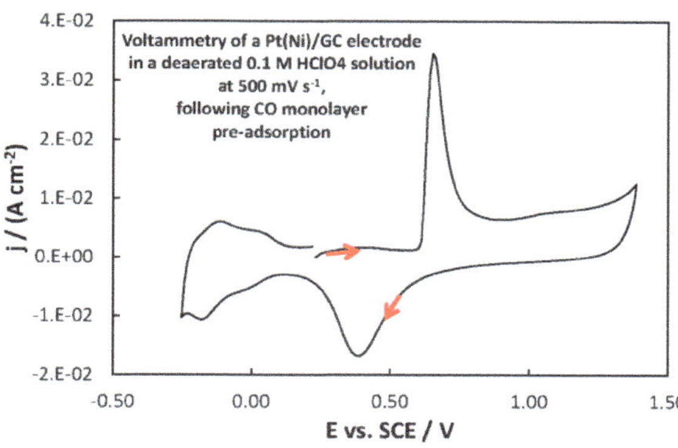

Figure 8. Voltammogram of a Pt(Ni)/GC electrode in a deaerated 0.1 M $HClO_4$solution at 500 mV s^{-1}, following the procedure for CO monolayer adsorption and subsequent oxidation described in section Electrochemical characterization of coatings. The scan was initiated at +0.20 V toward more positive potentials.

The roughness factor estimated by H desorption (taking the value of 210 μ C cm^{-2} for a complete H monolayer) was 13.9 whereas that based on CO oxidative $2e^-$ desorption (assuming 420 μ C $cm^{-2}$for a complete monolayer of linearly bonded CO) could be found as 13.1 (i.e., almost the same). If the effect of Ni on the adsorption properties of Pt in our samples was so strong as to affect the coverage of adsorbed H and CO it would be very unlikely that the extent of decrease in coverage would be identical for both species.

In the case of the Pt-Ru(Ni)/GC electrode, three features of the corresponding voltammogram of Figure 6 point to the co-existence of Ru at the catalyst surface (Gasteiger et al., 1993; Chu and Gilman, 1996; Tripkovic et al., 2002). First, the hydrogen adsorption/desorption peaks in the 0.00 to −0.30 V range are ill-defined, as expected if surface sites were also occupied by Ru, since that does not show significant H atom adsorption/desorption and the latter processes partially overlap with some Ru oxide reduction/formation. Second, the region of the voltammogram between 0.00 and +0.40 V (known as double layer region, where any current recorded in the case of pure Pt does not arise from a faradaic red/ox process but it is simply a capacitive, electrode charging current) is characterized by higher currents, indicative of Ru electrochemistry. Finally, the onset potential at which the formation of surface oxides occurs at the Pt-Ru(Ni) electrode (indicated by a current rise during the positive potential scan, beyond ca +0.40 V) is less positive than that for Pt(Ni). This could be explained by the presence of Ru on the surface whereby, in the presence of water, Ru surface oxides are formed easier/at lower potentials than Pt surface oxides. For the case of Pt-Ru(Ni)/GC too, the Pt electroactive area is similar to that of Pt(Ni)/GC. Since both EDS and XPS analysis confirmed that there was less Pt in the ternary catalyst, similar electroactive areas could only be explained if this decrease in Pt content is accompanied by an increase in the Pt-Ru(Ni) deposit roughness (as confirmed at microscopic level by the SEM pictures of Figure 1) It should be noted that the use of the charge under the hydrogen adsorption or desorption region in the case of PtRu electrodes will most likely lead to an overestimate of Pt electroactive area since it contains charge associated with Ru oxide processes too (Gasteiger et al., 1993); however if this overestimate is accounted for/corrected the electroactive area-specific activity of the electrode toward a Pt-catalyzed reaction (such as methanol oxidation) would further increase.

Methanol and CO Oxidation in Acid

The electrochemical oxidation of methanol at Pt-based electrodes was studied by near-steady state, slow linear potential sweep (at 5 mV s^{-1}) voltammetry, in a 0.5 M MeOH + 0.1 M $HClO_4$ solution. Figure 9 presents the corresponding voltammograms (positive-going potential sweeps) for Pt(Ni)/GC, Pt-Ru(Ni)/GC and pure Pt electrodes. The current density is reported per Pt electroactive area (as estimated from H monolayer adsorption-see above) so as to depict the inherent catalytic activity of the material and correct for mere increased surface area effects.

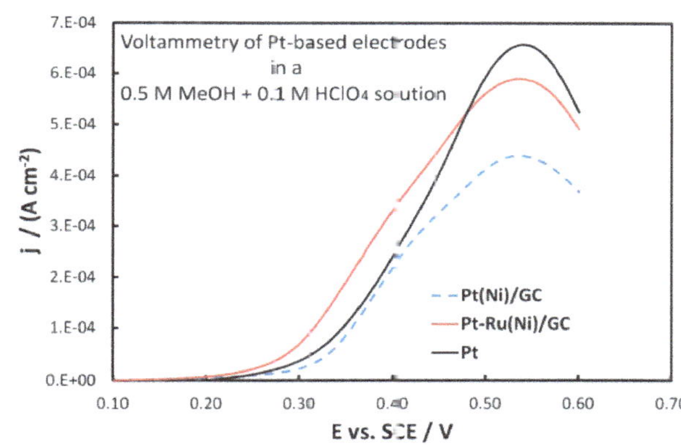

Figure 9. Voltammograms (third positive-going scan at 5 mV s^{-1} **potential scan rate) of** Pt(Ni)/GC, Pt-Ru(Ni)/GC, and bulk Pt electrodes (as indicated in the graph) in a deaerated 0.5 MeOH + 0.1 M $HClO_4$ **solution**. Current density is per electroactive surface area.

They all show the peak-shaped characteristics of methanol oxidation [(Papadimitriou et al., 2007,2010) and references therein]. Up to the peak potential of ca. +0.55 V chemisorbed methanol and the intermediate/poison CO are oxidized at an increasing rate on Pt free metal sites (Pt is only partially covered by Pt oxides in that potential range). At more positive potentials the higher coverage of Pt by its surface oxides limits the number of free Pt centers necessary for the initial reaction step of methanol chemisorption, thus decreasing the observed current.

Both Pt-Ru(Ni) and, mainly, Pt(Ni) catalysts show smaller peak currents than plain Pt. According to the above-mentioned, generally accepted, interpretation of the shape of methanol oxidation voltammogram, this may arise from a higher Pt surface oxide coverage at these potentials, on Ni containing systems The surface voltammetry of Pt(Ni) and Pt(Co) electrodes presented in Tegou et al. (2010) and the experimental findings of high OH species coverage at Pt-Fe electrodes (Wakisaka et al., 2008), do indeed support this idea. In the case of Pt-Ru(Ni) the effect is less pronounced since the oxides are preferentially formed on Ru than on Pt. At this point we would like to stress the fact that electrochemical measurements, due to their surface specific nature, can sometimes provide macroscopic evidence for subtle interactions (in this case, Pt-Ni interactions) that even surface sensitive spectroscopic techniques (in this case XPS) fail to provide.

From a fuel cell catalyst point of view, the potential range of interest for methanol oxidation is at low ovepotentials, i.e., in the potential range below +0.40 V vs. SCE. In that range, the superiority of the Pt-Ru(Ni) catalyst compared to the Pt(Ni) and plain Pt is clearly seen. This is in complete accordance with the well-established view (Gasteiger et al., 1993; Chu and Gilman, 1996; Tripkovic et al., 2002) that the presence of Ru on the surface of a Pt-based catalyst is beneficial for methanol oxidation since its oxides accelerate CO oxidative removal from nearby Pt sites, at low overpotentials where formation of Pt oxides is limited and where the applied potential is not high enough for direct CO electro-oxidation. This can be supported by the voltammograms of Figures 10,11 showing the oxidation of adsorbed CO at the three electrodes studied (recorded at a scan rate of 500 mV s^{-1}), whereby CO is oxidized at less positive potentials on the PtRu(Ni) electrode.

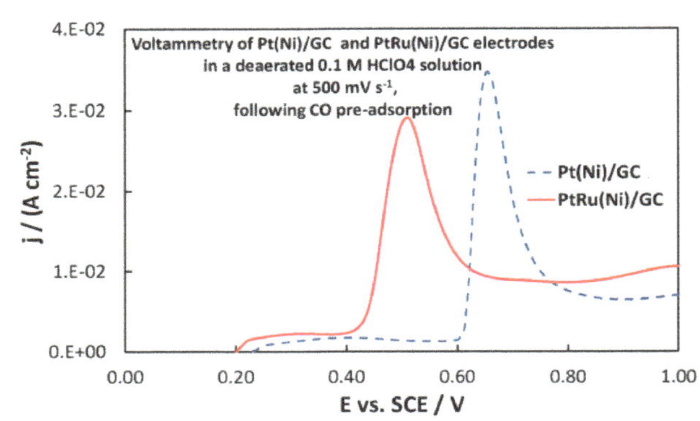

Figure 10. Voltammograms of PtRu(Ni)/GC and Pt(Ni) electrodes in a deaerated 0.1 M $HClO_4$ **solution at 500 mV** s^{-1}, following the procedure for CO monolayer adsorption and subsequent oxidation described in section Electrochemical characterization of coatings. The scan initiated at +0.20 V toward more positive potentials.

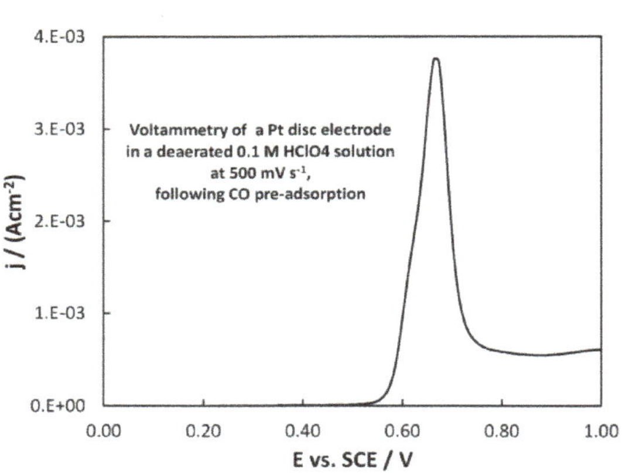

Figure 11. Voltammogram of a Pt disc electrode in a deaerated 0.1 M $HClO_4$solution at 500 mV s^{-1}, following the procedure for CO monolayer adsorption and subsequent oxidation described in section Electrochemical characterization of coatings. The scan initiated at +0.20 V toward more positive potentials.

On the other hand, the near-coincidence of the Pt(Ni) and Pt curves for methanol oxidation (Figure9) is in line with results reported in the literature for Pt-Ni and Pt-Co carbon supported catalysts (Yang et al., 2005; Antolini et al., 2006b) as well as with our own recent voltammetric results on Pt(Ni) electrodes prepared by transmetalation (Papadimitriou et al., 2010). The inefficiency of Pt(Ni) catalysts (at least in the short period timescale of voltammetric experiments) has been interpreted by the fact that the lowering of the d-band center energy, ε_d, of Pt and the associated decrease in its adsorption affinity for small atoms and molecules (Ruban et al., 1997; Kitchin et al., 2004; Nørskov et al., 2004), results in an interplay between decreased methanol chemisorption (decreasing the overall rate of oxidation) and decreased CO intermediate poison adsorption (increasing the rate of oxidation).

CONCLUSIONS

Ternary Pt-Ru-Ni catalyst layers have been prepared on a glassy carbon, GC, substrate by galvanic partial replacement of electrodeposited Ni layers, upon their immersion at room temperature in mixed Pt and Ru salt solutions.

Both the surface electrochemistry of the ternary catalyst as well as its XPS spectra point to the existence of a Pt-Ru shell and the recession of Ni in the interior of the layers.

The ternary catalyst exhibited high electrocatalytic activity toward methanol electrooxidation at low overpotentials, highlighting the further opportunities offered by the method for the preparation of efficient practical Pt-Ru catalysts on high surface area carbon supports for use in direct methanol fuel cells.

Conflict of Interest Statement

The authors declare that the research was conducted in the absence of any commercial or financial relationships that could be construed as a potential conflict of interest.

Acknowledgment

IKY, Greece for an MSc Fellowship to Athanasios Papaderakis. The Aristotle University of Thessaloniki for a "*Promoting Excellence in Basic Research*" grant.

REFERENCES

1. Ando, Y., Sasaki, K., and Adzic, R. (2009). Electrocatalysts for methanol oxidation with ultra low content of Pt and Ru.*Electrochem. Commun*. 11, 1135–1138. doi: 10.1016/j.elecom.2009.03.031
2. Antolini, E. (2003). Formation, microstructural characteristics and stability of carbon supported platinum catalysts for low temperature fuel cells. *J. Mater. Sci*. 38, 2995–3005. doi: 10.1023/A:1024771618027
3. Antolini, E. (2007). Platinum-based ternary catalysts for low temperature fuel cells: part I. Preparation methods and structural characteristics. *Appl. Catal. B Environ*. 74, 324–336. doi: 10.1016/j.apcatb.2007.03.002

4. Antolini, E., Lopes, T., and Gonzalez, E. R. (2008). An overview of platinum-based catalysts as methanol-resistant oxygen reduction materials for direct methanol fuel cells. *J. Alloys Comp*. 461, 253–262. doi: 10.1016/j.jallcom.2007.06.077

5. Antolini, E., Salgado, J. R. C., dos Santos, A. M., and Gonzalez, E. R. (2006b). The methanol oxidation reaction on platinum alloys with the first row transition metals: the case of Pt–Co and –Ni alloy electrocatalysts for DMFCs: a short review. *Appl. Catal. B. Environ*. 63, 137–149. doi: 10.1016/j.apcatb.2005.09.014

6. Antolini, E., Salgado, J. R. C., and Gonzalez, E. R. (2006a). The stability of Pt–M (M = first row transition metal) alloy catalysts and its effect on the activity in low temperature fuel cells: a literature review and tests on a Pt–Co catalyst. *J. Power Sources*160, 957–968. doi: 10.1016/j.jpowsour.2006.03.006

7. Beden, B., Kadirgan, F., Lamy, C., and Leger, J. M. (1981). Electrocatalytic oxidation of methanol on platinum-based binary electrodes. *J. Electroanal. Chemistry* 127, 75–85. doi: 10.1016/S0022-0728(81)80469-X

8. Bianchi, I., Guerrini, E., and Trasatti S. (2005). Electrocatalytic activation of Ni for H_2 evolution by spontaneous deposition of Ru. *Chem. Phys*. 319, 192–199. doi: 10.1016/j.chemphys.2005.06.042

9. Bonnemann, H., Brixoux, W., Brinkmann, R., Dinjus, E., Jouben, T., and Korall, B. (1991). Formation of colloidal transition metals in organic phases and their application in catalysis. *Angew. Chem. Int. Ed. Engl*. 30, 1312–1314. doi: 10.1002/anie.199113121

10. Boutonnet, M., Kizling, J., Stenius, P., and Maire, G. (1982). The preparation of monodisperse colloidal metal particles from microemulsions. *Colloids Surf*. 5, 209–225. doi: 10.1016/0166-6622(82)80079-6

11. Brankovic, S. R., McBreen, J., Adzic, R. R. (2001b). Spontaneous deposition of Pt on the Ru(0001) surface. *J. Electroanal. Chem*. 503, 99–104. doi: 10.1016/S0022-0728(01)00349-7

12. Brankovic, S. R., McBreen, J., and Adzic, R. R. (2001c). Spontaneous deposition of Pd on a Ru(0001) surface. *Surf. Sci*. 479, L363–L368. doi: 10.1016/S0039-6028(01)01006-8

13. Brankovic, S. R., Wang, J. X., and Adzic, R. R. (2001a). Metal monolayer deposition by replacement of metal adlayers on electrode surfaces. *Surf. Sci*. 474, L173–L179. doi: 10.1016/S0039-6028(00)01103-1

14. Brouzgou, A., Song, S. Q., and Tsiakaras, P. (2012). Low and non – platinum electrocatalysts for PEMFCs: current status, challenges and prospects. *Appl. Catal. B. Environ*. 127, 371–388. doi; 10.1016/j.apcatb.2012.08.031

15. Callister, W. D. Jr. (1997). *Materials Science and Engineering: An Introduction*. New York, NY: John Wiley and Sons Inc.

16. Chouhan, N., and Liu, R.-S. (2012). "Electrochemical technologies for energy storage and conversion," in *Electrochemical Technologies for Energy Storage and Conversion*, Vol. 1, eds R.-S. Liu, L. Zhang, X. Sun, X. Liu, and J. Zhang (Weinheim: Wiley-VCH Verlag GmbH & Co.), 1–43. doi: 10.1002/9783527639496.ch1

17. Chu, D., and Gilman, S. (1996). Methanol electro-oxidation on unsupported Pt–Ru alloys at different temperatures. *J. Electrochem. Soc*. 143, 1685–1690. doi: 10.1149/1.1836700

18. Cooper, J. S., and McGinn, P. J. (2006). Combinatorial screening of thin film electrocatalysts for a direct methanol fuel cell anode. *J. Power Sources* 163, 330–338. doi; 10.1016/j.jpowsour.2006.09.028

19. Demirci, U. B. (2007). Theoretical means for searching bimetallic alloys as anode electrocatalysts for direct liquid-feed fuel cells *J. Power Sources* 173, 11–18. doi: 10.1016/j.jpowsour.2007.04 069

20. Denno, K. (1988). *Power System Design and Applications for Alternative Energy Sources*. New York, NY: Prentice Hall.

21. Ehteshami, S. M. M., and Chan, S. H. (2013). A review of electrocatalysts with enhanced CO tolerance and stability for polymer electrolyte membarane fuel cells. *Electrochim. Acta*. 93, 334–345. doi: 10.1016/j.electacta.2013.01.086

22. Fowler, B., Lucas, C. A., Omer, A., Wang, G., Stamenković, V. R., and Markoviæ, N. M. (2008). Segregation and stability at $Pt_3Ni(111)$ surfaces and $Pt_{75}Ni_{25}$ nanoparticles. *Electrochim. Acta*. 53, 6076–6080. doi: 10.1016/j.electacta.2007.11.063

23. Gasteiger, H. A., Markovic, N., and Ross, P. N. Jr., and Cairns, E. J. (1993). Methanol electrooxidation on well-characterized platinum-ruthenium bulk alloys. *J. Phys. Chem*. 97, 12020–12029. doi: 10.1021/j100148a030

24. Hamnett, A. (2003). "Fundamentals technology and applications," in *Handbook of Fuel Cells*, Vol. 1, eds W. Vielstich, A. Lamm, and H. A. Gasteiger (Chichester: Wiley), 305.

25. Hamnett, A., and Kennedy, B. J. (1988). Bimetallic carbon supported anodes for the direct methanol-air fuel cell. *Electrochim. Acta*. 33, 1613–1618. doi: 10.1016/0013-4686(88)80233-0

26. Hoffmannová, H., Okube, M., Petrykin, V., Krtil, P., Mueller, J. E., and Jacob, T. (2013). Surface stability of Pt_3Ni nanoparticulate alloy electrocatalysts in hydrogen adsorption. *Langmuir* 29, 9046–9050. doi: 10.1021/la401562t

27. Pubmed Abstract | Pubmed Full Text | Huang, T., Liu, J., Li, R., Cai, W., and Yu, A. (2009). A novel route for preparation of PtRuMe (Me = Fe, Co, Ni) and their catalytic performance for methanol electrooxidation. *Electrochem. Commun*. 11, 643–646. doi: 10.1016/j.elecom.2009.01.008

28. Ishikawa, Y., Liao, M.-S., and Cabrera, C. R. (2004). "A theory guided design of bimetallic nanoparticle catalysts for fuel cell applications," in *Computational Materials Science, Theoretical and Computational Chemistry Series*, Vol. 15, ed J. Leszzynski (Amsterdam: Elsevier Science), 384–424.

29. Jackson, S. T., and Nuzzo, R. G. (1995). Determining hybridization differences for amorphous carbon from the XPS C 1s envelope. *Appl. Surf. Sci*. 90, 195–203. doi: 10.1016/0169-4332(95)00079-8

30. Kamarudin, S. K., Achmad, F., and Daud, W. R. W. (2009). Overview on the application of direct methanol fuel cell (DMFC) for portable electronic devices. *Int. J. Hydrogen Energy* 34, 6902–6916. doi: 10.1016/j.ijhydene.2009.06.013

31. Kibler, L. A. (2008). Dependence of electrocatalytic activity on film thickness for the hydrogen evolution reaction of Pd overlayers on Au(111). *Electrochim. Acta* 53, 6824–6828. doi: 10.1016/j.electacta.2008.01.097

32. Kim, J., Poliquit, B. Z., Nam, H.-S., Kim, S. H., Lee, Y. M., Kim, K. M., et al. (2012). Electrochemical activities of Pt–Ru–Co and Pt–Ru–Ni electrocatalysts supported by a carbon fiber web. *Curr. Appl. Phys.* 12, 254–258. doi: 10.1016/j.cap.2011.06.014

33. Kitchin, J. R., Nørskov, J. K., Barteau, M. A., and Chen, J. G. (2004). Modification of the surface electronic and chemical properties of Pt(111) by subsurface 3d transition metals. *J. Chem. Phys.* 120, 10240–10246. doi: 10.1063/1.1737365

34. Pubmed Abstract | Pubmed Full Text | Koffi, R. C., Coutanceau, C., Garnier, E., Léger, J.-M., and Lamy, C. (2005). Synthesis, characterization and electrocatalytic behaviour of non-alloyed PtCr methanol tolerant nanoelectrocatalysts for the oxygen reduction reaction (ORR). *Electrochim. Acta.* 50, 4117–4127. doi: 10.1016/j.electacta.2005.01.028

35. Kokkinidis, G., Stoychev, D., Lazarov, V., Papoutsis, A., and Milchev, A. (2001). Electroless deposition of Pt on Ti: part II. Catalytic activity for oxygen reduction. *J. Electroanal. Chem.* 511, 20–30. doi: 10.1016/S0022-0728(01)00505-8

36. Kreuer, K.-D. (2013). "Fuel cells, introduction," in *Fuel Cells, Selected Entries from the Encyclopedia of Sustainability Science and Technology*, ed K.-D. Kreuer (Berlin: Springer), 1–7.

37. Kristian, N., and Wang, X. (2008). Controlled deposition of Pt on Au nanorods and their catalytic activity towards formic acid oxidation. *Electrochem. Commun.* 10, 12. doi: 10.1016/j.elecom.2007.10.011

38. Lamy, C., Léger, J.-M., and Srinivasan, S. (2000). "Direct methanol fuel cells: from a twentieth century electrochemist's dream to a twenty-first century emerging technology," in *Modern Aspects of Electrochemistry*, eds J. O. M. Bockris and B. E. Conway, Vol. 34, 53, (New York: NY: Plenum Press). doi: 10.1007/0-306-46923-5_3

39. Lamy, C., Lima, A., LeRhun, V., Delime, F., Coutencau, C., and Léger, J.-M. (2002). Recent advances in the development of direct alcohol fuel cells (DAFC). *J. Power Sources* 105, 283–296. doi: 10.1016/S0378-7753(01)00954-5

40. La O', G. J., In, H. J., Crumlin, E., Barbastathis, G., and Shao-Horn, Y. (2007). Recent advances in microdevices for electrochemical energy conversion and storage. *Int. J. Energy Res.* 31, 548–575. doi: 10.1002/er.1280

41. Lee, S.-H., Kim, D.-J., and Yoon, Y.-S. (2013). Electrochemical characterization of hydrothermally synthesized Pt–Ru–Ni–P catalyst for direct methanol fuel cell Japanese. *J. Appl. Phys.* 52:035001. doi: 10.7567/JJAP.52.035001

42. Leiro, J. A., Heinonena, M. H., Laihoa, T., and Batirev, I. G. (2003). Core-level XPS spectra of fullerene, highly oriented pyrolitic graphite, and glassy carbon. *J. Electron Spectrosc. Relat. Phenom.* 128, 205–213. doi: 10.1016/S0368-2048(02)00284-0

43. Li, Q., Xu, Q., Zhou, X., and Li, J. (2013). Preparation and electrochemical research of anode catalyst PtRuNi/C for direct methanol fuel cell. *J. Biobased Mater. Bioenergy* 7, 525–528. doi: 10.1166/jbmb.2013.1351

44. Lima, A., Coutencau, C., Léger, J.-M., and Lamy, C. (2001). Investigation of ternary catalysts for methanol electrooxidation. *J. Appl. Electrochem.* 31, 379–386. doi: 10.1023/A:1017578918569

45. Liu, F., Lee, J. Y., and Zhou, W. J. (2006). Segmented Pt/Ru, Pt/Ni, and Pt/RuNi nanorods as model bifunctional catalysts for methanol oxidation. *Small* 2, 121–128. doi: 10.1002/smll.200500253

46. Pubmed Abstract | Pubmed Full Text | Lovrecek, B., Mekjavic, I., Metikos-Hukovic, F. M., Bard, A. J., Parsons, R., and Jordan, J. (eds.). (1985). *Standard Potentials in Aqueous Solution*. New York, NY; Basel: Marcel Dekker, Inc.

47. Martz, N., Roth, C., and Fueb. H. (2005). Characterization of different Pt/Metal/Complex catalysts as anode catalysts for the PEM fuel cell. *J. Appl. Electrochem.* 35, 85–90. doi: 10.1007/s10800-004-4059-9

48. Meier, J., Schiotz, J., Liu, P., Norskov, J. K., and Stimming, U. (2004). Nano-scale effects in electrochemistry. *Chem. Phys. Lett.* 390, 440–444. doi: 10.1016/j.cplett.2004.03.149

49. Mintsouli, I., Georgieva, J., Armyanov, S., Valova, E., Avdeev, G., Hubin, A., et al. (2013a). Pt-Cu electrocatalysts for methanol oxidation prepared by partial galvanic replacement of Cu/carbon powder precursors. *Appl. Catal. B Environ.* 136, 160–167. doi: 10.1016/j.apcatb.2013.01.059

50. Mintsouli, I., Georgieva, J., Valova, E., Armyanov, S., Kakaroglou, A., Hubin, A., et al. (2013b). Pt–Ni carbon-supported catalysts for methanol oxidation prepared by Ni electroless deposition and its galvanic replacement by Pt. *J. Solid State Electrochem.* 17, 435–443. doi: 10.1007/s10008-012-1915-0

51. Narayan, R., and Viswanathan, B. (1998). *Chemical and Electrochemical Energy Systems.* Hyderabad: Universities Press.

52. Newman, J. (1996). Resistance for flow of current to a disk. *J. Electrochem. Soc.* 113, 501–502. doi: 10.1149/1.2424003

53. Nørskov, J. K., Rossmeisi, J., Logadottir, A., Lindqvist, L., Kitchin, J. R., Bligaard, T., et al. (2004). Origin of the overpotential for oxygen reduction at a fuel-cell cathode *J. Phys. Chem. B* 108, 17886–17892. doi: 10.1021/jp047349j

54. Page, T., Johnson, R., Hormes, J., Noding, S., and Rambabu, B. (2000). A study of methanol electro-oxidation reactions in carbon membrane electrodes and structural properties of Pt alloy electro-catalysts by

EXAFS. *J. Electroanal. Chem.* 485, 34–41. doi: 10.1016/S0022-0728(00)00090-5

55. Papadimitriou, S., Armyanov, S., Valova, E., Hubin, A., Steenhaut, O., Pavlidou, E., et al. (2010). Methanol oxidation at Pt-Cu, Pt-Ni, and Pt-Co electrode coatings prepared by a Galvanic replacement process. *J. Phys. Chem. C* 114, 5217–5223. doi: 10.1021/jp911568g

56. Papadimitriou, S., Tegou, A., Pavlidou, E., Armyanov, S., Valova, E., Kokkinidis, G., et al. (2008). Preparation and characterization of platinum- and gold-coated copper, iron, cobalt and nickel deposits on glassy carbon substrates. *Electrochim. Acta.* 53, 6559–6567. doi: 10.1016/j.electacta.2008.04.015

57. Papadimitriou, S., Tegou, A., Pavlidou, E., Kokkinidis, G., and Sotiropoulos, S. (2007). Methanol oxidation at platinized lead coatings prepared by a two-step electrodeposition–electroless deposition process on glassy carbon and platinum substrates.*Electrochim. Acta.* 52, 6254–6260. doi: 10.1016/j.electacta.2007.04.020

58. Park, K.-W., Choi, J.-H., Kwon B.-K., Lee, S.-A., Sung, Y.-E., Ha, H.-Y., et al. (2002). Chemical and electronic effects of Ni in Pt/Ni and Pt/Ru/Ni alloy nanoparticles in methanol electrooxidation. *J. Phys. Chem. B* 106, 1869–1877. doi: 10.1021/jp013168v

59. Petry, O. A., Podlovchenko, B. I., Frumkin, A. N., and Lal, H. (1965). The behaviour of platinized-platinum and platinum-ruthenium electrodes in methanol solutions. *J. Electroanal. Chem.* 10, 253–269. doi: 10.1016/0022-0728(65)85060-4

60. Poh, C. K., Tian, Z., Gao, J., Liu, Z., Lin, J., Feng, Y. P., et al. (2012). Nanostructured trimetallic Pt/FeRuC, Pt/NiRuC, and Pt/CoRuC catalysts for methanol electrooxidation. *J. Mater. Chem.* 22, 13643–13652. doi: 10.1039/c2jm31956j

61. Rao, C. R. K., and Trivedi, D. C. (2005). Chemical and electrochemical depositions of platinum group metals and their applications. *Coord. Chem. Rev.* 249, 613–631. doi: 10.1016/j.ccr.2004.08.015

62. Royochowdhury, C., Matsumoto, F., Zeldovich, V. B., Warren, S. C., Mutolo, P. F., Ballesteros, M. J., et al. (2006). Synthesis, characterization, and electrocatalytic activity of PtBi and PtPb nanoparticles prepared by borohydride reduction in methanol.*Chem. Mater.* 18, 3365–3372. doi: 10.1021/cm060480e

63. Pubmed Abstract | Pubmed Full Text | Ruban, A., Hammer, B., Stoltze, P., Skriver, H. L., and Nørskov, J. K. (1997). Surface electronic structure and reactivity of transition and noble metals. *J. Mol. Catal. A Chem.* 115, 421–429. doi: 10.1016/S1381-1169(96)00348-2

64. Salgado, J. R. C., Antolini, E., and Gonzalez, E. R. (2005). Carbon supported Pt–Co alloys as methanol-resistant oxygen-reduction electrocatalysts for direct methanol fuel cells. *Appl. Catal. B* 57, 283–290. doi: 10.1016/j.apcatb.2004.11.009

65. Salgado, J. R. C., Antolini, E., Santos, L. G. R. A., Garcia, G., Pastor, E., Ticianelli, E. A., et al. (2006). Carbon supported Pt–Cr alloys as oxygen-reduction catalysts for direct methanol fuel cell. *J. Appl. Electrochem.* 36, 355–362. doi: 10.1007/s10800-005-9072-0

66. Shibata, M., and Motoo, S. (1986). Electrocatalysis by ad-atoms: part XX. Rate-determining step in methanol oxidation enhanced by oxygen-adsorbing ad-atoms. *J. Electroanal. Chem.* 209, 151–158. doi: 10.1016/0022-0728(86)80193-0

67. Shibata, M., and Motoo, S. (1987). Electrocatalysis by ad-atoms: part XXI. Catalytic effects on the elementary steps in methanol oxidation by non-oxygen-adsorbing ad-atoms. *J. Electroanal. Chem.* 229, 385–394. doi: 10.1016/0022-0728(87)85155-0

68. Srinivasan, S. (2006). *Fuel Cells: from Fundamentals to Applications*. Berlin: Springer.

69. Stamenkovic, V. R., Mun, B. S., Arenz, M., Mayrhofer, K. J. J., Lucas, C. A., Wang, G., et al. (2007). Trends in electrocatalysis on extended and nanoscale Pt-bimetallic alloy surfaces. *Nat. Mater* 6, 241–247. doi: 10.1038/nmat1840

70. Pubmed Abstract | Pubmed Full Text | Stamenkovic, V. R., Mun, B. S., Mayrhofer, K. J. J., Ross, P. N., and Markovic, N. M. (2006). Effect of surface composition on electronic structure, stability, and electrocatalytic properties of Pt-transition metal alloys: Pt-skin versus Pt-skeleton surfaces. *J. Am. Chem. Soc.* 128, 8813–8819. doi: 10.1021/ja0600476

71. Pubmed Abstract | Pubmed Full Text | Stassi, A., D'Urso, C., Baglio, V., Di Blasi, A., Antonucci, V., Arico, A. S., et al. (2006). Electrocatalytic behaviour for oxygen reduction reaction of small nanostructured crystalline bimetallic Pt–M supported catalysts. *J. Appl. Electrochem.* 36, 1143–1149. doi: 10.1007/s10800-006-9197-9

72. Tegou, A., Armyanov, S., Valova, E., Steenhaut, O., Hubin, A., Kokkinidis, G., et al. (2009). Mixed platinum–gold electrocatalysts for borohydride oxidation prepared by the galvanic replacement of nickel deposits. *J. Electroanal. Chem.* 634, 104–110. doi: 10.1016/j.jelechem.2009.07.016

73. Tegou, A., Papadimitriou, S., Armyanov, S., Valova, E., Kokkinidis, G., and Sotiropoulos, S. (2008). Oxygen reduction at platinum- and gold-coated iron, cobalt, nickel and lead deposits on glassy carbon substrates. *J. Electroanal. Chem.* 623, 187–196. doi: 10.1016/j.jelechem.2008.07.012

74. Tegou, A., Papadimitriou, S., Kokkinidis, G., and Sotiropoulos, S. (2010). A rotating disc electrode study of oxygen reduction at platinized nickel and cobalt coatings. *J. Solid State Electrochem.* 14, 175–184. doi: 10.1007/s10008-009-0879-1

75. Tegou, A., Papadimitriou, S., Mintsouli, I., Armyanov, S., Valova, E., Kokkinidis, G., et al. (2011). Rotating disc electrode studies of borohydride oxidation at Pt and bimetallic Pt–Ni and Pt–Co electrodes. *Catal. Today* 170, 126–133. doi: 10.1016/j.cattod.2011.01.003

76. Tegou, A., Papadimitriou, S., Pavlidou, E., Kokkinidis, G., and Sotiropoulos, S. (2007). Oxygen reduction at platinum- and gold-coated copper deposits on glassy carbon substrates. *J. Electroanal. Chem.* 608, 67–77. doi: 10.1016/j.jelechem.2007.05.008

77. Tiwari, J. N., Tiwari, R. N., Singh, G., and Kim, K. S. (2013). Recent progress in the development of anode and cathode catalysts for direct methanol fuel cells. *Nano Energy* 2, 553–578. doi: 10.1016/j.nanoen.2013.06.009

78. Tripkovic, A. V., Popovic, K. D., Grgur, B. N., Blizanac, B., Ross, P. N., and Markovic, N. M. (2002). Methanol electrooxidationon supported Pt and PtRu catalysts in acid and alkaline solutions. *Electrochim. Acta*. 47, 3707–3714. doi: 10.1016/S0013-4686(02)00340-7

79. Tusi, M. M., Brandalise M., Polanco, N. S. D. O., Correa, O. V., Da Silva, A. C., Villalba, J. C., et al. (2013). Ni/Carbon hybrid prepared by hydrothermal carbonization and thermal treatment as support for PtRu nanoparticles for direct methanol fuel cell.*J. Mater. Sci. Technol*. 29, 747–751. doi: 10.1016/j.jmst.2013.05.004

80. Van Brussel, M., Kokkinidis, G., Hubin, A., and Buess-Herman, C. (2003). Oxygen reduction at platinum modified gold electrodes. *Electrochim. Acta*. 48, 3909–3919. doi: 10.1016/S0013-4686(03)00529-2

81. Van Brussel, M., Kokkinidis, G., Vandendael, I., and Buess-Herman, C. (2002). High performance gold-supported platinum electrocatalyst for oxygen reduction. *Electrochem. Commun*. 4, 808–813. doi: 10.1016/S1388-2481(02)00437-X

82. Viswanathan, B., and Scibioh, M. A. (2007). *Fuel Cells: Principles and Applications*. Oxford: Taylor & Francis Group.

83. Wakisaka, M., Mitsui, S., Hirose, Y., Kawashima, K., Uchida, H., and Watanabe, M. (2006). Electronic structures of Pt-Co and Pt-Ru alloys for CO-tolerant anode catalysts in polymer electrolyte fuel cells studied by EC-XPS. *J. Phys. Chem. B* 110, 23489–23496. doi: 10.1021/jp0653510

84. Pubmed Abstract | Pubmed Full Text | Wakisaka, M., Suzuki, H., Mitsui, S., Uchida, H., and Watanabe, M. (2008). Increased oxygen coverage at Pt-Fe alloy cathode for the enhanced oxygen reduction reaction studied by EC-XPS. *J. Phys. Chem. C* 112, 2750–2755. doi: 10.1021/jp0766499

85. Wang, Z. B., Yin, G. P., Shi, P. F., and Sun, Y. C. (2006). Novel Pt–Ru–Ni/C catalysts for methanol electro-oxidation in acid medium. *Electrochem. Solid-State Lett*. 9, A13–A15. doi: 10.1149/1.2133722

86. Xu, C., Wang, L., Mu, X., and Ding, Y. (2010). Nanoporous PtRu alloys for electrocatalysis. *Langmuir* 26, 7437–7443. doi: 10.1021/la9041474

87. Pubmed Abstract | Pubmed Full Text | Yang, H., Coutanceau, C., Léger, J.-M., Alonso-Vante, N., and Lamy, C. (2005). Methanol tolerant oxygen reduction on carbon-supported Pt–Ni alloy nanoparticles. *J. Electroanal. Chem*. 576, 305–313. doi: 10.1016/j.jelechem.2004.10.026

88. Yeom, J., Mozsgai, G. Z., Flachsbart, B. R., Choban, E. R., Asthana, A., Shannon, M. A., et al. (2005). Microfabrication and characterization of a silicon-based millimeter scale, PEM fuel cell operating with hydrogen, methanol, or formic acid. *Sens. Actuators B Chem*. 107, 882–891. doi: 10.1016/j.snb.2004.12.050

89. Yong, F., Ruiying, M., Tongtao, W., Xindong, W., and Shibi, F. (2009). Research progress of polymer proton exchange membranes for direct methanol fuel cells. *Acta Polymerica Sinica* 10, 992.

90. Yuan, W., Scott, K., and Cheng, H. (2006). Fabrication and evaluation of Pt–Fe alloys as methanol tolerant cathode materials for direct methanol fuel cells. *J. Power Sources* 163, 323–329. doi: 10.1016/j.jpowsour.2006.09.005

91. Zhang, X., Tsang, K. Y., and Chan, K. Y. (2004). Electrocatalytic properties of supported platinum–cobalt nanoparticles with uniform and controlled composition. *J. Electroanal. Chem*. 573, 1–9. doi: 10.1016/j.jelechem.2004.06.023

92. Zhang, X., Zhang, F., Guan, R.-F., and Chan, K.-Y. (2007). Preparation of Pt-Ru-Ni ternary nanoparticles by microemulsion and electrocatalytic activity for methanol oxidation. *Mater. Res. Bull*. 42, 327–333. doi: 10.1016/j.materresbull.2006.05.021

93. Zhang, Y., and McGinn, P. J. (2012). Combinatorial screening for methanol oxidation catalysts in alloys of Pt, Cr, Co and V. *J. Power Sources* 206, 29–36. doi: 10.1016/j.jpowsour.2012.01.016

94. Zhao, H., Li, L., Yang, J., and Zhang, Y. (2008). Co@Pt–Ru core-shell nanoparticles supported on multiwalled carbon nanotube for methanol oxidation. *Electrochem. Commun*. 10, 1527–1529. doi: 10.1016/j.elecom.2008.07.047

95. Zhou, W.-P., Lewera, A., Bagus, P. S., and Wieckowski, A. (2007). Electrochemical and electronic properties of platinum deposits on Ru(0001): combined XPS and cyclic voltammetric study. *J. Phys. Chem.C* 111, 13490–13496. doi: 10.1021/jp072993u

CHAPTER 8

FORMIC ACID OXIDATION AT PLATINUM–BISMUTH CATALYSTS

Ksenija Đ. Popović and Jelena D. Lović

INTRODUCTION

For many years, much attention has been focused on low-temperature fuel cells with energy conversion based on the electrocatalytic oxidation of small organic molecules, such as methanol, ethanol and formic acid. The fuel cell industry has focused the majority of its research on the development of cost-effective, reactive, and durable catalysts. Electrochemical oxidation of formic acid is under comprehensive investigation for two main reasons: formic acid can be used as a fuel in direct formic acid fuel cell (DFAFC) and can serve as a model reaction that provides a simplified example of the oxidation of more complex organic molecules that can also be used for this purpose.[1] A direct formic acid–oxygen fuel cell with a polymer electrolyte membrane (PEM) has some advantages over a direct methanol fuel cell. Oxidation of formic acid commences at a lower positive potential than methanol oxidation and the crossover of formic acid through the polymer membrane is lower than that of methanol.[2,3] Moreover, formic acid is a relatively benign and non-explosive fuel, which makes it facile in handling and distribution, as compared to hydrogen. On the other hand, it has a lower energy content with respect to hydrogen or methanol. Recent data showed, however, that formic acid fuel cells are attractive alternatives for small portable fuel cell applications.[4,5]

Electrochemical oxidation of formic acid has been widely studied on different metal electrodes. Among them, platinum was shown to exhibit the highest catalytic activity of all the pure metals. This reaction has been investigated at platinum since the early work of Breiter[6] and the results were reviewed by Parsons and Van der Noot[1], and Jarvi and Stuve.[7] Many results have been reported concerning the electrocatalytic oxidation of formic acid from the fundamental viewpoint[8–12] in contrast to the limited information on the properties of formic acid as a fuel. However, in the last few years, this reaction has attracted more attention.[13,14]

As already mentioned above, formic acid oxidation on platinum is considered as a model reaction in electrocatalysis, because the oxidation of formic acid is a simple structure-sensitive process. In spite of the apparent simplicity of the process, that is, the oxidation only requires the elimination of two hydrogen atoms in the form of protons and the transfer of the two corresponding electrons, the oxidation mechanism has a double pathway. Thus, the oxidation of formic acid on Pt electrodes follows the dual path mechanism introduced by Capon and Parsons,[15,16] involving a direct path (dehydrogenation) and an indirect path (dehydratation), both generating CO_2 as the final reaction product. This finding was later confirmed by differential electrochemical mass spectrometry (DEMS) measurements.[3] Both routes are structure sensitive and there is a clear dependence of the reactivity on the surface structure, as experiments with single crystal electrodes revealed:[17–21]

$$HCOOH \xrightarrow{\text{Dehydrogenation}} CO_2 + 2H^+ + 2e^-$$

$$HCOOH \xrightarrow{\text{Dehydration}} CO_{ads} + H_2O \xrightarrow{k} CO_2 + 2H^+ + 2e^-$$

In the direct path, formic acid is oxidized through a reactive intermediate to CO_2, while in the indirect path, formic acid is first dehydrated to the adsorbed CO intermediate, as a poisoning species that hinders the direct reaction path, followed by oxidation of the adsorbed CO by OH formed at higher potentials.

The direct path that proceeds through an active intermediate is the simplest one. Formic acid adsorbs on the surface, probably transferring one electron, to form an active intermediate, and then this intermediate is oxidized to CO_2. Regarding the nature of the active intermediate, the question is not yet fully resolved.[22–24] Adsorbed formate (HCOO), rather than the formic acid fragment (COOH), was proposed as the reactive intermediate[23,25–27] and this assumption was confirmed by direct surface-enhanced infrared absorption spectroscopy (SEIRAS).[28] In the indirect pathway, the first step is a dehydration step with the loss of an oxygen atom to form adsorbed CO, which was detected by IR spectroscopy.[10,29]

Citation: Ksenija Đ. Popović and Jelena D. Lović, Formic Acid Oxidation at Platinum–Bismuth Catalysts, doi: 10.2298/JSC150318044P.

CO adsorbs strongly on the surface, and for this reason, the route is also known as the poisoning route. Thus, adsorbed CO blocks the active sites on the surface and prevents the reaction from proceeding. Accordingly, the catalytic performance of Pt is significantly reduced at low potentials due to CO poisoning.

However, besides being a poisoning species, CO may act as a reactive intermediate, whereby some fraction of COads can be oxidized with OHads to produce CO_2. 12 Nevertheless, platinum is an unavoidable material and considered as one of the most efficient catalysts for the oxidation of small organic molecules but, on the other hand, has several significant disadvantages: high cost and extreme susceptibility to poisoning due to strongly adsorbed intermediates, which are formed during the oxidation processes.30 Taking this into account, the ideal electrocatalysts would be one that accelerates the direct route and prevents the formation of CO. It is now well known that these requirements are fulfilled by bimetallic catalysts, which often show electronic and chemical properties that are distinct from those of their constituent metals and offer the chance to obtain new catalysts with enhanced selectivity, activity and stability.

Several approaches have been taken to achieve these goals, but in general, they consist of the modification of platinum with a second element. This modification is usually realized by alloying or by modification of the Pt surface with adsorbed foreign metals in an amount less than a full monolayer. 31,32 The presence of a foreign metal alters the properties of Pt in the bimetallic surfaces.12,33 The effects of these atoms can be classified in three main categories: electronic, bifunctional and third body effects. The first one, which involves ligand and strain effects, relates to a change in the electronic properties of the catalytically active material. The bifunctional effect is present when the second metal becomes the source of the oxygen required for the oxidation of the fuel. The third body effect implies a change in the distribution of the active adsorption sites due to dilution of the catalytically active material. Generally, more than one of these factors controls the enhanced properties of bimetallic surfaces making separation of the individual contributions difficult. In order to improve Pt electrocatalytic activity towards HCOOH oxidation and tolerance to CO, addition of metals such as Ru, Pb, Os, Li, Pd, Fe, Bi etc. 12,34–39 were applied. Especially, bismuth has received significant attention as a Pt-modifier,18,31,32,40,41 and different systems, such as PtBi intermetallics,42–46 PtBi alloys,39,47,48 electrochemically co-deposited carbon supported PtBi (PtBi/C),49 or Pt modified by Bi either by underpotential deposition (UPD) or irreversible adsorption50,51 were proposed as good catalysts for formic acid oxidation.

Besides the many methods for the synthesis of bimetallic catalysts, a new method for the preparation of noble metal coatings was recently proposed. This procedure includes the replacement of the surface layer of a less precious metal (Ru, Cu, Pb and Ti) with a more noble metal (Pt and Pd) by spontaneous electroless exchange upon immersion into a complex solution of Pt or Pd ions.52,53

An additional approach for the formation of low dimensional systems is electrodeposition of mono or multilayer metals on different substrates. 54–56 This concept of bimetallic mono and multilayer catalysts has received much attention regarding its possibility to reduce the noble metal quantity and maintain the activity by replacing the under-layer (bulk of the catalyst) with a less noble metal. Moreover, unlike other bimetallic catalysts where the second metal is either in the form of an adatom or as a component of a surface alloy, this type of catalyst allows the study of the electronic effect of the second metal under-layer on the noble catalyst over-layer, as the only operating factor. In addition, from a practical point of view, long-term stability of the investigated catalysts for formic acid oxidation is very important. Therefore, it is necessary to determine which of the factors mostly affected the improvement of the formic acid oxidation rate and the stability of the catalyst. In this paper, recent advances in HCOOH oxidation research are presented with focus on the progresses that have been made on Pt–Bi catalysts for the possible use in DFAFCs. Different preparation methods were employed to adapt the properties of Pt. Each method will be emphasized for its advantage and discussed in terms of its limitations, based on the physicochemical and electrochemical characterizations of the catalysts in order to explain the mechanism of action of bismuth added to platinum, the importance of surface composition and surface morphology for the reaction of formic acid oxidation.

BULK CATALYSTS FOR ELECTRO-OXIDATION OF FORMIC ACID

Activity of Bulk Catalysts

Platinum Electrodes

Platinum, the most studied catalyst for formic acid oxidation, is very susceptible to poisoning species, which significantly reduces its catalytic performance at low potentials, as is well known from the literature.12 Traditional single crystals represent an ideal model surfaces for the oxidation reaction of small organic molecule and are suitable for surface characterization methods both in situ and ex situ. Studies on single crystal Pt samples showed that formic acid oxidation is a strongly structure sensitive reaction. The most complete study on formic acid structure sensitivity was realized by the Motoo group,57 using a complete series of stepped surfaces around the stereographic triangle. The studies of this reaction on stepped Pt surfaces were performed in order to clarify how exactly the step density influences this reaction, i.e. higher index or stepped surfaces were used to verify the active site assumption, whether low coordination sites are particularly active. The less poisoned surface was Pt(111), as indicated by the low observed hysteresis, but the activity toward formic acid oxidation was low. It should be noted that the presence of defects, especially on (100) steps, considerably increased the activity of a Pt(111) electrode. For example, in order to achieve higher rates at moderate poisoning, electrodes having 5–6 atoms wide (111) terraces were the best under the experimental conditions used.

The oxidation of formic acid is a structure sensitive reaction, which implies the existence of adsorption steps in the process. In an electrochemical environment, the adsorption processes have to be considered as complex steps since they always involve the competitive adsorption of anions, water and/or hydrogen, which can have different dependences on concentration, and affects the relative rates of dissociative adsorption in a complex way.21 These kind of studies are relevant not only from a fundamental point of view but also from a practical one, because in practical applications, the stepped surfaces may be considered as models for surface defects always present on polycrystalline electrodes. Remembering that the cyclic voltammogram for an as-prepared polycrystalline Pt electrode (Fig. 1a) is described by a region of hydrogen adsorption/ /desorption (E < 0.05 V vs. SCE), separated by a double layer from the region of surface oxide formation (E > 0.45 V vs. SCE). The absence of well-developed peaks at an as-prepared Pt polycrystalline electrode in the hydrogen adsorption/ /desorption region is caused by the employed preparation procedure.

The activity of Pt electrode towards formic acid oxidation is given in Fig. 2a. The cyclic voltammogram shows a well-established feature for formic acid oxidation.7 In the forward scan, the current slowly increases reaching a plateau at ≈0.25 V vs. SCE followed by an ascending current starting at 0.5 V vs. SCE, which attains a maximum at ≈0.62 V vs. SCE. Such behavior could be explained considering the dual path mechanism, i.e., dehydrogenation assigned as the direct path, based on the oxidation of formate,28 and dehydration, indirect path, assumes the formation of CO_{ads}, both generate CO2 as the final reaction product. At low potentials, HCOOH oxidizes through the direct path with the simultaneous formation of CO_{ads}. Increasing coverage with CO_{ads} reduces the Pt sites available for the direct path and current slowly increases reaching a plateau. Subsequent formation of oxygen-containing species on Pt enables the oxidative removal of CO_{ads}, more Pt sites become available for HCOOH oxidation and current increases until Pt oxide, inactive for HCOOH oxidation, is formed, which results in a current peak at ≈0.62 V vs. SCE. In the backward scan, the sharp increase in the HCOOH oxidation current coincides with the reduction of Pt oxide. The currents are much higher than in the forward sweep, because the Pt surface is freed of CO_{ads}.

Effect of Bi Adatoms

The addition of foreign metals to Pt surfaces in amounts less than a full monolayer results in modified surface catalytic properties. Therefore, surfaces of bimetallic electrodes often show improved electrocatalytic behavior. It was reported in earlier studies that Bi modification of platinum electrodes could exceptionally increase their reactivity toward HCOOH oxidation, depending on the Bi surface coverage.40,47 Most current studies exploring structure/property/activity relationships based on experimental approaches either use well-defined single crystals as a model or are based on modified noble single crystal surfaces. Feliu and co-workers58,59 as well as Abruna and co-workers60,61 reported that bismuth-modified platinum low and high-index single crystal surfaces, which were prepared via the under-potential deposition (UPD) process, exhibited extraordinary enhancement in reactivity towards formic acid oxidation.

Irreversibly adsorbed Bi40,47,51,62 inhibits poison formation simultaneously enhancing dehydrogenation,51 i.e., this modification is an efficient way to hinder the dehydration path (CO-intermediate pathway) in favor of the direct path.63 This increased selectivity for dehydrogenation was proposed to be an "ensemble effect"64,65 in which the adsorbed Bi divides the Pt surface into small domains where only dehydrogenation can occur. A correlation between ensemble size and formic acid oxidation activity was also established.66 According to literature data, the activity of Pt catalysts modified with Bi depends on the shape of the Pt nanocrystals,67 and varies with the size of the particles68 and the loading of the Pt catalyst.63

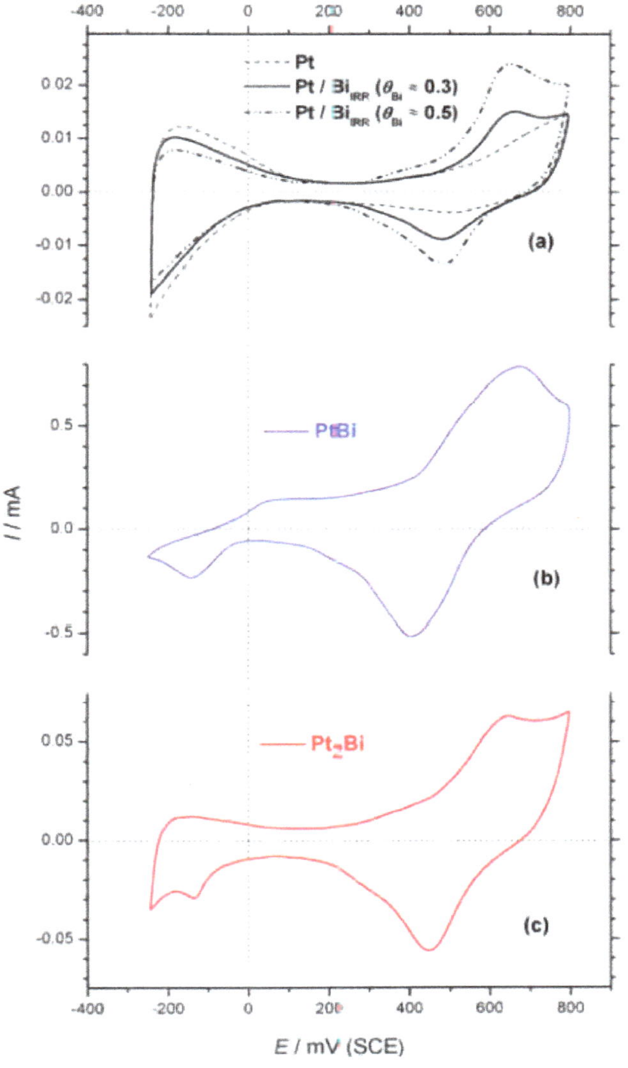

Figure 1. Basic voltammograms for Pt and Pt/BiIRR (a), PtBi alloy (b) and Pt2Bi (c) bulk electrodes in 0.1 M H_2SO_4. Scan rate: 50 mV s^{-1}. ω = 1500 rpm. T = 295 K.

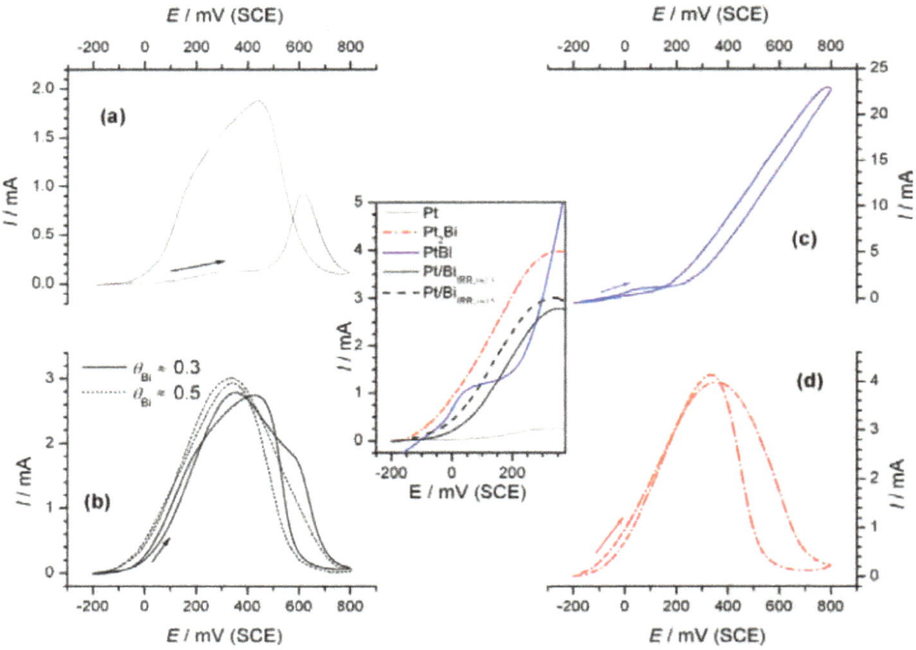

Figure 2. Cyclic voltammograms for the oxidation of 0.125 M HCOOH in 0.1 M H_2SO_4 solution on: a) Pt, b) Pt/BiIRR, c) PtBi and d) Pt2Bi catalysts. Inset: magnification of the onset potential region. Scan rate: 50 mV s^{-1}. ω = 1500 rpm. T = 295 K.

The beneficial effect of Bi on Pt for this reaction could be due to changes in the Pt–Pt distance that favor the direct route in formic acid oxidation,69 or to the formation of surface Bi oxides that participate in the oxidation of intermediates,39,70 or to electronic effects by lowering the electron density of the 5d orbitals, resulting in a considerable decrease of the CO binding strength to Pt,33,71 or to the ensemble effect creating an appropriate size of Pt domains and thereby providing direct oxidation of HCOOH to CO2. 48,64 Depending on the preparation of the catalysts and their resulting surface composition, the contribution of the above effects may vary.

In the research performed by our group,72 the oxidation of formic acid was studied on polycrystalline Pt modified by irreversibly adsorbed Bi (Pt/Bi_{IRR}) (Fig. 2b). The experiments were realized without Bi cations in solution, as opposed to modification by an UPD metal, thereby avoiding competition between adsorption/reaction steps of the reactant and modifier. The results were contrasted to pure Pt (Fig. 2a). Modification of the Pt electrode was performed at the open circuit potential, as described elsewhere.65 After modification, the electrode was rinsed with water and transferred into a cell containing supporting electrolyte. The fraction of the sites covered by Bi (denoted $Pt/Bi_{IRR;\theta}$) was estimated from the decrease in the charge for desorption of hydrogen, due to the fact that hydrogen does not adsorb on Bi,73 i.e., only the Pt sites not blocked by Bi were available for hydrogen adsorption. In order to avoid Bi dissolution, the anodic potential limit was set at 0.5 V vs. SCE, since it was established that the initial amount of Bi was almost completely retained when the upper potential limit was fixed below 0.75 V vs. RHE.74 This finding is in agreement with that reported using electrochemical quartz crystal microbalance (EQCM) analysis of the Bi oxidation mechanism on smooth Pt electrodes.75

The activities of Pt/BiIRR and Pt electrodes towards formic acid oxidation are compared in Fig. 2a and b and the results showed that the onset potential for the reaction on the Pt/Bi_{IRR} ($\theta_{Bi} \approx 0.3$) was about 0.1 V less positive than on the Pt electrode. The current reached a peak that corresponded to the oxidation of HCOOH to CO2 via the direct path, occurring on Pt sites that were not blocked by the poisoning COads species. On the descending part of the curve, a shoulder appeared at almost the same potential as the peak on the curve for the pure Pt electrode, which arises from the HCOOH oxidation on the sites being freed by CO_{ads} oxidation. Thus, HCOOH oxidation on $Pt/Bi_{IRR;\theta \approx 0.3}$ proceeded predominantly by the dehydrogenation path with some minor degree of dehydratation also occurring. HCOOH oxidation was also tested on a Pt/Bi_{IRR} electrode with larger coverage by Bi ($\theta_{Bi} \approx 0.5$). Bell-shaped voltammogram clearly suggests that oxidation of HCOOH on this electrode proceeds through dehydrogenation path. It is noticeable that the dehydration path is completely suppressed. The increased selectivity toward the dehydrogenation path on Pt/BiIRR compared to Pt was mainly the result of an ensemble effect caused by Bi reducing the continuous Pt sites necessary for dehydration. Nevertheless, the ensemble effect on the Pt/Bi_{IRR} catalyst was enabled by adsorbed Bi, which practically had no influence on the neighboring free Pt atoms.

PtBi Ordered Intermetallic Compounds

Recently, Abruna and co-workers42,43,45,69 studied the electro-oxidation of formic acid on the surface of ordered intermetallic compounds PtBi and $PtBi_2$. The choice of PtBi was based on extensive earlier studies in which the enhanced activity of Pt surfaces modified with irreversibly adsorbed Bi adlayers toward the oxidation of formic acid was established.40,60–62

Intermetallics are binary or multi-elemental metallic compounds that, since they have well-defined crystalline structures, offer predictable control over structural, geometric, and electronic effects in a manner that is not available when disordered alloys are used. In principle, the electronic and atomic structures, both of which are well known to be important parameters for electrocatalytic activity, can be significantly controlled. As the order in intermetallic phases arises from the high enthalpy of mixing, a high chemical and structural stability could be expected. Therefore, in contrast to disordered alloys, all Pt (and Bi) atoms on the surface of an ordered intermetallic phase have the same local geometry and thus, the same activity. The results obtained in these studies relating to formic acid oxidation indicated that the PtBi ordered intermetallic phase has properties and reactivity that are dramatically different from those of bare platinum surfaces. Especially, the onset potential for the electrocatalytic oxidation of formic acid is significantly shifted (by over 300 mV) to more negative values and the current density (at a given potential) is significantly enhanced when compared to pure Pt. Moreover, PtBi displayed virtual immunity to CO poisoning.69 Oana et al. 76 found for the intermetallic structures that the susceptibility for CO adsorption on Pt was drastically reduced on PtBi2 and PtBi surfaces, with respect to Pt, due to an increase in the Fermi level of the system induced by Bi.

According to Abruna and co-workers,45,69 the origin of the catalytic activity was related to electronic effects enhancing the affinity of PtBi for formic acid adsorption and producing surface oxides at low potentials, as well as to geometric effects that reduces the affinity for CO poisoning. The shift in the onset potential and the increase in the current density are due to electronic effects. In essence, the formation of the PtBi ordered intermetallic results in a charge redistribution (as a first approximation arising from work function differences), which enhances the affinity of PtBi towards formic acid and further gives rise to the formation of surface oxides at much lower potentials. Regarding the geometric effect, contrary to any Pt-based alloy where the Pt–Pt distance for nearest-neighbor Pt atoms is essentially the same as in Pt metal (2.78 Å), in ordered intermetallic compounds, the Pt–Pt distances can be modulated over a range of a factor of 2. For example, in PtBi the Pt–Pt distance in the (001) plane is 4.32 Å. Such distances were expected to diminish significantly CO poisoning, by reducing bridge sites and eliminating 3-fold hollow adsorption sites. However, some electronic effects could also be involved. Therefore, ordered intermetallic compounds, which were proposed as powerful catalysts for formic acid oxidation,42–46 not only exhibited greatly enhanced electrocatalytic activity (especially relative to Pt), but they could also serve as model systems to explore structure/composition/property/activity relationships.

PtBi Alloys

The fact that formic acid is a good candidate for DFAFCs initiated the study of its oxidation on so-called real catalysts, such as bulk PtBi alloy and two-phase Pt_2Bi catalysts.39,48,72 In one of our previous works, formic acid oxidation was investigated on PtBi alloy samples obtained according to the Bi–Pt phase diagram39 by melting the pure elements under an inert atmosphere in the proportion of Bi to Pt of 1:1, and the alloy was characterized by X-ray photoelectron spectroscopy (XPS) and by X-ray diffraction (XRD) analysis.39,48 XPS analysis revealed three chemical states of Bi, i.e., PtBi or Bi(0), Bi_2O_3 and BiO(OH) on the catalyst surface (Table I).39 These results suggested a model in which the PtBi alloy was covered by a layer of Bi_2O_3 and the very top of this layer contained BiO(OH) species.

Table i. XPS analysis of the surface composition of differently treated PtBi 1:1 alloy; BE – the binding energies of the Bi $4f_{7/2}$ excitation; FWHM – the full width at half maximum; takeoff angle (90 or 15°) is next to the sample name in brackets

Sample	BE Bi $4f_{7/2}$, eV	FWHM, eV	Approximate composition, %	Species
Polished electrode	157.3	0.9	34.3	Bi–Pt
	158.8	1.42	65.7	Bi^{3+}
Equilibrated at OCP (90°)	157.4	0.9	63.2	Bi–Pt
	158.3	1.6	31.8	Bi^{3+}
	159.6	1.9	4.9	BiO(OH)
Equilibrated at OCP (15°)	157.5	1.0	64.7	Bi–Pt
	158.5	1.6	22.0	Bi^{3+}
	159.5	1.7	13.3	BiO(OH)
Oxidized at 0.8 V *vs.* SCE (90°)	157.5	0.9	43.2	Bi–Pt
	158.5	1.3	44.1	Bi^{3+}
	159.3	1.8	12.6	BiO(OH)
Oxidized at 0.8 V *vs.* SCE (15°)	157.8	1.1	34.4	Bi–Pt
	158.8	1.2	51.8	Bi^{3+}
	159.6	1.3	13.8	BiO(OH)

XRD characterization of the PtBi alloy was performed to determine its phase composition. The diffraction pattern reveals peaks characteristic for hexagonal structure of the PtBi alloy and very small additional maxima that were assigned to traces of platinum cubic phase (≈0.7 wt. %).48 Electrochemical characterization showed that PtBi followed the behavior of its constituents.39 In the potential range up to 0.05 V vs. SCE, the electrode activity and the processes involved were determined by the behavior of pure Bi. It was established that the activity of PtBi was highly dependent on the reduction/oxidation of Bi species. Dissolution of Bi, i.e., leaching from the alloy matrix, proceeded all along the anodic potential scan.39,42 Comparison of the basic voltammograms for PtBi and Pt (Fig. 1a and b) showed that the deposition/dissolution of Bi completely suppressed hydrogen adsorption/desorption on Pt, as well as that the surface oxidation on PtBi was initiated at significantly lower positive potentials. The fact that adsorption/desorption of hydrogen was completely suppressed on the PtBi surface, as well as absolute inactivity of both the Bi and PtBi surfaces for the adsorption of CO, made determination of the real surface area impossible. Therefore, comparison of activity was given by the geometric surface area.

The voltammograms for the oxidation of formic acid on Pt and PtBi alloy electrodes, given in Fig. 2a and c, respectively, clearly indicated a dependence of the activity on the reduction/oxidation processes of Bi. Oxidation of formic acid does not occur on pure Bi39,77 and, consequently, does not occur on PtBi covered by Bi. Hence, the beginning of the reaction must be linked to free Pt sites on the PtBi. Relative to Pt, the onset potential on the PtBi electrode was shifted towards negative potentials by more than 0.25 V and the current densities at 0.05 V vs. SCE were higher by about two and half orders of magnitude. The exceptional activity of PtBi is caused by UPD phenomena of Bi on Pt, which was electrochemically detected. Namely, after recording the voltammogram for PtBi, this electrode was replaced with Pt and voltammograms were taken in the supporting solution with and without HCOOH. The voltammetric profiles obtained in the supporting solution indicated an underpotential deposition of Bi on Pt, evidencing that dissolved Bi could be adsorbed on Pt sites as an UPD layer. In the presence of formic acid, voltammogram displayed typical features for formic acid oxidation on a Bi-modified Pt surface,12 clearly suggesting that this reaction on the PtBi alloy occurs on the Bi modified Pt sites on the PtBi surface and the huge increase in catalytic activity relative to polycrystalline Pt was attributed to UPD phenomena of Bi leached from the alloy matrix and readsorbed on Pt. In addition, based on XPS analysis, it is proposed that some contribution of a bifunctional action, enabled by the presence of hydroxylated Bi species, should be taken into consideration. Our studies of Pt_2Bi electrode,48,72 a two-phase material consisting of PtBi alloy and pure Pt, revealed that this is a powerful catalyst for formic acid oxidation. Characterization of the catalyst was realized by XRD spectroscopy (phase composition), STM (surface morphology) and CO_{ads} stripping voltammetry (surface composition).

Comparison of the basic voltammograms of Pt_2Bi electrode and Pt in the potential range up to 0.05 V vs. SCE revealed that the presence of Bi suppressed hydrogen adsorption/desorption to a large extent (Fig. 1c). The cathodic peak at approximately –0.1 V vs. SCE could be correlated to the reduction of Bi oxide species and adsorbed Bi^{3+}, species formed in the positive going scan.39,43 Comprehensive oxide formation/reduction was presented by the anodic and cathodic peaks that are superimposed over those of Pt oxide formation and reduction, although the peak position that corresponds to oxide reduction on Pt_2Bi was slightly shifted towards negative potentials with respect to Pt, indicating some electronic interaction between Pt and Bi. A similar behavior was reported for smooth polycrystalline Pt electrodes in the presence of Bi(III) ions78 and for PtBi (1:1) alloy.39 The cyclic voltammogram for formic acid oxidation on Pt_2Bi is presented in Fig. 2d. On Pt2Bi catalyst, formic acid oxidation commenced more than 0.2 V earlier than on Pt. The currents increased reaching a peak with a current ≈30 times higher than the plateau on Pt and than diminished up to the positive potential limit. As Bi does not adsorb formic acid,39,77 the oxidation of formic acid occurred on pure Pt domains and on Pt atoms on PtBi domains. The bell-shaped voltammogram for formic acid oxidation suggests that the reaction on Pt_2Bi proceeded through the dehydrogenation path with the dehydration path being completely suppressed. Compared to Pt, the high activity of the Pt_2Bi catalyst could be explained by increased selectivity toward the dehydrogenation path caused by an ensemble effect originating from the interruption of continuous Pt sites by Bi atoms.

CO_{ads} stripping voltammetry recorded at Pt_2Bi and Pt48 demonstrated that onset potential and the peak position at Pt2Bi were slightly shifted to more negative potentials relative to Pt, indicating the presence of some electronic modification of the Pt surface atoms, capable for CO adsorption, by Bi. Since Bi74 and PtBi69 are inactive for CO adsorption, the oxidation of CO occurs only on the Pt domains. Therefore, the charge under the CO_{ads} peak at $P_{t2}Bi$ reflecting a process at the surface of the Pt phase was used for determining the contribution of pure Pt in the surface composition of the Pt_2Bi catalyst. The estimated contribution of pure Pt on the $P_{t2}Bi$ surface corresponded closely to bulk composition, thus indicating that adsorbed CO also prevents leaching of Bi.

In order to test whether the surface morphology of Pt_2Bi changes during formic acid oxidation, STM measurements were performed before and after the reaction and insignificant changes in the surface morphology and roughness were found. Consequently, it appears that the Pt_2Bi surface became kinetically stabilized due to the competition between the oxidation of formic acid at the electrode/ /solution interface and Bi leaching, i.e., corrosion/oxidation processes of the electrode surface itself.79 Accordingly, the main reasons for high stability of Pt_2Bi catalyst is the suppression of Bi leaching, as well as inhibition of dehydration path in the reaction of formic acid oxidation. On the other hand, the lower onset potential for HCOOH oxidation and higher reaction currents on Pt_2Bi alloy compared to both Pt/Bi_{IRR} electrodes were the result of ensembles that were created by alloyed Bi atoms incorporated into the Pt lattice, causing a shift in the d-band center of the adjacent Pt atoms. Therefore, Bi in the alloy also exhibited an electronic effect, which could enhance the affinity towards HCOOH adsorption and thus increasing the interaction of HCOOH molecules with the catalyst surface. It was found that all the investigated bimetallic catalysts were more active than Pt with the onset potentials shifted to more negative values and the currents at 0.0 V vs. SCE (under steady state conditions) improved by up to two order of magnitude (Table II).

Table II. Activity of the respective catalysts at E = 0.0 V (SCE) determined under steady- -state conditions

Activity parameter	Catalyst			
	Pt poly	$Pt/Bi_{IRR,\theta\approx0.3}$	$Pt/Bi_{IRR,\theta\approx0.5}$	Pt_2Bi
j / mA cm^{-2}	0.0038	0.058	0.36	0.48
$j_{Pt_2Bi}/j_{respective\ catalyst}$	127	8.3	1.33	1

Comparison of the results obtained for these different bulk Pt–Bi catalysts indicated that Bi in the alloy and irreversibly adsorbed Bi exhibited different effects on the catalytic activity. This enables distinguishing between the role of the ensemble and electronic effects in the oxidation of formic acid on Pt–Bi electrodes. The electronic effect, existing only on the alloy, contributes to an earlier start of the reaction, while the maximum current originates from an ensemble effect. During potential cycling of the Pt/Bi_{IRR} electrode, Bi was leached from the electrode surface and the ensemble effect was reduced over time, or lost.

Stability of Bulk Catalysts

From a practical point of view, long-term stability of the investigated catalysts for the oxidation of formic acid is very important. The stabilities of the Pt– Bi catalysts were tested by chronoamperometric measurements and by prolonged potential cycling in the supporting solution as well as in the supporting solution containing formic acid. The aim of the study was to establish which factors mostly affected the improvement of the oxidation rate of formic acid and the stability of the catalyst.

Additionally, a parallel study on a Pt electrode was performed to verify the promotional role of Bi. Cyclic voltammograms recorded on Pt_2Bi catalyst in the formic acid containing electrolyte are shown in Fig. 3. Over the potential cycling up to 0.8 V vs. SCE, the activity of Pt_2Bi electrode slowly decreased during the first 5–7 cycles reaching values of ≈85 % of the initial currents. After these first few sweeps, the currents remain unchanged with further cycling. On the contrary, the cycling of Pt_2Bi in supporting electrolyte led to enhancement of the currents related to the oxidation of Bi species, indicating some surface decomposition caused by Bi leaching/dissolution process (Fig. 3, inset A). It appears, consequently, that stability of Pt_2Bi during oxidation of formic acid could be induced by the presence of formic acid in the electrolyte.

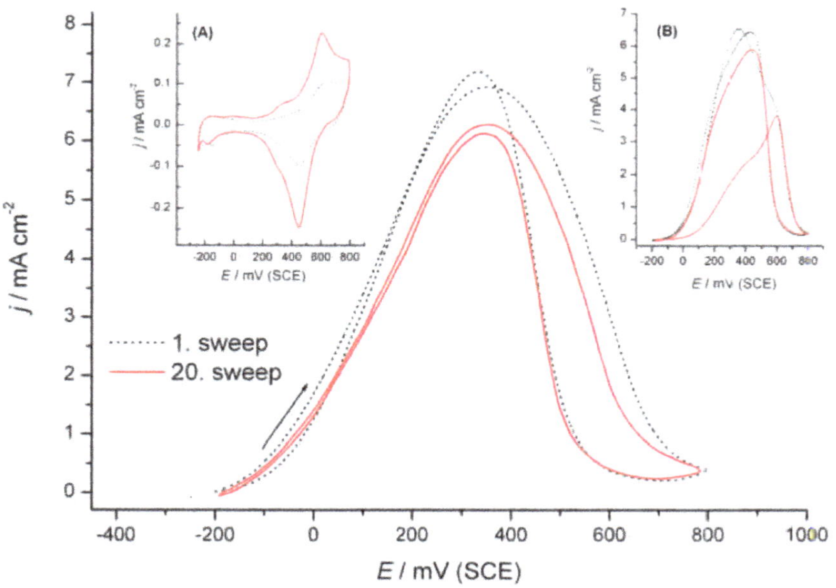

Figure 3. Cyclic voltammograms (1_{st} and 20_{th} sweep) for the oxidation of 0.125 M HCOOH in 0.1 M H_2SO_4 solution at a Pt_2Bi catalyst. Insets: A) basic voltammograms (1st and 20th sweep) for Pt_2Bi electrode in 0.1 M H_2SO_4 solution and B) cyclic voltammograms (1st and 20th sweep) for the oxidation of 0.125 M HCOOH in 0.1 M H_2SO_4 solution at $Pt/Bi_{IRR;\theta\approx0.3}$ electrode. Scan rate: 50 mV s-1. ω = 1500 rpm. T = 295 K.

Contrary to the Pt_2Bi catalyst, the Pt/BiIRR electrode shows significant changes with continuous cycling in the solution containing formic acid (Fig. 3, inset B).72 Repetitive cycling up to 0.8 V vs. SCE shifted the onset potential for formic acid oxidation to more positive values, decreased the reaction currents, while anodic peak diminished and a new peak starts to emerge and grow at ≈0.6 V vs. SCE. This transformation of the cyclic voltammograms indicated modification of the surface composition due to continuous Bi dissolution. Apparently, re- -adsorption of Bi species from the solution was rather low, so the initial voltammogram was never restored, which is in accordance with results obtained for formic acid oxidation on bismuth-coated mesoporous Pt microelectrodes.80 To test the assumption that the presence of formic acid stabilizes the catalyst, a Pt_2Bi electrode was subjected to potential cycling in the supporting electrolyte and after 20 cycles the electrode was replaced by a Pt electrode (Fig. 4a). The results of this experiment showed a slightly reduced charge of hydrogen adsorption/desorption indicating underpotential deposition of the previously leached/dissolved Bi. The same procedure was repeated in the electrolyte containing formic acid. The voltammogram of formic acid oxidation recorded on Pt in this experiment almost retraced the characteristic profile of pure Pt, suggesting that leaching of Bi was suppressed in the presence of formic acid. This was confirmed by STM measurements performed before and after the oxidation of formic acid, indicating a small difference in roughness and an insignificant change in the surface morphology.48

It should be noted that the experiment performed after replacing the PtBi alloy with Pt revealed significant Bi leaching under the same experimental conditions, meaning that $P_{t2}Bi$ was more resistant to Bi leaching than PtBi (Fig. 4b). The same experiment conducted with the Pt/Bi_{IRR} electrode showed considerable changes in surface composition due to Bi dissolution. Upon prolonged cycling, the electrode surface became enriched in platinum and exhibited a Pt-like electrochemical behavior in acid electrolyte containing formic acid (Fig 4c). Although Pt/Bi_{IRR} shows remarkable initial activity compared to pure Pt, this electrode was not stabilized by formic acid oxidation, since the desorption of Bi was not suppressed in the presence of formic acid. In addition, the poisoning effect induced by the dehydration path was not avoided and Bi_{IRR} did not provoke any significant modification of the electronic environment. Therefore, Pt/Bi_{IRR} catalysts are less active than the corresponding alloy. Chronoamperometric experiments were performed to prove the activity and stability of the investigated catalysts (Fig. 5). Insight into the chronoamperometric curves confirmed the advantage of alloys, i.e., the necessity of alloying Pt with Bi to obtain a catalyst with stable activity.

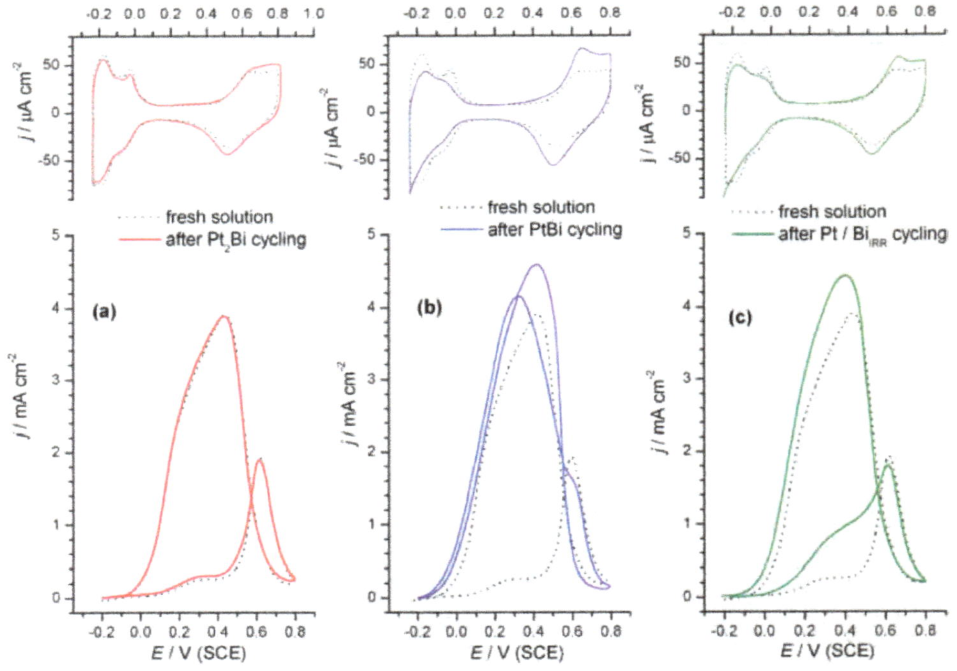

Figure 4. Comparison of cyclic voltammograms recorded after replacement of: a) Pt_2Bi, b) PtBi and c) Pt/Bi_{IRR} with polycrystalline Pt in 0.1 M H_2SO_4 solution and in 0.125 M HCOOH solution. Scan rate: 50 mV s^{-1}. ω = 1500 rpm. T = 295 K.

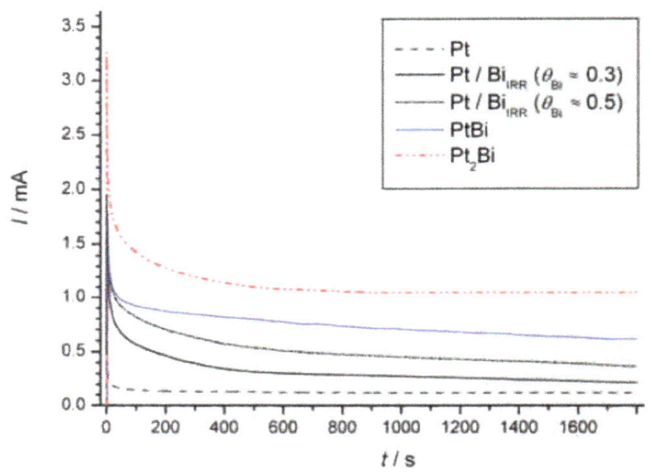

Figure 5. Chronoamperometric curves for the oxidation of 0.125 M HCOOH at 0.2 V in 0.1 M H_2SO_4 solution on PtBi, Pt_2Bi, Pt/Bi_{IRR} and Pt catalysts. ω = 1500 rpm. T = 295 K.

In summary, the main reason for the high stability of the Pt_2Bi catalyst is the inhibition of the dehydration path in the reaction, as well as suppression of Bi leaching in the presence of formic acid, which is specified by a minor change in the surface morphology and roughness.48,72 Comparing the results obtained for the two types of Pt–Bi catalysts, polycrystalline Pt modified by irreversible adsorbed Bi and for Pt_2Bi catalyst, the role of the ensemble effect and electronic effect in the oxidation of formic acid was distinguished.48 The electronic effect contributes to a lower onset potential of the reaction, while the maximum current comes from the ensemble effect. During the potential cycling treatment of the Pt/Bi_{IRR} electrode, Bi is dissolved from the electrode surface and the ensemble effect is reduced or lost over time. On the other hand, the high stability of the Pt_2Bi catalyst, confirmed by chronoamperometric experiments, proves an advantage of alloys, i.e., the necessity of alloying Pt with Bi to obtain a corrosion resistant catalyst. According to Liu et al.,79 the high stability of the PtBi intermetallic is due to suppressed leaching of Bi in the presence of formic acid because of the effective competition between oxidation of the organic molecule at the electrode surface and the corrosion/oxidation of the electrode surface itself.

Activity of Low-Loading Catalysts

The concept of bimetallic mono and multilayer catalysts has received much attention regarding its possibility to reduce the noble metal loading and maintain the activity by replacing the under-layer (or bulk of the catalyst) with a less noble metal. In addition, unlike other bimetallic catalysts where the second metal is either in the form of an adatom or as a surface alloy component, this type of catalyst allows the study of the electronic effect of the second metal under-layer to the noble catalyst over-layer, as the only operating factor.

Pt/GC Catalyst

Pt was deposited onto a glassy carbon substrate (Pt/GC) using chronocoulometry at the potential corresponding to Pt limiting current plateau.70 For the sake of comparison, a Pt/GC electrode was prepared using the same electrochemical procedure and the quantity analogous to one for a bimetallic electrode. AFM imaging of the Pt deposit on the GC substrate showed randomly distributed clusters (agglomerates), which consisted of spherical nanoparticles with a regular size distribution of 5.7±1.5 nm, as revealed by STM measurements.70 Catalytic activity of Pt/GC electrode for formic acid oxidation was examined by potentiodynamic and quasi steady-state measurements. The reaction proceeds through both paths featured by higher first and lower second anodic, indicating lower poisoning of this Pt surface compared to the Pt bulk electrode and a shift in the reaction towards the direct path. STM analysis of this electrode revealed rather small, loosely packed particles with a diameter of ≈5 nm, which should have a lower number of defects and smaller Pt ensembles exposed to the reaction. Such morphology of the particles should lead to a more pronounced direct path in formic acid oxidation. These results are in accordance with a previously reported conclusion81 that the particle structure, i.e., morphology, rather than the particle size plays a predominant role in the activity of Pt catalysts for formic acid oxidation. The particle structure is directed by particle growth, which is influenced by the support morphology and saturation of the active centers of the support by a metal precursor.

Pt@Bi/GC Clusters

Formic acid oxidation was studied on a Pt–Bi catalyst obtained using an unusual approach for the preparation, i.e., modification of the Bi deposit with a Pt overlayer.70 Briefly, platinum–bismuth deposits on a glassy carbon (GC) rotating disk electrode were prepared by a two-step process. Electrochemical deposition of a controlled amount of Bi was performed at –0.1 V vs. SCE onto a mirror-like polished GC substrate. Subsequently, the electrodes were rinsed and transferred to the solution for deposition of Pt. It should be stressed that Pt deposition was realized at a potential of –0.1 V vs. SCE, which corresponded to the Pt limiting current plateau, in order to avoid any displacement reaction between Pt and the less noble Bi and/or GC substrate. For the same reason, the GC and GC/Bi electrodes were immersed and pulled out from the solution for Pt deposition at –0.1 V vs. SCE. The electrodes prepared in such manner, denoted as Pt@Bi/GC, were characterized by AFM spectroscopy, which indicated that Pt crystallized preferentially onto the previously formed Bi particles.

Analysis of the current vs. time transient responses demonstrated compliance with theoretical curves for progressive 3D nucleation82 in all three cases, i.e., for Bi deposition on a GC substrate, for Pt deposition on Bi/GC surface and Pt deposited alone on a GC support. The density of nuclei at saturation imply not only higher coverage of GC surface by Bi in comparison to Pt, but also that Pt could be better spread over Bi than over GC. The issue of Bi leaching (dissolution) from PtBi catalysts, and their catalytic effect alongside the HCOOH oxidation is rather unresolved. In order to control Bi dissolution, as prepared electrode were subjected to electrochemical oxidation by slow sweep in the supporting electrolyte within the relevant potential range. Such oxidized electrodes are denoted as "treated Pt@Bi/GC". This procedure led to quantitative oxidation of Bi partially occluded by Pt, but simultaneously to the formation of Bi oxide, accordingly creating a surface composed of Pt and Bi oxide.

The SEM micrographs of the as-prepared and treated Pt@Bi/GC electrodes are shown in Fig. 6a and b, respectively. The SEM image of the as-prepared sample shows well separated, randomly distributed clusters (agglomerates), with size of about 700 nm, which were formed from smaller particles. After electrochemical treatment of Pt@Bi agglomerates, the SEM image showed a decrease in the number and size of the isolated clusters, indicating some dissolution of uncovered/unprotected Bi deposit. Anodic striping charges indicated that along oxidation procedure about 70% of the quantity of deposited Bi was oxidized (Fig. 6c). The shape of the stripping peak implies the possibility of its deconvolution into two peaks, a sharp one that corresponds to Bi dissolution and a broader one at more positive potential that suits Bi oxide formation.83 ICP mass spectroscopic analysis of the electrolyte after this electrochemical treatment revealed that Bi was only partially dissolved, which confirmed the possibility of the formation of some Bi oxide species. These assumptions were confirmed by EDX spectra, which evidenced the presence of Pt, Bi and O, showing a decrease in Bi quantity in the Pt@Bi/GC electrode after the slow sweep and significant increase in the oxygen content, which may be attributed to Bi oxide formation (Fig. 6d). Thus, the composition of the treated Pt@Bi/GC obtained by EDX analysis corresponded qualitatively well to the results obtained by the deconvolution of the stripping peak.

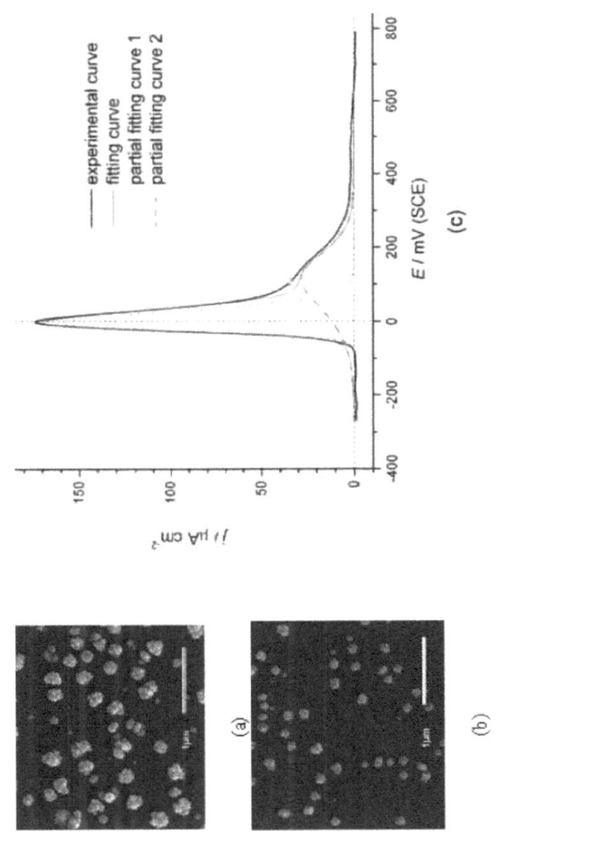

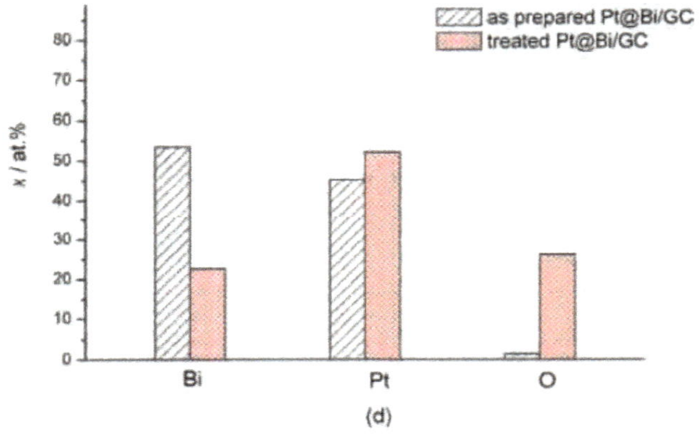

Figure 6. SEM images for Pt@Bi/GC electrode a) before and b) after electrochemical treatment; c) deconvoluted anodic stripping voltammogram for Bi oxidation in 0.1 M H_2SO_4 solution and d) element content in the catalyst before and after electrochemical treatment, as analyzed by EDX.

In this way, the prepared electrode exhibited significantly high activity and exceptional stability in comparison to the Pt/GC electrode. Formic acid oxidation proceeded predominantly through the dehydrogenation path on the treated Pt@Bi/GC electrode, resulting in its high activity. The onset potential was shifted by 150 mV to more negative values and the currents were about 10 times higher than those at the same potential on Pt/GC, as revealed both by potentiodynamic as well as by quasi-steady state measurements (Fig. 7).

This high activity and increased selectivity toward dehydrogenation is the result of well-balanced ensemble effect originating from the interruption of continuous Pt sites by Bi-oxide domains. The possibility of some electronic effect of non-oxidized Bi under the Pt on the activity of the Pt@Bi/GC electrode could not be excluded. Prolonged cycling and chronoamperometry tests revealed exceptional stability of the catalyst during formic acid oxidation.70 This low loading Pt-based electrode exhibited activity for the oxidation of formic acid similar to that of bulk Pt_2Bi alloy, which has been shown to be one of the best Pt–Bi bimetallic catalysts for the oxidation of formic acid.

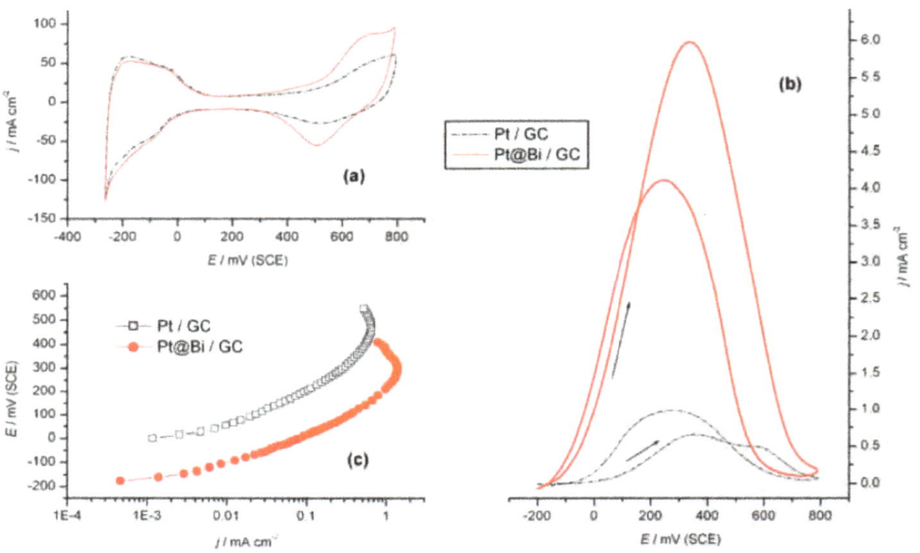

Figure 7. Initial cyclic voltammograms: a) for treated Pt/GC and treated Pt@Bi/GC electrodes in 0.1 M H_2SO_4 solution and b) for the oxidation of 0.125 M HCOOH in 0.1 M H_2SO_4 solution (scan rate 50 mV s^{-1}); c) corresponding Tafel plots (scan rate 1 mV s^{-1}). ω = 1500 rpm. T = 295 K.

Stability of Low-Loading Catalysts

As already stated, upon treatment of as prepared Pt@Bi/GC clusters by a slow anodic sweep, a unique bimetallic structure consisting of a Bi core occluded by Pt and a Bi-oxide was obtained. Consideration of the stability of the Pt@Bi/ /GC catalyst was realized by applying prolonged potential cycling up to 0.8 V vs. SCE (hereinafter referred to as a cycling protocol) in supporting electrolyte or in supporting electrolyte containing formic acid and the results were compared with data obtained at a Pt/GC electrode treated in the same manner. An attempt was made to correlate the electrochemical response with the structural features of these two catalysts and to signify the effects that determine the electrode stability in formic acid oxidation.84 The difference in activity between Pt@Bi/GC and Pt/GC catalyst after the application of cycling protocol in appropriate electrolyte revealed a complex phenomenology, suggesting that the interplay of several factors, such as nanoparticle size, surface morphology, influence of formic acid and its intermediate CO_{ads}, determines the performance of the catalysts in formic acid oxidation. Oxidation of HCOOH at the Pt@Bi/GC electrode occurred directly through dehydrogenation to CO_2 enabled by the well-balanced Bi-oxide domains and small Pt ensembles, which resulted in remarkable stability of this catalyst. Serious formic acid oxidation effectively competes with oxidation of the electrode surface, i.e., dissolution of Bi.79 Leaching of Bi from Pt@Bi/GC in the presence of formic acid was not prevented but highly suppressed due to the source of Bi, i.e., Bi core. Since this electrode is composed of a Bi core occluded by Pt and Bi oxide, the morphology of the surface slightly changes and facilitates high activity and stability of this catalyst.

Pt@Bi/GC catalyst exhibited excellent stability during prolonged oxidation of formic acid even above the potentials of Bi dissolution, as revealed by the negligible decrease in activity (Fig. 8a). Such stability of the catalyst was confirmed by quasi-steady state measurements performed on the surface previously treated by potential cycling in supporting electrolyte containing formic acid. The data obtained under slow sweep conditions corroborated the difference in the activities of investigated catalysts, as was found by the potentiodynamic measurements. These measurements also showed high selectivity of this surface toward the dehydrogenation path. Tafel slops of 120 mV obtained at Pt@Bi/GC electrode after 1, 20 and 100 cycles in the supporting electrolyte containing formic acid indicate that the dehydrogenation path in formic acid oxidation proceeded on a surface free of adsorbed CO species. Contrary to the Pt@Bi/GC electrode, which possessed high stability and unusual cycling performance in formic acid oxidation, the activity of the Pt/GC electrode during the cycling protocol in the presence of formic acid in the supporting electrolyte continuously decreased (Fig. 8b). Even more, while at Pt@Bi/GC electrode, the reaction proceeded continuously through the direct path, in the case of Pt/GC, the reaction mechanism changes during cycling. As can be seen from Fig. 8b, in the first cycle, the current raises and reaches a peak that indicates predominant direct oxidation of HCOOH, while the appearance of a well-defined shoulder on the descending part of the curve signifies the indirect path in the reaction as well. As the number of cycles increases, the currents related to dehydrogenation diminish and simultaneously, the well-defined shoulder transforms into a peak, indicating the increased role of the indirect path in the reaction.

This decrease in the activity of Pt/GC electrode could not only be due to a gradual accumulation of reaction residues, i.e., poisoning, on the electrode surface, but also to structural adjustments of the platinum nanoparticles as a result of the changes in the potentials during the scanning in acidic solutions, especially in the presence of the organic compound. The Tafel slope obtained for Pt/GC electrode of 140 mV (Fig. 8b) indicates that the reaction occurred on the surface partially covered by CO_{ads}, not only through the dehydrogenation path, but also via the dehydration path occurring in parallel. Significant decreases in the currents and increases in the Tafel slope after 20 and 100 cycles indicate retardation of the reaction due to adsorption and accumulation of CO on the surface.

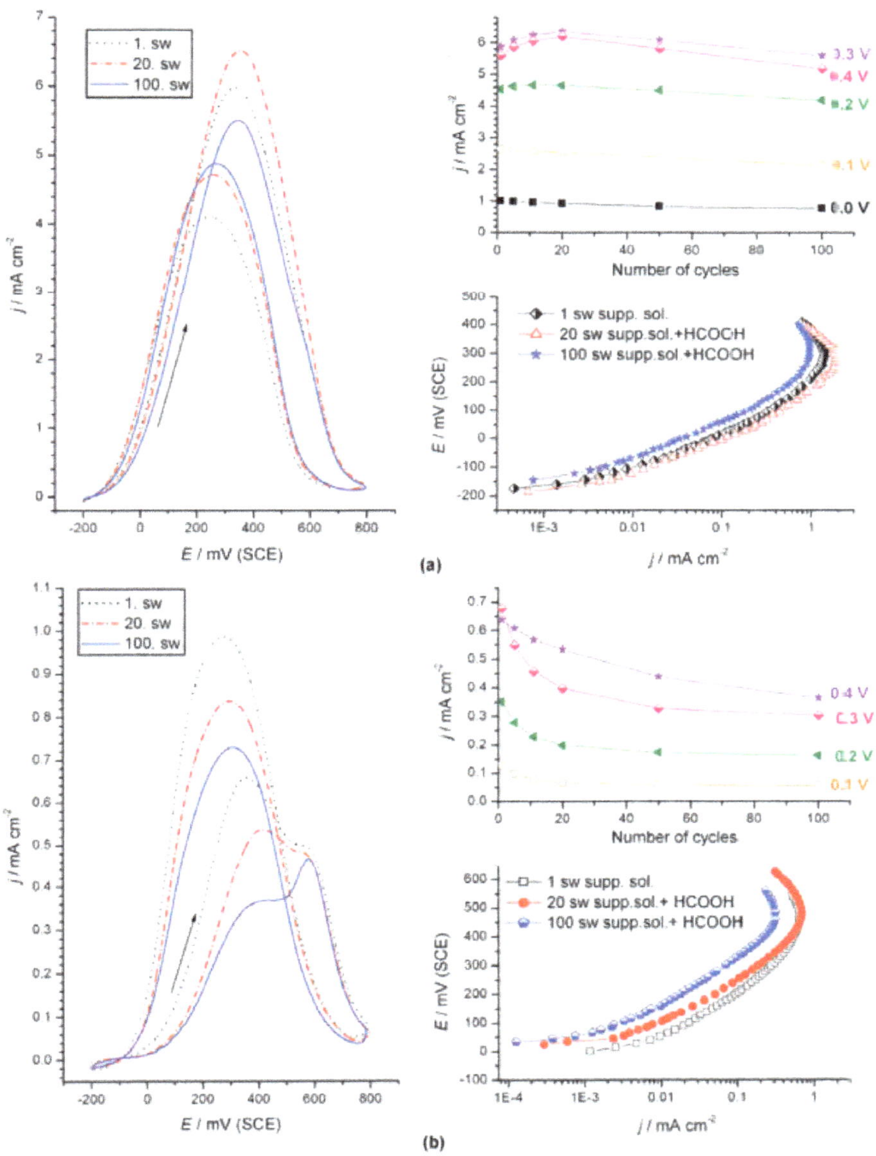

Figure 8. Cyclic voltammograms for the oxidation of 0.125 M HCOOH in 0.1 M H^2SO^4 (1st, 20th and 100th sweep at a rate of 50 mV s^{-1}); effect of cycling – plots of current density vs. number of cycles and corresponding Tafel plots (scan rate 1 mV s^{-1}) obtained a) on a Pt@Bi/GC electrode and b) on a Pt/GC electrode. ω = 1500 rpm. T = 295 K.

The oxidation of adsorbed CO, usually used for surface characterization, was also employed to determine the electrochemically active surface area (ECSA) of a catalyst. The values calculated for Pt@Bi/GC and Pt/GC electrodes revealed that, regardless of the presence or absence of HCOOH in the supporting electrolyte, the ECSA of the Pt@Bi/GC electrode increased during the cycling protocol, while the ECSA decreased for the Pt/GC electrode (Fig. 9). However, the degree of ECSA change upon cycling for both electrodes significantly depended on whether the supporting electrolyte contained HCOOH. The ECSA of the Pt@Bi/ /GC electrode slightly increased during this treatment in presence of formic acid due to some Bi dissolution and was confirmed in the experiments with polycrystalline Pt (Fig. 10b).

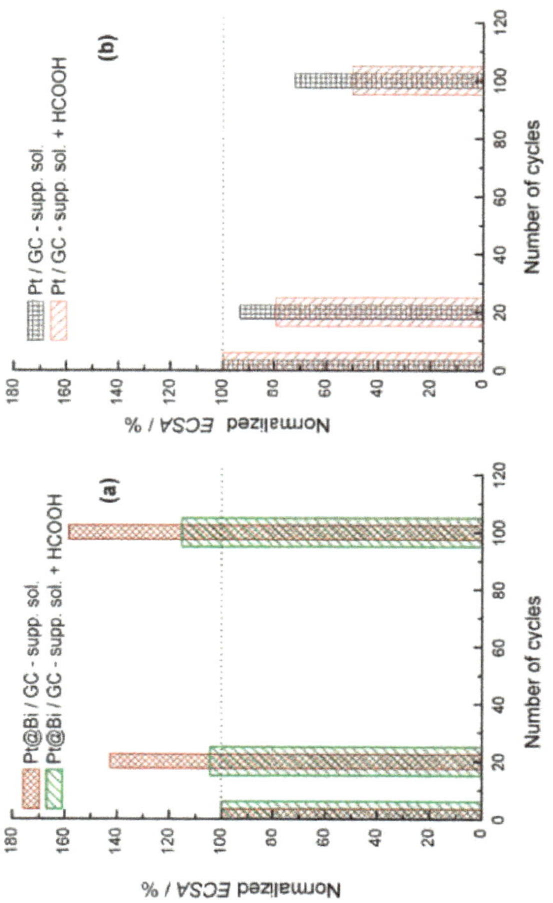

Figure 9. Normalzed ECSA values calculated with CVs upon potential cycling in supporting solution and in supporting solution containing 0.125 M HCOOH on: a) Pt@Bi/GC and b) Pt/GC electrodes. The normalized ECSA values were calculated by dividing the ECSA value after a certain number of potential cycles by that of the first cycle.

However, when this electrode was cycled in pure supporting electrolyte, the ECSA value increased significantly during the treatment (Fig. 9) because of intense leaching of Bi from the electrode, which was confirmed by the experiment with polycrystalline Pt (Figs. 10a and b). On the other hand, the Pt/GC electrode exhibited completely opposite properties upon similar treatment in the supporting electrolyte with or without formic acid. When this electrode was subjected to potential cycling in the absence of HCOOH, the ECSA value decreased slightly (Fig. 9), which was primarily due to the coalescence and agglomeration of the particles. Some negligible dissolution of Pt could also be possible.85 The result of these phenomena was the formation of defects on the surface that led to some negative shift of the onset and the peak potential of the oxidation of CO_{ads}, which indicates a slight change in the surface morphology had occurred. The consequence of these minor surface alternations was a small decrease in the activity for the oxidation of formic acid.

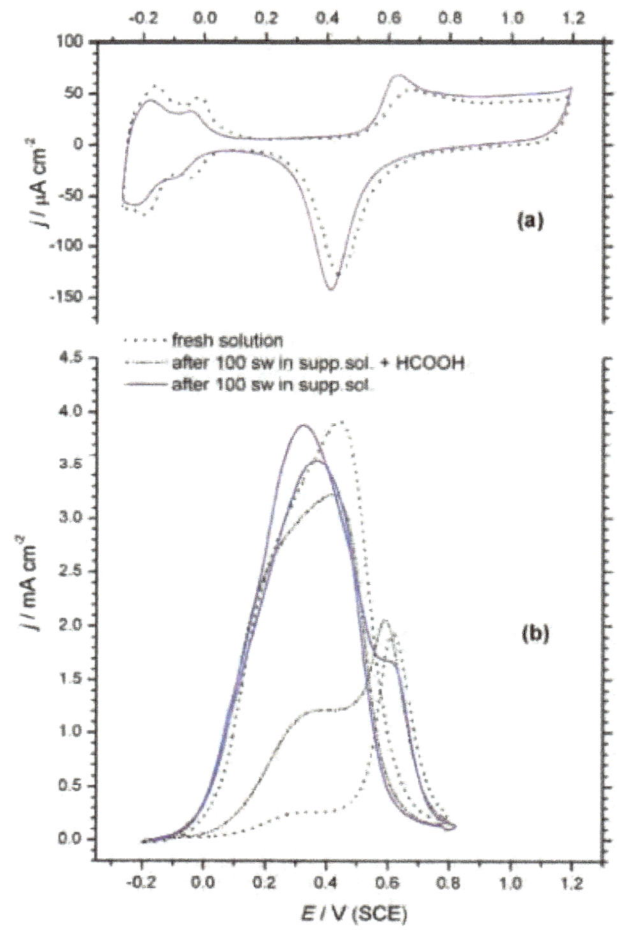

Figure 10. Cyclic voltammograms recorded a) after replacement of Pt@Bi/GC that has been exposed to cycling protocol with a polycrystalline Pt electrode in 0.1 M H_2SO_4 solution and b) for the oxidation of 0.125 M HCOOH in 0.1 M H_2SO_4. Scan rate: 50 mV s^{-1}. ω = 1500 rpm. T = 295 K.

Furthermore, when the Pt/GC electrode was similarly treated in the presence of HCOOH, the ECSA decreased much more (Fig. 9) and a larger shift of the CO_{ads} stripping peaks in the negative direction was observed, indicating to a considerable perturbation in the surface morphology. The consequence of these changes was a significantly lower activity of these surfaces. Considering that formic acid oxidation proceeds on a Pt/GC electrode through both the direct dehydrogenation and the indirect dehydration path, and that during cycling the reaction turns more to the latter one in which CO adsorbs on the surface, it seems that adsorption of CO and its oxidation contribute not only to particle agglomeration but even more to Pt dissolution.

High stability of Pt@Bi/GC electrode is confirmed in chronoamperometric experiment (Fig. 11). Current density recorded on treated Pt@Bi/GC during 1800 s at a constant potential of 0.2 V was again significantly higher than on the Pt/GC electrode. At Pt/GC electrode, the current decayed rapidly reaching a low steady state value within a few minutes. Contrarily, the current decreased slowly at the Pt@Bi/GC electrode and stabilized at a value that was more than 10 times higher than for the other electrode. This experiment also demonstrated the higher stability of the Pt@Bi/GC electrode in comparison to the Pt_2Bi catalyst48 since under the same conditions, the decrease in the currents for HCOOH oxidation at the Pt@Bi/GC electrode was much lower compared to the decrease registered for the Pt_2Bi catalyst.

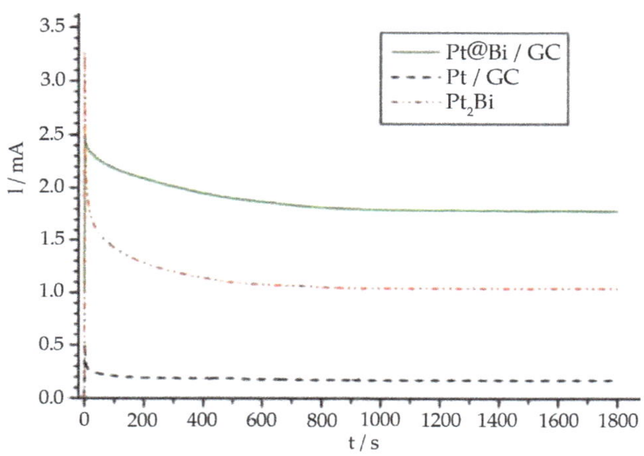

Figure 11. Chronoamperometric curves for the oxidation of 0.125 M HCOOH in 0.1 M H_2SO_4 solution at 0.2 V on Pt@Bi/GC and Pt/GC electrodes. ω = 1500 rpm. T = = 295 K.

Pt(Bi)/GC Shell–Core Catalyst

The bimetallic PtBi electrodes, as catalysts for formic acid oxidation, were examined in terms of obtaining a steady state electrode surface. The catalyst, denoted as Pt(Bi)/GC was produced using a similar methodology of preparation as previously described for the Pt@Bi/GC catalyst, i.e., by sequential deposition of Bi followed by deposition of Pt. In contrast to the experiments to test the stability of the Pt@Bi/GC catalyst, when its activation was performed by two slow sweeps up to 0.8 V vs. SCE, the Pt(Bi)/GC electrodes after metal deposition were activated by cycling the potential at a scan rate of 50 mV s–1 between hydrogen and oxygen evolution (–0.27 up to 1.2 V vs. SCE) in 0.1 M H_2SO_4 solution, prior to use as catalysts for the oxidation of formic acid. Generally, potential cycling is an electrochemical treatment that determines the degree of surface reconstruction and the size of the electrochemically active area. In the case of the Pt(Bi)/GC electrodes, this treatment was applied to quantify the amount of remaining Bi and Bi oxide in order to explain the importance of the composition and morphology of the surface for the reaction of formic acid oxidation.86

EDX and ICP-MS analysis, and AFM and electrochemical characterization, revealed that initially unfinished core–shell structures were formed. AFM characterization of the electrode surface indicated that Pt was preferentially deposited on the previously formed Bi particles, but cyclic voltammetry revealed leaching of Bi, meaning that Bi was not completely occluded by Pt. The Pt(Bi)/GC catalysts were not stable at potentials beyond 0.4 V vs. SCE due to Bi leaching/dissolution from the surface, which occurred through the oxidation of the less-noble metal. Electrochemical treatment by potential cycling of the as-prepared electrode (Fig. 12a) led to quantitative oxidation of Bi from the

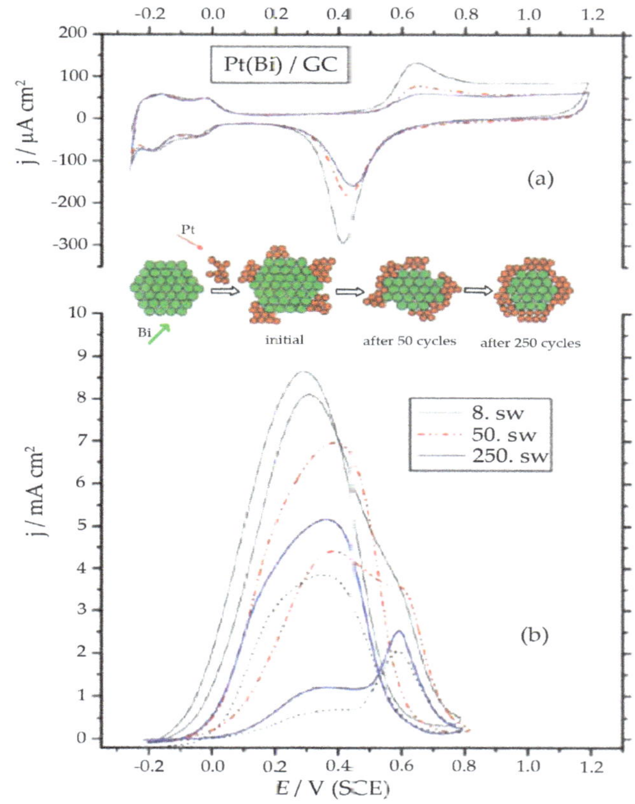

Figure 12. Cyclic voltammograms recorded on the Pt(Bi)/GC electrode: a) in 0.1 M H_2SO_4 and b) for the oxidation of 0.125 M HCOOH in 0.1 M H2SO4 after 8, 50 and 250 cycles. Scan rate: 50 mV s^{-1}. ω = 1500 rpm. T = 295 K.

unprotected core and the simultaneous formation of Bi oxide, consequently creating a shell composed of Pt and Bi-oxide. By prolonged cycling, the amount of surface oxides diminished creating finally a Pt@Bi shell–core structure. These electrodes exhibit enhanced electrocatalytic activity in formic acid oxidation in comparison to Pt/GC electrode treated in the same manner (Fig. 12b). This behavior could be explained primarily by an ensemble effect induced by surface Bi oxides interrupting the Pt domains, but to some extent, could also be attributed to the influence of the under-lying Bi onto the Pt surface layer, affecting the extent of poison adsorption on the Pt. Electronic modification of Pt both by surface and sub-surface Bi can play some role as well. Significantly prolonged potential cycling in supporting electrolyte of Pt(Bi)/GC electrodes previously stabilized by Bi oxide led to considerably lower Bi leaching and was accompanied by dissolution and redeposition of Pt, resulting in a Pt shell over a Bi core. The Pt@Bi/GC (shell–core) catalysts, with only Pt in the surface layer exhibit somewhat enhanced activity due to the electronic effect of the remaining under-lying Bi, which depended on the thickness of the Pt layer determined by the quantity of Bi in the core. This observation was confirmed by results obtained at three catalysts with different ratio of Pt and Bi (different underlying thickness of Bi).86 In this way, by controlling the thickness of the Bi and Pt layers using electrochemical techniques, it was possible to improve the electrocatalytical properties of Pt(Bi)/GC in HCOOH oxidation and to create a low-loaded Pt-based catalyst with comparable activity to that of the bulk metal alloy.

CONCLUSIONS

According to the results presented in this work dealing with the effects influencing the overall formic acid oxidation, it was found that both types of Pt–Bi bimetallic catalysts (bulk and low-loaded deposits on GC) showed superior catalytic activity compared to Pt, in terms of the lower onset potential and higher oxidation current density. Both types of Pt–Bi catalysts were investigated in order to establish how Bi atoms affect the adsorption characteristics of Pt towards formic acid. It was found that among all the tested Pt–Bi bimetallic bulk catalysts, Pt2Bi is the most powerful for formic acid oxidation, exhibiting high activity and stability. High activities of the bulk bimetallic catalysts result from the fact that formic acid oxidation proceeds completely through the dehydrogenation path. Increased selectivity toward dehydrogenation is caused by an ensemble effect. The high stability of Pt2Bi surfaces are induced by the suppression of Bi leaching, as was evidenced by the insignificant changes in the morphology and roughness of the surfaces before and after electrochemical treatment in formic acid containing solution. The results presented indicate that Bi in alloy and irreversibly adsorbed Bi exhibit different effects on the catalytic activity. Bi in the alloy not only induces an ensemble effect, but also has an electronic effect, which could be the reason for better performance of this catalyst resulting in higher currents and a lower onset potential. The activity of PtBi alloy is caused by UPD phenomena of Bi on Pt, which was electrochemically detected. In addition, based on XPS analysis, it is proposed that the activity of PtBi could also be caused by the bifunctional action of hydroxylated Bi species.

Comparing the results obtained for these bulk Pt–Bi catalysts, the role of the ensemble effect and electronic effect in the oxidation of formic acid could be distinguished. The electronic effect, existing only on the alloy, contributed to the earlier start of the reaction, while the enhanced maximum current originates from the ensemble effect. Thus, the stability of the catalytic activity of Pt–Bi bimetallic electrodes is strongly related to the leaching tolerance of the electrode surface during formic acid oxidation. The leaching tolerance of anodic catalysts was greatly enhanced by the formation of alloy between Pt and Bi, i.e., it is necessary to alloying Pt with Bi to obtain a corrosion stable bulk catalyst. Electrochemical deposition of low loading Pt layer over Bi deposits on a GC electrode resulted in the formation of approximately spherical clusters of Bi covered by Pt. Treatment of the as-prepared electrode in the relevant potential range and supporting electrolyte leads to quantitative oxidation of Bi partially occluded by Pt, and simultaneously to the formation of Bi oxide, thus creating a surface composed of Pt and Bi-oxide. The so-prepared electrode exhibits higher activity and exceptional stability in comparison to a pure Pt/GC electrode. Formic acid oxidation proceeds predominantly through dehydrogenation path on treated Pt@Bi/GC electrode resulting in its high activity. The increased selectivity toward dehydrogenation is caused by an ensemble effect originating from the interruption of continuous Pt sites by Bi-oxide domains. The possibility of some electronic effect of non-oxidized Bi under the Pt on the activity of Pt@Bi/ /GC cannot be excluded. Such stability is induced by the stability of the Bi-oxide formed during electrode pre-treatment. This low loading Pt-based electrode exhibits activity for the oxidation of formic acid similar to the activity of bulk Pt_2Bi alloy, which has been shown to be one of the best Pt–Bi bimetallic catalysts for the oxidation of formic acid.

The use of bimetallic compounds as anode catalysts is an effective solution to overcome the problems with stability of the formic acid oxidation current for long-term applications. In the future, the tolerance of both CO poisoning and electrochemical leaching should be considered as the key factors in the development of electrocatalysts for anodic reactions.

Acknowledgement

This work was financially supported by the Ministry of Education, Science and Technological Development of the Republic of Serbia, Contract No. H-172060.

REFERENCES

1. R. Parsons, T. Van der Noot, J. Electroanal. Chem. 257 (1988) 9
2. X. Wang, J.-M. Hu, I.-M. Hsing, J. Electroanal. Chem. 562 (2004) 73
3. J. Willsau, J. Heitbaum, Electrochim. Acta 31 (1986) 943
4. X. Yu, P. G. Pickup, J. Power Sources 182 (2008) 124
5. C. Rice, S. Ha, R. I. Masel, A. Wieckowski, J. Power Sources 111 (2002) 83
6. M. W. Breiter, Electrochim. Acta 8 (1963) 447
7. T. D. Jarvi, E. M. Stuve, in Electrocatalysis, J. Lipkovski, P. N. Ross, Eds., Wiley–VCH, New York, 1998, pp. 75–154
8. T. J. Schmidt, R. J. Behm, B. N. Grgur, N. M. Marković, P. N. Ross, Langmuir 16 (2000) 8159
9. J. M. Feliu, E. Herrero, in: Handbook of Fuel Cells-Fundamentals Technology and Applications, Vol. 2: Electrocatalysis, W. Vielstich, A. Lamm, H.A. Gasteiger (Eds.), John Wiley & Sons Ltd., Chichister, 2003, pp. 625–634
10. S. Park, Y. Xie, M. J. Weaver, Langmuir 18 (2002) 5792
11. E. Levia, T. Iwasita, E. Herrero, J. M. Feliu, Langmuir 13 (1997) 6287
12. N. M. Marković, P. N. Ross Jr., Surf. Sci. Rep. 45 (2002) 117
13. A. Boronat-González, E. Herrero, J. M. Feliu, J. Solid State Electrochem. 18 (2014) 1181
14. N. V. Rees, R. G. Compton, J. Solid State Electrochem. 15 (2011) 2095
15. A. Capon, R. Parsons, J. Electroanal. Chem. 44 (1973) 1
16. A. Capon, R. Parsons, J. Electroanal. Chem. 45 (1973) 205
17. J. Clavilier, R. Parsons, R. Durand, C. Lamy, J. M. Leger, J. Electroanal. Chem. 124 (1981) 321
18. R. R. Adžić, A. Tripković, V. B. Vešović, J. Electroanal. Chem. 204 (1986) 329
19. A. Tripković, K. Popović, J. Serb. Chem. Soc. 60 (1995)

297

20. A. Tripković, K. Popović, R. Adžić, J. Chim. Phys. 88 (1991) 1635
21. T. Iwasita, X. Xia, E. Herrero, H. D. Liess, Langmuir 12 (1996) 4260
22. G. Samjeske, A. Miki, S. Ye, M. Osawa, J. Phys. Chem., B 110 (2006) 16559
23. M. F. Mrozek, H. Luo, M. J. Weaver, Langmuir 16 (2000) 8463
24. A. Cuesta, G. Cabello, C. Gutierrez, M. Osawa, Phys. Chem. Chem. Phys. 13 (2011) 20091
25. C. Lamy, J. M. Leger, J. Chim. Phys. 88 (1991) 1649
26. V. Grozovski, F. J. Vidal-Iglesias, E. Herrero, J. M. Feliu, ChemPhysChem 12 (2011) 1641
27. H. Okamoto, Y. Numata, T. Gojuki, Y. Mukouyama, Electrochim. Acta 116 (2014) 263
28. A. Miki, S. Ye, M. Osawa, Chem. Commun. (2002) 1500
29. S.-C. Chang, L.-W. H. Leung, M. J. Weaver, J. Phys. Chem. B 94 (1990) 6013
30. P. K. Babu, H. S. Kim, J. H. Chung, E. Oldfield, A. Wieckowski, J. Phys. Chem. B 108 (2004) 20228
31. R. Adžić, A. V. Tripković, N. Marković, J. Electroanal. Chem. 150 (1983) 79
32. R. R. Adžić, in Advances in Electrochemistry and Electrochemical Engineering, H. Gerischer, C. W. Tobias, Eds., Wiley, New York, 1984, p. 159
33. N. de-los-Santos-Alvarez, L. R. Alden, E. Rus, H. Wang, F. J. Di Salvo, H. D. Abruna, J. Electroanal. Chem. 626 (2009) 14
34. L. J. Zhang, Z. Y. Wang, D. G. Xia, J. Alloy. Compd. 426 (2006) 268
35. W. Liu, J. Huang, J. Power Sources 189 (2009) 1012
36. Z. Awaludin, T. Okajima, T. Ohsak, Electrochem. Comm. 31 (2013) 100
37. O. Winjobi, Z. Zhang, C. Liang, W. Li, Electrochim. Acta 55 (2010) 4217
38. W. Chen, J. Kim, S. Sun, S. Chen, Langmuir 23 (2007) 11303
39. A. V. Tripković, K. Dj. Popović, R. M. Stevanović, R. Socha, A. Kowal, Electrochem. Comm. 8 (2006) 1492
40. J. Clavilier, A. Fernandez-Vega, J. M. Feliu, A. Aldaz, J. Electroanal. Chem. 258 (1989) 89
41. M. Ball, C. A. Lucas, N. M. Marković, B. M. Murphy, P. Steadman, T. J. Schmidt, V. Stamenković, P. N. Ross, Langmuir 17 (2001) 5943
42. D. Volpe, E. Casado-Rivera, L. Alden, C. Lind, K. Hagerdon, C. Downie, C. Korzniewski, F. J. Di Salvo, H. D. Abruna, J. Electrochem. Soc. 151 (2004) A971
43. H. Wang, L. Alden, F. J. Di Salvo, H. D. Abruna, Phys. Chem. Chem. Phys. 10 (2008) 3739
44. C. Roychowdhury, F. Matsumoto, V. B. Zeldovich, S. C. Warren, P. F. Mutolo, M. J. Ballesteros, U. Wiesner, H. D. Abruna, F. J. Di Salvo, Chem. Mater. 18 (2006) 3365
45. E. Casado-Rivera, D. J. Volpe, L. Alden, C. Lind, C. Downie, T. Vazquez-Alvarez, A. C. D. Angelo, F. J. Di Salvo, H. D. Abruna, J. Am. Chem. Soc. 126 (2004) 4043
46. J. Sanabria-Chinchilla, H. Abe, F. J. Di Salvo, H. D. Abruna, Surf. Sci. 602 (2008) 1830
47. E. Herrero, A. Fernandez-Vega, J. M. Feliu, A. Aldaz, J. Electroanal. Chem. 350 (1993) 73
48. J. D. Lović, M. D. Obradović, D. V. Tripković, K. Dj. Popović, V. M. Jovanović, S. Lj. Gojković, A. V. Tripković, Electrocatalysis 3 (2012) 346
49. X. Yu, P. G. Pickup, Electrochim. Acta 56 (2011) 4037
50. S. Daniele, S. Bergamin, Electrochem. Comm. 9 (2007) 1388
51. B.-J. Kim, K. Kwon, C. K. Rhee, J. Han, T.-H. Lim, Electrochim. Acta 53 (2008) 7744
52. S. R. Branković, J. X. Wang, R. R. Adžić, Surf. Sci. Lett. 747 (2001) L173
53. S. Papadimitriou, S. Armyanov, E. Valova, A. Hubin, O. Steenhaut, E. Pavlidou, G. Kokkinidis, S. Sotiropoulos, J. Phys. Chem., C 114 (2010) 5217
54. R. G. Freitas, E. C. Pereira, Electrochim. Acta 55 (2010) 7622
55. R. G. Freitas, E. P. Antunes, E. C. Pereira, Electrochim. Acta 54 (2009) 1999
56. A. Tegou, S. Papadimitriou, G. Kokkinidis, S. Sotiropoulos, J. Solid State Electrochem. 14 (2010) 175
57. S. Motoo, N. Furuya, Ber. Bunsenges. Phys. Chem. 91 (1987) 457
58. R. Carbo, R. Albalat, J. Claret, J. M. Feliu, J. Electroanal. Chem. 446 (1998) 79
59. J. M. Feliu, A. Fernandez-Vega, J. M. Orts, A. Aldaz, J. Chim. Phys. 88 (1991) 1493
60. S. P. E. Smith, H. D. Abruna, J. Electroanal. Chem. 467 (1999) 43
61. S. P. E. Smith, K. F. Ben-Dor, H. D. Abruna, Langmuir 15 (1999) 7325
62. J. Clavilier, A. Fernandez-Vega, J. M. Feliu, A. Aldaz, J. Electroanal. Chem. 261 (1989) 113
63. A. Sáez, E. Expósito, J. Solla-Gullón, V. Montiel, A. Aldaz, Electrochim. Acta 63 (2012) 105
64. A. Lopez-Cudero, F. J. Vidal-Iglesias, J. Solla-Gullon, E. Herrero, A. Aldaz, J. M. Feliu, Phys. Chem. Chem. Phys. 11 (2009) 416
65. J. Clavilier, J. M. Feliu, A. Aldaz, J. Electroanal. Chem. 243 (1988) 419
66. J. Kim, C. K. Rhee, Electrochem. Comm. 12 (2010) 1731
67. Q.-S. Chen, Z.-Y. Zhou, F. J. Vidal-Iglesias, J. Solla-Gullon, J. M. Feliu, S.-G. Sun, J. Am. Chem. Soc. 133 (2011) 12930
68. C. Jung, T. Zhang, B.-J. Kim, J. Kim, C. K. Rhee, T.-H. Lim, Bull. Korean Chem. Soc. 31 (2010) 1543
69. E. Casado-Rivera, Z. Gal, A. C. D. Angelo, C. Lind, F. J. Di Salvo, H. D. Abruna, ChemPhysChem. 4 (2003) 193
70. J. D. Lović, S. I. Stevanović, D. V. Tripković, V. V. Tripković, R. M. Stevanović, K. Dj. Popović, V. M. Jovanović, J. Electrochem. Soc. 161 (2014) H547
71. L. R. Alden, D. K. Han, F. Matsumoto, H. D. Abruna, F.

J. Di Salvo, Chem. Mater. 18 (2006) 5591

72. J. D. Lović, D. V. Tripković, K. Dj. Popović, V. M. Jovanović, A. V. Tripković, J. Serb. Chem. Soc. 78 (2013) 1189
73. R. Gomez, J. M. Feliu, A. Aldaz, Electrochim. Acta 42 (1997) 1675
74. T. J. Schmidt, B. N. Grgur, R. J. Behm, N. M. Marković, P. N. Ross Jr., Phys. Chem. Chem. Phys. 2 (2000) 4379
75. S. Uhm, Y. Yun, Y. Tak, J. Lee, Electrochem. Comm. 7 (2005) 1375
76. M. Oana, R. Hoffmann, H. D. Abruna, F. J. Di Salvo, Surf. Sci. 574 (2005) 1
77. N. Kapur, B. Shan, J. Hyun, L. Wang, S. Yang, J. B. Nicholas, K. Cho, Mol. Simulat. 37 (2011) 648
78. V. Pautienienė, L. Tamašauskaitė-Tamašiūnaitė, A. Sudavičius, G. Stalnionis, Z. Jusys, J. Solid State Electrochem. 14 (2010) 1675.
79. Y. Liu, M. A. Lowe, F. J. Di Salvo, H. D. Abruna, J. Phys. Chem., C 114 (2010) 14929
80. S. Daniele, C. Bragato, D. Battistel, Electroanalysis 24 (2012) 759
81. D. Tripković, S. Stevanović, A. Tripković, A. Kowal, V. M. Jovanović, J. Electrochem. Soc. 155 (2008) B281
82. F. Gloaguen, J. M. Leger, C. Lamy, A. Marmanna, U. Stimming, R. Vogel, Electrochim. Acta 44 (1999) 1805
83. W. S. Li, X. M. Long, J. H. Yan, J. M. Nan, H. Y. Chen, Y. M. Wu, J. Power Sources 158 (2006) 1096.
84. J. D. Lović, S. I. Stevanović, D. V. Tripković, A. V. Tripković, R. M. Stevanović, V. M. Jovanović, K. Dj. Popović, J. Solid State Electrochem 19 (2015) 2223
85. L. Tang, B. Han, K. Persson, C. Friesen, T. He, K. Sieradzki, G. Ceder, J. Am. Chem. Soc. 132 (2010) 596
86. J. D. Lović, S. I. Stevanović, D. V. Tripković, V. M. Jovanović, R. M. Stevanović, A. V. Tripković, K. Dj. Popović, J. Electroanal. Chem. 735 (2014) 1.

CHAPTER 9

FUNDAMENTAL MECHANISTIC UNDERSTANDING OF ELECTROCATALYSIS OF OXYGEN REDUCTION ON PT AND NON-PT SURFACES: ACID VERSUS ALKALINE MEDIA

Nagappan Ramaswamy and Sanjeev Mukerjee

INTRODUCTION

Oxygen reduction reaction (ORR) on noble and nonnoble metal surfaces remains as one of the well-investigated electrochemical processes. While under acidic conditions Pt-based systems remain the mainstay as catalyst materials for ORR, in alkaline electrolyte a wide range of nonnoble metals and their oxides are stable enough for practical applications [1, 2]. ORR pathway rather than ORR mechanism has typically been addressed in the literature due to the easy accessibility of the former from rotating ring-disk electrode (RRDE) studies, and the complexity in understanding the latter [3]. ORR pathway is found to be similar in both acid and alkaline media on Pt-based materials [3, 4]. Based on the initial propositions by Damjanovic et al. [4–6], rate determining step (rds) on Pt electrodes is widely agreed to be the first electron transfer step to the adsorbed molecular O_2 with or without rapid proton transfer [4–6]. A major alternative viewpoint to rds in ORR was proposed by Yeager et al. [7], wherein it was proposed that ORR on Pt surfaces is likely to involve dissociative chemisorption of molecular O_2 with the initial adsorption of O_2 (with or without an electron transfer) as therds. ORR on Pt-based catalysts is understood to proceed via "parallel" routes with the $4e^-$ "direct" or "series" pathway as the predominant route and a minor route involving $2e^-$ pathway to peroxide.

Citation: Nagappan Ramaswamy and Sanjeev Mukerjee, "Fundamental Mechanistic Understanding of Electrocatalysis of Oxygen Reduction on Pt and Non-Pt Surfaces: Acid versus Alkaline Media," Advances in Physical Chemistry, vol. 2012, Article ID 491604, 17 pages, 2012. doi:10.1155/2012/491604.

Various kinetic models for ORR have been developed to understand the reaction pathways. The first model was developed by Damjanovic et al. [8] following which Wroblowa et al. [9], and Anastasijević et al. [10, 11], proposed extensive models.

For electrocatalytic reactions proceeding via inner-sphere electron transfer mechanism, it is typically assumed that either molecular adsorption of reactant species (dissociatively or nondissociatively) or an electron transfer is the first step [12]. For neutral, nonpolar species like molecular O_2, direct molecular O_2 adsorption is likely to be inhibited relative to, for example, the adsorption of charged superoxide radical anion ($O_2^{\bullet-}$) unless the free energy of adsorption of O_2 molecule is very exothermic on a specific catalyst surface. This is especially true under fuel cell conditions, where the cathodic reaction typically occurs at potentials well positive of the potential of zero charge (pzc). Multistep, multielectron transfer processes like ORR that involves many adsorbed intermediates are undoubtedly classified as an inner-sphere electron transfer reaction. However, among the many elementary reaction steps involved in ORR, there could be a surface-independent outer-sphere electron transfer component in the overall electrocatalytic $4e^-$ inner-sphere electron transfer reaction. In that perspective, O_2 reduction by one-electron transfer to superoxide ($O_2^{\bullet-}$) is observed at $E^\circ = -0.3\pm0.03$ V versus SHE corresponding to $\Delta G^\circ = 30\pm 2$ kJ mol^{-1} with both O_2 and $O_2^{\bullet-}$ remaining in the aqueous phase [13, 14]. Given the pH independence of this redox couple ($O_2/O_2^{\bullet-}$), the potential of this reaction does not change as the pH is varied from zero to fourteen [15]. Due to the occurrence of four proton transfer steps in O2 reduction to H2O/OH−, its standard reduction potential changes by 0.828 V from 1.229 V to 0.401 V versus SHE as the pH value changes from zero to fourteen. This causes the overpotential for the first electron transfer step ($O_2/O_2^{\bullet-}$) to decrease from 1.53 V at pH = 0 to 0.7 V at pH = 14 indicating a sharp decrease in overpotential at alkaline pH conditions. Blizanac et al. [15] argued based on a modified Pourbaix diagram approach that the above-mentioned decrease in overpotential is the primary thermodynamic reason for the applicability of a wide range of nonnoble materials in alkaline media. Due to the high overpotential required for $O_2/O_2^{\bullet-}$ redox couple in acidic media, only certain specific catalyst surfaces such as platinum that offer high free energy of adsorption for O^2 can catalyze ORR in acidic media. While the decrease in overpotential for the first electron transfer in alkaline media is certainly significant, this argument is primarily of thermodynamic origin. The concept of involving the possibilities of outer-sphere electron transfer during ORR in alkaline media bears importance, and it was pointed out earlier by Bockris [16] and Appleby [17] that the exchange current density values in alkaline media exhibit near independence on a large number of electrode materials including silver, gold, manganese oxides, perovskites, and various carbon surfaces. So certain steps in the overall ORR process in alkaline media could proceed via a non-electrocatalytic mechanism [17].

The drive to replace expensive and scarce Pt-based catalysts for ORR has led to a class of electrocatalysts composed of first-row transition-metal ions stabilized by surface nitrogen functionalities on graphitic surfaces [2, 18–21]. The nature of the active site (obtained after heat treatment) in terms of its location on the carbon support (edge versus basal plane) [22], coordination number (Fe-N_4 versus non-Fe-N_4 environment) [23], and chemical identity of the nitrogen functional groups (pyridinic, pyrrolic, and quaternary) [24] have remained a key aspect of intense discussion. Several theories exist to explain the nature of the active site such as those proposed by van Veen et al. [25–27], McBreen et al. [28], Schulenburg et al. [29], Yeager et al. [2, 30], Scherson et al. [31–33], and Dodelet et al. [22, 34–45]. Although some authors observed that ORR is conducted by sites comprised of surface nitrogen groups devoid of any metal ion centers [46, 47], it is now widely accepted that the transition-metal ion centers coordinated to four nitrogen groups (Me-N_4) on graphitic surfaces constitute the active site [22, 23, 30, 48], whereas chelation primarily serves to prevent the metal center from passivation/corrosion under electrochemical conditions [49]. Besides the exact structure of the active site, the low active site density or the metal loading that is obtained in these catalysts eludes clear understanding. A maximum in catalytic activity is obtained at a very low metal loading (~5000 ppm by weight for inorganic precursors and ~2 wt% for macrocycle precursors) [35]. Dodelet et al. [50] suggested that only the microporosity generated during pyrolysis upon gasification of disordered carbon content acts as a host for active sites. Irrespective of whatever the exact nature of the active site is, this does not necessarily explain the fundamental origin of ORR activity in this class of heat-treated catalysts. It was recently pointed out by Bezerra et al. [51] and Garsuch et al. [52] that most studies on this class of materials have focused on the optimal synthesis conditions and structure necessary for maximum activity, whereas a more fundamental understanding will be of great help in developing innovative routes for new catalyst synthesis.

In this paper, we present a systematic study of ORR reaction mechanisms on Pt and non-Pt surfaces. Fundamental reasons for the so-called kinetic facility of ORR in alkaline media are investigated from the perspectives of the changes in the double-layer structure and the reaction mechanisms in transitioning from acidic to alkaline conditions. New evidence unraveling the influences of inner-sphere and outer-sphere electron transfer mechanisms during ORR is elucidated. Besides the efforts to understand the nature of the active site in heat-treated metal macrocycle-based catalysts, a combination of electrochemical and advanced X-ray absorption spectroscopy techniques has been utilized to understand (1) the fundamental origin of electrocatalytic ORR activity upon heat treatment and (2) the reasons for the low density of active metal sites. To this end, a comparative study of ORR on Pt and non-Pt surfaces in both acid and alkaline media has been performed to (1) analyze the influence of both inner-sphere and outersphere electron transfer mechanisms, (2) elucidate the ORR reaction mechanisms, and (3) understand the importance of stabilization of the reaction intermediates on the active site.

EXPERIMENTAL

Electrochemical Characterization

All electrochemical measurements were made at room temperature using a rotating ring-disk electrode (RRDE) setup from Pine Instruments connected to an Autolab (Ecochemie Inc., model-PGSTAT 30) bipotentiostat. Alkaline (0.1 M NaOH) and acidic (0.1 M $HClO_4$) electrolytes were prepared using sodium hydroxide pellets (semiconductor grade, 99.99%, Sigma-Aldrich) and double-distilled 70% perchloric acid (GFS Chemicals), respectively. 30% Pt/C catalyst from BASF-ETEK (Somerset, NJ) was used as received. Ru/C catalyst was synthesized in-house via aqueous route at a metal loading of 20% by weight [53]. Iron(III) meso-tetraphenylporphyrin chloride (FeTPPCl) was procured from Alfa Aesar. FeTPPCl was mixed with Black Pearl carbon in the mass ratio 1:4 and ball milled for 2 hours at 400 rpm followed by pyrolysis at 800°C for 2 hours under argon atmosphere. Catalyst inks were prepared by dispersing 25 mg of the catalyst in 10 mL of 1:1 millipore H_2O:isopropyl alcohol mixture along with 100 μL of 5 wt% Nafion(R) solution as a binder. 10 μL aliquot of the catalyst ink was dispensed on Glassy Carbon (GC) disk of 5.61 mm dia. Gold ring electrode was held at 1.1 V versus RHE in alkaline electrolyte and at 1.3 V versus RHE in acidic electrolyte to detect stable peroxide intermediate. Collection efficiency of the disk-ring electrode was 37.5%. All potentials are referred to reversible hydrogen electrode (RHE) scale prepared from the same solution as the bulk electrolyte unless otherwise stated.

X-Ray Absorption Spectroscopic (XAS) Measurements

The in situ XAS studies at Fe K-edge (7112 eV) were performed at X19A beamline of National Synchrotron Light Source (NSLS, Brookhaven National Laboratory, NY). Detailed information on the spectroelectrochemical cell design are given elsewhere [54]. Spectra at Fe K-edge were collected in fluorescence mode using a PIPS detector. Argon- or oxygen-saturated 0.1 M NaOH was used as the electrolyte. Complete details on data analysis of X-ray absorption near edge structure (XANES) and extended X-ray absorption fine structure (EXAFS) are available elsewhere [54]. Typical k-range window during EXAFS fit was 2.500–12.500 Å^{-1} (Kaiser-Bessel). Data analysis for Delta-mu ($\Delta\mu$) studies at Fe K-edge involved specific normalization procedures detailed elsewhere [54, 55]. Briefly, this involves careful calibration of edge energy (Fe K-edge 7112 eV), alignment to standard reference scan to account for any drift in the beam energy. A postedge normalization procedure was then applied to the aligned scans via a cubic spline function which normalizes the oscillations over a specific energy range (typically 25 to 200 eV with respect to E_0) on a per-atom basis. Difference spectra were obtained using the equation $\Delta\mu = \mu(V) - \mu(0.1\,V)$, where $\mu(V)$ is the XANES spectra of the catalyst at various potentials and $\mu(0.1\,V)$ is the reference XANES signal at 0.1 V at which potential no evidence for electrochemical adsorbates (H_{upd}, O_{ads}, OH_{ads}) were found on iron-based catalysts. Theoretical delta-mu curves ($\Delta\mu_t$) were constructed using the FEFF 8.0 code [56]. This was accomplished using the relationship $\Delta\mu_t = \mu(O_{ads}\text{-Fe-N}_x\text{-C}) - \mu(\text{Fe-N}_x\text{-C})$, where the oxide species (O_{ads} or OH_{ads}) is in a specific binding site on Fe.

RESULTS AND DISCUSSIONS

ORR on Carbon-Supported Platinum Nanoparticles: Acid versus Alkaline Media

Figure 1 provides a snapshot of the cyclic voltammetry (CV) and ORR on 30% Pt/C in 0.1 M NaOH and 0.1 M $HClO_4$ electrolytes. As shown in Figure 1(a), CV of 30% Pt/C catalyst features the typical hydrogen underpotential deposition/stripping region below 0.5 V versus RHE in both of the electrolytes followed by oxide formation on Pt at potentials above 0.7 V versus RHE. Oxide formation on Pt in acidic media is due to oxidation of the solvent water molecules (water activation) [57] and in alkaline media is due to specific adsorption of hydroxide anions from the supporting electrolyte [58, 59]. The onset potential of Pt-OH formation is similar in both electrolytes, although in alkaline media oxide formation current exhibits a characteristic peak shape whereas in acidic media oxide formation current is relatively more flat. Half-wave potential ($E_{1/2}$) of Pt-OH in 0.1 M NaOH is 0.775 V whereas in 0.1 M $HClO_4$ $E_{1/2}$ of Pt-OH formation is shifted slightly more positive to 0.810 V. Figure 1(b) shows the ring current measured during ORR on Pt/C at 900 rpm in 0.1 M $HClO_4$ and 0.1 M NaOH. The corresponding ORR polarization curves are shown in the inset of Figure 1(b). The ORR onset potential in both of the electrolytes is ~1 V versus RHE which is followed by mixed kinetic-diffusion region between the potentials 0.7 V and 1 V. Well-defined diffusion-limited current density region is observed below 0.6 V versus RHE. The marginally higher Pt-OH coverage in alkaline media (between the potential region of 0.9 V to 0.8 V versus RHE) causes a penalty of ~25 mV higher overpotential at a kinetic current density of i_k=1 mA cm^{-2}_{geo} for ORR in 0.1 M NaOH relative to 0.1 M $HClO_4$. The most interesting observation in this experiment using Pt/C is in the ring currents measured during ORR. As seen in Figure 1(b), in 0.1 M $HClO_4$ the ring current due to peroxide oxidation is lower in the entire potential region and does not show any significant increase until the disk potential enters the H UPD region. On the contrary, the ring current due to peroxide oxidation in 0.1 M NaOH electrolyte shows a sharp increase at 0.9 V in the cathodic scan.

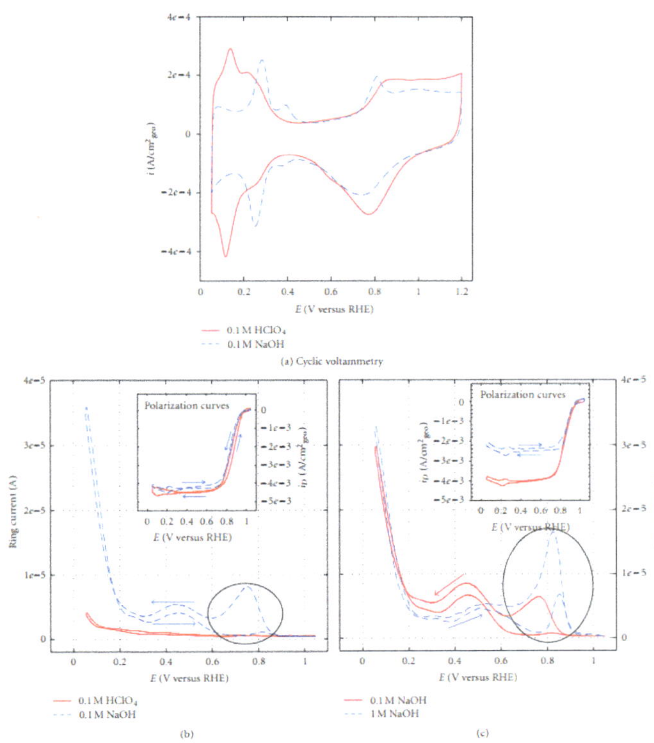

Figure 1: Comparison of electrochemical characteristics of BASF-ETEK 30% Pt/C in 0.1 M $HClO_4$, 0.1 M NaOH, and 1.0 M NaOH electrolytes. (a) CV in argon-saturated electrolytes, (b, c) Ring currents measured during ORR study at 900 rpm in O_2-saturated electrolytes. Insets in (b, c) show the corresponding ORR polarization curves at 900 rpm. Scan rate: 20 mV/s. E_{Ring} = 1.1 V versus RHE in 0.1 M NaOH and 1.3 V versus RHE in 0.1 M $HClO_4$. Catalyst loading: 15 μg_{Pt}15/cm^2 on 5.61 mm Glassy Carbon disk electrode.

The sharp increase in ring current at 0.9 V during ORR in alkaline media is closely related to Pt-OH formation from specific adsorption of hydroxide anions as seen from the CV in 0.1 M NaOH. As shown in Figure 1(a), in 0.1 M NaOH electrolyte, Pt-OH formation commences at ~0.7 V versus RHE and reaches a peak current at 0.81 V and a plateau at ~0.9 V. On the cathodic direction, the peak potential for Pt-OH reduction is ~0.75 V versus RHE. As seen in Figure 1(b), on the cathodic ORR scan, the increase in ring current commences at 0.9 V and reaches a peak potential of 0.75 V versus RHE. This clearly establishes the interaction between molecular oxygen and the hydroxyl species on the surface. This peroxide formation in the mixed kinetic-diffusion region in alkaline media is more evident at a higher electrolyte concentration of 1.0 M NaOH as shown in Figure 1(c). At a higher concentration of alkaline electrolyte, that is, at 1 M NaOH, peroxide is detected at the ring electrode in both the anodic and the cathodic scan [60].

In order to explain the link between the specifically adsorbed hydroxide anions and the peroxide detected in the mixed kinetic-diffusion region in alkaline media, it is important to understand the double-layer structure of the electrode-electrolyte interface during ORR at high-pH environments. A schematic illustration of the double-layer structure is shown in Figure 2. In an operating fuel cell, cathodic ORR process typically occurs at potentials well positive of the potential of zero charge (pzc). In alkaline media, water molecules not only act as solvent but also serve as the source of protons required in ORR. At these conditions in alkaline media, inner Helmholtz plane (IHP) is populated by specifically adsorbed hydroxyl species, solvent water dipoles (oriented with the oxygen atom towards the electrode surface), and chemisorbed O_2. Alkali metal ions are typically well solvated and are classically expected to populate the outer Helmholtz plane (OHP). In ORR, the case of a typical electrocatalytic inner-sphere electron transfer mechanism involves strong chemisorption of molecular O_2 (with or without bond breaking) on oxide-free Pt active sites followed by $4e^-/4H^+$ transfer. This inner-sphere mechanism depicted in Figure 2 (inset (a)) is common to both acidic and alkaline media, wherein all the reaction intermediates remain adsorbed on the active site until the final product is desorbed into the bulk electrolyte. Also shown in Figure 2 (inset (b)) is the outer-sphere electron transfer mechanism that appears to be unique to the alkaline conditions. The solvated molecular O_2, represented here as a cluster $O_2{\cdot}(H_2O)_n$, could interact with the surface hydroxyl species (OH_{ads}) via a hydrogen bond between H atom in OH_{ads} and O atom in the solvent water molecule. Such hydrogen bond energies (<35 kJ mol^{-1}) are typically much lower than the energy associated with covalent bond strength such as in the case of direct chemisorption of O_2 on Pt (>300 kJ mol^{-1}) [61]. This hydrogen bond formation stabilizes the solvated molecular oxygen $O_2{\cdot}(H_2O)$ cluster in the OHP and promotes an outer-sphere electron transfer to form the superoxide species.

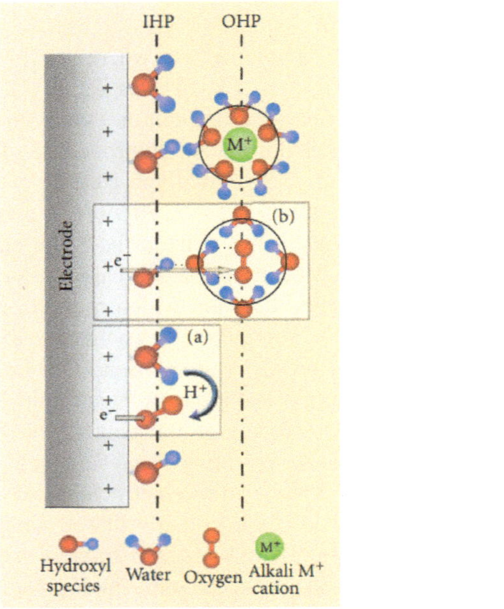

Figure 2: Schematic illustration of the double-layer structure during ORR in alkaline media. Insets (a, b) illustrate the inner- and outer-sphere electron transfer processes.

Following this new evidence on the involvement of outer-sphere electron transfer in alkaline media, two ORR mechanisms at high-pH environment are proposed here. The first is the well-known electrocatalytic inner-sphere electron transfer mechanism, where molecular O_2 undergoes direct chemisorption on oxide-free Pt site leading to a direct/series 4e $^-$ pathway without the desorption of reaction intermediates (such as peroxide) from the surface according to the following well-known reaction scheme (also depicted in Figure 3) [3]:

$$O_2 \longrightarrow O_{2,ads}, \tag{1a}$$

$$O_{2,ads} + H_2O + 2e^- \longrightarrow (HO_2^-)_{ads} + OH^-, \tag{1b}$$

$$(HO_2^-)_{ads} + H_2O + 2e^- \longrightarrow 3OH^-. \tag{2}$$

The second is the outer-sphere electron transfer mechanism, where solvated molecular O_2 cluster $O_2 \cdot (H_2O)$ n weakly interacts with adsorbed hydroxyl species to promote a $2e^-$ reaction pathway to HO_2^- anion as a reaction product which desorbs from the surface and is eventually detected at the ring electrode. This reaction is formulated as follows:

$$\begin{aligned} &\text{M-OH} + [O_2 \cdot (H_2O)_n]_{aq} + e^- \\ &\quad \longrightarrow \text{M-OH} + (HO_2^{\bullet})_{ads} + OH^- + (H_2O)_{n-1}, \end{aligned} \tag{3}$$

$$(HO_2^{\bullet})_{ads} + e^- \longrightarrow (HO_2^-)_{ads} \tag{4}$$

$$(HO_2^-)_{ads} \longrightarrow (HO_2^-)_{aq}. \tag{5}$$

The first step in the previous reaction shown in (3) involves electron transfer (or tunneling) from the electrode surface across a thin oxide film and at least one layer of solvation shell to solvated O_2. Equation (3) involves several elementary steps as written in the following:

$$\text{M-OH} + [\text{O}_2 \cdot (\text{H}_2\text{O})_n]_{\text{aq}} + e^- \longrightarrow \text{M-OH} + [\text{O}_2^{\bullet-} \cdot (\text{H}_2\text{O})_n]_{\text{aq}}, \quad (6a)$$

$$[\text{O}_2^{\bullet-} \cdot (\text{H}_2\text{O})_n]_{\text{aq}} \longrightarrow (\text{O}_2^{\bullet-})_{\text{ads}} + n\text{H}_2\text{O}, \quad (6b)$$

$$(\text{O}_2^{\bullet-})_{\text{ads}} + \text{H}_2\text{O} \longrightarrow (\text{HO}_2^{\bullet})_{\text{ads}} + \text{OH}^-. \quad (6c)$$

First electron transfer to $O_{2,aq}$ forms ($O_2^{\bullet-}$)aq which then undergoes desolvation and subsequent adsorption on the oxide substructure of the Pt surface to form ($O_2^{\bullet-}$)ads, followed by proton transfer to form adsorbed hydroperoxyl radical, ($HO_2^{\bullet}$)ads. Second electron transfer to $(HO_2^{\bullet})_{ads}$ yields $(HO_2^{\bullet-})_{ads}$. The binding energy of $(HO_2^{\bullet-})_{ads}$ on the oxide substructure of Pt is likely to be lower than that on oxide-free Pt site. This leads to the facile desorption of HO_2^- anion into the electrolyte which is eventually detected at the ring electrode. Also hydrogen peroxide anion formed via the homogeneous chemical disproportionation of superoxide anion would also be detected at the ring electrode. The interaction between the O_2·(H_2O)n cluster and the surface hydroxyl species causes certain nonspecificity to the identity of the underlying electrode metal. This nonspecificity opens the gate to use a wide-range of nonnoble metals and their oxides as electrode materials for ORR in alkaline media. On noble metals such as Pt, this nonspecificity is observed only in the oxide formation region, and depending on the extent of −OH coverage both inner- and outer-sphere electron transfer mechanisms coexist in alkaline media. On nonnoble metals such as the first-row transition elements that are completely passivated by a layer oxide film upon immersion into the aqueous electrolyte, this outer-sphere electron transfer process is likely to be the dominant mechanism. Furthermore as identified in Figures 1(b) and 1(c), we consider the peakshaped ring current in the potential range of 0.6 V to 0.9 V in alkaline media to be a characteristic signature of the outer-sphere electron transfer reaction mechanism. In acidic media, the adsorbed OH_{ads} species from water activation primarily serve only to block/inhibit the adsorption of molecular O2 and other reaction intermediates via the wellknown site-blocking effect [62, 63]. However, as shown here in alkaline media, the OH_{ads} species not only block the direct adsorption of O2 but also serve to promote the 2e$^-$ outer-sphere electron transfer reaction to peroxide. There are several precedents for electrochemical reactions which are mediated and/or promoted by specifically adsorbed anions and surface functional groups [64–67]. The anions or the surface groups act as an outer-sphere bridge between the reactant and the electrode surface. Furthermore as seen in Figures 1(b) and 1(c), peroxide intermediate is detected at the ring electrode at 0.5 V and also below 0.3 V versus RHE, and the reason for this was explained in detail in an earlier publication [53, 60]. Briefly, in alkaline media, the peak in ring current at 0.50 V is due to the carbon support and the increase in ring current below 0.3 V is due to reorientation of the water molecules at the electrode surface [53, 60].

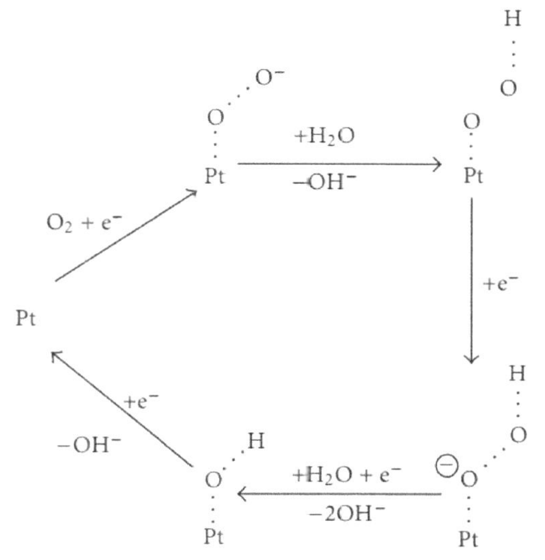

Figure 3: Electrocatalytic inner-sphere electron transfer mechanism for ORR on Pt in alkaline media assuming end-on adsorption of O_2 on a single Pt active site.

In light of this new evidence of involvement of outer-sphere electron transfer mechanism during ORR in alkaline media, it is noted that, while this mechanism allows the possibility of nonnoble metals to be used as electrode materials, it predominantly seems to yield the peroxide intermediate as the final product. So, for efficient conversion of O_2, it is important to facilitate direct adsorption of desolvated molecular O_2 on the OH-free active site and avoid the precipitous reaction of solvated molecular O_2 with the OH-covered active site. In agreement with Bard [12], this implies the necessity to promote the electrocatalytic inner-sphere electron transfer mechanism for efficient reduction of oxygen.

ORR on Heat-Treated Carbon-Supported Ru Nanoparticles

Figure 4 shows the ORR on Ru/C catalyst in O_2- saturated 0.1 M NaOH and 0.1 M HClO4 electrolytes recorded at 900 rpm. The ORR onset potential for Ru/C in 0.1 M HClO4 is 0.78 V versus RHE, whereas in 0.1 M NaOH it is 0.9 V versus RHE. In the alkaline media, mixed kineticdiffusion region between 0.9 V and 0.5 V is ensued by a welldefined limiting current region below 0.5 V versus RHE. In the acidic electrolyte, ORR is kinetically controlled as indicated by the absence of any clear diffusion-limited current region even at very high overpotentials. As shown in Figure 4(b) at a kinetic current density of 0.1 mA cm^{-2}_{geo}, Ru/C exhibits ∼125 mV lower overpotential in 0.1 M NaOH than in 0.1 M $HClO_4$. Primary reasons to this lower overpotential in alkaline media are delineated here. As seen in Figure 4(c) between the potential regions of 0.7 V to 0.8 V, the shoulder in ring current in 0.1 M NaOH electrolyte is characteristic of the outer-sphere electron transfer mechanism on oxidecovered Ru/C catalyst in alkaline media. This outer-sphere reaction mechanism was depicted in (3)–(5) giving rise to the peroxide intermediate (HO_2) as the final product at oxide-covered Ru sites. This (HO_2^-) aq formed is detected at the ring electrode and appears as a shoulder in ring current between 0.7 V and 0.8 V as shown in Figure 4(c). In acidic media, peroxide detected is very minimal and about an order of magnitude lower than that in 0.1 M NaOH. It should be noted that, in 0.1 M NaOH, this outer-sphere electron transfer occurs only at oxide-covered Ru sites, whereas at oxide-free Ru sites direct molecular O_2 adsorption should take place leading to efficient $4e^-$ reduction of O_2 to OH^- via an inner-sphere electrocatalytic pathway. So, in alkaline media, a combination of both inner-sphere and outer-sphere electron transfer mechanisms is operative. The consequence of this outer-sphere electron transfer in alkaline media is that this mechanism leads to higher concentration of HO_2^- to be generated near the electrode surface, that is, the double layer; however, no evidence for such an outer-sphere reaction is observed in acidic media. Higher activity of HO_2^- effectively shifts the potential of the electrode from that of the O_2/HO_2^- couple to that of the HO_2^-/OH^- redox couple by carrying out HO_2^- reduction to OH^- on oxide-free Ru sites. This is shown in Figure 4(d), where ORR and hydrogen peroxide reduction reaction (HPRR) are shown on Ru/C catalyst in both 0.1 M NaOH and 0.1 M $HClO_4$. $2e^-$ reduction of hydrogen peroxide in acidic (H_2O_2/H_2O) and alkaline media (HO_2^-/OH^-) are written as follows [13]:

$$\text{Acidic Media: } H_2O_2 + 2H^+ + 2e^- \longrightarrow 2H_2O \qquad E^\circ = 1.763\text{ V versus SHE}, \qquad (7)$$

$$\text{Alkaline Media: } HO_2^- + H_2O + 2e^- \longrightarrow 3OH^- \qquad E^\circ = 0.867\text{ V versus SHE}. \qquad (8)$$

As is well known, the standard reduction potentials of the above reactions are well positive of the reduction of molecular O_2 in both acidic and alkaline electrolytes. So, from a thermodynamic perspective, any peroxide intermediate formed should be immediately reduced further. This is also kinetically true on ruthenium as shown in Figure 4(d). Half-wave potential ($E_{1/2}$) of HPRR on Ru/C in 0.1 M NaOH is 40 mV positive compared to that of the $E_{1/2}$ of ORR [68]. So the kinetics of the reaction in alkaline media favors the immediate reduction of any peroxide intermediate generated during ORR. This is also true in acidic media although the shape of the HPRR profile in acidic media on Ru/C requires more explanation. The onset potential of HPRR on Ru/C in 0.1 M $HClO_4$ is as high as 0.88 V which is only 20 mV lower than ORR and HPRR onset potentials in 0.1 M NaOH electrolyte. As is known previously [69], H_2O_2 undergoes decomposition to O_2 and H_2O in acidic electrolyte at Ru/C surface. This decomposition reaction that generates O_2 near the electrode surface skews the HPRR profile in acidic media to higher overpotentials characteristic of ORR in acidic media. However, the aspect of relevance to the discussion here is that the kinetics of the system favors further reduction of hydrogen peroxide intermediate in both acidic and alkaline media. Once the H_2O_2/ HO_2^- stable intermediate is generated, this species undergoes adsorption at oxide-free Ru sites and further reduces to H_2O/OH^- according to (7)/(9). So, any in situ generation of hydrogen peroxide intermediate should shift the potential to more positive values, and indeed this is what is observed in alkaline media.

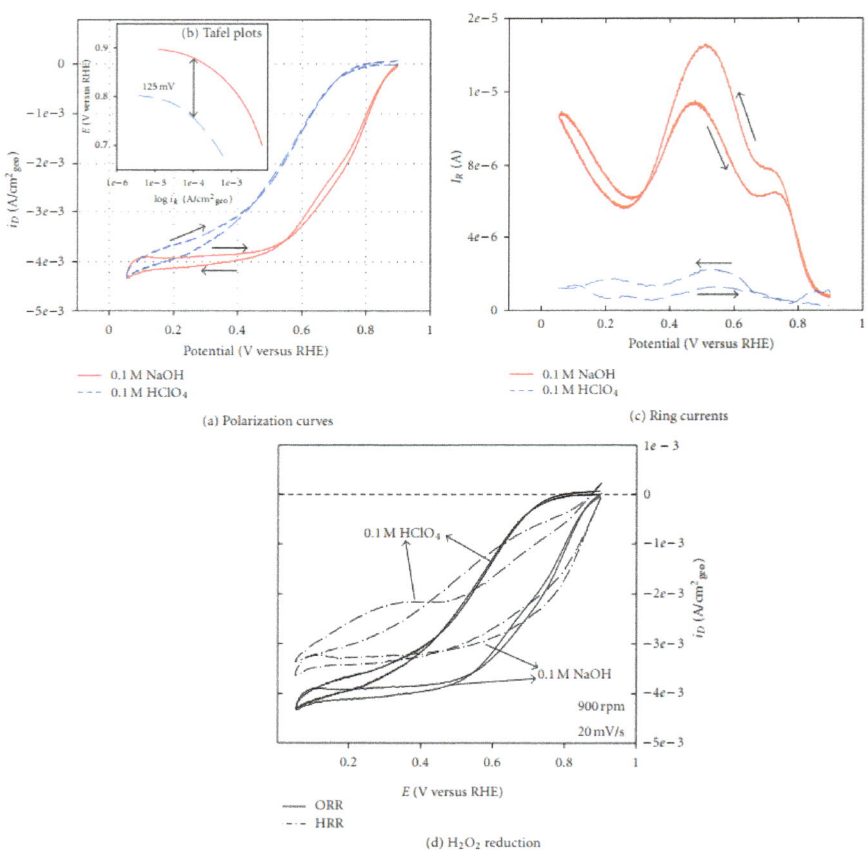

Figure 4: Oxygen reduction reaction on Ru/C heat-treated (HT) catalyst in 0.1 M NaOH and 0.1 M $HClO_4$ electrolytes. (a) ORR polarization curves at 900 rpm and 20 mV/s, (b) mass-transport-corrected Tafel plots, (c) ring current measured during ORR at 900 rpm. E_{Ring} = 1.1 V versus RHE in 0.1 M NaOH and E_{Ring} =1.3 V versus RHE in 0.1 M $HClO_4$, and (d) hydrogen peroxide reduction reaction (HPRR) in comparison to ORR in 0.1 M $HClO_4$ and 0.1 M NaOH. HPRR is shown in oxygen-free electrolytes containing externally added H_2O_2 at a concentration of 3.5 mM.

The in situ parallel generation of HO_2^- anion intermediate via the outer-sphere electron transfer reaction scheme shown in (3)–(5) serves to shift the ORR potential to more positive values in alkaline media. On the contrary, this excess parallel generation of H_2O_2 via the outer-sphere reaction mechanism is not observed in acidic media and, hence, the higher overpotential for ORR in acidic media. In general, this can also be further extended to other catalyst systems that exhibit lower ORR overpotential in alkaline media compared to that in acidic electrolytes. The prospects of modifying the carbon-supported ruthenium nanoparticle catalyst by chalcogen groups (S/Se) in order to suppress oxide formation and promote the inner-sphere electrocatalytic process by facilitating direct O_2 adsorption was discussed recently as a separate treatise [70].

Pyrolyzed Metal Macrocycle for Promoting Inner-Sphere Electron Transfer

Figure 5(a) shows a comparison in dilute acidic and alkaline electrolytes of the ORR activity of FeTPP/C catalyst pyrolyzed at 800°C. As observed in Figure 5(a), the onset potential for ORR in 0.1 M NaOH is 0.95 V versus RHE, whereas in 0.1 M $HClO_4$ electrolyte it is 0.80 V versus RHE. This 150 mV lower overpotential in alkaline media is clearly reflected over the entire mixed kinetic-diffusion region. In 0.1 M NaOH electrolyte, the mixed kinetic-diffusion region is ensued by a well-defined diffusion-limited region. In 0.1 M $HClO_4$, no clear diffusion-limited region could be discerned, which is indicative of kinetic control in acidic media even at high overpotentials. At a potential of 0.80 V versus RHE, the ORR kinetic current density of FeTPP/C (pyrolyzed at 800°C) is clearly four orders of magnitude higher in 0.1 M NaOH electrolyte than that in 0.1 M $HClO_4$. For a given catalyst, the four orders of magnitude difference in kinetic activity between acidic and alkaline media are intriguing. Figure 5(b) shows the HPRR activity of FeTPP/C catalyst (pyrolyzed at 800°C) in both acidic and alkaline media in comparison to the corresponding ORR polarization curves. The onset potential for peroxide reduction in 0.1 M $HClO_4$ is 0.84 V versus RHE, whereas in 0.1 M NaOH it is 1.01 V versus RHE. Besides this onset potential difference, in 0.1 M NaOH electrolyte, the mixed kinetic-diffusion region for peroxide reduction is more anodic compared to that of ORR in the same electrolyte which is then followed by a reasonably discernable diffusion-limited current density region. This clearly indicates that peroxide reduction in alkaline media is kinetically favored such that any peroxide intermediate formed during ORR in 0.1 M NaOH will be immediately reduced to the $4e^-$ product. On the contrary, the reduction of hydrogen peroxide in acidic media is kinetically unfavorable due to weak binding of H_2O_2 on the active site leading to its desorption into the bulk electrolyte and/or catalytic decomposition to molecular O_2.

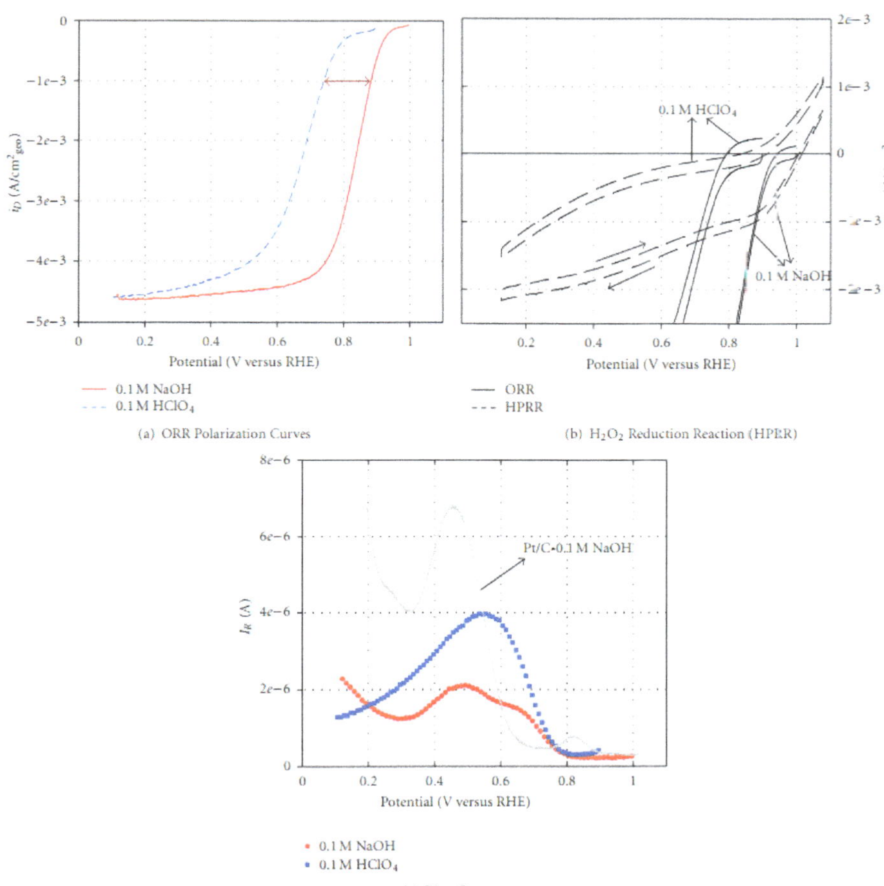

Figure 5: ORR activity of FeTPP/C catalyst (pyrolyzed at 800°C) in O_2 saturated acidic and alkaline electrolytes (a) ORR polarization curves, (b) hydrogen peroxide reduction reaction (HPRR) in comparison to ORR, and (c) ring current. All measurements were performed at 900 rpm rotation rate and 20 mV/s scan rate. E_{Ring} = 1.1 V versus RHE in 0.1 M NaOH and E_{Ring} = 1.3 V versus RHE in 0.1 M $HClO_4$. HPRR is shown in oxygen-free electrolytes containing externally added H_2O_2 at a concentration of 3.5 mM. Also shown in (c) is the ring current profile of 30% Pt/C in 0.1 M NaOH for comparison.

This clearly demonstrates that stabilizing the peroxide intermediate on the active site is important in effectively carrying out $4e^-$ ORR process, and the higher stability of peroxide intermediate on the active site leads to the so-called kinetic facility in alkaline media but not in acidic media. Given the pK_a values for the first and second ionization of H_2O_2 at 25°C (pK_1 = 11.69 and pK_2 = ~20), the predominant peroxide species for pH<12 is H_2O_2 whereas at pH>12 it is HO_2^- [58]. This leads to the formation of a stable Lewis acid-base adduct in alkaline media (Fe^{2+}- HO_2^-) that enables complete $4e^-$ transfer. On the contrary, the lack of stabilization of (Fe^{2+}-H_2O_2) adducts in acidic media leads to desorption of H_2O_2 into the bulk electrolyte. Figure 5(c) shows the ring current due to peroxide oxidation measured during ORR in both acidic and alkaline electrolytes. Clearly, the peroxide yield during ORR on FeTPP/C (pyrolyzed at 800°C) is higher in acidic media by an order of magnitude compared to that in alkaline media. This observation is opposite to what was observed on the Pt/C and Ru/C catalysts discussed previously. The onset potential for ORR in 0.1 M NaOH is 0.95 V versus RHE, whereas the corresponding peroxide oxidation current does not begin until 0.8 V. In 0.1 M $HClO_4$, the onset potential for both ORR and peroxide oxidation is 0.8 V. This is further proof for the instability of peroxide intermediate on the active site in acidic media because the weak binding of the H_2O_2 intermediate on the Fe^{2+} active site facilitates its desorption and/or decomposition into the bulk electrolyte. This desorption of H_2O_2 into the bulk electrolyte due to its weak binding on the active site is the primary source of peroxide detected at the ring in acidic media. At very high overpotentials (<0.3 V), the ORR process in acidic electrolyte becomes very efficient and consequently the peroxide detected decreases [71, 72]. The ring current profile in 0.1 M NaOH is more complex. Briefly, in alkaline media, the peak in ring current at 0.50 V is due to the carbon support and the increase in ring current below 0.3 V is due to reorientation of the water molecules [60, 73]. More importantly, a weak shoulder in the ring current in 0.1 M NaOH is observed between 0.6 V and 0.7 V. Also shown in Figure 5(c) is the ring current profile of Pt/C for comparison. As discussed previously, the ring current peak due to Pt/C around 0.80 to 0.90 V is the signature for the peroxide formation via outer-sphere electron transfer mechanism at –OH-covered Pt sites. On Fe-N_x-based catalyst, only a weak shoulder in ring current at more negative potentials of 0.60 to 0.70 V is observed indicating that the outer-sphere electron transfer mechanism is significantly muted. This indicates the facilitation of direct adsorption of O_2 on the Fe^{2+}active site. This is because the specifically adsorbed OH^- species on the Fe^{2+} site acts as a labile ligand that is easily displaced by the molecular O_2. This is ensued by the successful $4e^-$ reduction of O_2 with minimal occurrence of outer-sphere electron transfer mechanism.

The nature of the active site obtained upon pyrolysis and the fundamental origin of the activity is investigated here in detail. Figure 6(a) shows the square wave voltammetry (SWV) of non-heat-treated Fe(III)TPPCl/C in argon-saturated 0.1 M NaOH and 0.1 M $HClO_4$ electrolytes, and Figure 6(b) shows the corresponding CV. As seen in both the SWV and the CV profiles, the redox transition involving the metal center $Fe^{2+/3+}$ is observed at 0.314 V versus RHE in 0.1 M NaOH and at 0.155 V versus RHE in 0.1 M $HClO_4$ [30, 31, 74]. As seen in the SWV profile, the peaks at 1.260 V in 0.1 M $HClO_4$ and 1.508 V in 0.1 M NaOH correspond to the one-electron redox transition involving the delocalized π-electron system in the macrocycle ligand [74]. As observed in the literature [30], CV of the heat-treated catalysts does not yield useful information, since in most cases after heat treatment the double-layer charging current overwhelms the Faradaic currents. In order to overcome this limitation, SWV was performed as shown in Figure 7 in order to understand the evolution of the active site in these catalysts with increasing heat treatment temperatures. Figures 7(a) and 7(b) are divided into two panels each due to different potential ranges that are required for baseline correction of SWV profiles. As seen in Figure 7(a) in alkaline media, after 300°C pyrolysis, the $Fe^{2+/3+}$ redox couple at 0.31 V and the ligand oxidation peak at 1.5 V are similar to the case of non-heat-treated catalyst. After heat treatment at 500°C and 600°C, the magnitude of the peak currents decrease significantly due to the sublimation of some fraction of the metal macrocycle complex. More importantly, the $Fe^{2+/3+}$ peak potentials have shifted to more anodic potentials. As seen in Panel 1 of Figure 7(a), after pyrolysis at 500°C and 600°C, the peak potential (E_p) of the $Fe^{2+/3+}$ couple has shifted to 0.405 V and 0.427 V, respectively. Correspondingly, the ligand oxidation peak at ~1.5 V, shown in Panel 2 of Figure 7(a), decreases in magnitude significantly due to the sublimation and/or destruction of the macrocycle. The ligand oxidation peak is not observed after heat treatment at temperatures greater than 600°C indicating that the delocalized π-electron system of the macrocycle does not survive such high temperatures on the carbon support. After 800°C pyrolysis, $Fe^{2+/3+}$ couple has shifted more anodically to E_p=0.48 V. As shown in Figure 7(a) (Panel 2), the most interesting observation is that after pyrolysis at 600°C a new shoulder begins to emerge at ~1.2 V. This shoulder resolves into a clear peak after 800°C pyrolysis with a peak potential of 1.25 V versus RHE in alkaline media. This new peak is attributed to an anodic shift in the $Fe^{2+/3+}$ redox couple based on the in situ XANES experimental results shown later (vide infra). This indicates the presence of two $Fe^{2+/3+}$ redox couples, one at a low potential (~0.48 V) and another at a high potential (1.25 V). The anodic shift in $Fe^{2+/3+}$ redox peak potentials shown in Figure 7(a) (inset) clearly indicates that after 600°C pyrolysis the metal center exists in two redox environments. Qualitatively, similar behavior was observed in acidic media as shown in Figure 7(b). (Note that the ligand oxidation peak in acidic media that occurs above 1 V versus RHE is not shown in Panel 2 of Figure 7(b).) It is interesting to note that after heat treatment at 600°C the $Fe^{2+/3+}$redox peak potential in 0.1 M $HClO_4$ was shifted to an anodic potential of only 0.80 V versus RHE compared to the 1.25 V in alkaline media. Thus, there are two reasons for the lower ORR overpotential for FeTPP/C catalyst in alkaline media: (1) the higher redox potential of the $Fe^{2+/3+}$ metal center in alkaline media and (2) the improved stability of the peroxide intermediate on the active site. This translates into efficient $4e^-$ ORR reaction with lower overpotential in alkaline media compared to acidic media.

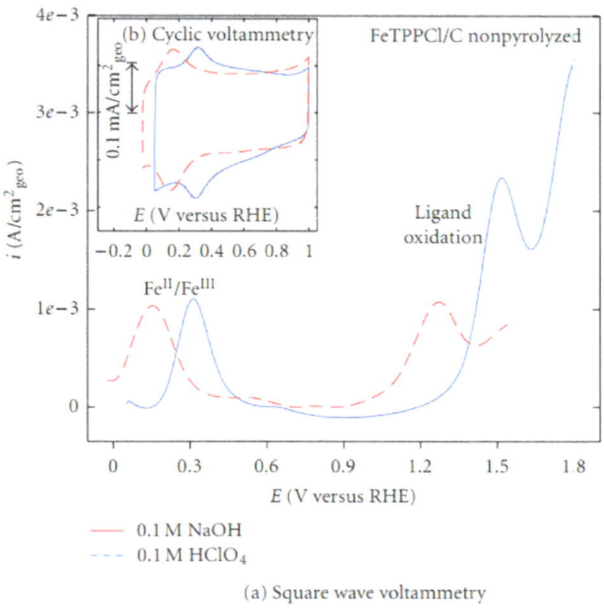

Figure 6: (a) Square wave voltammetry (SWV) and (b) cyclic voltammetry (CV) of as-received iron(III) meso-tetraphenylporphyrin chloride (FeTPPCl/C) supported on Black Pearl carbon. Experiments were performed in argon-saturated 0.1 M NaOH and 0.1 M $HClO_4$ electrolytes. SWV parameters: 5 mV step potential, 20 mV amplitude, and 10 Hz frequency. CV: 20 mV/s.

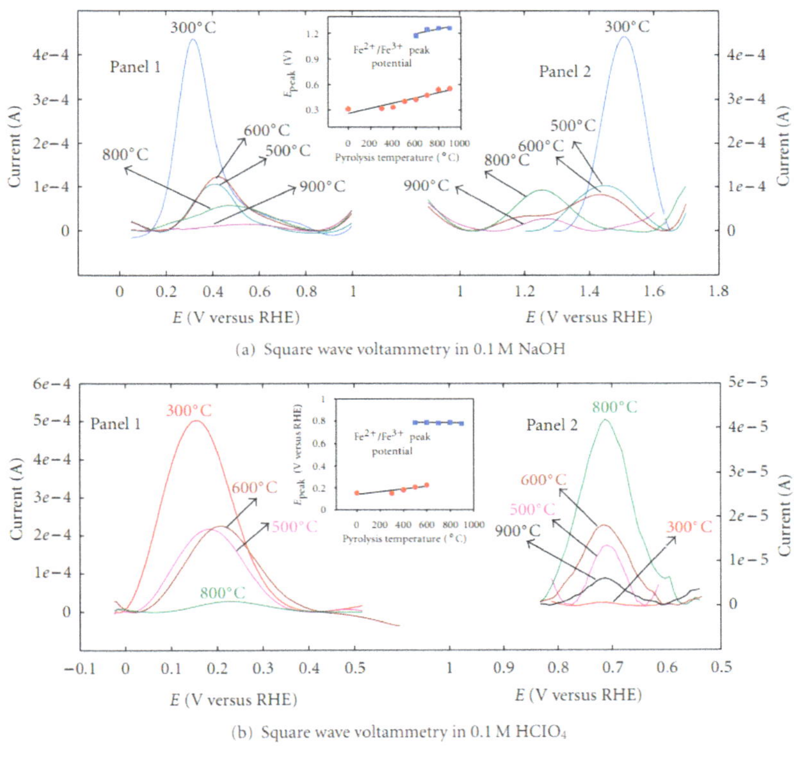

Figure 7: SWV profiles of FeTPP/C catalyst as a function of heat treatment temperature in (a) 0.1 M NaOH and (b) 0.1 M $HClO_4$ electrolytes. Insets show the peak potential of the $Fe^{2+/3+}$ redox couple as a function of the heat treatment temperature. SWV experiment parameters: step potential −5 mV, amplitude −20 mV, and scan frequency −10 Hz.

Delta-Mu (Δμ) Studies

Both the extended X-ray absorption fine structure (EXAFS) and X-ray absorption near edge structure (XANES) being bulk averaged techniques overlook the critical electrochemical reactions occurring on the catalyst surface. Delta-mu (Δμ) is a surface sensitive, spectral subtraction technique, where the bulk structure of the catalyst is effectively removed leading to information on the nature and site specificity of the surface adsorbates [54, 55, 75, 76]. Figure 8(a) shows the XANES region of FeTPP/C catalyst pyrolyzed at 300°C taken at two different potentials of 0.1 V and 0.9 V versus RHE. At 0.1 V, the metal center exists in the reduced Fe^{2+}state with no adsorbates (neither hydrides nor oxides) at the axial position, and immediate coordination environment is reminiscent of the iron(II) phthalocyanine complex, where the preedge peak is muted. At 0.9 V, the metal center is oxidized to Fe^{3+} state with an oxygen atom at the axial position, and the metal coordination environment is similar to that of the original porphyrin complex, where the preedge Fe(1s→3d) forbidden transition at 7112.5 eV is observed. The delta-mu spectra is obtained by subtracting the XANES regions according to the equation $\Delta\mu = \mu(0.90\,V) - \mu(0.10\,V)$. In the delta-mu spectra of Figure 8(a), the positive peak feature (boxed portion) indicates the difference in absorption probability at the preedge energy (7112.5 eV). This positive peak feature could be safely assigned as a signature for the existence of the metal center in a centrosymmetric environment undergoing a transition from Fe^{2+}-N_4 coordination geometry at 0.10 V to O-Fe^{3+}-N_4 coordination geometry at 0.90 V. Ensuing this positive peak is a steep negative dip featuring a split peak below 20 eV. This negative dip characterizes charge transfer from the metal center to the adsorbed oxygen species [77]. As shown in Figure 8(b), the XANES spectra of the 800°C pyrolyzed catalyst is predominantly characteristic of metallic iron at both 0.10 V and 1.10 V that precludes proper analysis of the active site. However, careful analysis of the corresponding delta-mu spectra clearly indicates the positive peak feature at the preedge energy indicative of the fact that the active site is Fe^{2+}-N_4 where the metal center is in a centrosymmetric environment which is mildly disrupted by the presence of an axial oxygen atom. This clearly indicates that the Fe^{2+} metal center surrounded by four nitrogen atoms is the active site that binds oxygen in the axial position and the redox transition from Fe^{3+} to Fe^{2+} triggers oxygen adsorption according to the redox mechanism [78]. Figure 9 shows the theoretical delta-mu ($\Delta\mu_t$) spectra obtained using FEFF8.0 code [56]. The structural models used are shown in the insets of Figure 9. Only the atoms encircled were used in the theoretical FEFF8.0 modeling. These theoretical spectra were calculated from Fe-N_4-C_x models derived from prior crystallographic data adjusted to the EXAFS fitting results according to the relation $\Delta\mu_t = \mu(O_{ads}\text{-Fe-}N_4\text{-C}) - \mu(\text{Fe-}N_4\text{-C})$, where the oxide species (O_{ads} or OH_{ads}) is in a specific binding site. In all cases, the positive peak feature at the preedge energy was observed only when the adsorbed oxygen atom was placed in the axial position of the metal center. No successful theoretical delta-mu fits could be obtained for Fe-N coordination numbers less than four or for oxygen adsorption modes other than at the axial position. As shown in the inset of Figure 9(a), after 300°C pyrolysis the immediate coordination environment of the original precursor porphyrin macrocycle is clearly retained. This corresponds to an FeN_4C_{12} cluster, where the metal center is coordinated to four nitrogen atoms and each nitrogen atom in turn bonded to two carbon atoms. Finally, inclusion of the four methine carbon bridges gives a tally of C_{12}. Figure 9(b) shows the theoretical delta-mu spectrum that likely mimics the line shape of the experimental delta-mu shown in Figure 8(b) for the 800°C heat-treated catalyst. As shown in the inset of Figure 9(b), the molecular cluster used to simulate the theoretical delta-mu spectrum consisted of FeN_4C_{10}. While compositionally this cluster is not very different from the 300°C pyrolyzed sample, the immediate coordination environment of the metal active site after 800°C heat treatment is found to be reminiscent of the crystallographic atomic defects such as the divacancy on the graphitic surfaces. This is schematically depicted in Figure 9(c) [79, 80]. Atomic defects such as monovacancy and divacancy on microporous carbon and carbon nanotubes are known to thermodynamically exist or can be induced via various chemical or physical processes [81–85]. The presence of a monovacancy in carbon atom creates three dangling bonds whereas divacancies create four dangling bonds. These dangling bonds give rise to unsaturated valences that then become favorable for nitrogen doping [80]. Consequently, these nitrogen-doped sites constitute defective pockets for metal coordination. Such atomic vacancies are either already present on the graphite surfaces or can be created during the heat treatment step [81, 83]. During heat treatment under inert atmosphere, carbothermic reaction causes desorption of oxygen functional groups along with creation of vacancy defects [83]. It has also been observed earlier that heat-treated Fe-N_x catalysts showed higher activity when supported on carbon that was previously treated in concentrated inorganic acids [52]. Since such acid treatment steps lead to oxygen functional groups on carbon support, it is likely that this yields higher number of defective sites during subsequent heat treatment. These defective pockets are likely the favorable zones for anchoring FeN_4 active sites. Presumably, the low concentration of the defective sites limits the active site density. Therefore, increasing the defect density likely holds key to increasing the metal loading in this class of catalysts.

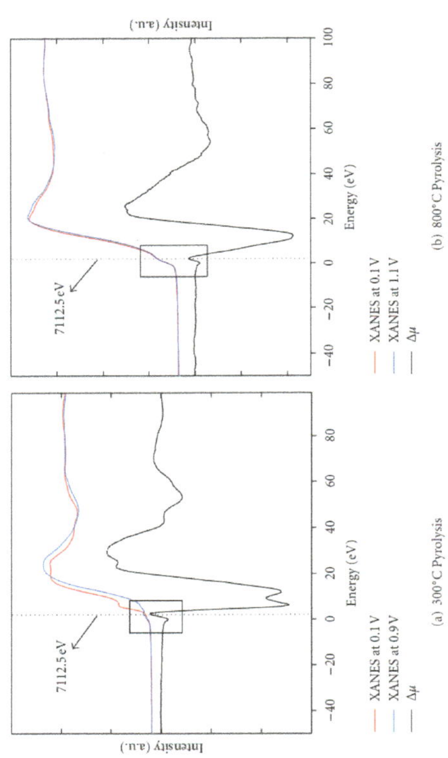

Figure 8: Experimental XANES and delta-mu signatures of FeTPP/C catalyst heat treated at (a) 300°C and (b) 800°C. Delta-mu signatures were obtained by subtracting the XANES signatures according to $\Delta\mu = \mu(0.90\text{ (or) }1.10\,V) - \mu(0.10\,V)$. Experiments were conducted at Fe K-edge under in situ conditions in argon-saturated 0.1 M NaOH electrolyte. Vertical dotted line indicates the preedge position at 7112.5 eV and the boxed region focuses the preedge region. Delta-mu spectra have been multiplied by a factor of 5 for visual comparison of the line shapes.

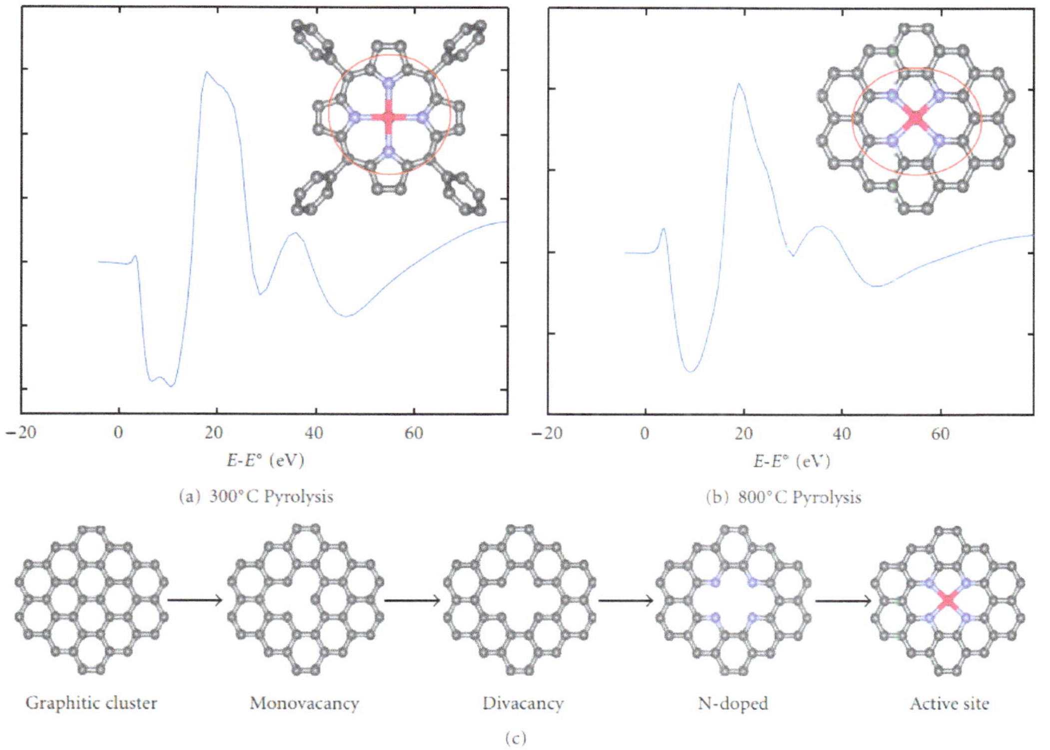

Figure 9: Theoretical FEFF8 $\Delta\mu = \mu(Fe\text{-}N_4\text{-}C_x\text{-}O_{ads}) - \mu(Fe\text{-}N_4\text{-}C_x)$ signatures obtained for (a) 300°C and (b) 800°C pyrolysis conditions. The insets in (a, b) show the corresponding structural models utilized. Only the atoms encircled in these structural models were used for FEFF8 simulation. (c) Schematic illustration of the mono- and divacant defective pockets in amorphous carbon acting as anchors for active site formation during pyrolysis. Color codes: pink: Fe, blue: N, and black: C.

The following observations are made in an attempt to correlate the SWV and the delta-mu results. The low-potential $Fe^{2+/3+}$ redox couple is characteristic of the original FeN_4 porphyrin moiety, whereas the high-potential $Fe^{2+/3+}$ redox couple is characteristic of the FeN_4 cluster found in the microporous carbon defect sites. At a pyrolysis temperature of 600°C, both of these sites seem to coexist and the higher ORR activity obtained for heat treatment temperatures ≥600°C is due to the high-potential $Fe^{2+/3+}$ redox couple seated in the defective pockets. Based on the above experimental results, the following reaction scheme is proposed for ORR in dilute alkaline media on heat-treated FeTPP/C catalyst. Similar set of reactions can be developed,mutatis mutandis, for dilute acidic media [22, 71, 72, 86]. Equation (9) shows the redox reaction involving the metal center that is a prerequisite for adsorption of molecular oxygen on the active site $[N_4\text{-}Fe^{II}\text{-}OH^-]$:

$$\left[N_4\text{-}Fe^{III}\text{-}OH\right] + e^- \longrightarrow \left[N_4\text{-}Fe^{II}\right] + OH^- \tag{9}$$

While the Fe^{II} valence state favors a square-planar tetracoordinate environment, the high potential of this redox reaction causes the OH^- species to poison the active site at the axial position. This poisonous OH^- species prevents direct molecular adsorption of O_2 on the active site. Further, the adsorbed OH^- species mediates the $2e^-$ outer-sphere electron transfer reduction of solvated O_2 molecule as shown in the reaction schemes (3)–(5) although this outer-sphere mechanism appears to be extremely weak in this class of pyrolyzed catalysts as explained previously. It should be noted that the case of outer-sphere mechanism does not arise in acidic media for reasons explained previously.

Finally, the $4e^-$ electrocatalytic inner-sphere electron transfer mechanism is shown in Figure 10, where the molecular O_2 displaces the OH^- species and chemisorbs directly on the Fe^{II} active site. Based on our experimental results, the electrocatalytic process in Figure 10 is shown to take place via a redox mechanism involving $Fe^{II/III}$ couple [78]. Once molecular O_2 adsorbs on the Fe^{II} active site, the reaction proceeds to the ferrous-hydroperoxyl adduct via the superoxo and the ferric-hydroperoxyl states. The ferrous-hydroperoxyl adduct is very critical since its stability determines the product distribution. Peroxide anion (HO_2^-) is a stable intermediate. So, weak binding of HO_2^- on the active site will lead to its desorption into the bulk electrolyte. As mentioned previously, the stable acid-base adduct formed in alkaline media stabilizes the ferrous-hydroperoxyl adduct. This ensures that the catalytic cycle shown in Figure 10 regenerates the active site via the formation of ferric-hydroxyl species. However, in acidic media the analogous ferrous-hydrogen peroxide adduct Fe^{II} -(OHOH) is unstable and causes desorption and/or disproportionation of the peroxide intermediate (H_2O_2) into the bulk electrolyte. This leads to higher peroxide yield in acidic media as shown previously.

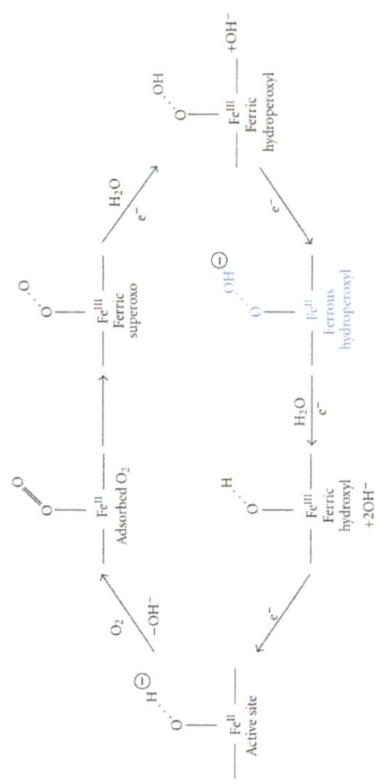

Figure 10: Catalytic cycle showing the redox mechanism involved in ORR on heat-treated iron porphyrin macrocycles in dilute alkaline medium. Nitrogen atoms in the square planar positions have been omitted for clarity.

CONCLUSIONS

The case of ORR in alkaline media shows interesting mechanistic differences compared to acidic media. In acidic media, the presence of OH_{ad} due to water activation blocks the direct adsorption of molecular O_2 on the active site. In alkaline media, the presence of OH_{ad} due to specific adsorption of hydroxide species not only blocks the O_2 adsorption but also promotes outer-sphere electron transfer mechanism to yield the $2e^-$ peroxide intermediate as the product. The interaction between the $O_2{\cdot}(H_2O)_n$ cluster and the surface hydroxyl species causes certain nonspecificity to the identity of the underlying electrode metal. This nonspecificity is a result of kinetic differences in alkaline media that opens the gate to use a wide-range of nonnoble metals and their oxides as electrode materials, while in acidic media the absence of this outer-sphere mechanism necessitates the use of highly catalytic surfaces based on Pt. On noble metals such as Pt, this nonspecificity in alkaline media is observed only in the oxide formation region, and depending on the extent of OH_{ad} coverage both inner- and outer-sphere electron transfer mechanisms coexist in alkaline media. On nonnoble metals such as the first-row transition elements that are completely passivated by a layer oxide film upon immersion into the aqueous electrolyte, this outer-sphere electron transfer process is likely to be the dominant mechanism. While the outer-sphere mechanism primarily seems to yield peroxide intermediate as the final product, this could be beneficial in cases like Ru/C catalyst, where the peroxide intermediate generated in the outer-sphere process could be readsorbed on the oxide-free Ru site to reduce further to form hydroxide as the final product. However, in cases like nonnoble metal surfaces, where the electrode surface is incapable of reducing peroxide intermediate any further, only $2e^-$ reduction products are likely to predominate in alkaline media. So, it is important to promote the inner-sphere electron transfer mechanism in alkaline media by facilitating the direct molecular O_2 adsorption and promoting complete $4e^-$ reduction process.

On catalyst surfaces such as the $Fe\text{-}N_4$ systems, the operation of redox mechanism allows the direct molecular O_2 adsorption on the $Fe^{2+}\text{-}N_4$ active site and prevents the precipitous outer-sphere reaction of solvated O_2molecule with the OH_{ad}-covered metal site. This is due to the OH_{ad} species acting as a labile ligand in the $Fe\text{-}N_4$-based catalysts that helps to promote the $4e^-$ inner-sphere electron transfer mechanism by efficiently reducing O_2 to OH^- in alkaline media. Further, in the class of $Fe\text{-}N_4$-based catalysts, it is important to perform ORR studies with a thin film of catalyst surface ($\leq 25\ \mu m$) rather than thick catalyst films that are typically used in the literature, where the peroxide formed is completely quenched within the catalyst layer and hence not being detected at the ring electrode. This is especially true while Koutecky-Levich plots for calculating the number of electrons transferred since such calculations were derived for smooth electrode surfaces. Also, it is noted that if a catalyst is considered to perform $4e^-$ oxygen reduction, then it is imperative to show the reduction of externally added hydrogen peroxide in the absence of O_2. Finally, in conclusion, the presence of the outer-sphere electron transfer mechanism in alkaline media is a kinetic effect. The only thermodynamic advantage of performing ORR at high pH is the improved stability of the hydroperoxide anion intermediate on the active site that enables complete $4e^-$ transfer. So, in order to stabilize the H_2O_2 intermediate in acidic media, the use of dimeric active sites that simultaneously bind both oxygen atoms are required.

Acknowledgments

The authors deeply appreciate financial assistance from the Army Research Office under the Single Investigator Grant. The authors also gratefully acknowledge the supply of platinum electrocatalysts from BASF fuel cells (Somerset, NJ, USA). Use of the National Synchrotron Light Source (NSLS) at Brookhaven National Laboratory (BNL) was supported by the US Department of Energy, Office of Basic Energy Sciences. Support from NSLS beamline personnel Drs. Kaumudi Pandya (X11A), Syed Khalid (X19A), and Nebojsa Marinkovic (X18B) is gratefully acknowledged.

REFERENCES

1. J. S. Spendelow and A. Wieckowski, "Electrocatalysis of oxygen reduction and small alcohol oxidation in alkaline media," Physical Chemistry Chemical Physics, vol. 9, no. 21, pp. 2654–2675, 2007.
2. E. Yeager, "Electrocatalysts for O_2 reduction," Electrochimica Acta, vol. 29, no. 11, pp. 1527–1537, 1984. ·
3. R. Adzic, "Frontiers in electrochemistry," in Electrocatalysis, J. Lipkowski and P. N. Ross, Eds., vol. 197, Wiley-VCH, New York, NY, USA, 1998.
4. D. B. Sepa, M. V. Vojnovic, and A. Damjanovic,

"Kinetics and mechanism of O_2 reduction at Pt in alkaline solutions," Electrochimica Acta, vol. 25, no. 11, pp. 1491–1496, 1980.

5. D. B. Sepa, M. V. Vojnovic, L. M. Vracar, and A. Damjanovic, "Different views regarding the kinetics and mechanisms of oxygen reduction at Pt and Pd electrodes," Electrochimica Acta, vol. 32, no. 1, pp. 129–134, 1987.

6. A. Damjanovic, A. Dey, and J. O. M. Bockris, "Kinetics of oxygen evolution and dissolution on platinum electrodes," Electrochimica Acta, vol. 11, no. 7, pp. 791–814, 1966.

7. E. Yeager, M. Razaq, D. Gervasio, A. Razaq, and D. Tryk, "The electrolyte factor in O_2 reduction electrocatalysis," Proceedings of the Electrochemical Society, vol. 92-11, pp. 440–473, 1992.

8. A. Damjanovic, M. A. Genshaw, and J. O'M Bockris, "Distinction between intermediates produced in main and side electrodic reactions," The Journal of Chemical Physics, vol. 45, no. 11, pp. 4057–4059, 1966.

9. H. S. Wroblowa, Y.-C. Pan, and G. Razumney, "Electroreduction of oxygen a new mechanistic criterion,"Journal of Electroanalytical Chemistry, vol. 69, no. 2, pp. 195–201, 1976.

10. N. A. Anastasijević, V. Vesović, and R. R. Adžić, "Determination of the kinetic parameters of the oxygen reduction reaction using the rotating ring-disk electrode. Part I. Theory," Journal of Electroanalytical Chemistry, vol. 229, no. 1-2, pp. 305–316, 1987.

11. N. A. Anastasijević, V. Vesović, and R. R. Adžić, "Determination of the kinetic parameters of the oxygen reduction reaction using the rotating ring-disk electrode. Part II. Applications," Journal of Electroanalytical Chemistry, vol. 229, no. 1-2, pp. 317–325, 1987.

12. A. J. Bard, "Inner-sphere heterogeneous electrode reactions. Electrocatalysis and photocatalysis: the challenge," Journal of the American Chemical Society, vol. 132, no. 22, pp. 7559–7567, 2010.

13. A. J. Bard, R. Parsons, and J. Jordan, Eds., Standard Potentials in Aqueous Solution, 1985.

14. H. H. Yang and R. L. McCreery, "Elucidation of the mechanism of dioxygen reduction on metal-free carbon electrodes," Journal of the Electrochemical Society, vol. 147, no. 9, pp. 3420–3428, 2000.

15. B. B. Blizanac, P. N. Ross, and N. M. Markovic, "Oxygen electroreduction on Ag(111): the pH effect,"Electrochimica Acta, vol. 52, no. 6, pp. 2264–2271, 2007.

16. J. O. Bockris and J. Appleby, "Alkaline Fuel Cells," in Assessment of Research Needs for Advanced Fuel Cells, S. S. Penner, Ed., vol. 11, p. 95, 1986.

17. A. J. Appleby, "Electrocatalysis," in Comprehensive Treatise of Electrochemistry, B. E. Conway, Ed., vol. 7, pp. 173–239, Plenum, New York, NY, USA, 1983.

18. R. Jasinski, "A new fuel cell cathode catalyst," Nature, vol. 201, no. 4925, pp. 1212–1213, 1964.

19. J. H. Zagal, "Metallophthalocyanines as catalysts in electrochemical reactions," Coordination Chemistry Reviews, vol. 119, pp. 89–136, 1992.

20. "Macrocycles," in Handbook of Fuel Cells-Fundamentals, J. H. Zagal, Ed., pp. 544–554, John Wiley & Sons, Chichester, UK, 2003.

21. B. Wang, "Recent development of non-platinum catalysts for oxygen reduction reaction," Journal of Power Sources, vol. 152, no 1-2, pp. 1–15, 2005.

22. J.-P. Dodelet, "Oxygen reduction in pem fuel cell conditions: heat- treated non-precious metal-N4 macrocycles and beyond," in N4-Macrocyclic Metal Complexes, J. H. Zagal, F. Bedioui, and J.-P. Dodelet, Eds., p. 83, Springer, Berlin, Germany, 2006.

23. J. M. Ziegelbauer, T. S. Olson, S. Pylypenko et al., "Direct spectroscopic observation of the structural origin of peroxide generation from Co-based pyrolyzed porphyrins for ORR applications," Journal of Physical Chemistry C, vol. 112, no. 24, pp. 8839–8849, 2008.

24. K. Artyushkova, S. Levendosky, P. Atanassov, and J. Fulghum, "XPS Structural studies of nano-composite non-platinum electrocatalysts for polymer electrolyte fuel cells," Topics in Catalysis, vol. 46, no. 3-4, pp. 263–275, 2007.

25. A. L. Bouwkamp-Wijnoltz, W. Visscher, and J. A. R. van Veen, "The selectivity of oxygen reduction by pyrolysed iron porphyrin supported on carbon," Electrochimica Acta, vol. 43, no. 21-22, pp. 3141–3152, 1998.

26. A. L. Bouwkamp-Wijnoltz, W. Visscher, J. A. R. van Veen, E. Boellaard, A. M. van der Kraan, and S. C. Tang, "On active-site heterogeneity in pyrolyzed carbon-supported iron porphyrin catalysts for the electrochemical reduction of oxygen: an in situ mössbauer study," Journal of Physical Chemistry B, vol. 106, no. 50, pp. 12993–13001, 2002.

27. A. L. Bouwkamp-Wijnoltz, W. Visscher, J. A. R. van Veen, and S. C. Tang, "Electrochemical reduction of oxygen: an alternative method to prepare active CoN_4 catalysts," Electrochimica Acta, vol. 45, no. 3, pp. 379–386, 1999.

28. J. McBreen, W. E. O'Grady, D. E. Sayers, C. Y. Yang, and K. I. Pandya, "An EXAFS study of pyrolyzed metal macrocyclic electrocatalysts," S. Srinivasan, S. Wagner, and H. Wroblowa, Eds., vol. 87-12, p. 182, The Electrochemical Society, Pennington, NJ, USA.

29. H. Schulenburg, S. Stankov, V. Schünemann et al., "Catalysts for the oxygen reduction from heat-treated iron(III) tetramethoxyphenylporphyrin chloride: structure and stability and active sites," Journal of Physical Chemistry B, vol. 107, no. 34, pp. 9034–9041, 2003.

30. A. A. Tanaka, S. L. Gupta, D. Tryk, C. Fierro, and E. B. Yeager, "Electrochemical and spectroscopic aspects of heat-treated transition metal macrocycles as electrocatalysts for oxygen reduction," inStructural Effects in Electrocatalysis and Oxygen Electrochemistry, D. Scherson, D. Tryk, M. Daroux, and X. Xing, Eds., vol. 92-11, p. 555, Proceedings of the Electrochemical Society, 1992.

31. I. T. Bae, D. A. Tryk, and D. A. Scherson, "Effect of heat treatment on the redox properties of iron porphyrins

adsorbed on high area carbon in acid electrolytes: an in situ Fe K-edge X-ray absorption near-edge structure study," Journal of Physical Chemistry B, vol. 102, no. 21, pp. 4114–4117, 1998. ·

32. S. Kim, I. T. Bae, M. Sandifer et al., "In situ XANES of an iron porphyrin irreversibly adsorbed on an electrode surface," Journal of the American Chemical Society, vol. 113, no. 24, pp. 9063–9066, 1991. ·

33. S. Kim, D. A. Tryk, I. T. Bae et al., "In situ extended X-ray absorption fine structure of an iron porphyrin irreversibly adsorbed on an electrode surface," Journal of Physical Chemistry, vol. 99, no. 25, pp. 10359–10364, 1995. ·

34. M. Lefèvre, J. P. Dodelet, and P. Bertrand, "Molecular oxygen reduction in PEM fuel cells: evidence for the simultaneous presence of two active sites in Fe-based catalysts," Journal of Physical Chemistry B, vol. 106, no. 34, pp. 8705–8713, 2002.

35. M. Lefevre, J. P. Dodelet, and P. Bertrand, "Oi reduction in PEM fuel cells: activity and active site structural information for catalysts obtained by the pyrolysis at high temperature of Fe precursors,"Journal of Physical Chemistry B, vol. 104, no. 47, pp. 11238–11247, 2000. ·

36. M. Lefèvre and J. P. Dodelet, "Fe-based catalysts for the reduction of oxygen in polymer electrolyte membrane fuel cell conditions: determination of the amount of peroxide released during electroreduction and its influence on the stability of the catalysts," Electrochimica Acta, vol. 48, no. 19, pp. 2749–2760, 2003.

37. M. Lefèvre, E. Proietti, F. Jaouen, and J. P. Dodelet, "Iron-Based catalysts with improved oxygen reduction activity in polymer electrolyte fuel cells," Science, vol. 324, no. 5923, pp. 71–74, 2009.

38. G. Lalande, R. Côté, D. Guay, J. P. Dodelet, L. T. Weng, and P. Bertrand, "Is nitrogen important in the formulation of Fe-based catalysts for oxygen reduction in solid polymer fuel cells?" Electrochimica Acta, vol. 42, no. 9, pp. 1379–1388, 1997. ·

39. G. Lalande, G. Faubert, R. Côté et al., "Catalytic activity and stability of heat-treated iron phthalocyanines for the electroreduction of oxygen in polymer electrolyte fuel cells," Journal of Power Sources, vol. 61, no. 1-2, pp. 227–237, 1996. ·

40. M. C. M. Alves, J. P. Dodelet, D. Guay, M. Ladouceur, and G. Tourillon, "Origin of the electrocatalytic properties for O_2 reduction of some heat-treated polyacrylonitrile and phthalocyanine cobalt compounds adsorbed on carbon black as probed by electrochemistry and X-ray absorption spectroscopy," Journal of Physical Chemistry, vol. 96, no. 26, pp. 10898–10905, 1992. ·

41. G. Faubert, R. Côté, J. P. Dodelet, M. Lefèvre, and P. Bertrand, "Oxygen reduction catalysts for polymer electrolyte fuel cells from the pyrolysis of FeII acetate adsorbed on 3,4,9,10-perylenetetracarboxylic dianhydride," Electrochimica Acta, vol. 44, no. 15, pp. 2589–2603, 1999. ·

42. G. Faubert, G. Lalande, R. Côté et al., "Heat-treated iron and cobalt tetraphenylporphyrins adsorbed on carbon black: physical characterization and catalytic properties of these materials for the reduction of oxygen in polymer electrolyte fuel cells," Electrochimica Acta, vol. 41, no. 10, pp. 1689–1701, 1996.

43. P. He, M. Lefèvre, G. Faubert, and J. P. Dodelet, "Oxygen reduction catalysts for polymer electrolyte fuel cells from the pyrolysis of various transition metal acetates adsorbed on 3,4,9,10-perylenetetracarboxylic dianhydride," Journal of New Materials for Electrochemical Systems, vol. 2, no. 4, pp. 243–251, 1999. ·

44. C. Médard, M. Lefèvre, J. P. Dodelet, F. Jaouen, and G. Lindbergh, "Oxygen reduction by Fe-based catalysts in PEM fuel cell conditions: activity and selectivity of the catalysts obtained with two Fe precursors and various carbon supports," Electrochimica Acta, vol. 51, no. 16, pp. 3202–3213, 2006. ·

45. F. Jaouen, E. Proietti, M. Lefèvre et al., "Recent advances in non-precious metal catalysis for oxygen-reduction reaction in polymer electrolyte fuel cells," Energy and Environmental Science, vol. 4, no. 1, pp. 114–130, 2011. ·

46. D. P. Amalnerkar, S. Radhakrishnan, H. Minoura, T. Sugiura, and Y. Ueno, "Origin of the photocathodic effect at CdS + CuCl electrodes," Journal of Electroanalytical Chemistry and Interfacial Electrochemistry, vol. 260, no. 2, pp. 433–440, 1989.

47. G. Gruenig, K. Wiesener, S. Gamburzev, I. Iliev, and A. Kaisheva, "Investigations of catalysts from the pyrolyzates of cobalt-containing and metal-free dibenzotetraazaannulenes on active carbon for oxygen electrodes in an acid medium," Journal of Electroanalytical Chemistry, vol. 159, no. 1, pp. 155–162, 1983. ·

48. J. A. R. van Veen, J. F. van Baar, and K. J. Kroese, "Effect of heat treatment on the performance of carbon-supported transition-metal chelates in the electrochemical reduction of oxygen," Journal of the Chemical Society, Faraday Transactions 1, vol. 77, no. 11, pp. 2827–2843, 1981.

49. P. Vasudevan, Santosh, N. Mann, and S. Tyagi, "Transition metal complexes of porphyrins and phthalocyanines as electrocatalysts for dioxygen reduction," Transition Metal Chemistry, vol. 15, no. 2, pp. 81–90, 1990.

50. F. Jaouen, M. Lefèvre, J. P. Dodelet, and M. Cai, "Heat-treated Fe/N/C catalysts for O_2 electroreduction: are active sites hosted in micropores?" Journal of Physical Chemistry B, vol. 110, no. 11, pp. 5553–5558, 2006.

51. C. W. B. Bezerra, L. Zhang, K. Lee et al., "A review of Fe–N/C and Co–N/C catalysts for the oxygen reduction reaction," Electrochimica Acta, vol. 53, no. 15, pp. 4937–4951, 2008.

52. A. Garsuch, A. Bonakdarpour, G. Liu, R. Yang, and J. R. Dahn, "Time to move beyond transition metal–N–C catalysts for oxygen reduction," in Handbook of Fuel Cells, vol. 5, Springer, New York, NY, USA, 2009.

53. N. Ramaswamy, R. J. Allen, and S. Mukerjee, "Electrochemical kinetics and X-ray absorption spectroscopic investigations of oxygen reduction on chalcogen-modified ruthenium catalysts in alkaline

media," Journal of Physical Chemistry C, vol. 115, no. 25, pp. 12650–12664, 2011. ·

54. T. M. Arruda, B. Shyam, J. S. Lawton et al., "Fundamental aspects of spontaneous cathodic deposition of Ru onto Pt/C electrocatalysts and membranes under direct methanol fuel cell operating conditions: an in situ X-ray absorption spectroscopy and electron spin resonance study," Journal of Physical Chemistry C, vol. 114, no. 2, pp. 1028–1040, 2010.

55. T. M. Arruda, B. Shyam, J. M. Ziegelbauer, S. Mukerjee, and D. E. Ramaker, "Investigation into the competitive and site-specific nature of anion adsorption on Pt using in situ X-ray absorption spectroscopy," Journal of Physical Chemistry C, vol. 112, no. 46, pp. 18087–18097, 2008. · ·

56. A. L. Ankudinov, B. Ravel, J. J. Rehr, and S. D. Conradson, "Real-space multiple-scattering calculation and interpretation of x-ray-absorption near-edge structure," Physical Review B, vol. 58, no. 12, pp. 7565–7576, 1998. ·

57. V. S. Murthi, R. C. Urian, and S. Mukerjee, "Oxygen reduction kinetics in low and medium temperature acid environment: correlation of water activation and surface properties in supported Pt and Pt alloy electrocatalysts," Journal of Physical Chemistry B, vol. 108, no. 30, pp. 11011–11023, 2004.

58. M. R. Tarasevich, A. Sadkowski, and E. Yeager, "Kinetics and mechanisms of electrode processes," inComprehensive Treatise of Electrochemistry, B. E. Conway, J. O. M. Bockris, E. Yeager, S. U. M. Khanand, and R. E. White, Eds., vol. 7, pp. 301–398, Plenum Press, New York, NY, USA, 1983.

59. N. M. Marković, H. A. Gasteiger, and P. N. Ross, "Oxygen reduction on platinum low-index single-crystal surfaces in alkaline solution: rotating ring diskPt(hkl) studies," Journal of Physical Chemistry, vol. 100, no. 16, pp. 6715–6721, 1996. ·

60. N. Ramaswamy and S. Mukerjee, "Influence of inner- and outer-sphere electron transfer mechanisms during electrocatalysis of oxygen reduction in alkaline media," Journal of Physical Chemistry C, vol. 115, no. 36, pp. 18015–18026, 2011.

61. E. Janin, H. Von Schenck, M. Göthelid, U. O. Karlsson, and M. Svensson, "Bridge-bonded atomic oxygen on Pt(110)," Physical Review B, vol. 61, no. 19, pp. 13144–13149, 2000. ·

62. G. Jerkiewicz, G. Vatankhah, J. Lessard, M. P. Soriaga, and Y. S. Park, "Surface-oxide growth at platinum electrodes in aqueous H_2SO_4 Reexamination of its mechanism through combined cyclic-voltammetry, electrochemical quartz-crystal nanobalance, and Auger electron spectroscopy measurements,"Electrochimica Acta, vol. 49, no. 9-10, pp. 1451–1459, 2004. · ·

63. N. M. Markoví and P. N. Ross, "Surface science studies of model fuel cell electrocatalysts," Surface Science Reports, vol. 45, no. 4–6, pp. 117–229, 2002. ·

64. P. Chen and R. L. McCreery, "Control of electron transfer kinetics at glassy carbon electrodes by specific surface modification," Analytical Chemistry, vol. 68, no. 22, pp. 3958–3965, 1996. ·

65. P. Chen, M. A. Fryling, and R. L. McCreery, "Electron transfer kinetics at modified carbon electrode surfaces: the role of specific surface sites," Analytical Chemistry, vol. 67, no. 18, pp. 3115–3122, 1995. ·

66. N. Wakabayashi, F. Kitamura, T. Ohsaka, and K. Tokuda, "Effect of adsorbed anions on the outer-sphere electron-transfer reactions of cobalt complexes at platinum single-crystal electrodes," Journal of Electroanalytical Chemistry, vol. 499, no. 1, pp. 161–168, 2001.

67. T. Wandlowski and R. de Levie, "Double-layer dynamics in the adsorption of tetrabutylammonium ions at the mercury—water interface Part 4. The reduction of hexammine-cobalt(III) through tetrabutylammonium films," Journal of Electroanalytical Chemistry, vol. 380, no. 1-2, pp. 201–207, 1995. ·

68. N. A. Anastasijevic, Z. M. Dimitrijevic, and R. R. Adzic, "Oxygen reduction on a ruthenium electrode in alkaline electrolytes," Journal of Electroanalytical Chemistry and Interfacial Electrochemistry, vol. 199, no. 2, pp. 351–364, 1986.

69. N. A. Anastasijevic, Z. M. Dimitrijevic, and R. R. Adzic, "Oxygen reduction on a ruthenium electrode in acid electrolytes," Electrochimica Acta, vol. 31, no. 9, pp. 1125–1130, 1986.

70. N. Ramaswamy, R. J. Allen, and S. Mukerjee, "Electrochemical kinetics and X-ray absorption spectroscopic investigations of oxygen reduction on chalcogen-modified ruthenium catalysts in alkaline media," The Journal of Physical Chemistry C, vol. 115, no. 25, pp. 12650–12664, 2011.

71. R. Boulatov, "Billion-Year-Old Oxygen Cathode that Actually Works: Respiratory Oxygen Reduction and Its Biomimetic Analogs," in N4-Macrocyclic Metal Complexes, J. H. Zagal, F. Bedioui, and J.-P Dodelet, Eds., p. 1, Springer, Berlin, Germany, 2006.

72. R. Boulatov, "Metalloporphyrin Catalysts of Oxygen Reduction," in Fuel Cell Catalysis, M. T. M. Koper, Ed., p. 637, John Wiley & Sons, New York, NY, USA, 2008.

73. N. Ramaswamy and S. Mukerjee, "Electrocatalysis of oxygen reduction on nonprecious metallic centers at high pH environments," ECS Transactions, vol. 33, no. 1, pp. 1777–1785, 2010.

74. K. M. Kadish, E. van Caemelbecke, and R. Royal, "Electrochemistry of metalloporphyrins in nonaqueous media," in The Porphyrin Handbook, K. M. Kadish, K. M. Smith, and R. Guilard, Eds., vol. 8, chapter 55, pp. 1–97, Academic Press, San Diego, Calif, USA, 2000.

75. M. Teliska, V. S. Murthi, S. Mukerjee, and D. E. Ramaker, "Correlation of water activation, surface properties, and oxygen reduction reactivity of supported Pt-M/C bimetallic electrocatalysts using XAS,"Journal of the Electrochemical Society, vol. 152, no. 11, pp. A2159–A2169, 2005.

76. M. Teliska, W. E. O'Grady, and D. E. Ramaker, "Determination of O and OH adsorption sites and coverage in situ on Pt electrodes from Pt L23 X-ray absorption spectroscopy," Journal of Physical Chemistry B, vol. 109, no. 16, pp. 8076–8084, 2005.

77. E. A. Lewis, C. U. Segre, and E. S. Smotkin, "Embedded cluster Δ-XANES modeling of adsorption processes on Pt," Electrochimica Acta, vol. 54, no. 28, pp. 7181–7185, 2009.

78. F. Beck, "The redox mechanism of the chelate-catalysed oxygen cathode," Journal of Applied Electrochemistry, vol. 7, no. 3, pp. 239–245, 1977. ·

79. S. R. P. Silva, "Properties of Amorphous Carbon," in Properties of Amorphous Carbon, EMIS Datareviews Series 29, 2003.

80. Y. J. Cho, H. S. Kim, S. Y. Baik et al., "Selective nitrogen-doping structure of nanosize graphitic layers,"Journal of Physical Chemistry C, vol. 115, no. 9, pp. 3737–3744, 2011. ·

81. J. R. Hahn and H. Kang, "Vacancy and interstitial defects at graphite surfaces: scanning tunneling microscopic study of the structure, electronic property, and yield for ion-induced defect creation,"Physical Review B, vol. 60, no. 8, pp. 6007–6017, 1999. ·

82. S. L. Mielke, D. Troya, S. Zhang et al., "The role of vacancy defects and holes in the fracture of carbon nanotubes," Chemical Physics Letters, vol. 390, no. 4–6, pp. 413–420, 2004.

83. Y. Luo, Y. Heng, X. Dai, W. Chen, and J. Li, "Preparation and photocatalytic ability of highly defective carbon nanotubes," Journal of Solid State Chemistry, vol. 182, no. 9, pp. 2521–2525, 2009.

84. T. Kondo, T. Suzuki, and J. Nakamura, "Nitrogen doping of graphite for enhancement of durability of supported platinum clusters," Journal of Physical Chemistry Letters, vol. 2, no. 6, pp. 577–580, 2011. ·

85. P. Nemes-Incze, Z. Koínya, I. Kiricsi et al., "Mapping of functionalized regions on carbon nanotubes by scanning tunneling microscopy," Journal of Physical Chemistry C, vol. 115, no. 8, pp. 3229–3235, 2011. ·

86. A. B. Anderson and R. A. Sidik, "Oxygen Electroreduction on FeII and FeIII Coordinated to N4 Chelates. Reversible Potentials for the Intermediate Steps from Quantum Theory," Journal of Physical Chemistry B, vol. 108, no. 16, pp. 5031–5035, 2004. ·

CHAPTER 10

ELECTRO-CATALYSIS SYSTEM FOR BIODIESEL SYNTHESIS FROM PALM OIL OVER DIELECTRIC-BARRIER DISCHARGE PLASMA REACTOR

I. Istadi, Ardian D. Yudhistira, Didi D. Anggoro, Luqman Buchori

Keywords: Plasma reactor, electro-catalysis, biodiesel, transesterification, cracking, energetic electrons

INTRODUCTION

Recent worldwide energy policies and increased environmental concerns are driving industry to come up with an alternative fuels, such as biodiesel. Biodiesel is an alternative fuel consisting of fatty acid methyl esters (FAME) produced by transesterification of triglycerides, the main constituent of various kinds of vegetable oil. This renewable fuel can be used in any compression ignition engine without major modifications. Biodiesel fuel is defined as the alkyl esters of long chain fatty acids, which fulfills certain standards. The biodiesel chemical is obtained by the transesterification or alcoholysis of natural triacylglycerols or triglyceride with short-chain alcohols, usually methanol or ethanol. The triglyceride was obtained from various crude such as vegetable oils, animal fats, waste fats and greases, waste cooking oils or side-stream products of refined edible oil production.

The transesterification reaction requires the presence of a catalyst, which can be a base, an acid, or an enzyme to occur at an acceptable rate, although it can be conducted in noncatalytic but supercritical conditions [1,2]. The base and acid catalysts, most commonly used in biodiesel production, include both homogeneous and heterogeneous catalysts. Until now, homogeneous base catalysts (KOH or NaOH) have been well known used in industry, since the process has fast reaction rate at mild reaction conditions. Unfortunately, their utilization in vegetable oils transesterification has big problems including forming soaps as undesirable by-products in which large amounts of wastewater was resulted, corrosiveness, and needs longer time for separation process. Therefore, the use of heterogeneous catalysts allows a more environmentally benign process to be used for biodiesel production. The use of heterogeneous catalysts could enhance efficiency of reactor design, enable continuous process, and improve the economics of biodiesel production. The use of heterogeneous catalysts does not produce soaps through free fatty acid neutralization, which simplifies the separation and purification processes. Due to the potential interest of heterogeneous catalyst usage, researches on the transesterification reaction using the solid catalysts for biodiesel production has increased over the past decade. Heterogeneous catalysts could overcome the mentioned problems of homogeneous catalysts usage. In fact, solid acid catalysts have been used in many industrial processes because they contain a variety of acid sites with different strength of Brönsted or Lewis acidity, compared to the homogenous acid catalysts.

Many kinds of solid catalysts for biodiesel production have been investigated by previous researchers including base and acid solid catalysts. The solid base catalyst includes CaObased oxides [3,4], lithium-based [5], and zinc oxide-based catalysts [6-8]. Heterogeneous solid acid catalysts have been used as an alternative to basic and enzyme catalysts, since they can be used in both transesterification and esterification reactions [9]. However, the acid catalysts require high alcohol-to-oil molar ratios and catalyst concentrations to achieve satisfactory transesterification conversions. The solid acid catalysts

Citation: Istadi, I., Yudhistira, A.D., Anggoro, D.D., Buchori, L. (2014). Electro-Catalysis System for Biodiesel Synthesis from Palm Oil over Dielectric-Barrier Discharge Plasma Reactor. Bulletin of Chemical Reaction Engineering & Catalysis, 9 (2): 111-120. (doi:10.9767/bcrec.9.2.6090.111-120).

include sulphated zirconia [9-12]. However, the use of solid acid and basic catalysts still have some problems, especially longer duration of reaction, mass transfer resistance between liquid oil and solid catalyst, and separation between phases. Transesterification of vegetable oils to yield biodiesel can also be performed by microwave irradiation with acid, base, and heterogeneous catalysts [13,14], as well as ultrasonic-assisted catalysis [15,16]. Production of biodiesel by transesterification reaction under microwave irradiation is effective in accelerating the reaction rate. Microwave effect on the transesterification reaction includes enhancement of reaction by a microwave thermal effect. The microwave heating shows a promising technique for biodiesel production in a short reaction time with high product yield. In term of ultrasound technology in biodiesel processing, the use of sonochemistry is also applied in biodiesel production as the ultrasonic field is known to produce chemical and physical effects that arise from the collapse of cavitation bubbles. Low frequency sonication can be used to produce emulsions from immiscible liquids and since oil and alcohol are immiscible, this effect could be employed for biodiesel preparation. Production of biodiesel under ultrasonic processing possesses the following advantages: reduction in processing time, amount of alcohol and catalyst required, separation time and reaction temperature [15].

Another promising method is the use of plasma technology, whether plasma electrocatalysis or using integrated plasma assisted – catalyst [17-21] that had been applied for other synthesis. Over the past decade, non-thermal plasmas have been investigated as a way to catalyze the conversion processes of various hydrocarbon fuels into syngas. The nonthermal plasmas are electrically-driven ionized gases that are known to generate chemically active species, such as O and OH, which can act as chain-branching radicals that assist with the propagation of fuel conversion reactions and can lower the temperature required for a reaction to proceed. For this reason, some researchers believed that the non-thermal plasma has a catalytic effect assisting the fuel conversion processes. In plasma processes for the biodiesel production by methanolysis of vegetable oils, electro-chemical reaction is expected to split carbon-carbon bonds within the triglyceride reactants, unlike conventional processes that break weaker carbon-oxygen bonds. Carbon-to-carbon bonds are much stronger and require higher catalytic energy than available in conventional homogeneous and heterogeneous catalysis. The higher electro-chemical catalytic energy also enables addition reaction, attacking the unsaturated double bonds of triglycerides through oxygenation [17,22-24]. Oxygen molecules can be supplied by methanol during reaction process, which replace each carbon-carbon double bond to yield a saturated molecule that having better thermal storage stability. A high voltage power supply is required to store more high energetic electrons that help catalysing the methanolysis of triglyceride. However, the detailed mechanisms of the plasma catalysis effect are still unknown. Possible reaction mechanism during the plasma treatment is very important to be known in order to obtain the optimum operating conditions. Therefore, further study on high voltage plasma treatment of vegetable oil to produce biodiesel is importantly required. The aim of this paper is to study biodiesel synthesis reaction routes from palm oil using plasma electro-catalysis process over DielectricBarrier Discharge (DBD) plasma reactor. The study was focused on finding possible reaction mechanism route during plasma electrocatalysis process. The prediction was performed based on the changes of GC-MS and FT-IR analyses to the biodiesel products with respect to time length of plasma treatment.

MATERIALS AND METHODS

Materials

Commercial refine palm oil was purchased from local market (Tropicana ®) to be used as raw material of triglyceride, while methanol (Merck, 99.9%) were used as reactant.

Plasma Electro-Catalysis Process

Plasma electro-catalysis reaction of triglyceride with methanol was performed in a Dielectric Barrier Discharge plasma reactor. The DBD plasma reactor was composed of a high voltage electrode, a ground electrode, and a barrier of glass material. The glass barrier, like a dielectric, was placed adhere to ground electrode. The reactor was equipped with DC type high voltage power supply (support up to 15 kV), a high voltage probe (SEW PD-28), and multimeter (Sanwa) to measure the voltage during the reaction process. The high energetic electrons flowed from high voltage electrode to ground electrode through the glass barrier. The glass barrier was purposed to distribute the electrons flowed toward ground electrode.

The plasma electro-catalysis was conducted by filling the reactor with palm oil and methanol with methanol to oil mole ratio of 6:1. Thus, the reactor was heated up to 65 oC, which in turn a DC 10 kV high voltage was applied on the electrode. In this reactor, distance between both electrodes was 1.5 cm, while the ground electrode was connected to earth ground. Effect of discharge duration was studied by varying the reaction or contact time as 30, 120, and 300 seconds, while the methanol to oil mole ratio, voltage, and gap of electrodes were kept constant at 6:1, 10 kV, and 1.5 cm, respectively. Effect of high voltage was investigated by varying voltage as 6, 8, and 10 kV, while the methanol to oil mole ratio, reaction duration, and gap of electrodes were kept constant at 6:1, 120 seconds, and 1.5 cm, respectively. Effect of distance between electrode was investigated by varying the electrodes gap as 1.5, 2.5, and 3.5 cm, while the methanol to oil mole ratio, voltage, and reaction duration were kept constant at 6:1, 10 kV, and 120 seconds, respectively.

Biodiesel Product Characterization

In order to predict the possible reaction routes during plasma electro-catalysis process, the biodiesel product was analyzed by using GC-MS for identification of FAME components (ASTM E202), and FT-IR for identification of chemical structures (ASTM E168) analysis. Beside the components and chemical structure identification, other analysis of important biodiesel characteristics were also performed, they are: density (ASTM D1298), viscosity (ASTM D445), acid value (FBI-A01-03), cetane number (ASTM D613), and heating value (ASTM D1405).

RESULTS AND DISCUSSION

Prediction of Reaction Routes within Plasma Electro-Catalysis System

Possible reaction routes within plasma process is essentially different with other conventional catalytic reactions. Initially, the energetic electrons collide with the liquid bulk molecules resulting in excitation, dissociation, and/or ionization, and the formation of atoms and metastable compounds. When the electric field is high enough to lead breakdowns, a large number of micro-discharges are produced. Thus, these active atoms and metastable compounds subsequently collide with reactant molecules. According to this mechanism, various reactions may occur and mixture of compounds with various compositions may be consequently produced [20,21].

In this plasma electro-catalysis experiments, the reaction products were analyzed using GC-MS and was confirmed with FT-IR spectra. In fact, reaction between triglyceride and methanol with the presence of high energetic electrons from plasma system at mild duration (120 seconds) can produce various chemicals, i.e. fatty acid methyl ester, aldehydes, alkynes, alcohols, esters, and carboxylic acids as presented in Figure 1 and Table 1. From Figure 1 and Table 1, various fatty acid methyl esters were detected, i.e. pentadecanoic acid methyl ester ($C_{17}H_{34}O_2$); 9,12-octadecadienoic acid methyl ester ($C_{19}H_{34}O_2$); 9-octadecenoic acid methyl ester ($C_{19}H_{36}O_2$); pentadecanoic acid methyl ester ($C_{17}H_{34}O_2$); and 9-octadecenoic acid hexyl ester ($C_{24}H_{46}O_2$). Meanwhile, other chemicals were also resulted for 30 seconds reaction time by plasma electro-catalysis process, they were: 1- heptyne (alkyne), octadecanoic acid (carboxylic acid), 1-hexadecanol, and 9-octadecenal (aldehyde). At reaction time of 120 seconds, the following chemicals were also detected, i.e.: 1- heptyne (alkynes), isooctanol (alcohol), hexadecenoic acid (carboxylic acid), octadecanal and 9-octadecenal (aldehydes). Other chemicals, such as gases, were also produced as bubble within reactor and indicated by increasing reactor pressure.

The detected chemicals as revealed by the GC-MS were confirmed with FT-IR from chemical structure point of view as depicted in Bulletin of Chemical Reaction Engineering & Catalysis, 9 (2), 2014, 114 Figure 2 and Table 2. From Figure 2 and Table 2, the detected chemical bondings include: C-H, $-CH_3$, C=O, C-O, C-C, dan O-H. From the facts of IR spectra and confirmed with GC-MS, the following chemicals were predicted, i.e. fatty acid methyl esters, aldehydes, alkynes, alcohols, esters, and carboxylic acids. Fatty acid methyl esters were formed during reaction implied from the following IR spectra as listed in Table 2, i.e.: stretching mode of C-H alkane and C-H alkene, bending mode of C-H alkane, stretching mode of C=O, stretching mode of C-C and C=O. Alkyne compounds were indicated from stretching mode of C=O, stretching mode of C-H, stretching mode of C≡C, and bending mode of C-H. Alcohol compounds were suggested from detected C-H stretching, C-H bending, C-O stretching, C-C stretching, and OH stretching. In term of carboxylic acid possibility, stretching and bending modes of CH, stretching mode of C-O and C=O, and stretching mode of C-C and O-H indicate the presence of carboxylic acids.

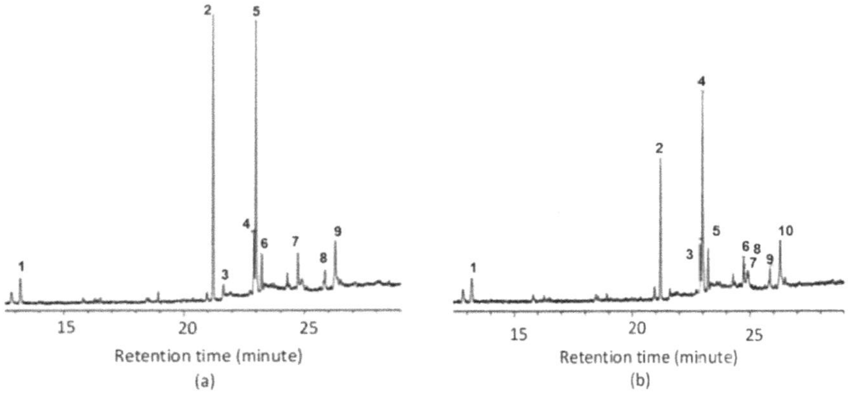

Figure 1. GC-MS Chromatogram of identified chemical components (mainly fatty acid methyl ester). (a) reaction time 30 seconds, (b) reaction time 120 seconds.

According to the GC-MS and FT-IR identifications as aforementioned, the possible reaction routes occurred in the plasma electrocatalysis system is non-thermal cracking or pyrolysis due to collision between high energetic electrons from plasma source and the reaction mixture. The high energetic electrons supplied from high voltage power supply are able to affect the electrons pair of covalent bondings to be excited or dissociated even ionized depending on their energy. The vibration of electron pairs due to high energetic electrons effects is dependent on chemical bonding energy (listed in Table 3) [24,25] as compared to energy brought by the energetic electrons. Therefore, the plasma electro-catalysis could able to break the chemical bonding to produce the various chemicals. Types and composition of resulted compounds are dependent on the power of plasma or energetic electrons and reactor configuration/design. Another possible reaction mechanism is transesterification reaction which is catalyzed by the energetic electrons. However, with respect to resulted compounds, non-thermal cracking mechanism is dominant rather than transesterification mechanism. The resulted compound groups, such as fatty acid methyl esters, aldehydes, alkynes, alcohols, carboxylic acids, and esters, were formed due to breaking and reforming of chemical bonding during the reaction time. Various chemicals were formed during the plasma electro-catalysis system at 30 and 120 seconds contact time which may be due to continuing chain chemical reactions. The drawbacks of this plasma electro-catalysis are how to control the reaction mechanism, how to drive which bonds will be excited or ionized, and when to stop the plasma electro-catalysis process. In order to avoid the chain reactions, the plasma electro-catalysis should be terminated at very short time of reaction, but the electrons should be having high energy to do the expected reaction mechanism. The higher energetic electrons could be produced by applying higher voltage by power supply and low current.

Table 1. Identification and composition of GC-MS chromatogram at respected reaction time

Reaction Time (s)	Peak	Identified Chemical Components		Composition (wt%)
		Chemical Formula	Chemical Compound	
30	1	C_7H_{12}	1-Heptyne	4.91
	2	$C_{17}H_{34}O_2$	Pentadecanoic Acid Methyl ester	30.60
	3	$C_{18}H_{36}O_2$	Octadecanoic Acid	2.15
	4	$C_{19}H_{34}O_2$	9,12-Octadecadienoic Acid Methyl ester	7.13
	5	$C_{19}H_{36}O_2$	9-Octadecenoic Acid Methyl ester	33.12
	6	$C_{17}H_{34}O_2$	Pentadecanoic Acid Methyl ester	4.15
	7	$C_{16}H_{34}O$	1-Hexadecanol	4.66
	8	$C_{24}H_{46}O_2$	9-Octadecenoic Acid Hexyl ester	3.48
	9	$C_{18}H_{34}O$	9-Octadecenal	9.80
			Total	100.00
120	1	$C_{12}H_{12}$	1-Heptyne	7.17
	2	$C_{17}H_{34}O_2$	Pentadecanoic Acid Methyl ester	22.00
	3	$C_{19}H_{34}O_2$	9,12-Octadecadienoic Acid Methyl ester	8.06
	4	$C_{19}H_{36}O_2$	11-Octadecenoic Acid Methyl ester	36.43
	5	$C_{17}H_{34}O_2$	Pentadecanoic Acid Methyl ester	7.05
	6	$C_{16}H_{18}O$	Isooctanol	4.91
	7	$C_{10}H_{18}O_2$	7-Nonenoic Acid Methyl ester	2.11
	8	$C_{16}H_{30}O_2$	9-Hexadecenoic Acid	2.53
	9	$C_{16}H_{36}O$	Octadecanal	2.22
	10	$C_{18}H_{34}O$	9-Octadecenal	751
			Total	100.00

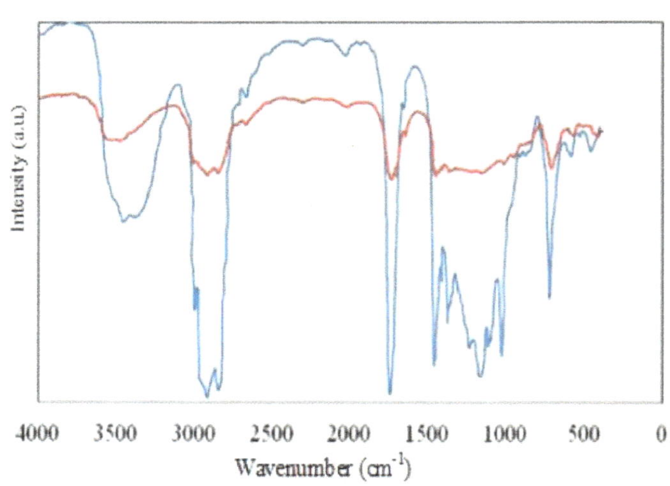

Figure 2. FT-IR spectra of plasma electrocatalysis products

Table 2. Identification of vibrational mode of atomic bonding in reaction products

No	Detected peaks at wavenumber (cm^{-1})		Vibrational mode of IR detection	Type of possible chemical compounds
	30 seconds Reaction Time	120 seconds Reaction Time		
1.	2853, 2922	2853, 2922	C-H Alkane (Stretching)	Ester, FAME, alkane, alkyne, alcohol, aldehyde, carboxylic Acid
2.	1377, 1418	1415, 1438	C-H Alkane (Bending)	Ester, FAME, alkane, alkyne, alcohol, aldehyde, carboxylic acid
3.	3005	3005	C-H Alkane (Stretching)	Ester, FAME, alkane, alkyne, alcohol, aldehyde, carboxylic acid
4.	2166	2160	C≡C (Stretching)	Alkyne
5.	1748	1740	C=O (Stretching)	FAME, ester, carboxylic acid, aldehyde
6.	1032, 1098, 1117, 1165, 1238	1032, 1119, 1171, 1233	C-O (Stretching)	Alcohol, ester, FAME, carboxylic acid
7.	3381, 3466	3472	O-H (Stretching)	Alcohol, carboxylic acid
8.	1456	1455	C-C (Stretching)	Ester, FAME, alkane, alkyne, alcohol, aldehyde, carboxylic acid

Table 3. Chemical bondings energy (23,24)

No.	Bonding types	Chemical bonding energy (eV)
1	C-C	3.6
2	C-O	3.7
3	C-H	4.3
4	H-H	4.5
5	O-CO	4.8
6	C=C	6.3
7	C=O	7.7
8	C≡C	8.7

As mentioned before, the possible dominant reaction routes in this plasma electro-catalysis system was non-thermal cracking or pyrolysis. The possible reaction mechanism in this system was predominantly controlled by higher electrons energy from high voltage electrode than chemical bonding within reaction mixture. The energetic electrons from high voltage electrode have internal energy about 1-10 eV [24] leading to the chemical bonding energy. This electrons energy should be higher than chemical bonding energy to make sure that the covalent bonding to be cracked. Unfortunately, which covalent bonding would be cracked cannot be controlled easily, like in conventional catalysis system. Atomic covalent bonding with lower bonding energy is easier to be cracked than the higher bonding energy.

The following are predicted reactions route and mechanism during the plasma electrocatalysis system of palm oil and methanol:

1. Methanol was collide with high energetic electrons to form excited and dissociated groups or atoms (CH_3^*, OH^*, H^*, CH_3O^*) as depicted in Equation (1).

$$CH_3\text{-}OH + e \rightarrow CH_3^* + OH^* + H^* + CH_3O^* \quad (1)$$

2. High energetic electrons cut the C-C bonding (3.3 eV) within triglyceride which in turn react with H* from methanol to form methyl ester as described in Equations (2) and (3).
3. Due to long duration of electrons collision, the fatty acid methyl ester may undergo continuing cracking of C-C and C-O bonding leading to shorter C-C bonding. In this case, formed FAME may undergo continuing cracking or pyrolysis due to too long duration of electrons collision leading to other detected chemicals such as aldehydes, alcohols, esters, and carboxylic acids. The continuing electrons cracking may be caused by higher energy content of electrons and the avalanche electrons. From this phenomena, control of contact time between high energetic electrons and the reactants is very important to favor which chemicals would be produced.
4. The predicted aldehydes formation was implied from scission of C-O bonding in alkyl ester groups and reacted with H^* from methanol to form aldehydes compounds as described in Equations (4) and (5). This reaction mechanism prediction is also confirmed by detected long-chain aldehydes. Therefore, formation of aldehyde compounds is due to scission of C-O in methyl ester.
5. Alkyne compounds (1-heptyne) may be formed from C-C scission in end section of methyl ester chain and reacts with CH^* from methanol as predicted in Equations (6) and (7).
6. Alcohols (1-hexadecanol) may be formed due to continuing reaction of Equation (6) with OH^* radical as predicted in Equation (8).
7. Some esters (9-octadecenoic acid hexyl ester) are detected due to continuing reaction between result of scission of C-O with product of Equation (6) as predicted in Equations (9) and (10).
8. Carboxylic acid (octadecanoic acid) may be formed in the plasma electro-catalysis system due to reaction between result of Equation (9) with H^* from methanol. This route was predicted in Equations (11) and (12). The acid was also formed due to reaction between result of Equation (4) with OH^* from methanol.

Potential of Plasma Electro-Catalysis System for Biodiesel Production

Various plasma sources are currently used for plasma-assisted reactions. Among those, DBD was exclusively used because of its several benefits. The DBD plasma reactor has a

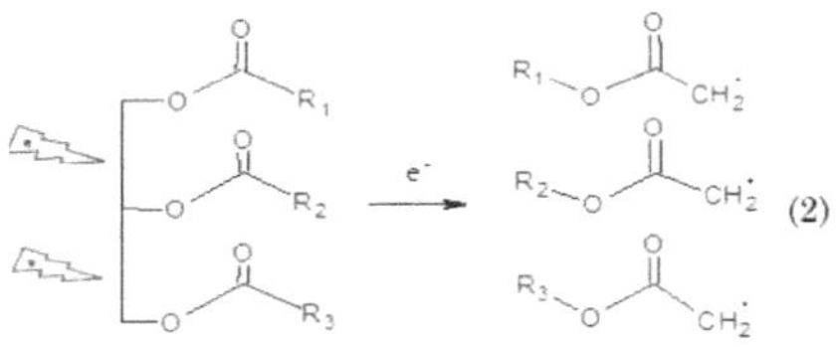

(2)

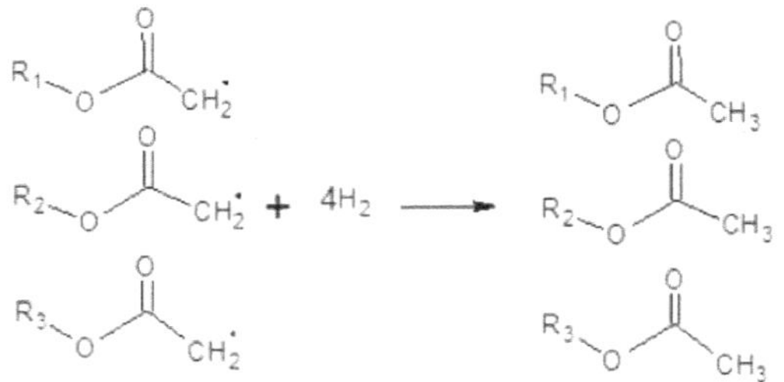

(3)

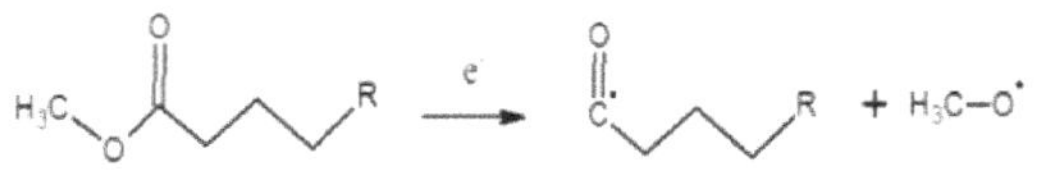

(4)

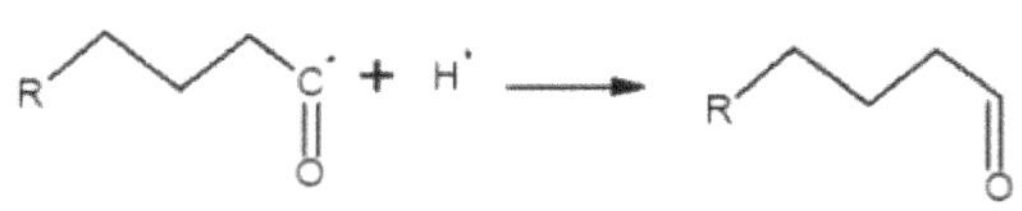

(5)

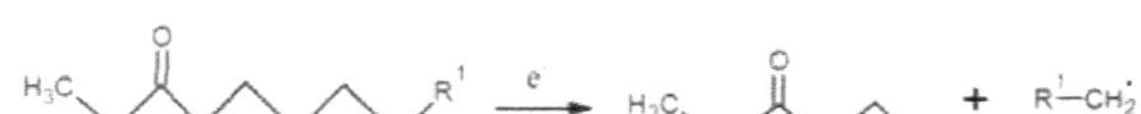

(6)

(7)

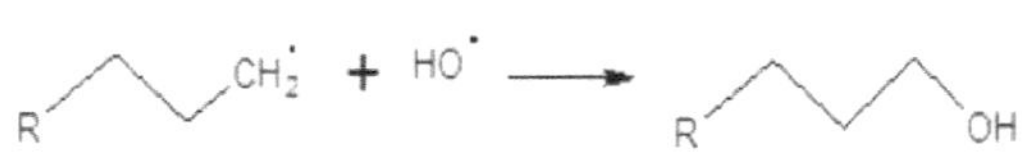

(8)

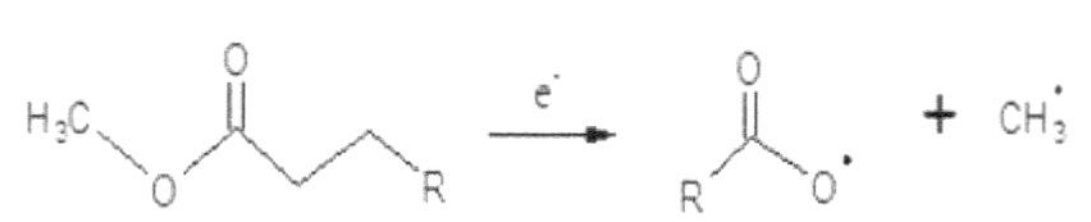

(9)

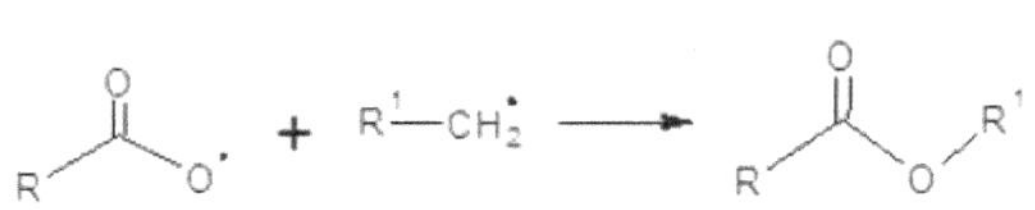

(10)

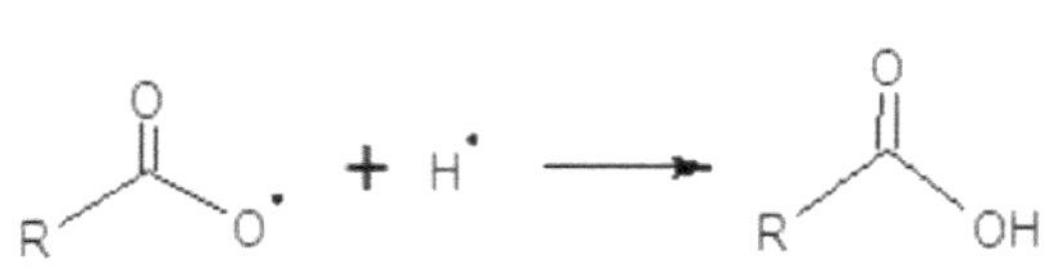

(11)

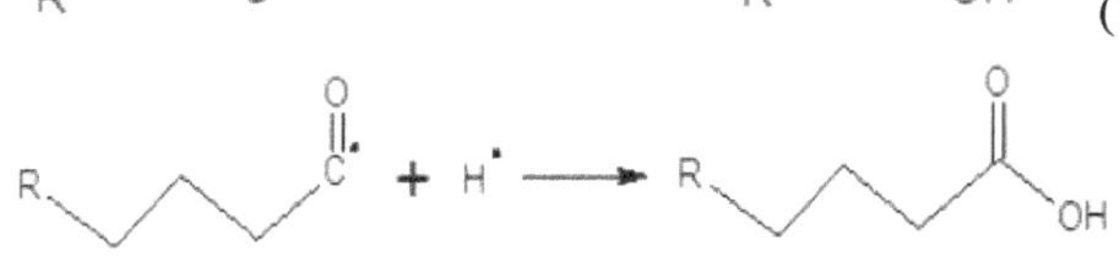

(12)

long and proven history originating from ozone synthesis [21]. Its fundamental of physical and chemical properties were well understood. This fact indicates that DBD is scalable technology to treat large amounts of gaseous materials. The DBD only slightly transits to unstable spark or arc plasma because of the presence of a dielectric barrier between the gaps. Great flexibility also exists in terms of electrode and reactor configurations. Therefore, the combination of DBD plasma reactor and catalysis is promising in term of plasmaassisted catalysis [17,18,20]. Moreover, the reactor can be constructed using inexpensive materials, such as glass and polymers. In terms of gas processing, partially excited, dissociated, and ionized gas, in which the gas temperature remains near room temperature, is readily created using simple electrode configuration and inexpensive power sources. High-energetic electrons produced by high voltage electric fields with low current collide against neutral molecules, producing ions, radicals, and electrically/vibrationally excited species almost independently of the reaction temperature [24]. These species are responsible for triggering fuel conversion at low temperatures.

The electro-catalysis phenomena within methanolysis reaction of triglyceride over DBD plasma reactor was investigated in this paper. The energetic electrons supplied by DC type - high voltage power supply collide with bulk liquid of mixed triglyceride - methanol to produce other valuable chemicals. In this nonthermal system, the bulk liquid only heated up to about 65 oC which is very far from electrons temperature of about 104 K. Consequently, this electro-chemical reaction is expected to split carbon-carbon bonds within the triglyceride reactants, unlike conventional processes that break weaker carbon-oxygen bonds. In fact, carbon-to-carbon bonds are much stronger and require higher energy to cut than that in conventional homogeneous and heterogeneous catalysis. The strong electro-chemical catalytic energy also enables and addition reaction, attacking the unsaturated double bonds of triglycerides through oxygenation. Oxygen molecules can be supplied by methanol during the reaction process, which replace each carbon -carbon double bond to yield a saturated molecule that having better thermal storage stability.

The plasma electro-catalysis system is promising for biodiesel synthesis from vegetable oils due to very short time reaction (main feature of plasma technology), even no catalyst required, no soap formation, and no glycerol by product. From the experimental works, it could produce fatty acid methyl ester yield of 75.65 % at 120 seconds and there was no glycerol and soap resulted. However, control of reaction mechanism due to the high energetic electrons within the plasma was very difficult to be done. The important think is how to avoid continuing reactions due the action of high energetic electrons. The recommendation for operating the plasma reactor is short duration contact between the energetic electrons and the reaction mixture, but the voltage should be very high at lower current. This suggestion is purposed to avoid continuing sustained chain reactions by applying very high energetic electrons at very short contact time.

Characterization of Biodiesel Product

Resulted biodiesel, mainly composed of fatty acid methyl esters as well as other chemical formed, was characterized in terms of density, kinematic viscosity, acid number, cetane number, iod value, saponification number, and calorific value as listed in Table 4. All parameters fulfil the requirement of biodiesel as Indonesian National Standard (Standar Nasional Indonesia - SNI-04-7182-2006) except for kinematic viscosity and density. The high viscosity of biodiesel product is due to formation of 1-hexadecanol, 9-octadecenal, and octadecanal that having high melting point. The unexcpected compounds are soluble in the biodiesel product that cause high viscosity and density. It is suggested that continuing reaction during plasma electro-catalysis should be avoided. The recommendation for applying the plasma electro-catalysis reactor is shorter duration of contact between the energetic electrons and the reactions mixtures at very high voltage and lower current.

Table 4. Characterization of biodiesel product and its comparison with Indonesian National Standard (SNI-04-7182-2006)

Parameters	Experimentals	SNI Standard
Density (g/ml)	0.90	0.85 – 0.89
Kinematic Viscosity (cSt)	20	2.3 – 6.0
Acid Number (-)	0.671	max 0.8
Cetane Number (-)	51.73	min 51
Iod Value (-)	97.65	max 115
Saponification Number (-)	199.44	261.26
Calorific Value (cal/gr)	9105	-

CONCLUSION

The plasma electro-catalysis system was promising for biodiesel synthesis from vegetable oils due to only very short time reaction, even no catalyst required, no soap formation, and no glycerol by product. From the plasma electro-catalysis experiments, it can produce fatty acid methyl ester yield of 75.65 % at short reaction time of 120 seconds and there was no gliserol and no soap resulted. However, other chemicals may be resulted such as alkanes, alkynes, aldehydes, esters, and carboxylic acids. Control of reaction mechanism in plasma electro-catalysis was very important due to the high energetic electrons within plasma. The important idea was how to avoid continuing reactions due the action of high energetic electrons. Short duration contact between the energetic electrons and the liquid reactants was recommended, but the plasma electro-catalysis reactor should be operated at very high voltage and lower current. The purpose of the suggestion was to avoid continuing chain reactions by applying very high energetic electrons at very short contact time.

Acknowledgments

The authors would like to express their sincere gratitude to the Directorate General of Higher Education (DIKTI) for the financial support received under the project of Hibah Kompetensi Year 2012-2014 by Contract No.: 314b-5/UN7.5/PG/2013.

REFERENCES

1. Macario, A., Giordano, G., Onida, B., Cocina, D., Tagarelli, A., Giuffrè, A.M. (2010). Biodiesel production process by homogeneous/heterogeneous catalytic system using an acid–base catalyst. Appl. Catal. A Gen. 378(2): 160-168.
2. Yusuf, N.N., Kamarudin, S.K., Yaakub, Z. (2011). Overview on the current trends in biodiesel production. Energy Convers. Manag. 52(7): 2741-2751.
3. Kouzu, M., Hidaka, J. (2012). Transesterification of vegetable oil into biodiesel catalyzed by CaO: A review. Fuel. 93: 1-12.
4. Taufiq-Yap, Y.H., Lee, H.V., Hussein, M.Z., Yunus, R. (2011). Calcium-based mixed oxide catalysts for methanolysis of Jatropha curcas oil to biodiesel. Biomass Bioenergy 35(2): 827- 834.
5. Istadi, I., Pramudono, B., Suherman, S., Priyanto, S. (2010). Potential of LiNO3/Al2O3 C a t a l y s t f o r H e t e r o g e n e o u s Transesterification of Palm Oil to Biodiesel. Bull. Chem. React. Eng. Catal., 5(1): 51-56
6. Islam, A., Taufiq-Yap, Y.H., Chu, C.M., Chan, E.S., Ravindra, P. (2013). Studies on design of heterogeneous catalysts for biodiesel production. Process Saf. Environ. Prot. 91(1- 2): 131-144.
7. Kesić, Ž., Lukić, I., Brkić, D., Rogan, J., Zdujić, M., Liu, H., et al. (2012). Mechanochemical preparation and characterization of CaO·ZnO used as catalyst for biodiesel synthesis. Appl. Catal. A Gen. 427-428: 58-65.
8. Lukić, I., Kesić, Ž., Maksimović, S., Zdujić, M., Liu, H., Krstić, J., et al. (2013). Kinetics of sunflower and used vegetable oil methanolysis catalyzed by CaO·ZnO. Fuel. 113: 367-378.
9. Nandagopal, J., Subramanian, S. (2010). Synthesis of Biodiesel from Neem Oil using Sulfated Zirconia via Transesterification. Brazilian J. Chem. Eng. 27(04): 601-608.
10. Lam, M.K., Lee, K.T. (2010). Accelerating transesterification reaction with biodiesel as co-solvent: A case study for solid acid sulfated tin oxide catalyst. Fuel. 89(12): 3866-3870.
11. Rattanaphra, D., Harvey, A.P., Thanapimmetha, A., Srinophakun, P. (2012). Simultaneous transesterification and esterification for biodiesel production with and without a sulphated zirconia catalyst. Fuel. 97: 467-475.
12. Reddy, B.M., Sreekanth, P.M., Lakshmanan, P. (2005). Sulfated zirconia as an efficient catalyst for organic synthesis and transformation reactions. J. Mol. Catal. A: Chem. 237: 93-100.
13. Azcan, N., Yilmaz, O. (2012). Microwave Irradiation Application in Biodiesel Production from Promising Biodiesel Feedsto ck : Microalgae (Chlorella

prototthecoides). Proc. World Congr. Eng. Comput. Sci. San Fransisco, USA.

14. Gude, V., Patil, P., Martinez-Guerra, E., Deng, S., Nirmalakhandan, N. (2013). Microwave energy potential for biodiesel production. Sustain. Chem. Process. 1(5): 1- 31.
15. Babajide, O., Petrik, L., Amigun, B., Ameer, F. (2010). Low-Cost Feedstock Conversion to Biodiesel via Ultrasound Technology. Energies. 3(10): 1691-1703.
16. Veljković, V.B., Avramović, J.M., Stamenković, O.S. (2012). Biodiesel production by ultrasound-assisted transesterification: State of the art and the perspectives. Renew. Sustain. Energy Rev. 16(2): 1193-1209.
17. Lee, D.H., Kim, T. (2013). Plasma-catalyst hybrid methanol-steam reforming for hydrogen production. Int. J. Hydrogen Energy. 38(14): 6039-6043.
18. Gallagher, M.J., Geiger, R., Polevich, A., Rabinovich, A., Gutsol, A., Fridman, A. (2010). On-board plasma-assisted conversion of heavy hydrocarbons into synthesis gas. Fuel. 89(6): 1187-1192.
19. Rafiq, M.H., Hustad, J.E. (2011). Biosyngas production by autothermal reforming of waste cooking oil with propane using a plasmaassisted gliding arc reactor. Int. J. Hydrogen Energy. 36(14): 8221-8233.
20. Nozaki, T., Okazaki, K. (2013). Non-thermal plasma catalysis of methane: Principles, energy efficiency, and applications. Catal. Today. 211: 29-38.
21. Istadi, I., Saidina Amin, N.A. (2006). Cogeneration of synthesis gas and C2+ hydrocarbons from methane and carbon dioxide in a hybrid catalytic-plasma reactor: A review. Fuel. 85(5-6): 577-592.
22. Motasemi, F., Afzal, M.T. (2013). A review on the microwave-assisted pyrolysis technique. Renew. Sustain. Energy Rev. 28: 317-330.
23. Rahimpour, M.R., Jahanmiri, A., Mohamadzadeh Shirazi, M., Hooshmand, N., Taghvaei, H. (2013). Combination of nonthermal plasma and heterogeneous catalysis for methane and hexadecane co-cracking: Effect of voltage and catalyst configuration. Chem. Eng. J. 219: 245-253.
24. Fridman, A. (2008). Plasma Chemistry. Cambridge: Cambridge University Press; p. 978.
25. Sanderson, R.T. (1983). Polar Covalence. London: Academic Press; p. 240.

CHAPTER 11

A REVIEW ON DIRECT ELECTROCHEMISTRY OF CATALASE FOR ELECTROCHEMICAL SENSORS

Periasamy Arun Prakash, Umasankar Yogeswaran and Shen-Ming Chen

Keywords: Direct electrochemistry, catalase, electrocatalysis, hydrogen peroxide, electrochemical sensors, nanomaterials

INTRODUCTION

The study of the direct electron transfer pathways of redox proteins or enzymes is very significant in understanding the redox proteins, as well as in development of enzyme biosensors, biofuel cells and biomedical devices [1]. An in-depth study of the direct electrochemistry (DET) of heme enzymes provides satisfactory information regarding these bioelectrocatalysts (enzymes) and their major roles in electrocatalysis [2–6]. In the past decades, numerous works have been done to study the DET of various redox proteins and enzymes at modified electrodes. In 1977, a new gateway for protein direct electrochemistry was opened, when two groups, Yeh *et al.* and Eddowes *et al.* [7,8], independently reported the phenomenon of reversible electron transfer at a cytochrome c (cyt c) modified electrode. Since then, a large number of proteins and enzymes were successfully immobilized on a variety of biological and chemical matrices. Recently, Wolenberger *et al.* have reviewed direct protein electrochemistry using isolated enzymes and enzyme-protein couples [9]. The studies of redox proteins' direct electrochemistry are thus important in understanding the mechanism of electron transfer between immobilized proteins and the electrodes.

On the other hand, an electrochemical sensor is a device in which a biochemical recognition process is coupled to an appropriate electrochemical transducer where one of the electrochemical transducer components is the electrode. To enhance the efficiency of the transducer, the electrode surfaces have been modified either using redox active compounds or bio-components like enzymes and proteins. Among enzymes, nearly 3,000 wild type enzymes are known and 1,060 of them are oxidoreductases. The oxidoreductase enzymes family are very important since they catalyze H_2O_2 reduction within the cells. During this reduction process, H_2O_2 is reduced to H_2O and molecular O_2 without forming free radicals. This reaction is mainly catalyzed by catalase (CAT) present in the red blood cells of mammals. CAT is a heme-containing redox protein with ferritoprotoporphyrin IX at its redox centre which belongs to the oxidoreductase family class [10]. It possesses a heme prosthetic group at its active site with metallic iron (Fe^{III}). The catalytic ability of CAT to reduce H_2O_2 has been applied in the development of H_2O_2 sensors [11]. In order to investigate this catalytic ability of CAT in detail, the direct electrochemistry of CAT at the transducer surface has to be examined, but earlier studies showed that immobilization of CAT directly on the bare electrode surface lead to poor electron transfer [12]. This may be due to the fact that the $Fe^{(III/II)}$group gets deeply buried inside the huge structure of CAT. In order to achieve better electron transfer, CAT has been immobilized on various modified electrode surfaces. The electrode surfaces were modified either with surfactants, biopolymers, organic polymers, hydrogels, sol-gels or dendrimers, respectively [15–25]. Though noticeable electron transfer was observed with a few modified electrodes, the majority of them fail to promote the electron transfer process. This might be due to the diminished contact between CAT and such modified matrices. To overcome these problems, the bioelectric contact

Citation: Periasamy Arun Prakash, Umasankar Yogeswaran and Shen-Ming Chen, A Review on Direct Electrochemistry of Catalase for Electrochemical Sensors, doi:10.3390/s90301821.

between CAT and the bare electrode was improved with the implementation of nanomaterials [29]. It is a well known fact that, with large surface area and highly porous network, nanomaterials are well suited for the entrapment of biological molecules [49–51]. Further, proteins or enzymes can be easily immobilized at these nanomaterials without any damage. In recent years, many researchers have attempted to immobilize CAT on various nanomaterial modified electrode surfaces. Their results ultimately illustrate that CAT immobilized on nanomaterial matrices possesses enhanced electron transfer and excellent catalytic activity. However, a survey of the literature shows that no one has presented a review on the direct electrochemistry of CAT and the influence of nanomaterials on the direct electron transfer of CAT. Consequently, in the present review we have discussed the direct electrochemistry of CAT on various modified electrode surfaces, both in the presence and absence of nanomaterials. This review mainly highlights the various electrode fabrication methods, their electrochemical characterizations, miscellaneous enzyme immobilization strategies used by researchers in the present and past decades, along with the benefits of using nanomaterials as enzyme immobilization matrices.

NANOMATERIAL FREE MATRICES FOR CAT IMMOBILIZATION - VERSATILE APPROACHES

Polyelectrolyte Encapsulated CAT

Earlier attempts were made by researchers to immobilize CAT onto various nanomaterial free matrices. Caruso and co-workers carried out a simple yet versatile approach to encapsulate CAT crystals within polymer multilayer capsules via the stepwise regular assembly of oppositely charged polyelectrolytes, poly(styrenesulfonate) (PSS) (negatively charged) and poly(allylamine hydrochloride) (PAH) (positively charged), using biocrystals as templates [13]. The schematic representation of the process of enzyme encapsulation into the polymer films is shown in Figure 1. They examined the activity and stability of CAT present inside the polymer multilayer films through proteolysis. They observed that the polymer encapsulated CAT retained 100% of its activity, even after 100 min incubation with protease, whereas polymer uncoated CAT loses more than 90% of its initial activity within 100 min under the same conditions.

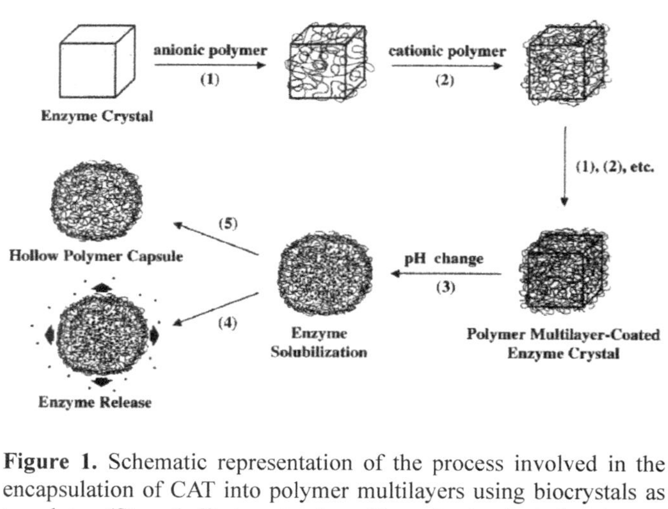

Figure 1. Schematic representation of the process involved in the encapsulation of CAT into polymer multilayers using biocrystals as templates. (Steps 1, 2) stepwise deposition of poly electrolyte layers; (Step 3) Exposure of the enzyme to solutions of pH > 6 or acidic solution (pH < 4), which results in the morphology change of the polymer capsule accompanied by the solubilisation of the enzyme; (Step 4) Exposure of the enzyme to solutions of pH > 11 leads to the release of the enzyme by rupturing of the polymer capsule; (Step 5) Exposure of the encapsulated enzyme to an oxidizing solution results in the decomposition of the enzyme and thus enzyme expelled from the interior of the capsule through the polymer walls, leaving behind the hollow polymer capsules (reproduced with permission from Caruso *et al. Langmuir* **2000**, *16*, 1485–1488).

Yu *et al.* employed a similar approach to encapsulate CAT into polyelectrolytes, PAH and PSS respectively [14]. They investigated the direct electron transfer behavior and electrocatalytic response of polyelectrolyte encapsulated CAT to H_2O_2reduction. The polyelectrolyte encapsulated CAT displays 4.5 times higher electrocatalytic response to H_2O_2 than non-encapsulated and solubilized CAT. However, the relative increase in catalytic response at the former was not proportional to the amount of enzyme deposited, since the entire CAT molecules were not found in their electrocatalytic active site. This might be due to the reason that the distance between the CAT molecules and the electrode surface gets significantly increased after polyelectrolyte encapsulation. Furthermore, they observed through their detailed studies that, with increases in precursor polyelectrolyte layer number, the current response to H_2O_2 decreased significantly, while the electrode response time increased. In contrast, implementation of fewer polyelectrolyte layers lead to sensitive and fast response in H_2O_2 determinations. This approach of encapsulating CAT micro crystals into polyelectrolyte layers for biosensing provides a versatile approach to prepare high enzyme content films with improved enzyme activities.

Surfactant Modified Matrices for CAT Immobilization

In an alternative approach, Chen *et al.* attempted to immobilize CAT onto cationic surfactant, (didodecyldimethylammonium bromide, DDAB) liquid crystal films [15]. They revealed through their results that the electron transfer of CAT is greatly improved in the presence of DDAB. The electron transfer rate constant (k_s) was observed to be 3.0 ± 0.4 s^{-1}. However, their circular dichroism (CD) results illustrated that the conformation of CAT gets slightly perturbed by the hydrophobic nature of DDAB (Figure 2). The CD spectrum of CAT exhibits a negative peak at 237 nm, whereas for CAT-DDAB film it was observed at 232 nm. The negative peak at 232 nm was regarded as the shift from 237 nm of CAT film alone.

Gebica *et al.* also studied the interaction of the surfactant sodium bis(2-ethylhexyl) sulfosuccinate (AOT) with CAT and horseradish peroxidase (HRP) [16]. They studied comparatively the properties of CAT and HRP entrapped into AOT/*n*-heptane with those in aqueous solutions of AOT. Their detailed studies illustrate that the secondary structure of CAT changed significantly in the presence of aqueous micellar solutions of AOT and the activity of CAT was completely lost. Interestingly, the secondary structure of CAT remains unchanged in the AOT/*n*-heptane reverse micelles.

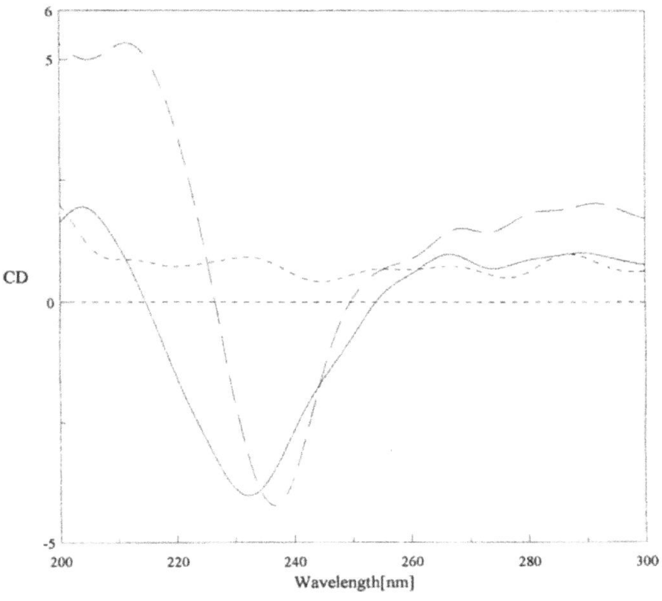

Figure 2. CD spectra of CAT (— —), DDAB (– – –), CAT-DDAB (—) films. CAT was prepared in pH 6.1, 10 mM phosphate buffer solution (PBS) with 50 mmol: l KCl. Temperature: 25°C (reproduced with permission from Chen *et al. Biosens. Bioelectron.* **2001**, *16*, 115–120).

Immobilization of CAT at Various Polymers and Gel Matrices

Other than surfactants, CAT has been immobilized in various matrices like biopolymers, organic polymers, hydrogels and sol-gels, respectively. Huang and co-workers attempted to immobilize heme proteins like myoglobin (Mb), hemoglobin (Hb), HRP and CAT into chitosan (CS) film modified pyrolytic graphite (PG) electrodes [17]. They obtained stable voltammogramms for MB, Hb and HRP films, whereas less stable voltammogramms were obtained in the case of CAT-CS films. Furthermore, compared to other protein-CS films, the electron transfer process also happens at a much slower rate in CAT-CS films and the stability was relatively very poor. Lu *et al.* have successfully immobilized CAT on polyacrylammide (PAM) hydrogel film-coated PG electrodes [18]. They investigated the direct voltammetric and electrocatalytic behavior of these CAT incorporated PAM films. They observed that the electron transfer of CAT was greatly facilitated in the microenvironment of PAM films. However, the electron transfer of these CAT-PAM films was much smaller and relatively less stable than CAT-CS films described above. Through UV–vis spectroscopy, they observed the position change of the Soret band for CAT-PAM films (Figure 3). They revealed through their results that CAT in dry PAM films retains a secondary structure that is the same as the native state of CAT in dry CAT films. Further, CAT in PAM films essentially retains its native state in buffers at medium pH. When the pH was more acidic or more basic, the Soret band showed a blue-shift with some distortion. As shown in Figure 3(f), substantial denaturation of CAT in PAM films takes place at pH 3.0. This illustrates the poor stability of CAT-PAM films at acidic pH values. Nevertheless, CAT-PAM shows good catalytic activity towards H_2O_2 reduction in the concentration range 0.4 – 0.8 mM (See Table 1).

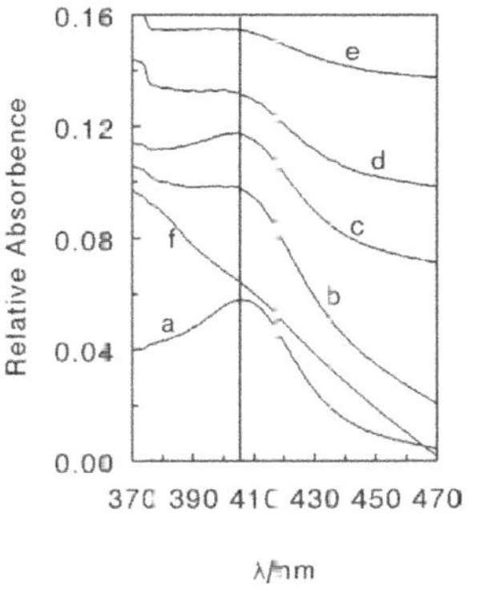

Figure 3. UV–vis absorption spectra of (a) dry CAT film, (b) dry CAT-PAM film, and CAT-PAM films in different pH buffer solutions: (c) pH 7.0; (d) pH 5.0; (e) pH 9.0 and (f) pH 3.0 (reproduced with permission from Lu *et al. Biophys. Chem.* **2003**, *104*, 623–632).

Li *et al.* investigated the direct voltammetric behavior of heme proteins CAT and HRP immobilized on methyl cellulose (MC) modified edge-plane pyrolytic graphite electrodes [19]. Both CAT-MC and HRP-MC films displayed a good response to H_2O_2 reduction in the concentration range of 0.02 to 0.12 mM and 0.02 to 0.11 mM, respectively (See Table 1). However, the electron transfer rate observed on CAT-MC films was relatively lower than that of CAT-CS and CAT-PAM films. Li *et al.* also successfully immobilized CAT and Hb onto collagen modified PG electrodes [20]. Though they noticed direct electron transfer in these protein-collagen films, the redox peak potential of CAT-collagen films was found to be more negative than those of Hb-collagen films. This may be likely due to the reason that the larger polypeptide chains of CAT needs more active energy than Hb to electrochemically reduce at the PG electrodes. Moreover, Hb-collagen films were found to be more stable than CAT-collagen films. The reason for the greater stability of Hb-collagen films may be due to the electrostatic interactions between the oppositely charged groups present in collagen and Hb films. In addition, the hydrophobic and hydrogen-bond interactions between the collagen and Hb films further help to maintain the stability of these films. However, the reasons for the poor stability of CAT-collagen films were not reported. Further results show that, a more facile electron transfer process occurs at a much faster rate with Hb-collagen films than CAT-collagen films, which is obvious from the reported k_s values, ($46 \pm 6\ s^{-1}$) and ($31 \pm 3\ s^{-1}$), respectively.

Wu *et al.* investigated the direct electrochemistry of heme proteins Mb, Hb, HRP and CAT entrapped in silk fibroin (SF) films [21]. Among various protein-SF films, CAT-SF films relatively possess lower k_s value ($0.34\ s^{-1}$). This indicates that, much slower electron transfer process occurs at CAT-SF films. In contrast, the amperometric i-t measurement results reveal that, both HRP-SF and CAT-SF films were more active than Mb-SF and Hb-SF films towards the catalytic reduction of H_2O_2(Figure 4).

Table 1. The electrochemical parameters of various CAT-modified electrodes calculated through CV and amperometric i-t studies.

Modified Electrode	Techniques	E_{pa}^{a} (or) E_{pc}^{b} (V)	Linear range (mM)	Sensitivity (μA mM^{-1} cm^{-2})	Detection limit (μM)	Ref
CAT-PAM	CV	-0.5^{b}	0.4 - 0.8	-	-	[18]
CAT-MC	CV	-0.4^{b}	0.02-0.12	-	-	[19]
CAT-SF	i-t	-0.2^{b}	0.003-0.158	-	-	[21]
CAT-PNM	i-t	-0.25^{b}	0.002-0.035	-	-	[22]
CAT/cysteine/Si sol–gel	i-t	0.1^{b}	0.001-0.03	-	0.4	[23]
CAT-agarose	CV	-0.24^{b}	0.001 - 0.818	-	-	[24]
CAT-SWCNTs	CV	-0.4^{a}	0.7-1.1	-	4.0	[29]
CAT-SWCNTs-CS	CV	-0.52^{b}	5–50	6.32	2.5	[35]
NF-MWCNTs-CAT–GNP	CV	-0.29^{a}	1-5	-	-	[36]
CAT-MWCNTs	i-t	-0.3^{b}	0.01-0.1	3.3	1.0	[40]
CAT-MWCNTs	CV	-0.56^{b}	1.0 - 4.8	-	-	[46]
MWCNTs-NF-(DDAB/CAT)	CV	-0.38^{b}	0.5 – 1.2	35.62	150	[47]
CAT-NiO	i-t	-0.3^{b}	0.001-1.0	15.9	0.60	[51]

[a] Anodic peak potential (Epa)
[b] cathodic peak potential (Epc)

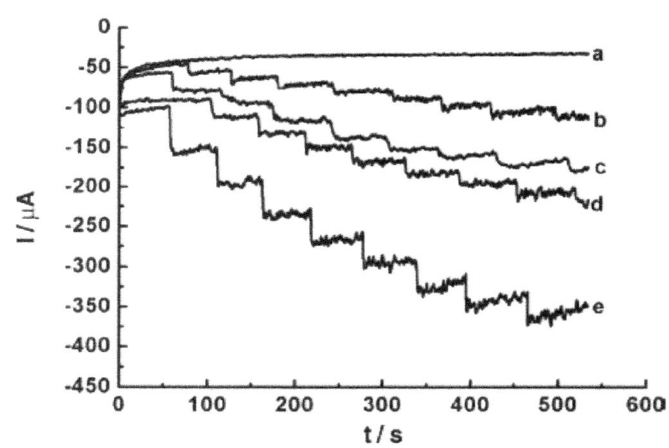

Figure 4. Amperometric i-t responses observed at different heme-proteins-SF/GCEs at − 0.2V in 10 ml of PBS pH 7.0 for 10μl injection of 22 mM H_2O_2 for every 80 s: (a) SF films; (b) Mb–SF films; (c) Hb–SF films; (d) CAT–SF films; and (e) HRP–SF films (reproduced with permission from Wu *et al. Anal. Chim. Acta* **2006**, *558*, 179–186).

From the calibration curves for H_2O_2 determination (not shown), the linear range of H_2O_2 was observed to be 0.004 to 0.085 mM for Mb–SF films, 0.003 to 0.091 mM for Hb–SF films, 0.002 to 0.354 mM for HRP–SF films and 0.003 to 0.158 mM for CAT–SF films, respectively.

In an alternative approach, Sun *et al.* immobilized heme proteins Hb, Mb and CAT into an organic polymer matrix, poly (N-isopropylacrylamide-co-3-methacryloxypropyltrimethoxysilane) (PNM) film coated glassy carbon electrode (GCE) [22]. All three protein-PNM films exhibit a facile electron transfer process, which is obvious from the obtained k_s values of 16, 17 and 20 s^{-1} for Hb-PNM, CAT-PNM and Mb-PNM films respectively. However, the stability of CAT-PNM films was much lower than Hb and Mb containing PNM films. CAT-PNM films show good catalytic activity towards H_2O_2 in the concentration range 0.002–0.035 mM (See Table 1).

The direct electrochemistry of CAT immobilized in silica sol-gel in the presence of cysteine on gold (Au) electrode was reported by Di *et al.* [23]. They confirmed the different surface morphology between silica sol-gel matrices with and without CAT through AFM studies. Figure 5 shows the AFM image of sol-gel matrix without CAT and with CAT. In their detailed study, they used this CAT/cysteine/Si sol–gel modified electrode for the determination of H_2O_2 in the range 0.001 to 0.03 mM with a detection limit of 0.4 μM, respectively (See Table 1). Further, they investigated the interaction of CAT and aluminium (Al^{3+}) ions. They observed that, the electrocatalytic activity of

CAT increased with the addition of Al^{3+} ions. Their detailed voltammetry and fluorescence spectroscopy results also confirms that, the addition of Al^{3+} ions leads to conformational changes in CAT molecule, which might be the reason for the improved catalytic activity of CAT.

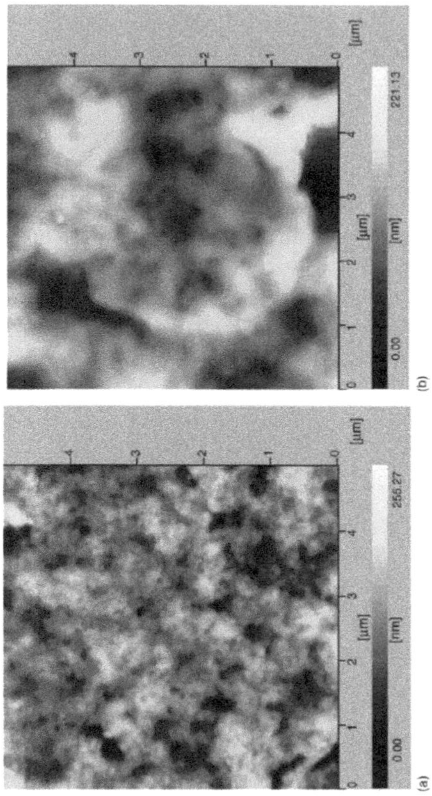

Figure 5. AFM images of: (a) Cysteine/Silica sol–gel and; (b) CAT/cysteine/Si sol–gel modified Au electrodes (reproduced with permission from Di *et al. Biosens. Bioelectron.***2006**, *22*, 247–252).

Recently Wang *et al.* immobilized the heme proteins Hb, Mb and CAT into agarose hydrogel films [24]. They also investigated the effects of 1-butyl-3-methylimidazolium tetrafluoroborate ([bmim][BF_4]) along with a small amount of water on the direct electrochemistry and bioelectrocatalysis of the entrapped heme proteins. Their CV results provide strong evidence that CAT entrapped in agarose films exhibits direct electron transfer and retains its electrochemical activity in [bmim][BF_4] solution. A more facile electron transfer process occurs at CAT-agarose films (k_s = 49.1±9.3) than Hb-agarose (k_s= 36.1±7.3) and Mb-agarose (k_s = 29.1±7.0) films respectively. Further, CAT-agarose films show good catalytic response towards H_2O_2 in the concentration range 0.001 – 0.818 mM (See Table 1). The Michaelis-Menten constant (K_m) for CAT-agarose films (317±77 μM) is two folds smaller than Hb and Mb, which indicates that the CAT entrapped in agarose film exhibits a higher affinity for H_2O_2.

Dendrimer Matrices for CAT Immobilization

Shen *et al.* successfully used a layer-by-layer assembly technique to immobilize Hb, Mb and CAT with polyamidoamine (PAMAM) dendrimers on different solid surfaces [25]. As well known, at pH 7 protonated PAMAM possesses positive surface charges, whereas the proteins have net negative surface charges. Thus, layer-by-layer $\{PAMAM\text{-}protein\}_n$ films were assembled with alternate adsorption of oppositely charged PAMAM and proteins from their aqueous solutions mainly by electrostatic interaction. The assembly process was monitored by QCM, UV-vis spectroscopy and CV. The growth of the protein multilayer films was regular and linear, whereas the electroactivity of the films extended only to a few bilayers. CVs of $\{PAMAM\text{-}protein\}_n$ films showed a pair of well-defined and nearly reversible peaks which are characteristic of the heme $Fe^{(III)/(II)}$ redox couples of the corresponding proteins. Although $\{PAMAM\text{-}Hb\}_n$ and $\{PAMAM\text{-}Mb\}_n$ films showed very similar properties, $\{PAMAM\text{-}CAT\}_n$ films displayed dissimilar properties. The substrates with biological or environmental significance, such as oxygen, H_2O_2, trichloroacetic acid, and nitrite, were catalytically reduced at $\{PAMAM\text{-}protein\}_n$ film electrodes. This shows the potential applicability of the $\{PAMAM\text{-}protein\}_n$ films as new types of biosensors or bioreactors. Both the electrochemical and electrocatalytic activity of $\{PAMAM\text{-}protein\}_n$ films can be tailored precisely by controlling the number of bilayers or the film thickness. Among various protein-PAMAM films, CAT incorporated PAMAM films showed poor stability and the catalytic response towards H_2O_2 was limited only to two layers. The previous experimental results showed that greater the distance between the enzyme and the electrode surface lesser will be the stability. The CAT with greater molecular dimension (diameter 10.5 nm) also experience weaker interactions with PAMAM, which could be partly responsible for the less stability of $\{PAMAM\text{-}CAT\}_n$ films.

As discussed in Section 2, CAT immobilized on various nanomaterial-free matrices possesses unique characteristics like facile electron transfer, good catalytic activity, fast response and appreciable stability. More immense evolution in the characteristics of CAT based sensors has been achieved in the recent years with the implementation of various nanomaterial matrices. The utilization of nanomaterials as CAT immobilization matrices narrowed the difficult tasks that existed over the past decades in the achievement of direct electrochemistry. Besides these, the long term stability of such nanomaterial-incorporated CAT films has promoted their wide applications in various fields. The significant advancements in the implementation of nanomaterial matrices for CAT immobilization towards the development of efficient H_2O_2 sensors are explained in detail in the subsequent section.

NANOMATERIAL MATRICES USED FOR CAT IMMOBILIZATION

The direct electrochemistry of CAT at various nanomaterial modified matrices is discussed in the following sections.

CAT Immobilized in Single-wall Carbon Nano-tubes (SWCNTs) Matrices

SWCNTs with high tensile strength and unique ability to grow on various electrode surfaces have advanced their applications in electrochemical sensor development. Recently, Salimi *et al.* reported the lowest ever detection limit of 10 pM determination of H_2O_2 with an excellent sensitivity (47.6 nA pM^{-1}) at an NAD^+ strongly adsorbed SWCNTs modified electrode [26]. This shows the excellent application of SWCNTs matrices towards the direct determination of H_2O_2. Moreover, the exceptional properties of SWCNTs have made them extremely suitable for enzyme immobilization and thus they are exclusively used for enzyme based biosensor development [27,28]. Interestingly, Wang and co-workers reported the direct electrochemistry of CAT at SWCNTs-Au electrode surface [29]. Their detailed electrode fabrication procedure involves the following steps: about 1 mg of purified SWCNTs was dissolved in 10 mL of *N, N*-dimethylformamide and the whole mixture was ultrasonicated until a uniform dispersion was obtained. About 15 μL of this SWCNTs suspension was casted on a clean Au electrode surface and dried under an infrared lamp. This SWCNTs-Au electrode was then immersed into 50 μM CAT in pH 5.9 PBS and kept for 20 min. For better performance, the CAT concentration was optimized as 8 μM to 80 μM respectively.

Their CV results illustrate that CAT adsorbed on bare Au electrode surface exhibits no significant redox peaks (Figure 6a). However, the SWCNTs-Au modified electrode with CAT exhibits a pair of well defined quasi reversible peaks at −0.414 V vs. SCE with a peak potential separation (ΔEp) value of 32 V (Figure 6b). Though in the absence of CAT, no noticeable peaks were observed for the SWCNTs modified Au electrode, considerable back ground current was found (Figure 6c). This provides evidence that the surface area of the Au electrode has been significantly increased in the presence of SWCNTs and this indeed enhanced the electron transfer process of CAT. Moreover, from their pH studies it was obvious that the catalytic peak current of CAT negatively shifted with pH 3.1 to 8.9 with a slope of 36 V. This reveals that the Fe $^{(III/II)}$ redox couple of CAT involves a one proton and two electron transfer process.

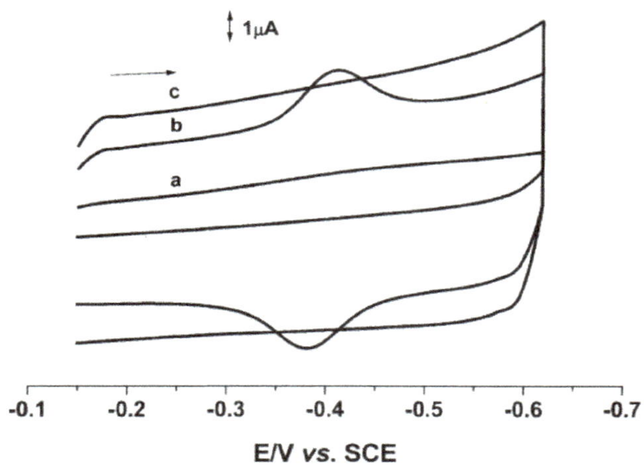

Figure 6. (a) CVs of CAT (adsorbed from a 50 μM CAT solution) at a bare Au electrode (b) SWCNTs-Au modified electrode with CAT and (c) SWCNTs-Au modified electrode without CAT. The electrolyte solution used was 0.05 M PBS pH 5.9 and the scan rate employed was 0.1 V s^{-1} (reproduced with permission from Wang *et al. Electroanalysis* **2004**, *16*, 627–632).

Moreover, their catalytic activity studies show that the CAT immobilized on SWCNT matrices exhibits a good response towards H_2O_2 in the concentration range 0.7 mM to 1.1 mM, respectively. All the above studies show that CAT has been successfully immobilized on SWCNTs, and the direct electron transfer and catalytic activity are greatly enhanced in the presence of SWCNTs. The analytical data have been given in Table 1.

CAT Immobilized in a SWCNTs-CS Matrix

CS is a multifunctional biopolymer which possesses exceptional properties like excellent film-forming ability, good adhesion, biocompatibility, high mechanical strength and has thus been used to immobilize biomolecules in recent years [30–33]. In order to increase the contact area between enzyme and SWCNTs, for better sensor sensitivity and stability, CS was widely employed to wrap SWCNTs. [34]. Jiang *et al.* observed that the incorporation of CS into SWCNTs lead to significant enhancements in the mechanical strength, tensile strength and conductivity of SWCNTs [35]. They have successfully immobilized CAT onto this SWCNTs-CS composite matrix and thereby observed the direct electrochemistry of CAT. In their detailed electrode fabrication process, SWCNTs were dispersed in 0.5% by weight CS solution and sonicated until a 0.5 mg mL^{-1} black suspension was obtained. About 5 μL of this SWCNT suspension was casted onto a cleaned GCE surface with a microsyringe, and dried at room temperature. The SWCNTs-GCE was immersed overnight in 10 mg mL^{-1}CAT solution at 4°C.

CV studies show that SWCNTs-CS-CAT-GCE exhibits a pair of well defined redox peaks at ca. −0.45 V and −0.5 V. The different scan rate studies explain that a surface confined process occurs at the modified electrode surface. From the average surface coverage concentration (Γ) value of CAT (16.4 nmol cm^{-2}) they confirmed that a monolayer of CAT has been formed over the SWCNTs modified electrode. The pH studies show that a one electron/one proton coupled reaction occurs at SWNTs-CS-CAT-GCE. These results clearly illustrate that direct CAT electrochemistry has occurred in the SWCNTs-CS matrix. They further observed through their CV studies that SWNTs-CS-CAT-GCE exhibits a steady increase in cathodic reduction peak current at −0.50 V for every substantial addition of H_2O_2 to the analyte solution (Figure 7). The linear concentration range was observed to be 5–50 mM H_2O_2. The sensitivity was 6.32 μA mM^{-1}cm^{-2} and detection limit was 2.5 μM, respectively (see Table 1). The following catalytic mechanism explains in detail about the electrocatalysis process that occurs at the electrode.:

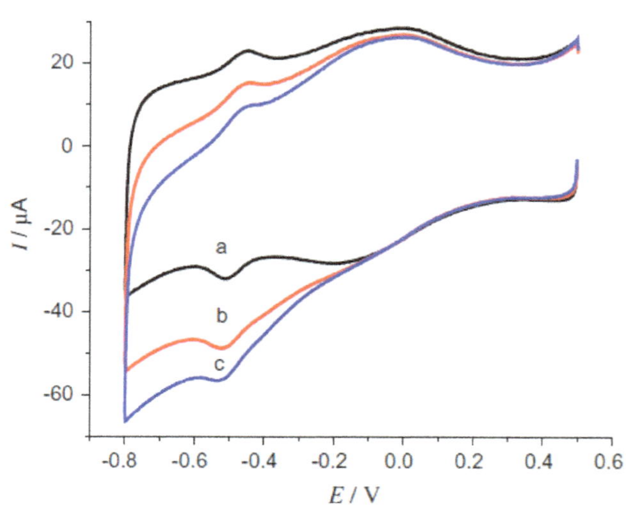

Figure 7. CVs of SWNTs-CS-CAT-GCE in PBS pH 7 at a scan rate of 0.1 V s^{-1} in the presence of (a) 0, (b) 1, and (c) 4 mM H_2O_2 (reproduced with permission from Jiang *et al. J. Electroanal. Chem.* **2008**, *623*, 181–186).

$$\text{Heme Fe}^{(III)} + H_2O_2 \longrightarrow \text{Compound I} + H_2O \quad (1)$$

$$\text{Compound I} + 2H^+ + 2e^- \longrightarrow \text{Heme Fe}^{(III)} \quad (2)$$

The electron transfer process becomes much more facile at SWCNTs platform, which was obvious from the very high k_s value of 118 s^{-1} respectively. The k_s value observed at SWCNTs-CS-CAT-GCE was relatively higher than the nanomaterial free matrix modified electrodes [15,20–22,24] reported in Section 2.

CAT Immobilized in Nafion-nano Gold-MWCNT Matrices

Zhou *et al.* reported the direct electrochemistry of CAT on multiwalled carbon nanotubes (MWCNTs) and gold nanoparticles (GNPs) modified pyrolitic graphite electrode (PGE) [36]. MWCNTs and GNPs provided an extremely stable microenvironment for the entrapped CAT molecule. Colloidal Au was prepared with slight modification according to the procedure of Frens *et al.* [37], and MWCNTs was pretreated to introduce carboxylic acid groups on the surface of CNTs [38]. The pretreated MWCNTs were dispersed in 1 mg mL^{-1} of 1% Nafion (NF) solution. The NF-CAT-GNP-MWCNTs-PGE was prepared by the following procedure: Initially a 6 mL mixture of MWCNTs and 1% NF solution was dropped on the surface of a cleaned PGE and dried at room temperature. Another 6 mL of a mixture of CAT (10 mg of CAT in 1 mL of 0.1M pH 6.98 PBS) and GNP (v = v ¼ 1:1) solution was dropped above the MWCNTs and NF surface and dried at ambient temperature without light. For stability, NF-CAT-GNP-MWCNTs-PGE was covered with a thin layer of 4 mL of 1% NF solution. Similarly NF-CAT-PGE and NF-CAT-MWCNTs-PGE were prepared as mentioned above.

The effect of GNPs on microstructures of CAT was examined through UV-vis spectroscopy results. An intense band at 280 nm with similar site/shape was observed for both CAT solution and CAT-GNP mixture. This in turn confirms that CAT in the mixture retains its native structure without the loss of biological activity. CV studies explain that NF-CAT-GNP-MWCNTs-PGE displays a pair of well-defined and nearly reversible redox peak in 6.98 PBS (Figure 8c). The anodic peak potential (Epa), cathodic peak potential (Epc) and formal potential (E^0) at NF-CAT-GNP-MWCNTs-PGE was observed to be −0.461, −0.495 V and −0.478 V, respectively. From the slope of cathodic peak currents vs. scan rate, Γ value of CAT was found to be about 2.4 nmol cm^{-2}. Electrochemical impedance spectroscopy (EIS) results show that both GNP-PGE and MWCNTs-PGE exhibits two obvious semicircles with smaller diameter, which provides strong evidence that GNP and MWCNTs promoted the electron transfer between CAT and PGE. As reported by Zhang *et al*, the interaction between the amino groups of enzyme and GNPs greatly facilitates the electron transfer process [39]. The transfer coefficient (α) and k_s of CAT was observed to be 0.49 and 1.387 s^{-1} (±0.1), respectively. This observed k_s value was relatively lower than that reported at nanomaterial-free matrices [20–22,24]. However, it was higher than CAT-SF films [21]. NF-CAT-GNP-MWCNTs-PGE also exhibits a high stability on storage. When stored at −20° C for 20 days the reduction peak current decreased only to 8%.

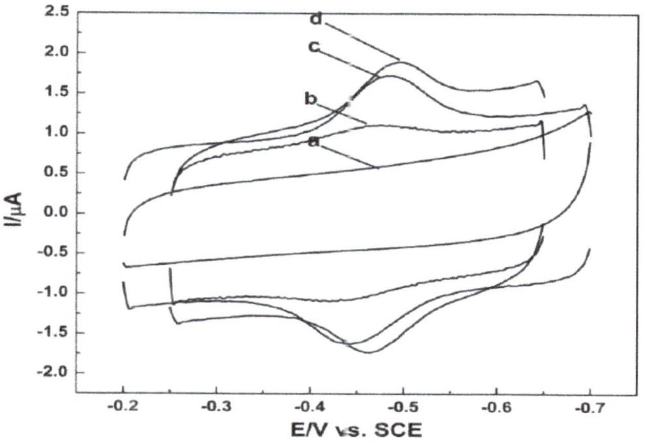

Figure 8. CVs of (a) NF-GNP-PGE, (b) NF-CAT-PGE, (c) NF-CAT-GNP-PGE (d) NF-CAT-GNP-MWCNTs-PGE electrode in 0.1 M pH 6.98 PBS at scan rate of 100 mV s^{-1}(reproduced with permission from Zhou *et al. Anal. Lett.* **2008**, *41*, 1832–1849).

Further, the NF-CAT-GNP-MWCNTs-PGE exhibits an enhanced catalytic activity towards H_2O_2 and the catalytic current increased with increase in H_2O_2 concentration. The linear response of the modified electrode was 1–5 mM (See Table 1). The K_m value was observed to be 1.71 mM (±0.05), which in turn implies that the modified PGE exhibits higher affinity towards H_2O_2.

CAT Immobilized on MWCNTs Incorporated Glassy Carbon Electrodes

Salimi *et al.* successfully immobilized CAT on a MWCNTs modified GCE [40]. The detailed electrode modification procedure of MWCNTs-CAT modified GCE is illustrated as follows: prior to each experiments, 15 cycle scans were performed on a the clean GCE in the potential range of −2.0 to +2.0 V vs. SCE in a solution of 1 M H_2SO_4 [41]. About 25 μL of 0.4 mg mL^{-1} of acetone-MWCNTs solution was casted on the GCE and dried in air. In order to obtain a well defined stable CV, the resulting MWCNTs-GCE was dipped in 0.5 mM of CAT in 0.1 M of PBS solution and subjected to consecutive potential cyclic scans in the potential range of −1.0 to +1.5 V (scan rate 100 mV s^{-1}). Finally, the modified GCE was removed from CAT solution and washed with double distilled water and stored in buffer at pH 7 at 4 °C before use.

The direct electrochemistry of CAT immobilized on MWCNTs modified GCE was investigated through CV. A pair of well defined and nearly reversible redox peaks were observed at −0.05 V in PBS 6.5 for MWCNTs-CAT-GCE. The formal potential $E^{0'}$ of MWCNTs-CAT-GCE was ultimately more positive than the other electrodes [42–44] and approximately between 350–400 mV. The surface coverage (Γ) value of CAT (24 nmol cm^{-2}) confirms the existence of a CAT monolayer on the MWCNTs electrode surface. A slope value of 61 mV pH^{-1} obtained through pH studies shows that the redox reaction of CAT was single electron coupled single proton transportation process [45]. MWCNTs-CAT-GCE also possesses a good stability on storage. When stored at 4 °C for 10 days, it shows a similar catalytic response and the reduction peak currents decreased only to 3%. The α and k_s values are 0.42 and 80 s^{-1}, respectively. Thus, the k_s value observed at MWCNTs-CAT-GCE was comparatively higher than that reported at nanomaterial free matrices [15,20–22,24] and lower than that reported at nanomaterial modified matrices [36]. With the addition of H_2O_2, MWCNTs-CAT-GCE exhibits an obvious catalytic reduction peak at −0.25V. The catalytic peak current increased linearly with increase in H_2O_2 concentration in the range of 0.01 to 1 mM. The correlation coefficient was 0.9889 and the detection limit was 50 μM at a signal-to-noise ratio of 3. Amperometric i-t measurements show that MWCNTs-CAT GCE shows steady state response to H_2O_2 at −0.3 V for a linear concentration range 0.01 to 0.1 mM. The sensitivity was 3.3 nA μM^{-1} cm^{-2}. The correlation coefficient was 0.9998 and the detection limit was 1 M, respectively (See Table 1). Moreover, MWCNTs-CAT-GCE exhibits greater affinity towards H_2O_2which was obvious from the observed K_m value, 1.7 mM.

Zhou and co-workers also succeeded in immobilizing CAT onto MWCNTs modified GCEs. They investigated the direct electrochemistry of CAT in 0.1M tris(hydroxymethyl) aminomethane (Tris)–HCl buffer solution [46]. Interestingly, they reported the effect of La^{3+} ions on the electrocatalytic activity of CAT. Their results demonstrate that the electrochemical reaction of the immobilized CAT was promoted by low concentration of La^{3+} (0.001 mM) and inhibited by the high concentration (0.01 mM) of La^{3+}. This might be due to the reason that low concentration of La^{3+} increases the non planarity and the exposure extent of CAT heme group and vise versa. Moreover, the CAT modified electrode retains its catalytic activity and exhibits a good response to H_2O_2 concentration in the concentration range 1 to 4.8 mM respectively. The $k_s$value was observed to be 1.0 7 s^{-1}. This was much lower than that reported at some nanomaterial modified matrices [36,40] and nanomaterial free matrices [20,22,24].

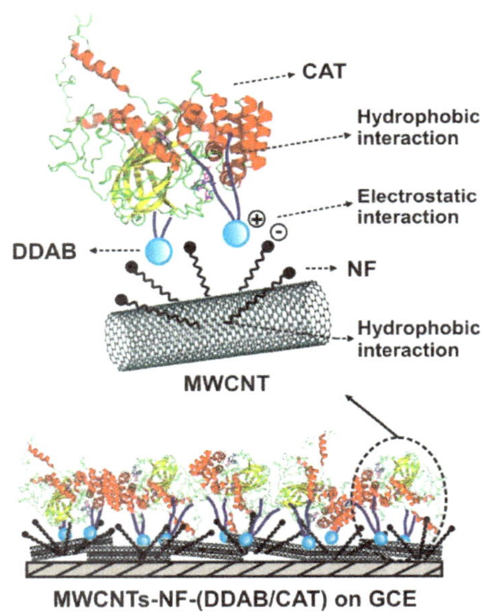

Figure 9. The possible interaction between MWCNTs, NF, DDAB and CAT in MWCNTs-NF-(DDAB/CAT) film modified electrodes. (Reproduced with permission from Prakash *et al. Talanta*, doi:10.1016/j.talanta.2009.02.033, article in press).

Recently we reported the direct electrochemistry of CAT on a NF dispersed MWCNTs modified GCE both in the presence and absence of DDAB [47]. We revealed through our studies that the presence of DDAB in the MWCNTs-NF-CAT film produced a 48% enhancement in the surface coverage concentration of CAT ($Fe^{III/II}$) and 66% enhancement in k_s value, whereas the CAT stability increased by 57 %. We also reported a maximum k_s value of 11 s^{-1} at MWCNTs-NF-(DDAB/CAT) film modified GCE. Thus the k_s value observed at MWCNTs-NF-(DDAB/CAT) film modified GCE was comparatively higher than that observed at nanomaterial free matrices [15,21] and nanomaterial modified matrices [36,46]. Moreover, the sensitivity values of MWCNTs-NF-(DDAB/CAT) film for H_2O_2 reduction was calculated using CV and i-t curve measurements and it was about 35.62 and 101.74 μA mM^{-1} cm^2 respectively (See Table 1). The possible interactions between MWCNTs-NF, DDAB and CAT at MWCNTs-NF-(DDAB/CAT) film on GCE have been given in Figure 9.

Covalent Immobilization of CAT on Conductive Composite Nanofiber Meshes

Wang et al. co-immobilized CAT on a neat poly(acrylonitrile-co-acrylic acid) (PANCAA) nanofiber mesh using 1-ethyl-3-(3-dimethylaminopropyl) carbodiimide hydrochloride (EDC) and *N*-hydroxy-succinimide (NHS) [48]. They also immobilized CAT on PANCAA and MWCNTs incorporated composite matrices. Their detailed enzyme immobilization procedure was described here. About 2 mg of the pretreated PANCAA/MWCNTs nanofiber mesh was soaked into 50 mM EDS/NHS (1:1) PBS and shaken gently for 6 h at room temp and thus activated. This activated nanofiber mesh was later mixed with CAT (0.10 mg mL^{-1} PBS) at room temp for a suitable time. Their results show that the activity of CAT on composite nanofiber matrix was 42 % higher than that of neat PANCAA nanofiber matrices. Furthermore, the electrical conductivity and viscosity of PANCAA/MWCNTs were superior to that of PANCAA alone. This may be due to the interaction between -COOH groups of etched MWCNTs and DMF and -COOH groups of PANCAA through hydrogen bond. The K_m value for free CAT, CAT/PANCAA and CAT/PANCAA/MWCNTs composite nanofiber mesh were observed to be about 34.07, 77.9 and 58.14 mM, respectively. The reasons for the higher K_m value of CAT at PANCAA and PANCAA/MWCNTs matrix might be due to conformational changes of CAT and the formation of a substrate-CAT complex. Furthermore, the increased limitation in diffusion also lessens the accessibility of the substrate to the active sites of immobilized CAT and thus promotes the K_m value.

Similarly, Wan *et al.* used poly(acrylonitrile-co-*N*-vinyl-2-pyrrolidone) (PANCNVP) and polyacrylonitrile (PAN) electrospun nanofibrous membranes with and without MWCNTs for the efficient immobilization of CAT [49]. The enzyme immobilization procedure involves two steps as explained in Figure 10. The first step involves the alkali treatment of the electrospun nanofibrous membranes, which results in the generation of -COOH groups. The second step was the activation step and it involves the treatment of nanofibrous membranes with EDC/NHS. The FESEM and TEM results confirm that presence of MWCNTs in the nanofibers increased the surface area, number of carboxylic groups and surface roughness. Further studies show that the presence of MWCNTs in the nanofibrous membranes increased the amount of immobilized CAT. The maximum amount of CAT was immobilized on PAN/MWCNTs (29.81±3.76 mg g^{-1} fibres) and PANCNVP/MWCNTs (25.77±4.10 mg g $^{-1}$ fibres) composite nanofibrous membranes, respectively. MWCNTs presence increased the activity retention and electrical conductivity of the nanofibres. Thus, CAT immobilized at PANCNVP/MWCNTs membrane possesses maximum activity retention (48.7 %), whereas PAN/MWCNTs possess a little lower activity retention (45.3%). Compared to free CAT, nanofibrous membrane immobilized CAT possesses lower, maximum reaction rates (Vmax) and higher K_m values. Their detailed stability studies also elucidate that nanofibrous membrane immobilized CAT possess improved stability. Further the immobilized CAT remains more biocompatible especially in the presence of MWCNTs. Thus among various nanofibrous membranes, MWCNTs containing nanofibrous membranes are most suited for CAT immobilization.

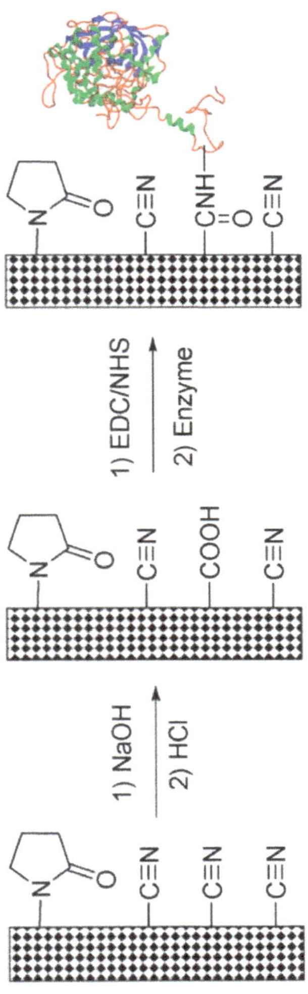

Figure 10. The two steps involved in the CAT immobilization are (Step 1) Activation of nanofibrous membrane through alkali treatment and (Step 2) Enzyme immobilization using EDC/NHS (reproduced with permission from Wan *et al. Enzyme Microb. Technol.* **2008**, *42*, 332–339).

In an alternative approach, Wan and co-workers covalently immobilized CAT onto electrospun nanofibres (dia = 180 nm) of acrylonitrile-based copolymers containing porphyrin pendants, either in the presence or absence of MWCNTs [50]. The terpolymer, PANAACoPP was prepared from acrylonitrile, acrylic acid and metalloporphyrin with Co^{2+}. Then nanofibers were fabricated from this terpolymer using an electrospining technique. The field emission scanning electron microscopy (FESEM) images obtained for various nanofibers both in the presence and absence of MWCNTs are shown in Figure 11. They clearly display that MWCNTs present in PAN/MWCNT and PANAACoPP/MWCNT nanofibres increased the roughness of the nanofibres. Moreover, due to the lower surface tension of the polymer solution, MWCNTs were mostly covered with thin polymer film and thus not much exposed to air.

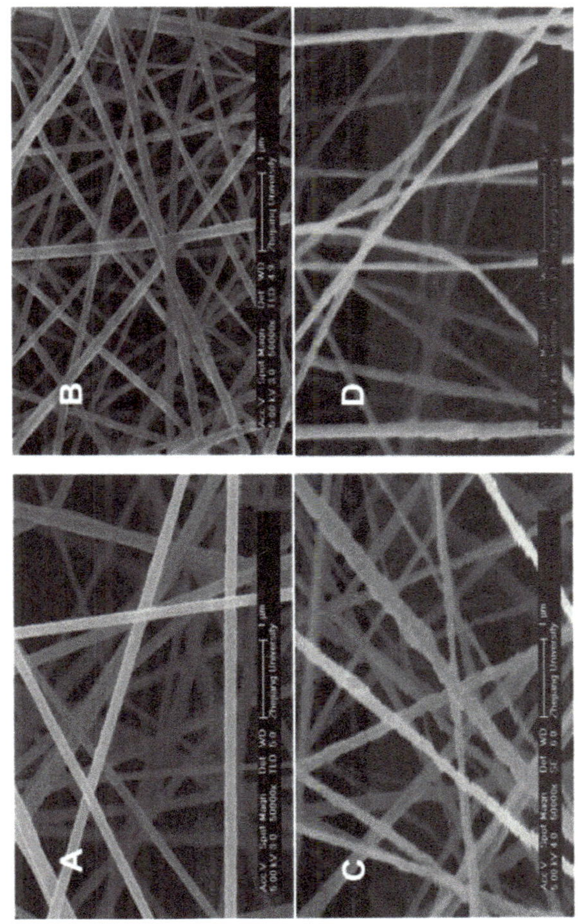

Figure 11. FESEM images of the electrospun nanofibers: (a) PAN (154±30 nm); (b) PANAACoPP (180±30 nm); (c) PAN/CNT (180±34 nm); (d) PANAACoPP/CNT (165±37 nm) (reproduced with permission from Wan *et al. J. Phys. Chem. C* **2007**, *111*, 14091–14097).

Their detailed studies show that the nanofibers with MWCNTs contain greater amounts of CAT than without MWCNTs. For instance, in the absence of MWCNTs, one gram of PAN fibers contains only 24.45±2.33 mg CAT, whereas they contain 29.81±3.76 mg CAT in the presence of MWCNTs. Similarly, one gram of PANAACoPP nanofibres has only 18.93±4.03 mg CAT, whereas PANAACoPP/MWCNTs contain more CAT (22.81±4.22 mg). The specific activity and retention activity also increased in the presence of MWCNTs, however, the *K*m value decreased and V_{max} increased in the presence of MWCNTs. This clearly shows that blending of nanofibres with MWCNTs has significantly increased the characteristics of CAT in the nanofiber matrices. Their further studies indicate that both the introduction of porphyrin pendants and MWCNTs obviously improve the activity and stabilities of the immobilized CAT. The reason for the improvement in activity and stability might be due to the enhanced electron transfer between the immobilized CAT molecules and the ability of both porphyrin pendants and MWCNTs to retain the active conformation of the CAT.

CAT Immobilized on Nickel Oxide Nanoparticles Modified Glassy Carbon Electrodes

Salimi and co-workers reported the direct electrochemistry of CAT on nickel oxide (NiO) nanoparticles modified GCE [51]. The NiO nanoparticles modified GC and indium tin oxide (ITO) electrodes were prepared as follows: initially Ni (−1.0 V, 5 min deposition time) was electrodeposited on a GCE using 1 mM nickel nitrate pH 4 acetate buffer solution. Then they repetitively cycled (30 scans) from 1.5 to −1.0 V at scan rate 100 mV s^{-1} in fresh PBS solution for electrodissolution and passivation of NiO layer. They determined the effective area of the modified electrode as 0.033 cm^2 from CVs of 10 mM $K_4[Fe(CN)_6]$ in buffer solution at pH 7. They employed the same procedure for electrodeposition of NiO nanoparticles on ITO.Figure 12 shows the scanning electron microscopy (SEM) image of NiO film electrochemically deposited at GCE. This shows that the particle size of NiO varied from 100 nm to 700 nm. The larger particles are shown at the bottom right. Similarly, the immobilization of CAT on GCE and ITO modified with nano-scale islands of NiO was carried out using the following procedures. CV was used for immobilization of CAT on NiO nanoparticles with a one step method.

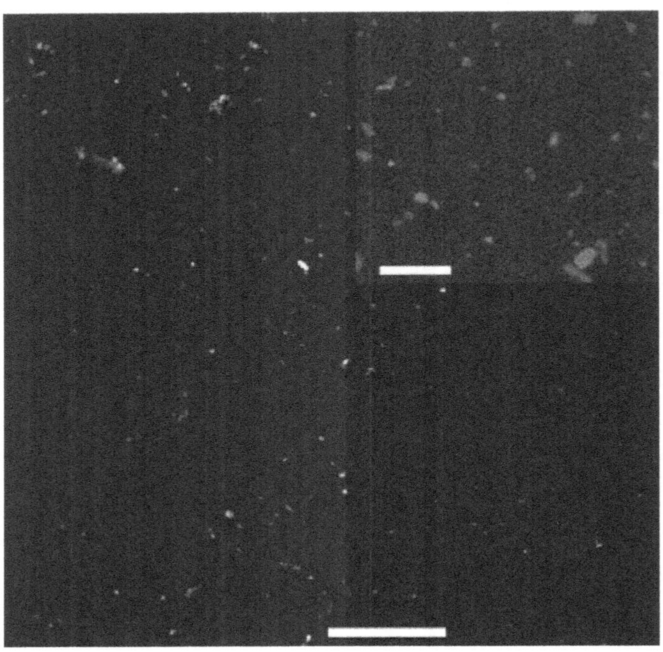

Figure 12. SEM image of the electrodeposited NiO on GCE, scale bare 20 μm. Inset is the SEM image with higher magnification for the same sample, scale bar is 2 μm (reproduced with permission from Salimi *et al. Biophys. Chem.* **2007**, *125*, 540–548).

After the deposition of metallic Ni on GCE, the electrode was immersed in fresh PBS pH 7 solution containing 5 mg mL^{-1}CAT and the potential was repetitively cycled (30 scans) from +1 to −0.5 V at scan rate 100 mV s^{-1} for electrodissolution and passivation of NiO layer and immobilization of CAT. They used the same procedure to immobilize CAT and NiO on an ITO electrode. The CAT films on NiO exhibits a pair of well defined, stable and nearly reversible CV peaks at about −0.05 V vs. SCE at pH 7. $E^{0\prime}$ of CAT in NiO film linearly varied in the range 1–12 with slope of 58.426 mV pH^{-1}, indicating that the electron transfer is accompanied by single proton transportation. Moreover, α and k_s values for NiO-CAT-GCE were observed to be 0.45 and 3.7 s^{-1} (±0.1), respectively. The reason for more facile electron transfer process at NiO-CAT-GCE was due to the larger surface area and more porous structure of NiO nanoparticles which helps for the better entrapment of CAT. Further, NiO acts as a molecular wire and enhances the direct electron transfer of CAT. Their detailed studies illustrate that the chemical stability of NiO film and the interaction between CAT and NiO NPs have enhanced the overall stability of modified electrode. NiO-CAT GCE on 10 days storage (4 °C) showed no shift in CV potentials and the reduction peak currents decreased only to 2%.

The CAT embedded in NiO nanoparticles shows excellent electrocatalytic activity towards H_2O_2 reduction with a linear range from 0. 1 to 5.0 mM. The detection limit was 10 μM with a signal to noise ratio of 3. NiO-CAT modified GCE also exhibits a fast amperometric response within 2 s to H_2O_2 in the linear concentration range 0.001 – 1.0 mM. The sensitivity was observed to be 15.9 nA μM^{-1} and the correlation coefficient was 0.9988 respectively. The detection limit was found to be 0.60 μM. Thus, NiO-CAT modified GCE possess an excellent stability, long life and good reproducibility towards H_2O_2determination. The apparent K_m value was 0.96 mM (±0.05), which shows the excellent catalytic activity of CAT incorporated NiO films towards H_2O_2. Moreover, exceptional electrochemical reversibility of redox couple, high stability, technical simplicity, lack of need for mediators and short preparation times are advantages of this NiO-CAT modified electrode.

BENEFITS OF NANOMATERIAL MATRICES ABOVE NANOMATERIAL FREE MATRICES FOR EFFICIENT CAT IMMOBILIZATION TOWARDS THE DEVELOPMENT OF CAT BASED ELECTROCHEMICAL SENSORS

Miscellaneous CAT immobilization techniques, detailed electrode fabrication procedures along with various electrochemical and morphological characterizations both in the absence and presence of nanomaterials have been presented in Sections 2 and 3. The electrochemical parameters obtained through CV and i-t curve measurements for various nanomaterial modified electrodes were compared with that of nanomaterial free electrodes and given in Table 1. It was obvious from this Table that excellent electrocatalytic activity of CAT with very high sentivity and low detection limit was more pronounced in nanomaterial matrices rather than nanomaterial free matrices. In addition, the results discussed in Sections 2 and 3 clearly show that nanomaterial modified matrices are superior to nanomaterial free matrices in several aspects. Being extremely miniaturized in size and having large surface areas, nanomaterials are an excellent choice for CAT immobilization. CAT immobilized on those nanomaterial matrices remains extremely compatible and undergoes facile electron transfer processes. With excellent mechanical strength and exceptional conductivity, SWCNTs are more appropriate for CAT immobilization. As discussed in Sections 3.1 and 3.2, the presence of SWCNTs greatly facilitates the electron transfer process. Moreover, the k_s value for CAT at SWCNT-CS matrix was observed to be 118 s^{-1}. It was comparatively higher than any of the nanomaterial modified and nanomaterial-free electrodes discussed before. Further, the detection limit, linear range and sentivity calculated at CAT-SWCNTs-CS modified electrode are given in Table 1. This in turn confirms the good catalytic ability of this SWCNTs modified electrode. Other than SWCNTs, the stability and activity of CAT greatly improved at MWCNTs matrices. This might be due to the biocompatible microenvironment of MWCNTs, which facilitates the immobilized CAT to retain its activity. Moreover, in the presence of MWCNTs the surface area and roughness largely increased. As described in Section 3.5, the activity retention and electrical conductivity extensively improved in the presence of MWCNTs. Moreover, the large surface area and porous nature of MWCNTs surfaces assists in the entrapment of more CAT. The electron transfer also becomes facile at MWCNTs matrices. It was obvious that the maximum peak current and lesser peak separation of CAT redox couples was observed at MWCNTs matrices. Furthermore, the electrocatalytic activity of CAT also considerably enhanced at MWCNTs matrices (see Table 1). MWCNTs-CAT modified GCE exhibits good catalytic activity towards H_2O_2. The lowest detection limit was observed to be 1 µM H_2O_2 and the sensitivity was 3.3 µA mM^{-1} cm^{-2}, respectively. Other than, MWCNTs, NiO nanoparticles modified GCE also possess promising catalytic activity. The NiO-CAT-GCE exhibits a very high sensitivity of 15.9 µA $mM^{-1}cm^{-2}$ with 0. 6 µM H_2O_2 as lowest detection limit. The linear range of H_2O_2 concentration was about 0.001–1.0 mM respectively. Thus among various nanomaterials, CNTs and CNT incorporated matrices are excellent choices for better CAT immobilization and for the achievement of facile direct electrochemistry.

CONCLUSIONS

The versatile approaches practiced in the immobilization of CAT onto various matrices for the achievement of direct electron transfer have been reviewed. For this purpose CAT has been immobilized on various modified electrode surfaces both in the presence and absence of nanomaterials. All the CAT modified films were electrochemically characterized and their catalytic efficiency towards H_2O_2 was also investigated. The results illustrate that CAT incorporated nanomaterial matrices show outstanding characteristics like facile electron transfer, high sensitivity, low detection limit and excellent stability. Thus they are superior to CAT incorporated nanomaterial free matrices in several aspects. Although some nanomaterial free matrices promoted the direct electrochemistry of CAT, the majority of them slow down the electron transfer process. This might be due to the lesser contact noticed between CAT and those nanomaterial free matrices. On the other hand, most of the nanomaterials acted as nanowires and helped for the achievement of better contact between CAT and the electrode surface. This ultimately facilitated the electron transfer process. In the midst of various nanomaterial matrices, more fascinating direct electron transfer process was observed at CNTs modified matrices. Moreover, with large surface area and numerous pores on their surfaces, CNTs matrices have excellent ability to trap large amount of CAT molecules. CNTs incorporated CAT also exhibits excellent catalytic activity towards H_2O_2 and ultimately helps for the CAT based electrochemical sensors development. The future aspects of CAT based sensors depend on the development of novel nanomaterials with some advantageous properties. This perhaps helps for the better immobilization and significant enhancement in the direct electrochemistry of CAT and other CAT based redox enzymes and protein direct electrochemistry.

Acknowledgments

National Taipei University of Technology assisted in meeting the publication costs of this article.

REFERENCES

1. Armstrong, F.A.; Wilson, G.S. Recent developments in faradaic bioelectrochemistry. *Electrochim. Acta* **2000**, *45*, 2623–2645.
2. Hill, H.A.O.; Hunt, N.I. Direct and indirect electrochemical investigations of metalloenzymes. *Methods Enzymol* **1993**, *227*, 501–522.
3. Bond, A.M. Chemical and electrochemical approaches to the investigation of redox reactions of simple electron transfer metalloproteins. *Inorg. Chim. Acta* **1994**, *226*, 293–340.
4. Ghindilis, A.L.; Atanasov, P.; Wilkins, E. Enzyme-catalyzed direct electron transfer: Fundamentals and Analytical Applications. *Electroanalysis* **1997**, *9*, 661–674.
5. Ferapontova, E.E.; Ruzgas, T.; Gorton, L. Direct electron transfer of heme and molybdopterin cofactor-containing chicken liver sulfite oxidase on alkanethiol-modified gold electrodes. *Anal. Chem* **2003**, *75*, 4841–4850.
6. Gorton, L.; Lindgren, A.; Larsson, T.; Munteanu, F. D.; Ruzgas, T.; Gazaryan, I. Direct electron transfer between heme-containing enzymes and electrodes as basis for third generation biosensors. *Anal. Chim. Acta* **1999**, *400*, 91–108.
7. Yeh, P.; Kuwana, T. Reversible electrode reaction of cytochrome C. *Chem. Lett* **1977**, *6*, 1145–1148.
8. Eddowes, M. J.; Hill, H.A.O. Novel method for the investigation of the electrochemistry of metalloproteins : cytochrome C. *J. Chem. Soc., Chem. Commun* **1977**, 771–772.
9. Wollenberger, U.; Spricigo, R.; Leimkühler, S.; Schröder, K. Protein electrodes with direct electrochemical communication. *Adv. Biochem. Engi. Biotechnol* **2008**, *109*, 19–64.
10. Nicholls, P.; Schonbaum, G.R.; Boyer, I.; Lardy, P.D.; Myrback, H. *The Enzymes*, Academic Press: Orlando, 1963; Volume 8, pp. 158–159.
11. Buleandra, M.; Racu, G.L; Tanase, I. Redox protein electroanalysis: metalloproteins. *Roum. Biotechnol. Lett* **2000**, *5*, 423–438.
12. Lai, M.E.; Bergel, A. Direct electrochemistry of catalase on glassy carbon electrodes. *Bioelectrochemistry* **2002**, *55*, 157–160.
13. Caruso, F.; Trau, D.; Mohwald, H.; Renneberg, R. Enzyme encapsulation in layer-by-layer engineered polymer multilayer capsules. *Langmuir* **2000**, *16*, 1485–1488.
14. Yu, A.; Caruso, F. Thin films of polyelectrolyte-encapsulated catalase microcrystals for biosensing. *Anal. Chem* **2003**, *75*, 3031–3037.

15. Chen, X.; Xie, H.; Kong, J.; Deng, J. Characterization for didodecyldimethylammonium bromide liquid crystal film entrapping catalase with enhanced direct electron transfer rate. *Biosens. Bioelectron* **2001**, *16*, 115–120.

16. Gebicka, L.; Gebicki, J.L. Interaction of sodium bis (2-ethylhexyl) sulfosuccinate (AOT) with catalase and horseradish peroxidase in an aqueous solution and in the reverse micelles of AOT/n-heptane. *Biochem. Mol. Biol. Int* **1998**, *45*, 805–811.

17. Huang, H.; Hu, N.; Zeng, Y.; Zhou, G. Electrochemistry and electrocatalysis with heme proteins in chitosan biopolymer films. *Anal. Biochem* **2002**, *308*, 141–151.

18. Lu, H.Y.; Li, Z.; Hu, N.F. Direct voltammetry and electrocatalytic properties of catalase incorporated in polyacrylamide hydrogel films. *Biophys. Chem* **2003**, *104*, 623–632.

19. Li, Y.M.; Chen, X.T.; Li, J.; Liu, H.H. Direct voltammetry and catalysis of hemoenzymes in methyl cellulose film. *Electrochim. Acta* **2004**, *49*, 3195–3200.

20. Li, M.; He, P.; Zhang, Y.; Hu, N. An electrochemical investigation of hemoglobin and catalase incorporated in collagen films. *Biochim. Biophys. Acta* **2005**, *1749*, 43–51.

21. Wu, Y.; Shen, Q.; Hu, S. Direct electrochemistry and electrocatalysis of heme-proteins in regenerated silk fibroin film.*Anal. Chim. Acta* **2006**, *558*, 179–186.

22. Sun, Y.X.; Wang, S.F. Direct electrochemistry and electrocatalytic characteristic of heme proteins immobilized in a new sol–gel polymer film. *Bioelectrochemistry* **2007**, *71*, 172–179.

23. Di, J.; Zhang, M.; Yao, K.; Bi, S. Direct voltammetry of catalase immobilized on silica sol–gel and cysteine modified gold electrode and its application. *Biosens. Bioelectron* **2006**, *22*, 247–252.

24. Wang, S.F.; Chen, T.; Zhang, Z.L.; Pang, D.W.; Wong, K.L. Effects of hydrophilic room-temperature ionic liquid 1-butyl-3-methylimidazolium tetrafluoroborate on direct electrochemistry and bioelectrocatalysis of heme proteins entrapped in agarose hydrogel films. *Electrochem. Commun* **2007**, *9*, 1709–1714.

25. Shen, L.; Hu, N. Heme protein films with polyamidoamine dendrimer: direct electrochemistry and electrocatalysis. *BBA-Bioenergetics* **2004**, *1608*, 23–33.

26. Salimi, A.; Miranzadeh, L.; Hallaj, R.; Mamkhezri, H. Picomolar detection of hydrogen peroxide at glassy carbon electrode modified with NAD^+ and single walled carbon nanotubes. *Electroanal* **2008**, *20*, 1760–1768.

27. Shi, L.; Liu, X.; Niu, W.; Li, H.; Han, S.; Chen, J.; Xu, G. Hydrogen peroxide biosensor based on direct electrochemistry of soybean peroxidase immobilized on single-walled carbon nanohorn modified electrode. *Biosens. Bioelectron* **2009**,*24*, 1159–1163.

28. Cosnier, S.; Ionescu, R.E.; Holzinger, M. Aqueous dispersions of SWCNTs using pyrrolic surfactants for the electro-generation of homogeneous nanotube composites. Application to the design of an amperometric biosensor. *J. Mater. Chem* **2008**, *18*, 5129–5133.

29. Wang, L.; Wang, J.; Zhou, F. Direct electrochemistry of catalase at a gold electrode modified with single-wall carbon nanotubes. *Electroanal* **2004**, *16*, 627–632.

30. Miao, Y.; Chia, L.S.; Goh, N.K.; Tan, S.N. Amperometric glucose biosensor based on immobilization of glucose oxidase in chitosan matrix cross-linked with glutaraldehyde. *Electroanal* **2001**, *13*, 347–349.

31. Okuma, H.; Watanabe, E. Flow system for fish freshness determination based on double multi enzyme reactor electrodes. *Biosens. Bioelectron* **2002**, *17*, 367–372.

32. Hikima, S.; Kakizaki, T.; Taga, M.; Hasebe, K. Enzyme sensor for L-lactate with a chitosan mercury film electrode. *J. Anal. Chem* **1993**, *345*, 607–609.

33. Chen, L.; Gorski, W. Bioinorganic composites for enzyme electrodes. *Anal. Chem* **2001**, *73*, 2862–2868.

34. Jiang, H.; Du, C.; Zou, Z.; Li, X.; Akins, D.; Yang, H. A biosensing platform based on horseradish peroxidase immobilized onto chitosan-wrapped single-walled carbon nanotubes. *J. Solid State Electrochem* **2009**, *13*, 791–798.

35. Jiang, H.J.; Yang, H.; Akins, D.L. Direct electrochemistry and electrocatalysis of catalase immobilized on a SWNT-nanocomposite film. *J. Electroanal. Chem* **2008**, *623*, 181–186.

36. Zhou, B.; Wang, J.; Gao, X.; Tian, Y. Attachment of nanoparticles to pyrolytic graphite electrode and its application for the direct electrochemistry and electrocatalytic behavior of catalase. *Anal. Lett* **2008**, *41*, 1832–1849.

37. Frens, G. Controlled nucleation for the regulation of the particle size in monodisperse gold suspensions. *Nat. Phys. Sci***1973**, *241*, 20–22.

38. Chattopadhyay, D.; Galeska, I.; Papadimitrakopoulos, F. Metal-assisted organization of shortened carbon nanotubes in monolayer and multilayer forest assemblies. *J. Am. Chem. Soc* **2001**, *123*, 9451–9452.

39. Zhang, B.; Zhang, Z.; Wang, B.; Yan, J.; Li, J.; Cai, S.M. Preparation of gold nano-arraied electrode on silicon substrate and its electrochemical properties: Probe into biosensor based on electroluminescence of porous silicon. *Acta Chimi. Sin* **2001**, *59*, 1932–1936.

40. Salimi, A.; Noorbakhsh, A.; Ghadermarz, M. Direct electrochemistry and electrocatalytic activity of catalase incorporated onto multiwall carbon nanotubes-modified glassy carbon electrode. *Anal. Biochem* **2005**, *344*, 16–24.

41. Zhao, G.C.; Zhang, L.; Wei, X.W.; Yang, Z.S. Myoglobin on multi-walled carbon nanotubes modified electrode: direct electrochemistry and electrocatalysis. *Electrochem. Commun* **2003**, *5*, 825–829.

42. Lu, H.; Li, Z.; Hu, N. Direct voltammetry and electrocatalytic properties of catalase incorporated in polyacrylamide hydrogel films. *Biophys. Chem* **2003**, *104*, 623–632.

43. Huang, H.; Hu, N.; Zeng, Y.; Zhou, G. Electrochemistry and electrocatalysis with heme proteins in chitosan biopolymer films. *Anal. Biochem* **2002**, *308*, 141–151.

44. Zhang, Z.; Chouchane, S.; Magiliozzo, R.S.; Rusling, J.F. Direct voltammetry and catalysis with mycobacterium

tuberculosis catalase-peroxidase, peroxidases, and catalase in lipid films. *Anal. Chem* **2002**, *74*, 163–170.

45. Bond, A.M. *Modern polarographic methods in analytical chemistry*; Marcel Dekker: New York, 1980.

46. Zhou, H.; Lu, T.H.; Shi, H.X.; Dai, Z.H.; Huang, X.H. Direct electrochemistry and electrocatalysis of catalase immobilized on multi-wall carbon nanotubes modified glassy carbon electrode and its application. *J. electroanal. Chem* **2008**, *612*, 173–178.

47. Prakash, P.A.; Yogeswaran, U.; Chen, S.M. Direct electrochemistry of catalase at multiwalled carbon nanotubes - nafion in presence of needle shaped DDAB for H_2O_2 sensor. *Talanta*.

48. Wang, Z.G.; Ke, B.B.; Xu, Z.K. Covalent immobilization of redox enzyme on electrospun nonwoven poly(acrylonitrile-co-acrylic acid) nanofiber mesh filled with carbon nanotubes: a comprehensive study. *Biotechnol. Bioeng* **2007**, *97*, 708–720.

49. Wan, L.S.; Ke, B.B.; Xu, Z K. Electrospun nanofibrous membranes filled with carbon nanotubes for redox enzyme immobilization. *Enzyme Microb. Technol* **2008**, *42*, 332–339.

50. Wan, L.S.; Ke, B.B.; Wu, J.; Xu, Z.K. Catalase immobilization on electrospun nanofibers: effects of porphyrin pendants and carbon nanotubes. *J. Phys. Chem. C* **2007**, *111*, 14091–14097.

51. Salimi, A.; Sharifi, E.; Noorbakhsh, A.; Soltanian, S. Direct electrochemistry and electrocatalytic activity of catalase immobilized onto electrodeposited nano-scale islands of nickel oxide. *Biophys. Chem* **2007**, *125*, 540–548.

CHAPTER 12

ELECTROCATALYSIS AND ELECTROCATALYSTS FOR LOW TEMPERATURE FUEL CELLS: FUNDAMENTALS, STATE OF THE ART, RESEARCH AND DEVELOPMENT

Hartmut Wendt, Estevam V. Spinacé; Almir Oliveira Neto; Marcelo Linardi

INTRODUCTION

The idea of electrochemical electricity generation from chemical energy in fuel cells originated already more the 150 years ago from the experiments of Grove and Schönbein[1]. However, the technical realisation had to wait for more than a hundred years as the relevant materials problems could not be solved for a very long time and the theoretical background of electrochemistry, not to speak of electrocatalysis, took so much time to be developed. It was apparent from the very beginning that the electrochemical combustion of any fuel (Equation 1) in an electrochemical cell brought about by separate anodic oxidation of a gaseous fuel, for instance hydrogen (Equation 1a), at the anode and cathodic reduction of oxygen (Equation 1b) needed the catalytic activation of the reactants, fuel as well as oxygen.

$$\text{Total reaction: fuel} + (x + y/_2)\ O_2 \rightarrow x\ CO_2 + y\ H_2O \quad (1)$$

$$\text{Anodic reaction: } H_2 + 2\ H_2O \rightarrow 2\ H_3O^+ + 2\ e^- \quad (1a)$$

$$\text{Cathodic reaction: } ½\ O_2 + 2\ H_3O^+ + 2\ e^- \rightarrow 3\ H_2O \quad (1b)$$

Citation: Hartmut Wendt, Estevam V. Spinacé, Almir Oliveira Neto, Marcelo Linardi, Electrocatalysis and electrocatalysts for low temperature fuel cells: fundamentals, state of the art, research and development, http://dx.doi.org/10.1590/S0100-40422005000600023.

Hydrogen (H_2) is a relatively stable molecule – and so are carbonaceous fuels. Due to the strength of the H-H bond or H_2-binding energy respectively of approximately 430 kJ/mol[2,3], only catalytic activation of this molecule would enhance its reaction at ambient temperature to a degree, which allows it to react with sizeable rate. For double-bonded and hence more strongly bonded oxygen (DH @ 500 kJ/mol) the catalytic activation is even more necessary. This statement holds for gas phase reactions (Equation 1) as well as for electrochemical conversion at electrode surfaces (Equation 1a and 1b). And as will be explained below, platinum in both cases is the catalyst of choice as already realized in the 19th century by Döbereiner[4]. Although this article deals almost exclusively with electrocatalysis in fuel cells, it is stressed that electrocatalysis is of high practical relevance in industrial electrochemistry. The present technology of chloroalkali electrolysis today depends essentially on electrocatalysis of the hydrogen and chlorine evolution reactions[5,6]. In the mid 80ties of the last century the phosphoric acid fuel cell technology gained shape. As documented by a number of important patents of that time it became evident that platinum and also a number of platinum alloys can be prepared on industrial scale in well-defined nanodisperse form, thus enabling fuel cell producing industries to make highest possible use of this expensive material[7-10]. In the same context also the use of active carbon supported nanodisperse platinum had been introduced into the practice of technical production of fuel cells[11].

The developments of modern fuel cells began with moderately high temperature alkaline cells[12] in the 20ies (working at 200 °C) and with high temperature molten carbonate cells[13] in the 50ties of the last century. A first technical application was the development of the alkaline cells for the Gemini space mission by United Technologies Corporation (UTC)[13,14]. The Varta-Siemens approach based on Raney-nickel electrodes in 30 wt% KOH did not lead to a convincing success[14]. Much more promising was the next step undertaken by UTC to introduce membrane fuel cells, which after disappointing experimentations with classical ion exchange membranes led to a technical brake through by introducing perfluorinated sulfonic acid membranes as so-called solid polymer electrolyte[15,16], which had been perfected by Siemens for powering submarines during the last years. Dupont initially had developed this type of membrane under the name of Nafion for the chloroalkali industry[17]. This eventually yielded into the now well-established polymer electrolyte membrane fuel cell (PEMFC) technology (Figure 1). The scanning electron microscopic cut through a so-called membrane/electrode assembly (MEA) of a PEMFC (Figure 2) shows the minute dimensions of only approximately 10 µm thick porous electrodes on both sides of the approximately 150 µm thick membrane. The membrane chemical composition is shown in Figure 3. The essential catalyst for hydrogen/oxygen fuel cells on the anode- and the cathode side is nano-dispersed platinum, supported on active carbon. Its preparation and morphology will be described below.

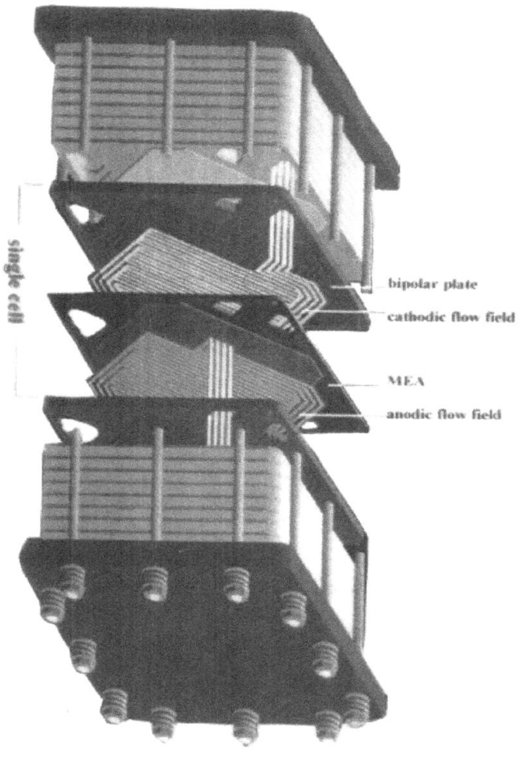

Figure 1. View of Proton Exchange Membrane Fuel Cell stack; included is the exploded view of a singular cell containing bipolar wall with flow field and at the centre the MEA (Membrane-Electrode-Assembly).

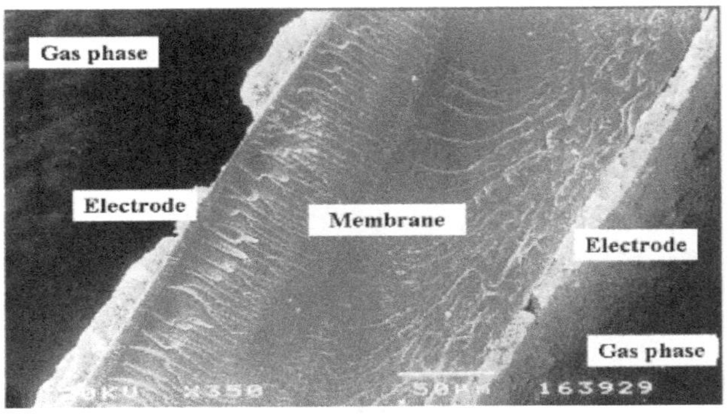

Figure 2. SEM micrograph of a typical MEA produced in the laboratory of the authors.

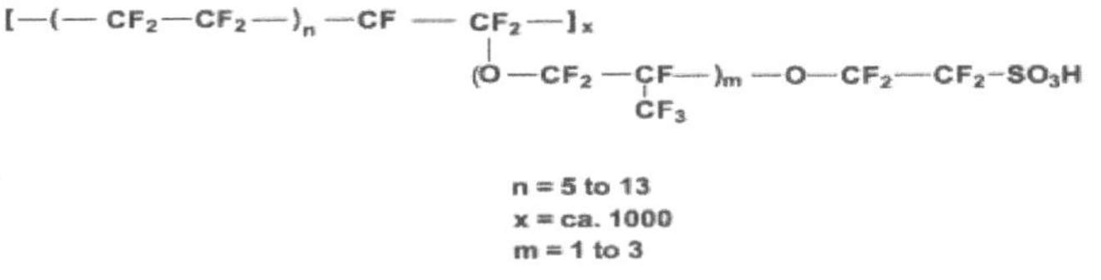

Figure 3. Schematic presentation of the chemical structure of Nafion.

The further advance of PEMFC-electrocatalyst development was due to the fact that for mobile or for stationary electricity generation with PEMFC using hydrogen, purified hydrogen in particular, seemed to be too expensive and therefore impractical. The relatively reactive and easily processable methanol but also natural gas, which demands higher processing efforts, are cheaper than hydrogen. Methanol additionally because it is a liquid at ambient temperature and pressure is easy to store and more appropriate as fuel for cars whereas natural gas is the primary fuel of choice for stationary electricity generation. These hydrogen carriers can be converted to hydrogen by the sequence of reforming and shift-conversion (Equations. 2 and 3) that produces gas mixtures containing hydrogen and carbon dioxide but also the catalyst poison carbon monoxide. The latter is found in concentrations of up to one percent because shift conversion is an equilibrium reaction, which favours CO at higher temperature and the equilibrium concentration of CO in reformate gas is that high at the operation temperature of the heterogeneously catalysed shift conversion process (200 °C).

Reformate from methane

Methane reforming:

$$CH_4 + H_2O \rightarrow CO + 3H_2 \text{ (800-900 °C)} \quad (2a)$$

Shift conversion:

$$CO + H_2O \rightarrow CO_2 + H_2 \text{ (beginning at 400 and ending at 200 °C)} \quad (2b)$$

Reformate from methanol

Methanol cracking and reforming:

$$CH_3OH \rightarrow CO + 2\,H_2 \text{ (300-400 °C)} \quad (3a)$$

Shift conversion:

$$CO + H_2O \rightarrow CO_2 + H_2 \text{ (beginning at 400 and ending at 200 °C)} \quad (3b)$$

The generation of hydrogen from carbonaceous hydrogen carriers therefore is highly problematic with respect to the CO

– sensitive platinum catalysts if the electrochemical reaction is to be performed at temperatures of between 60 and 80 °C, which is the case in PEMFCs. Their working temperature must be kept that low because of the necessary moisture in the membrane, which keeps the ionic conductivity of the membrane sufficiently high. At these low temperatures the catalytic activity of platinum is severely impaired by strongly adsorbed CO, as long as the CO concentration is not kept lower than only 10 ppm. Direct methanol fuel cells (DMFCs)[18] for automotive power generation and for so-called portables applications are even more demanding with respect to the anodic electrocatalysts. There the electrocatalyst has not only to provide for the anodic fission of O-H and C-H bonds by chemisorption of the respective fuel but the anodic oxidation of methanol ends up generating adsorbed CO in situ and thus the reaction on Pt is poisoning the catalyst. And this catalyst poison has to be removed steadily. Therefore, during the last years a couple of cocatalysts has been investigated and applied which enhance particularly the oxidation rate of CO in order to render anodic PEM-electrocatalysts more CO-tolerant. Among these cocatalysts ruthenium is the most prominent. It is technically applied since several years and it is very well documented and investigated[19]. Binary Pt/Ru catalysts supported on active carbon are commercial products. This article will therefore deal not only with dispersed platinum electrocatalysts, their preparation and characterization but will report briefly also on what we know today about composition, structure and nanomorphology of more elaborate nanodisperse, composite electrocatalyst systems. These more involved anodic catalyst systems, as far as they are used in PEMFCs, are almost exclusively supported on active carbon[20-22].

Phosphorous acid fuel cells (PAFCs) working at 200 °C are insensitive against CO and work on reformate with 1% of CO (without CO removal). But according to an announcement of UTC in 2003 (source: Internet, Hyweb) they will develop PEM fuel cells in preference to PAFCs for stationary electricity generation because the PAFC – technology does not promise any further cost reduction below the established system costs of 4,500 USD per kW.

FUNDAMENTALS: PLATINUM, THE CATALYST OF CHOICE FOR ANODIC HYDROGEN OXIDATION AND CATHODIC OXYGEN REDUCTION

Exchange Current Densities, an Indicator of Electrocatalysis

Current voltage correlations measured for the Butler-Volmer equation (Equation 4), describe mathematically the current-voltage correlation for a reversible electrochemical reaction, for instance, the anodic oxidation of hydrogen and cathodic hydrogen evolution. It comprises an anodic (*index a*) and a cathodic (*index c*) exponential terms with the overpotential h being the difference between actual electrode potential E and equilibrium potential E_o (h = E - E_o) and the charge transfer coefficients a_a and a_c in the exponent. Instead of current, I, the current density (current per unit surface), i, is chosen as the adequate quantity for this correlation, in order to obtain an equation, which is independent of the size of the electrode.

$$i = i_o \left(\exp\{\alpha_a F\eta / RT\} - \exp -\{\alpha_c F\eta / RT\}\right) \qquad (4)$$

In the Butler-Volmer equation the exchange current density, i_o, as the quantity describing charge exchange per unit surface at equilibrium potential in cathodic and anodic direction, is an indicator for electrocatalysis. Table 1collects the well-established charge exchange current densities observed at different electrode materials whose numerical values range from 10^{-3} down to 5 x 10^{-13} Ampere per square centimetre (A cm^{-2}) over more than twelve orders of magnitude. Obviously the metals contained in the group of the platinum metals (Pd, Pt, Rh, Ir and, not reported here, also Ru) at which the hydrogen reaction is recorded to have the highest exchange current densities, do catalyse this reaction. They are good electrocatalysts for this reaction, whereas mercury on which an exceedingly low exchange current density is reported is not an electrocatalyst at all[6].

Table 1. Exchange current densities of the hydrogen evolution/anodic oxidation reaction at different electrode materials in aqueous 1 N H_2SO_4 solution at ambient temperature

Table 1. Exchange current densities of the hydrogen evolution/anodic oxidation reaction at different electrode materials in aqueous 1 N H_2SO_4 solution at ambient temperature

Metal	i_o/A cm^{-2}
Palladium, Pd	1.0 x 10^{-3}
Platinum, Pt	8.0 x 10^{-4}
Rhodium, Rh	2.5 x 10^{-4}
Iridium, Ir	2.0 x 10^{-4}
Nickel, Ni	7.0 x 10^{-6}
Gold, Au	4.0 x 10^{-6}
Tungsten, W	1.3 x 10^{-6}
Niobium, Nb	1.5 x 10^{-7}
Titanium, Ti	7.0 x 10^{-8}
Cadmium, Cd	1.5 x 10^{-11}
Manganese, Mn	1.3 x 10^{-11}
Thallium, Tl	1.0 x 10^{-11}
Lead, Pb	1.0 x 10^{-12}
Mercury, Hg	0.5 x 10^{-13}

Cyclic Voltammetry at Smooth Platinum, an Indicator of H-Electrosorption and Electrocatalytic Activity for Hydrogen Evolution

The cyclic voltammogram of a platinum electrode immersed in acid electrolyte is depicted by broken line inFigure 4, which would be very similar in any electrolyte of any pH, provided that electrosorption of components of the electrolyte on platinum can safely be excluded[23]. One observes a relatively flat anodic and a very steep cathodic current increase at the right and left hand side of the plot, which are due to anodic oxygen and cathodic hydrogen evolution. The peaks and waves between electrochemical gas evolutions are due to electrochemical surface reactions at the platinum electrode. The voltammogram is observed by sweeping continuously the electrode potential with constant velocity of, for instance, 10 mV per second and thereby measuring the electrode current. Positive current peaks or waves do mean anodic oxidation processes and negative currents are correspondingly due to cathodic reactions at the platinum/electrolyte interface. A constant current between 0.3 and 0.6 V signify charging of the electrode/electrolyte interface, which forms a capacitance. The electrode potential is measured with reference to the Reversible Hydrogen

Electrode (RHE) in the applied electrolyte. Beginning at 0 V vs. RHE, where the observed current is zero, one observes two oxidation peaks one centred around +150 and the other around +300 mV, which are attributed to the anodic oxidation of adsorbed hydrogen ($H_{ad}{}^{I,\,II}$ ® $H^+ + e^-$), which due to the respective different adsorption sites are adsorbed with different strengths. At 150 mV the free energy of adsorption corresponds to –14.5 kJ/mol and correspondingly the second peak signifies a free adsorption enthalpy of approximately –29 kJ per mol of adsorbed hydrogen atoms (reaction: ½H_2 ® H_{ad}). These weak free adsorption enthalpies evidence that the binding energy of two H-atoms on platinum almost matches the dissociation/binding energy of the H-H bond. Neglecting for the moment the further anodic details of the voltammogram at more anodic potentials one observes on reversal of the voltage sweep cathodic currents and electrosorption peaks of hydrogen ($H^+ + e^-$ ® H_{ad}), which mirror the discussed electrodesorption peaks at +300 and +150 mV. This evidences that the electrochemical reaction $H^+ + e^-$ ® $H_{ad}$is a fast and reversible reaction on the time scale of the voltage sweep. These cathodic electrosorption and anodic electrodesorption peaks signify the pronounced adsorptive properties of platinum. Similar peaks are also observed at electrodes of other platinum metals, which are known to adsorb hydrogen[23].

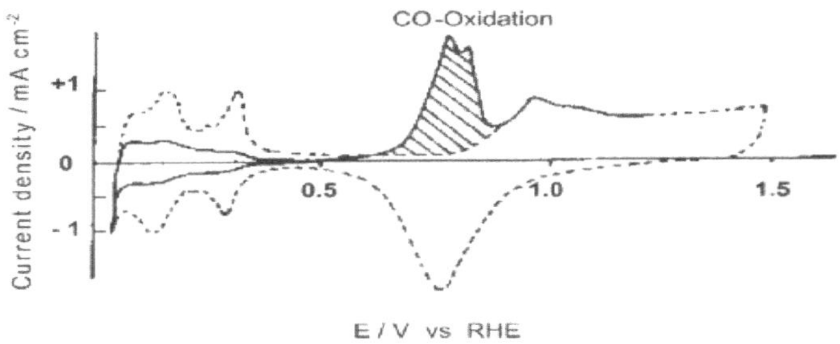

Figure 4. Cyclic voltammograms of a platinum electrode in acid solution (broken line) and CO-poisoned demonstrating anodic stripping of CO above 0.6V vs. RHE due to oxidation of adsorbed CO by anodically formed Pt-oxide.

Proceeding further with the sweep of the electrode potential to more positive potentials one passes through a region, where only a weak and constant anodic current is observed which signify only charging the capacity of the electrode double layer. Then a little bit above 600 mV the anodic current begins to rise approaching a current niveau above 800 mV. This increasing anodic current above 600 mV is connected to the beginning of surface oxidation of platinum (Pt + H_2O ® $PtOH_{surface}$ + $H^+ + e^-$), which forms a closed Pt oxide layer at approximately +800 mV. The almost constant anodic current, which is observed further, at still higher potential, serves then only to increase the thickness of the oxide layer and to increase the degree of oxidation of Pt (increasing the formal valency of the metal from +1 to higher values). Above 1.4 V, well above 1.2 V, the equilibrium potential of the oxygen electrode – the anodic current rises again and exponentially with electrode potential due to the onset of anodic oxygen evolution which shows a remarkable anodic overpotential because the reaction H_2O ® ½ O_2 + $2H^+ + 2e^-$ is slow, kinetically hindered and Pt – oxides, which cover the electrode surface, are not particularly active catalysts to enhance it. Reversing the sweep leads to fast decline of anodic oxygen evolution currents and surface oxide charging currents. A broad cathodic current peak shifted negatively with respect to 0.6 V, the onset potential of Pt oxidation demonstrates the retarded reduction of the Pt-oxides. They can only be reduced with some cathodic overpotential because these oxides are (due to aging and further oxidation at higher potential) no longer the same and more stable and chemically more inert than those, which had been formed initially between +600 and +800 mV. Then approaching closer to the hydrogen equilibrium potential (0 V vs. RHE) one observes the cathodic electrosorption peaks for cathodic deposition of adsorbed hydrogen in two different adsorption states, which had been discussed before. Almost exactly at 0 V the cathodic hydrogen evolution current commences to increase steeply and exponentially with increasing overpotential. This increase is much steeper than that of the anodic oxygen evolution. Electrochemical ad- and desorption and electrochemical hydrogen evolution can be slowed appreciably or completely suppressed by strongly adsorbed substances – compare for CO-adsorption in Figure 4. Three items are immediately extracted from this cyclic voltammogram:

i) Hydrogen is relatively strongly adsorbed on platinum and this voltammogram reveals different adsorption states with different adsorption enthalpies – both with a fixed stoichiometry. In contact with aqueous electrolyte at least two different adsorption sites on polycrystalline Pt can be identified with a free adsorption enthalpy of roughly –15 and –30 kJ per mol of adsorbed H. This adsorption is a stoichiometric reaction with Pt: H = 1:1, which allows coulometric determination (Integral [i dt]) of the free surface of Pt of disperded platinum.

ii) At platinum hydrogen evolution, and electroadsorption/electrodesorption are fast, strongly catalysed approximately reversible reactions demanding little, almost negligible overpotential. The assumption of microscopic reversibility for this reaction on Pt/electrolyte interfaces is justified, provided this interface is clean and not blocked by electrosorbed species of any kind.

iii) In contrast, the anodic oxygen evolution does not proceed at the metallic Pt/electrolyte interface because this Pt metal is almost completely covered by Pt oxides already at +800 mV vs. RHE that means already at a potential more negative than 400 mV compared to the oxygen equilibrium potential. On these Pt oxides, the oxygen evolution and reduction is only weakly catalysed and comparatively much slower than cathodic hydrogen evolution on metallic platinum. Since the double bonded oxygen molecule is relatively stable, neither its reductive splitting nor its anodic formation is fast, must be particularly catalysed and cannot be expected to be a microscopically reversible process.

Volcano plots correlating Me-H bond strengths and catalytic activities

An approach to explain the particular role of platinum as an electrocatalyst for anodic hydrogen oxidation and cathodic hydrogen evolution is correlating the catalytic activity of different metals for this reaction and the strength of the adsorption enthalpy of hydrogen on these metals in Figure 5 [6]. In this figure the current densities measured for cathodic release of hydrogen at a given overpotential are plotted vs. the strength of the Me-H bond (calculated in kJ/mol). The reader may be reminded, that free H–adsorption enthalpies for immersed electrodes, as determined from a voltammogram and Me-H bond strength are not identical but related to each other. Figure 5 reveals in the so-called volcano curve a maximum of obtained current densities and hence electrocatalytic activities for intermediate Me-H bond strengths. That means that with lower bond strengths the interaction between adsorbate and metal is too low to allow for an effective activation of the H-H bond whereas with high Me–H bond strengths the adsorbed hydrogen is bonded too strongly to the electrode surface, so that is inactivated with respect to further reaction – be it further anodic oxidation in a fuel cell or reaction to liberated H_2 in cathodic hydrogen evolution. Therefore the Me-H binding energy of two hydrogen atoms approximately matches the binding energy of the H_2-molecule at most of the platinum metal group. It must be underlined that the volcano curve of Figure 5 is by far not restricted to electrochemical hydrogen reactions but is also observed for chemical reactions, for instance, for hydrogenation reactions of different unsaturated organic compounds and therefore is an indication for chemical catalysis of hydrogen splitting rather than for the charge transfer, that is the electrochemical steps.

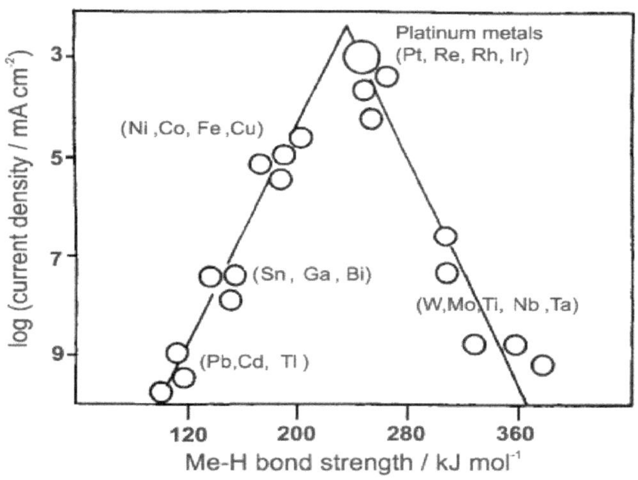

Figure 5. Volcano-type correlation of catalytic activity of different metals and the Me-H binding energy.

A different more fundamental approach to electrocatalysis of the anodic hydrogen oxidation consists in modelling hydrogen adsorption by theoretical chemical means, i.e., by calculations based on the functional density theory[24]. This approach aims at estimating binding energies of adsorbed hydrogen by calculating the surface specific number of empty d-electron states or d-orbitals respectively protruding from the metal surface towards the electrolytes (or for gas phase reactions towards the vacuum) at the metal surface. By taking into account the electronic binding energy of these d-orbitals of the metal (Fermi niveau) and the binding energy of electrons, coming from adsorbed molecules the energy of adsorption is calculated. This constitutes the strength of adsorption bonds of different adsorbates and allows calculating the binding energy of different adsorbed species, for instance hydrogen or carbon monoxide. For different platinum metal group and their alloys the results showed that hydrogen and carbon monoxide adsorption follow the same dependence that means, the stronger hydrogen is adsorbed, the stronger is also the adsorption of CO[25].

Electrocatalysis and cocatalysts for CO-oxidation

Carbon monoxide, CO, is strongly adsorbed on platinum and other metals of the platinum group and it is particularly inert against anodic oxidation, although thermodynamically its oxidation potential is calculated at almost the equilibrium potential of hydrogen. Figure 4 shows a typical voltammogram of platinum electrode immersed in aqueous acidic solution that had been exposed to CO and was recorded after thorough purging of the electrolyte with Ar or N_2 or another inert gas, which expels CO from the electrolyte completely. Even after removal of the carbon monoxide from the solution, the platinum surface remains completely blocked by carbon monoxide[26]. This is first of all documented by the absence of the anodic peaks for adsorbed hydrogen, because of complete displacement of adsorbed hydrogen by much more strongly adsorbed carbon monoxide. Additionally one observes at anodic potential of approximately 0.6 V a relatively broad peak (typical for strongly hindered and slow electrochemical processes), which is due to anodic oxidation of the adsorbed carbon monoxide

($CO_{ad} + H_2O$ ® $CO_2 + 2\,H^+ + 2e^-$). This process is called anodic stripping. And the anodic CO-stripping potential is kinetically – not thermodynamically – controlled proceeding with a high anodic overpotential. It is known, that this retarded oxidation is not a simple electrochemical reaction but that CO-oxidation is mediated by Pt-oxides, which are beginning to form at +0,6 V vs. RHE. Therefore the observed anodic CO-stripping can be formulated in two steps (Equations 5a and 5b):

$$Pt + H_2O \rightarrow PtOH + H^+ + e^- \quad (5a)$$

$$CO_{ad} + 2\,PtOH \rightarrow CO_2 + 2\,Pt + H_2O \quad (5b)$$

The platinum oxide is immediately recuperated by anodic oxidation of the platinum surface at the anodic stripping potential. Only at such high overpotential the platinum surface is cleared of adsorbed CO and only then it becomes available for electrocatalyzed hydrogen oxidation. This means that CO-contaminated hydrogen can only be oxidized at anodic overpotential close to and above 0.6 V. A similar situation does also prevail with anodic oxidation of methanol, ethanol or other energy alcohols, because by their anodic conversion always strongly adsorbed CO is generated which poisons the platinum catalyst after very short time. Therefore metal-additives as cocatalysts are necessary which are more easily oxidized then platinum and whose oxides are able to oxidise adsorbed CO at much lower potential than would platinum oxide.

In Figure 6 is shown the current-voltage curves using an active carbon supported platinum anode for oxidation of pure hydrogen and hydrogen with 150 ppm of CO but using a modified platinum catalyst which contains other metals as a cocatalyst[27]. Clearly the presence of 150 ppm of CO almost destroys the catalytic activity of platinum for anodic hydrogen oxidation due to adsorptive poisoning as almost no current can be drawn from the anode. The addition of other metals, which by themselves are no H_2-oxidizing catalyst, as a potent cocatalyst in a Pt:Me (atomic ratio of 1:1) restores the catalytic activity of platinum. The effect is due to oxygen-transfer or spillover from metal oxides, which are already formed at close to +0.2 V. Interesting enough recent investigations of Baltrushat and co-workers[28] had shown that this effect is produced even if relatively small amounts of molybdenum are adsorbed on platinum.

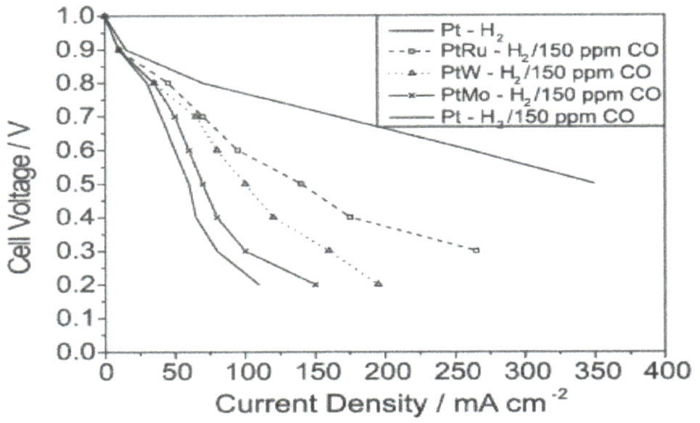

Figure 6. Current-cell voltage correlation of PEMFC with pure and CO contaminated hydrogen for supported pure Pt-electrocatalyst and binaryelectrocatalyst (Pt-loading 0.4 mg cm^{-2}, 75 o C, 1 bar).

The Bifunctional Electrocatalyst Formulation

The first and practically still most important binary catalyst formulation for anodic reformate and methanol conversion is ruthenium which had already been mentioned in the famous conference report edited by Sandstede in 1968[29]. The remarkable cocatalytic efficiency of ruthenium is based on the observation, that this metal is less noble than platinum and oxidizes at a potential at least 0.2 V more negative than platinum and that ruthenium surface oxides/hydroxides therefore can act as a chemical oxidant for CO adsorbed on Pt at a potential, which is between 0.3 and 0.4 V more negative than anodic stripping from pure platinum would demand – provided ruthenium surface oxides and platinum are directly neighboured as would be the case in a surface alloy or if at least Pt and Ru particles would touch each other and the surface diffusivity of CO on platinum is high enough[30], which is the case. Therefore in this case the composite catalyst is called to be "bifunctional", as the platinum due to its adsorptive properties cares for bond breaking in H-H, O-H and C-H groups, whereas ruthenium in the form of its surface oxides mediates the chemical oxidation of the strongly adsorbed poison CO. But ruthenium is not singular and quite a number of elements particularly transition elements like Mo, W and also Ni, all being much less noble than Pt can act as cocatalysts in exactly that sense. The total mechanism is very similar to the anodic stripping reaction of Equation 5, as given by Equations 6a and 6b,

$$Me + H_2O \rightarrow MeOH + H^+ + e^- \quad (6a)$$

$$CO_{\text{adsorbed on Pt}} + 2\ MeOH \rightarrow CO_2 + Me + H_2O \quad (6b)$$

where, Me could be any metal being adsorbed on or alloyed with platinum and which is oxidized at potentials between 0 and 0.6 V vs. RHE as, for instance, Mo, W or others. Table 2 collects some of the alloying elements that had been shown to possess cocatalytic properties in binary Pt-metal electrocatalyst formulations[31]. A particular case is tin. Tin forms a well defined alloy Pt_3Sn, which is an active and particularly efficient cocatalyst for CO-oxidation[32], whose use in membrane fuel cells might become state of the art within the next years.

Table 2. Composition of various CO-tolerant binary Pt-based electrocatalysts

Adittive	Composition
Ru	$Pt_{35}Ru_{65}$
	$Pt_{57}Ru_{43}$
Co	$Pt_{57}Co_{43}$
	$Pt_{82}Co_{18}$
Ni	$Pt_{58}Ni_{52}$
	$Pt_{76}Ni_{24}$
Fe	$Pt_{85}Fe_{15}$
	$Pt_{95}Fe_{0.5}$
Sn	Pt_3Sn
Mo	$Pt_{45}Mo_{55}$
	$Pt_{67}Mo_{33}$

Anodic oxygen evolution in electrolysis processes and cathodic oxygen reduction in fuel cells

Anodic oxygen evolution in aqueous solutions applied, for instance, in water electrolysis or metal electrowinning processes, Equation 7,

$$H_2O \rightarrow 2H^+ + \tfrac{1}{2}\ O_2 + 2\ e^- \quad (7)$$

proceeds always at metal oxide covered metal electrodes, the anodic overpotential usually exceeding 0.2 V and is supposed to proceed according to the so-called Krasilch'shikov mechanism[6], in which unstable, overoxidized metal oxide sites are self stabilizing by mutual redox- or disproportionation reactions by release of molecular oxygen (schematically described by Equation 7a and 7b) with regeneration of the lower valent metal oxide:

$$Me_xO_y + H_2O \rightarrow Me_xO_yOH + H^+ + e^- \quad (7a)$$

$$4\ Me_xO_yOH \rightarrow 4\ Me_xO_y + O_2 + 2\ H_2O \quad (7b)$$

This mechanistic interpretation is based on the observation, that searching for a volcano-like correlation the correct quantity for correlating different catalytic activities has little to do with adsorption of oxygen. It is simply the free enthalpy of formation of the overoxidized metal oxide sites – namely the free enthalpy of formation of the higher oxide from the stable metal oxide. The maximum, the tip of the volcano, is observed, where the equilibrium potential for the lower valent and higher valent metal oxides (which can be calculated from the Gibbs-enthalpy of the oxidation reaction of the metal oxide) matches that of the equilibrium potential of the oxygen electrode (+1.2 V vs. RHE)[33].

Mechanism of the Cathodic Oxygen Reduction

Because of non-existing microscopic reversibility the cathodic reduction of oxygen follows, however, quite a different route than oxygen evolution. By adsorptive interaction of O_2 with the noble platinum metals the molecule is activated towards bond-splitting and electron uptake. The best electrocatalyst for cathodic oxygen reduction hitherto known is platinum metal. Referring to the cyclic voltammogram of Figure 4, one observes the formation of surface oxides on a platinum electrode above +600 mV and the anodic current reaches a plateau above +800 mV as an indication of the formation of an almost closed platinum oxide layer that grows steadily. But even at +800 mV still a minute fraction of the metal surface is not covered by Pt oxide (PtOH) but it becomes less and less with increasing the electrode potential. The degree of coverage, Q, in a simplified correlation, follows a type of electrochemical Langmuir isotherm, Equation 8a and 8b

$$\Theta / (1-\Theta) = K_{ox} \quad (8a)$$

$$K_{ox} = \exp (E - E_{1/2}) F / RT \quad (8b)$$

where, Q is the degree of coverage of the surface by the oxide and K_{ox} is the potential-dependent equilibrium constant. $E_{1/2}$ is the potential at which half of the surface is covered by the oxide. Practically this is the potential at half of the surface oxidation current is measured (approximately 0.7 V). Equation 8a and 8b are only to be understood in a schematic way, but they demonstrate the most important fact, that increasing the anode potential still leaves a smaller and smaller but still finite rest of the platinum surface uncovered by oxide. Therefore oxygen adsorption and reduction is not a question of yes or no but of the degree of Pt-oxide coverage, which depends for a chemical equilibrium reaction on the applied electrode potential.

The adsorptive interaction of electron-donating double bonded oxygen and empty d-orbitals at the Pt-surface acting as acceptors with the ability of d_P - p_P^* back donation activates the oxygen molecule to an extent which allows for a fast transfer of four electrons, a fast sequence of several electron transfer and chemical steps, which are today not yet completely elucidated. This explains, why a relatively high cathodic overpotential of at least – 0.2 V is necessary in order to start with cathodic oxygen reduction at Pt–electrodes, as a rough calculation on the basis of Equatiion 8a clarifies, that the fraction of free metallic surface atoms amounts to roughly 10^{-3} at 1 V vs. RHE (– 0.2 V vs. oxygen equilibrium).

Electrocatalyzed oxygen reduction proceeds at Pt_ electrodes only on metal–sites in the adsorbed state and two parallel reaction pathways are observed in acid solution: the direct reduction of O_2 to H_2O (Equation 9a) and an indirect pathway with H_2O_2 (Equation 9b and 9c)[26,34]:

Direct reduction

$$O_2 + 4H^+ + 4e^- \rightarrow 2H_2O \quad E^0 = 1.23 \text{ V} \quad (9a)$$

Indirect reduction

$$O_2 + 2H^+ + 2e^- \rightarrow H_2O_2 \quad E^0 = 0.682 \text{ V} \quad (9b)$$

$$H_2O_2 + 2H^+ + 2e^- \rightarrow 2H_2O \quad E^0 = 1.77 \text{ V} \quad (9c)$$

In order to the oxygen reduction reaction to take place, the oxygen bond must be weakened, which in turn implies that a strong interaction with the surface of the catalyst will be necessary. If the adsorption of the O_2molecule is end-on, this does not weaken the O-O bond sufficiently, and the indirect mechanism is favoured, whereas bidentate bonding of the O_2 to a surface metal atom, or bridge bonding between two surface metal atoms will favour the direct process, since in this way the O-O bond is substantially weakened[26,34]. The 2-electron reduction of oxygen to hydroperoxide with still intact O-O bond ($O_2 + 2e^- + H_2O$ ® H_2O_2) is produced, for instance, by O_2–reduction to hydroperoxide (H_2O_2) on graphite electrodes in an industrial process in the pulp and paper industry as bleaching agent.

HIGHLY DISPERSE PLATINUM AND PLATINUM ALLOYS: THE ELECTROCATALYST FOR PHOSPHORIC ACID AND MEMBRANE FUEL CELLS

Dispersed Platinum Inverted For Phosphoric Acid Fuel Cells

Already in the late 70ies of the last century UTC (United Technology Corporation) developed highly disperse platinum supported on active carbon as anodic (H_2 oxidation) and cathodic (O_2 reduction) electrocatalyst for phosphoric acid fuel cells[10]. Much older had been the use of so-called platinum black as a form of – although not very well defined – disperse platinum metal for catalytic purposes for instance in the Döbereiner lighter (the oldest and very popular application) and for catalytic hydrogenations in synthetic organic chemistry. Already in this very early times of the 18th century existed the idea to increase the effective catalytic activity of the metal catalyst by dispersing the material as finely as possible. The relevant idea was to make highest possible use of the precious metal by dividing it into very small crystallites. Thus the relevant surface to volume ratio according to Equation 10 for a globular particle can be approximated by 3/r, with r being the particle radius. Thus for a particle of 1 nm (10^{-9} m or 10 Å) the surface to volume ratio amounts to 10^7 cm²/cm³ = 10^7 cm⁻¹. Even more impressive than this number is the knowledge, that in a globular platinum particle of approximately 2 nm diameter roughly 1/3 of all platinum atoms are exposed to the surface and only 2/3 are buried in the interior of this particle.

$$(\text{Surface/Volume})_{\text{spere}} = 4\Pi r^2 / 4/3\, \Pi r^3 = 3/r \quad (10)$$

Practically the utilisation of platinum cannot be expected to be much greater in a particle intended to be stable long term in an operating fuel cell. Highly dispersed materials are thermodynamically unstable. Due to their high surface energy they tend to agglomerate, which would decrease the excess surface energy.

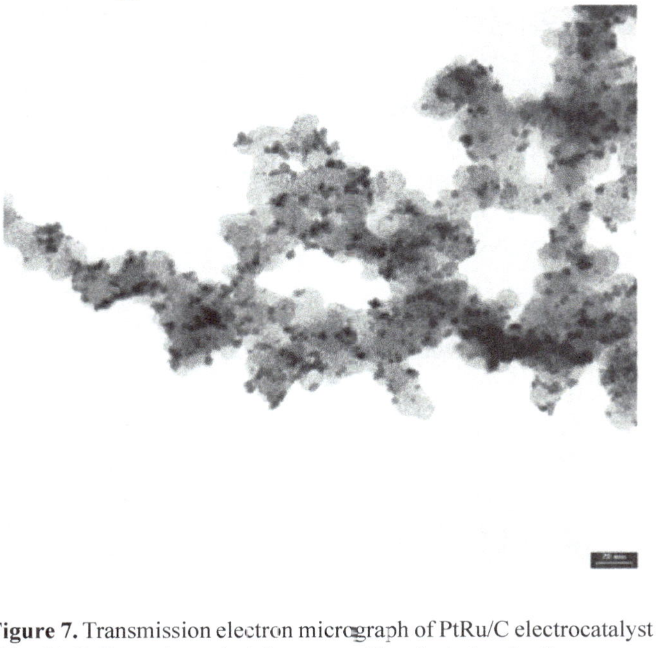

Figure 7. Transmission electron micrograph of PtRu/C electrocatalyst (20wt%, Pt:Ru molar ratio 1:1) prepared by alcohol-reduction process

Hence fast deactivation of the catalyst due to "sintering" would occur. Anchoring such particles by adsorption on the surface of active carbon allows separating these particles from each other stabilizing them long-term. Figure 7 shows a micrograph of such active carbon-supported platinum-ruthenium nanoparticles obtained by transmission electron microscopy[35]. The carbon supported Pt catalysts prevents the fast agglomeration of the nanoparticles, a process that would occur with platinum black, because the platinum particles are touching each other.

INDUSTRIAL ACTIVE CARBON SUPPORTED ELECTROCATALYSTS, THEIR PROPERTIES AND PREPARATION

Properties of Active Carbon

Active carbon is a technical mass product, a type of soot, obtained by incomplete combustion and gas phase pyrolysis of hydrocarbons, unsaturated hydrocarbons as ethene (ethylene) and ethyne (acetylene) and aromatic compounds in particular[11]. The production procedure secures reproducible qualities both with respect to specific surface and with respect to surface chemistry. Active carbons come in different qualities and most frequently are amorphous and possess an inner surface of the order of from several tens to several 100 m^2 per gram. Two of the most popular qualities are Black Pearls 2000® with approximately 1.300 m^2 per gram and Vulcan® XC-72 with 250 m^2 per gram both from Cabot Corp. Literally there are hundreds of different active carbon qualities available on the market. The surface of active carbons is to some extent oxidized, carrying hydroxyl-, carbonyl-, carboxylate-, ester-, ether- and quinone-groups and unsaturated groups _ sometimes also nitrogen. Upon oxidation by strong oxidants like chromic acid they obtain cation exchange properties[11]. Active carbons are, due to their high inner surface, thermodynamically unstable with respect to bulk graphite but non the less they are so inert, that they resist reductive and oxidative attack and form a material, which is long term stable in low temperature fuel cells even at the cathode side, where the working potential is oxidative (+800 mV vs. RHE). Their practical stability is accounting for life times of 20.000 h (vehicles) or even 40.000 h at 200 °C in phosphoric acid fuel cells. It should be kept in mind, that for instance in phosphoric acid fuel cells on the cathode, half of the supporting active carbon disappears by anodic oxidation within 40.000 h of service time or approximately within four years.

Preparation of Dispersed Platinum and Active Carbon Supported Platinum

The preparation of highly dispersed metals proceeds according to two principally different lines[36,37]:

1) On one hand the metal can be prepared separately as a sol or stabilized, solvated (not precipitated) nanodisperse material by deposition of the small metal particles from aqueous solutions of respective salts. Most commonly the precipitation is accomplished by strong reductants and assisted by the presence of surface-active substances, which help to prevent growing of the crystals beyond a certain size.

2) On the other hand the dispersed metal particles can be obtained on the active carbon surface after impregnating the material with noble metal salt solutions, slow drying with salt deposition on the active carbon followed by reduction. The reductive treatment may be most commonly performed by dissolved reductants as formaldehyde or hydrazine or may be performed by flowing hydrogen or other gaseous reductants at relatively low temperature (200-300 °C). As a reductant one may also use the supporting active carbon proper at higher temperatures exceeding 400 °C. By this way the so-called cabothermic reduction of the metal salts on active carbon can be performed.

Concerning these two lines, the preparation of highly dispersed platinum have been done by different approaches[37]:

i) Sols of the metal are obtained for instance by an initial formation of a metastable platinum – sulfito complex, which is inert at ambient temperature but decomposes and produces the small Pt-crystallites at temperatures in excess of 60 °C. Thus a relatively well-defined crystal size of from 2 to 6 nm can be obtained. Bringing then active carbon into contact with the sol "solution" results in anchoring of the metal crystallites on the carbon surface.

ii) Impregnating the active carbon by aqueous or nonaqueous, organic solvents containing platinum salts assisted by the addition of dissolved wetting agents, followed by drying the impregnated material leads to a relatively homogeneous distribution of platinum salts on the active carbon material. Successively reducing these salts, for instance with hydrogen at relatively low temperatures allows creating nanoscale platinum particles whose diameter vary, depending on reduction and preparation conditions, between 2 and 8 nm.

iii) Chemical oxidation of active carbon by strong oxidants as for instance by aqueous solution of chromic acid creates acidic groups (OH-groups, phenolic, and carboxylate groups) on the active carbon surface, which make the active carbon a cation exchange material. From aqueous solutions, which contain positively charged platinum complexes and entities (for instance, amino-complexes of Pt^{+2}), these moieties are absorbed and bound to the active carbon by ion exchange. Chemical or (less often used) electrochemical reduction produces the dispersed metal on active carbon.

Platinum Alloys and Composite Multi-Metal Catalysts for the Anodic Oxidation of Carbon Monoxide Containing Reformate, Of Methanol and Other Energy Alcohols

As documented by a considerable number of patents already in the late 70ies the developers of phosphoric acid fuel cells had detected that from highly dispersed platinum on active carbon, dispersed platinum alloys can be obtained by impregnating of the catalyst with aqueous solutions of transition metal salts and subsequent drying and heat treatment[36]. Subsequent leaching of the active carbon with hydrochloric acid removes excesses of non-alloyed metal oxide and leaves on the support the platinum-transition metal alloy. At that time alloy formation, which is still in habit for PAFC electrocatalysts had the main purpose of catalyst stabilisation. It is known today, that on Pt-transition metal alloys also adsorption of oxygen and its catalytic activation is enhanced compared to pure platinum. Therefore the alloy catalyst is not only more stable but also additionally more active because due to alloying the electronic structure of platinum is changed in a way that favours O_2-adsorption on platinum. In the context of reformate and alcohol consuming fuel cells alloy formation and catalyst formulations of binary and ternary composition became indispensable because cocatalysts are needed to oxidize carbon monoxide and other, strongly adsorbed oxidation- and oxidative degradation products of methanol or the higher alcohols together with hydrogen[38,39]. Ralph and Hogart[40], two scientists of Johnson Matthey published the history of particle stabilisation by alloying (for the cathode) and CO–resistance enhancement by alloying (for the anode) till 2001 and Thompson[41] published a respective article in the Handbook of Fuel Cells.

The Bönnemann Method[42]

Preparing nanodisperse electrocatalysts which are composed of two, three or even more elements, demands that the preparation method produces not a mixture of crystals of the different elements but a well defined molecular mixture – if possible well defined alloys of the elements in the electrocatalyst particles of nanometer size. By precipitation from aqueous solutions this cannot be attained. Therefore, recently the procedure of Bönnemann became very popular. It is essentially based on reductive precipitation of metals from their salt solutions in aprotic media by a borohydride, a very strong reductant. The intention is the forced codeposition of metals irrespective of their redox potential or different nobleness respectively. The tetrabutyl ammonium cation is a strong surfactant that adsorbs on the precipitated crystallites and prevents their growing beyond a size of between 2 and 3 nm. The result is a nanodisperse material in the form of a sol, which then is anchored by adsorption on active carbon as the usual support. Effectively the reductant is tetrabutyl-tridodecyl-borohydride $N(but)_4$ $HB(C_{12}H_{25})_3$. Precipitation for instance of a 1:1:1 mixture of Pt, Ru and W is performed in tetrahydrofuran solution containing dissolved H_2PtCl_6, $RuCl_3$ and WCl_3 in equimolar amounts[43]. By this method almost any desired mixture of elements in any desired molecular ratios can be produced. However, recent investigations show, that not always the desired mixture on molecular level can be obtained as the different reducibility of compounds of different elements occurs sometimes with different rate, so that in the precipitate a certain segregation of the elements is non-voluntarily obtained. Therefore it is clear, that mixed electrocatalysts produced by the Bönnemann method must always be investigated, whether alloys or segregated nanoparticles of variable compositions have been obtained and what is their microstructure and elemental composition.

In Brazil, the electrocatalysts produced by the Bönnemann method with binary, ternary and quaternary compositions have been studied at IPEN/CNEN-SP[44-50](Instituto de Pesquisas Energéticas e Nucleares). Other methodologies of electrocatalysts preparation have also been studied and developed at IPEN/CNEN-SP[51-55] and IQSC/USP[56-61] (Instituto de Química de São Carlos, Universidade de São Paulo).

Means and Methods Characterizing Dispersed Platinum and Platinum Alloy Catalysts

Cyclovoltammetry as a method to characterize the electrochemical properties and electrocatalytic potential of active carbon supported nanodisperse electrocatalyst can be applied in a similar manner as is usual for solid electrodes. It is possible to fix the catalyst on glassy carbon by a submicrometer layer of Nafion[62]. This, however, allows only a measurement on short term for screening the most potent formulations among a preselection of a number of different but similar catalyst formulations. The ultimate electrochemical characterization is the long-term performance test in real fuel cells under typical operation conditions. Here long-term means at least several hundred hours, if not more than 1000 h. For research and development, however, many questions concerning the nanostructure of the electrocatalyst must be asked and answered as:

i) Size and size distribution of the catalysts.

ii) Are the components in binary and ternary catalyst formulations attributed to Pt in a form of alloy?

iii) Are the additives evenly distributed on the support and in what chemical form and environment and what is their valency?

iv) If not alloyed, are the additives present in well-defined crystalline phases or are they X-ray amorphous? And how are they spatially distributes.

The most important method to answer these questions is by Transmission Electron Microscopy (TEM). This method fails with X-ray amorphous material but with well-crystallized catalysts it allows for size and size distribution and even determination of lattice parameters. X-ray diffraction according to Debye-Scherrer answers almost the same questions as widening of the refraction signals gives information about the size of the refracting particles. Another method allowing gross analytic identification of the elements and their relative amounts in the catalyst is XPS (X-ray Photoelectron Spectroscopy) and EDX (Energy Dispersive X-ray Spectroscopy). These methods are usually integrated in scanning electron microscopic equipment[63]. All these methods are typical ex-situ methods as they are performed in separate vacuum-apparatus. In-situ characterisation of the solid state properties of the electrocatalyst is only possible by X-ray absorption of "white" X-ray light which is only available in the form of synchrotron-radiation[64]. Although theory and practice of X-ray Absorption Spectroscopy (XAS) are complex and too much specialised to be treated in the context of this article it should be stressed that a great many questions in particular questions ii), iii) and iv) can be answered by the different forms of XAS. The second in-situ method is IR-Absorption Spectroscopy, which tells many things about specifically adsorbed species but almost nothing about the chemistry and physics of the adsorbent, the catalyst or parts of it.

FINAL CONSIDERATIONS

What will be the technological future of low temperature fuel cells and the respective Research and Development into electrocatalysts and electrocatalysis? For the time being, a certain deceleration of research and development is observed because the hopes for a fast evolving fuel cell automobile market could not be fulfilled. But the task of catalyst research, development and fabrication for low temperature fuel cells is far from being finished. The salient point with fuel cell cars was not technical feasibility but costs – which postponed fuel cell development for cars to the late twenties – and then it will be hydrogen/air not methanol reformate/air cells. Nevertheless with so-called portables running on direct methanol cells, APUs (auxiliary power sources for heavy duty vehicles, ships and planes) running on reformate from gasoline and domestic electricity and heat cogeneration based on natural gas there is ample demand for better and more stable anodic and cathodic electrocatalysts. Small portable fuel cells are operated on liquid methanol (more exactly approx. 1 mol L^{-1}aqueous solutions of methanol). Portables are small cells with electric power of 10 to 100 W, which supply camcorders, PCs and other electronic devices. Portables are in this specialised field with costs of from 15,000 to 18,000 USD/kW cost-effective enough to compete with high-valued batteries because their capacity is remarkably higher (charge lasts for approx. a day instead of 3 to 4 h). In the Internet there are some addresses as for instance Smart Fuel Cell® or Oksolar®, to name only two, which testify commercial activities and success.

Auxiliary Power Units (APU) are another interesting option for cars, ships and aeroplanes, which are operated on reformed gasoline or Diesel. They are supposed to substitute heavy and expensive lead acid batteries. For them the same or very similar catalyst formulations would be needed as for reformate powered cells. Finally domestic power sources, which provide electricity and heat for household fired with natural gas may become an option in the future.

For all these purposes PEM – cells have been and still are being developed. There are from the technical and fuel point of view two very different lines of development: PEMFCs for portables with aqueous methanol solutions as fuel which work at close to ambient temperature and PEMFCs cells for APUs, domestic and other stationary power and heat generation which work on reformate gas at enhanced (though low) temperature around 80 °C or somewhat higher. The demands with respect to the anodic electrocatalyst are very different for these two different technologies. Using as fuel aqueous methanol solution, the stability with respect to temperature needs not to be very high, but long-term stability together with stability against frequent current interruption, fuel starvation and voltage-cycling is very important. Additionally the cathode poses a particular problem. With liquid methanol at the anode, substantial amounts of methanol are leaking to the cathode whose performance is severely impaired by methanol poisoning. There are today interesting approaches to make the cathode less sensitive towards methanol poisoning by applying Ru and Pt containing Chevrel phases – but still the state of the art is rudimentary in this respect and the same holds with the anode catalyst, where still nothing better than Pt/Ru is being applied.

PEM cells for domestic use or for APUs working on reformate can still be improved by more sophisticated anodic electrocatalysts as discussed in this article. It is therefore expected, that intensive R&D research will continue on the still relatively young field of fuel cell electrocatalysis and that many innovations and advances are still ahead. Also production technologies must be developed for new catalysts.

Very important is it to take into account, what will be the advances in membrane-technology in the near future. Intensive work worldwide aims in finding a substitute for the usual Nafion membrane and to introduce less solvated ionomers into fuel cell technology in order to raise the working temperature of the gas-operated cell substantially above 100 °C (even to 200 °C). Clearly, higher operating temperatures demands new electrocatalysts. We do now know a lot about activity enhancement by alloying and this knowledge can also be accounted for in theoretical chemistry – but this has not yet predictive power, so that still experimentation, flanked by the marvellous extended analytical methods will be the usual tools of research and development.

Last but not least it should be mentioned that ethanol is a special and typical fuel option for Brazil. Be it reformate from ethanol for PEM-cells or direct anodic oxidation of ethanol in PEM-cells, for both purposes highly active and selective electrocatalysts are indispensable[65]. In Brazil, the direct anodic oxidation of ethanol has been studied at IPEN/CNEN-SP[53-55] and IQSC-USP[66,67], using electrocatalysts prepared by different methodologies. Pt/Sn electrocatalyst has been shown a performance superior to that of Pt/Ru electrocatalyst for direct anodic oxidation of ethanol (Figure 8)[68].

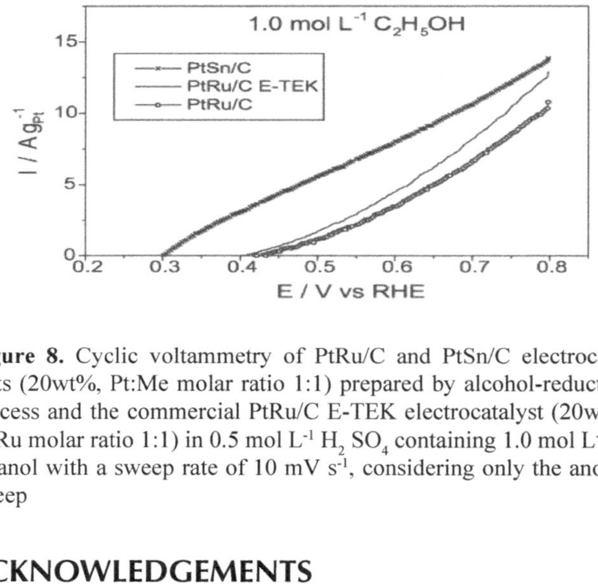

Figure 8. Cyclic voltammetry of PtRu/C and PtSn/C electrocatalysts (20wt%, Pt:Me molar ratio 1:1) prepared by alcohol-reduction process and the commercial PtRu/C E-TEK electrocatalyst (20wt%, Pt:Ru molar ratio 1:1) in 0.5 mol L^{-1} H_2SO_4 containing 1.0 mol L^{-1} of ethanol with a sweep rate of 10 mV s^{-1}, considering only the anodic sweep

ACKNOWLEDGEMENTS

H. Wendt is indebted to IPEN/CNEN-SP, to the Brazilian Ministry of Science and Technology and to FINEP for obtaining a scholarship at São Paulo in the year 2003, when he spent four weeks lecturing on *Electrocatalysis*and on *Modern Energy Technologies* as guest lecturer with the group of M. Linardi. His and his collaborators kind hospitability is gratefully acknowledged.

Financial supports of CTPETRO/FINEP and FAPESP (Fundação de Amparo a Pesquisa do Estado de São Paulo) for the research activities at IPEN/CNEN-SP are greatly acknowledged.

REFERENCES

1. Carrette, L.; Friedrich, K. A.; Stimming, U.; *Fuel Cells* 2001, *1*, 5.
2. Stull, D. R.; Prophet, H.; *Janaf Thermochemical Tables*, 2nd ed., US National Bureau of Standards: Washington,

1971.

3. Barin, I.; Knacke, O.; *Thermodynamical Properties of Inorganic Substances*, Springer Verlag: Berlin, 1973
4. Kauffmann, G. B.; *Platinum Metals Rev.* 1999, *43*, 122.
5. Wendt, H.; Rausch, S.; Borucinski, T.; *Adv. Catal.* 1994, *40*, 87.
6. Wendt, H.; Kreysa, G.; *Electrochemical Engineering, Science and Technology in Chemical and Other Industries,* Springer: Berlin, 1999, ch. 4 and 10.
7. Kemp, F.S.; George, M. A.; *US pat. 3,857,737* 1974.
8. Petrov, H. G.; Allen, R.J.; *US pat. 3,992,512* 1976.
9. Kinoshita, K.; Stonehart, P. In *Modern Aspects of Electrochemistry*; Bockris, J. O'M.; Conway, B. E., eds.; Plenum Press: New York, 1977, vol. 12.
10. Stonehart, P.; *Ber. Bunsen Phys. Chem.* 1990, *94*, 913.
11. Kinoshita, K.; *Carbon – Electrochemical and Physical Properties*, Wiley and Sons: New York, 1987.
12. Bacon, F. T.; *Electrochim. Acta* 1969, *14*, 569.
13. Ketelaar, J. A. A. In *Fuel Cell Systems;* Blomen, L. M. M. J.; Mugerwa, M. N., eds.; Plenum Press: New York, 1993.
14. Winsel, A.; Richter, G. J. In *Alkaline Fuel Cells in Electrochemical Hydrogen Technologies*; Wendt, H., ed.; Elsevier: Amsterdam, 1990
15. Watkins, D. S. In ref. 13.
16. Strasser, K.; *J. Electrochem. Soc.* 1980, *127*, 2172.
17. Steele, B. C. H.; Heinzel, A.; *Nature* 2001, *414*, 345.
18. Neergat, M.; Friedrich, K. A.; Stimming, U. In *Handbook of Fuel Cells – Fundamentals, Technology, Applications*; Vielstich, W.; Lamm, A.; Gasteiger, H. A., eds.; J. Wiley & Sons: New York, 2003, vol. 4, chap. 63
19. Acres, G. J. K.; Frost, J. C.; Hards, G. A.; Potter, R. J.; Ralph, T. R.; Thompsett, D.; Burstein, G. T.; Hutchings, G. J.; *Catal. Today* 1997, *38*, 393.
20. Roth, C.; Goetz, M.; Fuess, H.; *J. Appl. Electrochem.* 2001, *31*, 793.
21. Trasatti, S.; *Electrochim. Acta* 2003, *48*, 3729.
22. Debe, M. K. In ref. 18, vol. 3, chap. 45.
23. Vielstich, W. In ref. 18, vol. 2, chap. 14.
24. Liu, P.; Norskøv, J. K.; *Fuel Cells* 2001, *1*, 192.
25. Liu, P.; Logadottir, A.; Norskøv, J. K.; *Electrochim. Acta* 2003, *48*, 3731.
26. Hamann, C. H.; Hammett, A.; Vielstich, W.; *Electrochemistry*, Wiley-VCH: Weinheim, 1998.
27. Goetz, M.; Wendt, H.; *Electrochim. Acta* 1998, *43*, 3637.
28. Samjeske, G.; Wang, H.; Loftler, T.; Baltruschat, H.; *Electrochim. Acta* 2002, *47*, 3681.
29. Binder, H.; Köhling, G.; Sandstede, G.; *From Electrocatalysis to Fuel Cells*, University of Washington Press: Seattle, 1972.
30. Friedrich, K. A.; Geyzers, K-P.; Linke, U.; Stimming, U.; Stumper, J.; *J. Electroanal. Chem.* 1996, *402*, 123.
31. Watanabe, M. In ref. 18, vol. 2, chap. 28.
32. Gasteiger, H. A.; Marcovic, N. M.; Ross, P. N.; *J. Phys. Chem.* 1995, *99*, 8945.
33. Trasatti, S.; *Electrochim. Acta* 1984, *29*, 1503.
34. Tanata, A.; Kanamura, K.; Adzic, R.; Cahan, B.; Yeager, E.; *Proc.- Electrochem. Soc.* 1990, *90-1*, 969.
35. Spinacé, E. V.; Neto, A. O.; Vasconcellos, T. R. R.; Linardi, M.; *Br. Patent INPI-RJ* PI0304121-2,2003.
36. Jalan, V. M.; *US pat. 4,202,934* 1980.
37. Mukerjee, S.; Srinivasan, S. In ref. 18, vol. 2, chap. 34.
38. Torresi, R. M.; Wasmus, S. In ref. 18, vol. 2, chap. 15.
39. Iwasita, T.; *Electrochim. Acta* 2002, *47*, 3663.
40. Ralph, T. R.; Hogarth, M. P.; *Platinum Metals Rev.* 2002, *46*, 117.
41. Thomsett, D. In ref. 18, vol. 3, chap. 37.
42. Bönnemann, H.; Brijoux, W.; Brinkmann, R.; Fretzen, R.; Joussen, T.; Köppler, R.; Korall, B.; Neiterle, P.; Richter, J.; *J. Mol. Catal.* 1994, *86*, 129.
43. Goetz, M.; Wendt, H.; *J. Appl. Electrochem.* 2001, *31*, 811.
44. Neto, A. O.; Franco, E. G.; Aricó, E. M.; Linardi, M., Gonzalez, E. R.; *International Workshop on Ceramic & Metal Interfaces*, Oviedo, Spain, 2002.
45. Franco, E. G.; Oliveira Neto, A.; Linardi, M.; Aricó, E.; *J. Braz. Chem. Soc.* 2002, *13*, 516.
46. Neto, A. O.; Franco, E. G.; Aricó, E. M.; Raimundo, C. P.; Linardi, M.; *XV Congresso da Sociedade Ibero-Americana de Eletroquímica*, Évora, Portugal, 2002.
47. Oliveira Neto, A.; Franco, E. G.; Aricó, E.; Linardi, M.; Gonzalez, E. R.; *J. Eur. Ceram. Soc.* 2003, *23*, 2987.
48. Neto, A. O.; Franco, E. G.; Aricó, E. M.; Spinacé, E. V.; Linardi, M.; *Fuel Cells Science and Technology,* Amsterdam: Netherlands, 2002.
49. Franco, E. G.; Aricó, E.; Linardi, M.; Roth, C.; Martz, N.; Fuess, H.; *Mater. Sci. Forum* 2003, *416*, 4.
50. Franco, E. G.; Neto, A. O.; Raimundo, C. P.; Aricó, E. M.; Linardi, M.; *53rd Annual Meeting of the International Society of Electrochemistry*, Dusseldorf, Germany, 2002.
51. Spinacé, E. V.; Oliveira Neto, A.; Franco, E. G.; Linardi, M.; Gonzalez, E. R.; *Quim. Nova* 2004, *27*, 648.
52. Neto, A. O.; Franco, E. G.; Spinacé, E. V.; Linardi, M.; *54th Annual Meeting of the International Society of Electrochemistry*, São Pedro, Brasil, 2003.
53. Spinacé, E. V.; Neto, A. O.; Linardi, M.; *J. Power Sources* 2003, *124*, 426.
54. Spinacé, E. V.; Neto, A. O.; Linardi, M.; *J. Power Sources* 2004, *129*, 121.
55. Spinacé, E. V.; Neto, A. O.; Vasconcelos, T. R. R.; Linardi, M.; *J. Power Sources* 2004, *137*, 17.
56. Castro Luna, A. M.; Camara, G. A.; Paganin, V. A.; Ticianelli, E. A.; Gonzalez, E. R.; *Electrochem. Commun.*2000, *2*, 222.
57. Lizcano-Valbuena, W. H.; Souza, A.; Paganin, V. A.; Leite, C. A. P.; Galembeck, F.; Gonzalez, E. R.; *Fuel Cells* 2002, *2*, 159.
58. Colmati Jr., F.; Lizcano-Valbuena, W. H.; Camara, G. A.; Ticianelli, E. A.; Gonzalez, E. R.; *J. Braz. Chem.*

Soc. 2002, *13*, 474.

59. Camara, G. A.; Giz, M. J.; Paganin, V. A.; Ticianelli, E. A.; *J. Electroanal. Chem.* 2002, *537*, 21.

60. Lizcano-Valbuena, W. H.; Paganin, V. A.; Leite, C. A. P.; Galembeck, F.; Gonzalez, E. R.; *Electrochim. Acta*2003, *48*, 3869.

61. Lizcano-Valbuena, W. H.; Azevedo, D. C.; Gonzalez, E. R.; *Electrochim. Acta* 2004, *49*, 1289.

62. Schmidt, T. J. In ref. 18, vol. 2, chap. 22.

63. Hoster, H. E.; Gasteiger, H. A. In ref. 18, vol. 2, chap. 18.

64. Adzic, R. R.; Wang, J. X.; Ocko, B. M.; McBreen, J. In ref. 18, vol. 2, chap. 20.

65. Lamy, C.; Belgsir, E. M.; J-M. Leger, J-M.; *J. Appl. Electrochem.* 2001, *31*, 799.

66. Oliveira Neto, A.; Giz, M. J.; Perez, J.; Ticianelli, E. A.; Gonzalez, E. R.; *J. Electrochem. Soc.* 2002, *149*, A272.

67. Souza, J. P. I.; Queiroz, S. L.; Bergamaski, K.; Gonzalez, E. R.; Nart, F. C.; *J. Phys. Chem. B* 2002, *106*, 9825.

68. Spinacé, E. V.; Oliveira Neto, A.; Vasconcelos, T. R. R.; Linardi, M.; *Abstracts of the VI Encontro Regional de Catálise, Americana*, Brasil, 2004.

CHAPTER 13

ADVANCES AND RECENT TRENDS IN HETEROGENEOUS PHOTO (ELECTRO)-CATALYSIS FOR SOLAR FUELS AND CHEMICALS

James Highfield

Keywords: Solar fuels, heterogeneous photocatalysis, water splitting, renewable hydrogen, visible sensitization, hydrogen peroxide, bio-oxygenates photoreforming, electrocatalysis, carbon dioxide reduction

INTRODUCTION

In 2004 Nobel Laureate Richard Smalley, discoverer of fullerenes and pioneer of modern nanoscience, gave a talk on the 10 great challenges facing humanity in the 3rd millenium. Placing *energy* at the top of the list, he proposed divestiture of fossil fuels in favour of sustainable and environmentally benign alternatives [1]. He also envisioned that nanotechnology, *i.e.*, the design and assembly of nanometer scale structures, would play a key role in our future prosperity. In fact, it can be argued that nanomaterials have already been exploited for many years in the form of heterogeneous catalysts in industrial chemical processing, responsible for the rapid growth in civilized life during the 20th century. Much of this was done with only marginal understanding of their workings prior to the advent of *in-situ* surface characterization, high-resolution imaging tools, and theoretical (computational) modeling [2,3]. Nowadays, the development of nanocatalysts with functionality optimized "by rational design" is a popular theme [4,5,6,7,8] but it remains an uphill challenge due to the myriad complexity of catalytic phenomena [9]. Nonetheless, recent articles focusing on the prospects for such materials in renewable energy applications give some ground for optimism [10,11,12]. They also introduce the main topic of this report, viz., the prodigious growth in research into heterogeneous photocatalysis, as attested by the drastic increase in literature citations over the last 20 years. While many of these concern environmental or "advanced oxidation" applications [13], multitudinous examples related to energy topics can be located under key (search) phrases like solar fuels, photocatalytic hydrogen, photo-reforming, water photo-splitting, CO_2 photo-reduction, *etc.* A photocatalyst can be defined as "a solid material that accelerates a chemical reaction by light absorption while itself remaining unchanged" [14]. Just as in a thermal heterogeneous catalyst a chemical process is made faster due to a significant lowering of the energy barrier of the associated transition state, e.g., by providing surface sites that activate unique modes of adsorption. On a typical semiconductor photocatalyst like titanium dioxide (TiO_2), the adsorbed (dark) state is further activated (or entirely new states are created) by surface interaction with highly energetic photo-induced charges, viz., electrons (e^-) and holes (h^+). Band-gap excitation of TiO_2 ($\lambda \leq 400$ nm) creates photons with an energy $\geq$300 kJ/mol. In principle, the use of photocatalysts enables substitution of expensive process heat by cheap solar (photon) energy, leading to reduced operating costs. However, an added impetus for research is that they can also drive *endergonic* (thermodynamically uphill) processes by converting light energy into stored chemical energy (bond enthalpy). In other words, the reaction selectivity can be steered

Citation: James Highfield, Advances and Recent Trends in Heterogeneous Photo (Electro)-Catalysis for Solar Fuels and Chemicals, doi:10.3390/molecules20046739.

towards more useful products. By analogy with the natural process, when the reactants are water and carbon dioxide, this is sometimes referred to as *artificial photosynthesis* [15], in which H_2 and/or its reaction product with CO_2 is isolated as a *solar fuel*.

To better appreciate the future impact of any viable solar fuel technology, it is helpful to consider it in the broader context of renewable energy schemes and their current limitations. In principle, the ideal energy carrier from an energetic and environmental viewpoint is hydrogen. On a weight basis, it has the highest energy density of any fuel (143 MJ/kg, or 3× the value of gasoline) and it burns cleanly and efficiently to water, producing heat and/or electrical power in a fuel cell [16]. Unfortunately, H_2 also has the lowest volumetric energy density under ambient conditions (0.011 MJ/L), making its storage in physical form (compressed, liquefied, adsorbed, *etc.*) impractical and expensive, especially for applications in the transportation sector [17]. Any renewable fuel should retain the positive attributes of gasoline but offer the environmental benefits of H_2. It must be a liquid (for ease of handling) and have a practical energy density (≥20 MJ/L). It should also contain a substantial level of "incipient hydrogen" (≥12 wt %) and be carbon-neutral in the long term. The recent (transitional) strategy of fuel "decarbonization" aims to exploit the clean energy associated with the H-component in fossil fuels as these become depleted [18].

A previous review by this author [19] dealt almost exclusively with *thermal* heterogeneous catalytic processes in a future renewable hydrogen energy system schematized in Figure 1.

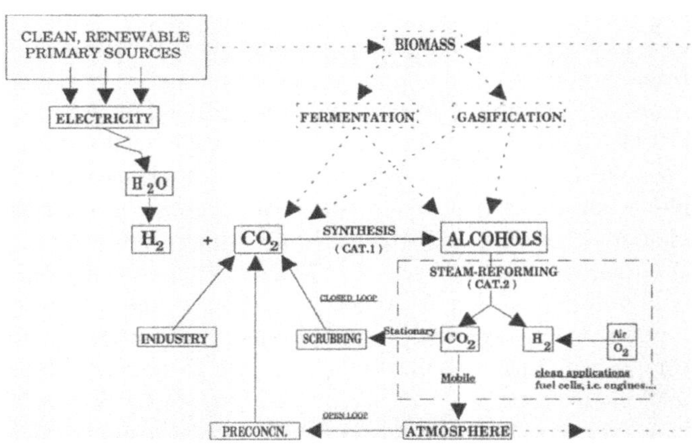

Figure 1. Idealized energy scheme based on renewable H_2 stored in simple alcohols (reproduced from [19] with permission).

In a gas phase catalytic process, H_2 derived from renewable primary sources (solar-electric, hydro-electric, *etc.*) is converted to alcohol(s) by a synthesis reaction with CO_2, itself recycled from industrial sources in concentrated form (closed loop). In the near future, the more challenging (open loop) process, *i.e.*, direct capture of CO_2 from the atmosphere, may become technical reality [20,21]. Simple alcohols are ideal H-carriers as they can be synthesized quite selectively while H_2can be released at relatively low temperature. Insofar as biomass can supply renewable H_2, alcohols and oxygenates (even sugars) by gasification, fermentation, carbohydrate hydrolysis, *etc.* [22], this natural renewable (CO_2-neutral) energy source is integrated into the overall scheme. Essentially the same scientific precepts have since been advanced in a monograph by Olah *et al.*, which focuses on storage solely as methanol since this is already technically feasible [23]. The current status of *The Methanol Economy* has been the subject of a recent review, which re-emphasizes that the price of the alcohol will be strongly dependent on the cost of renewable electricity for H_2 generation [24]. Cost analyses show that "renewable" methanol will probably be 2–3× more expensive than the methane-based commodity [25] (see also Section 7). As regards the main catalytic cycle, the alcohol(s) synthesis can be written in general form as:

$$n\,CO_2 + 3n\,H_2 \leftrightarrow C_nH_{2n+1}OH + (2n-1)\,H_2O \qquad (1)$$

Under high pressure, reaction (1) is favoured ($\Delta G < 0$) at ambient temperature but synthesis catalysts generally operate above 220 °C due to kinetic limitations. Product selectivity in ethanol synthesis is a major challenge [26,27]. Upon demand, H_2 is released catalytically from aqueous alcohol vapours to generate heat and power via steam-reforming (SR):

$$C_nH_{2n+1}OH + (2n-1)\,H_2O \leftrightarrow n\,CO_2 + 3n\,H_2 \qquad (2)$$

SR is an endothermic process ($\Delta H \geq 0$) but reaction (2) is theoretically feasible ($\Delta G \leq 0$) near ambient for methanol, and above 230 °C for ethanol due to the molar volume increase (entropy factor). Since SR is simply the reverse of synthesis, catalysts for both reactions are similar. Formulations based on Cu (modified by Co, Ni, Pd, Rh, *etc.*) on various oxide supports are quite effective. The main technical hurdle is the need for "fuel processing" because low levels of carbon monoxide in the reformate poison the fuel cell (Pt) anode and must be eliminated [28,29]. In addition, although it has been a growing area of research in the last 20 years, ethanol SR still suffers from catalyst deactivation due to carbon deposition, probably linked to the high temperatures necessary to establish good conversion rates [30,31]. Nevertheless, the scope for bio-hydrogen resources now encompasses a range of oxygenates and polyols, e.g., glycerol (a waste product from bio-diesel synthesis [32,33]), in which coking can be minimized via aqueous-phase-reforming (APR) under mild conditions [34].

For maximum technical impact, photocatalysis should logically be applied to the most energy-demanding steps in the scheme under consideration. Artificial photosynthesis to create a fuel such as methanol from aqueous CO_2:

$$CO_2 + 2\,H_2O \leftrightarrow CH_3OH + 1.5\,O_2 \qquad (3)$$

is highly endothermic ($\Delta H = +727$ kJ/mol) but can be driven by solar photons in the visible/near IR region because it is 6 e^-process. Since reaction (3) involves water splitting (H abstraction) implicitly, it effectively couples two stages in the above scheme (H_2 generation and methanol synthesis), offering a potential process simplification. Supplying the energy needed for (endothermic) steam-reforming of alcohols ($\Delta H° \approx +130$ kJ/mol CH_3OH or $+175$ kJ/mol C_2H_5OH) by means of photons is another obvious prospect, giving rise to intensive recent interest in *photo-reforming* [35]. In the review that follows, space limitations and a plethora of recent review articles [10,11,36,37,38,39,40,41,42,43,44,45,46,47,48,49,50,51,52,53,54,55,56] act as major con-straints on detail and impose a necessarily high degree of selectivity in given examples. Similarly, the emerging field of "renewable fuels from CO_2 and H_2O by solar-*thermal* processes" is beyond its scope [57]. As the title implies, this review deals only with materials that respond to (absorb) a significant fraction of the solar power spectrum, over 90% of which lies in the visible and near infrared region. Thus, recent advances in "sensitization" methods for TiO_2, the benchmark photocatalyst, are covered in some depth along with exploration of stable and non-toxic semiconductors of more suitable bandgap, either for independent use or in tandem (composite) arrangement for improved efficiency. The author has endeavoured to strike a balance between topicality and novelty, and apologizes in advance to the many authors who are not cited directly. Advances in selected materials are given regardless of their testing configuration, be it as dispersed nanoparticles in suspension or as "wired" electrodes in a photoelectrochemical cell (PEC). The question of the degree of complexity of any solar fuel "device", its future amenability to scale-up, and ultimate impact on process techno-economics (fuel cost) is still at an early stage of evaluation [58,59,60,61,62]. This is considered near the end of the review, along with the inevitable trend towards devices composed of "earth-abundant" elements [62,63,64,65,66,67,68], as necessitated by the ultimate (global) scale of fossil fuels replacement (see Section 7).

WATER SPLITTING AND TARGET EFFICIENCY IN SOLAR HYDROGEN GENERATION

The fundamental groundwork in evaluating the maximum efficiency achievable in a solar photonic device was done by Bolton and co-workers [69,70,71], taking water splitting as the model reaction:

$$H_2O \rightarrow H_2 + 0.5\ O_2 \quad (4)$$

This two-electron, or effectively two-photon (one photon per H atom), reduction process is highly energetic and thermodynamically uphill by 237 kJ/mol H_2 or 119 kJ/mol photons, corresponding to a wavelength of 1008 nm (1.23 eV—all redox potentials are given *vs.* the normal hydrogen electrode at pH = 0). Inspection of the solar spectrum in Figure 2 shows that photons of greater energy ($\lambda \leq 1008$ nm) constitute only one half of the total incident solar flux.

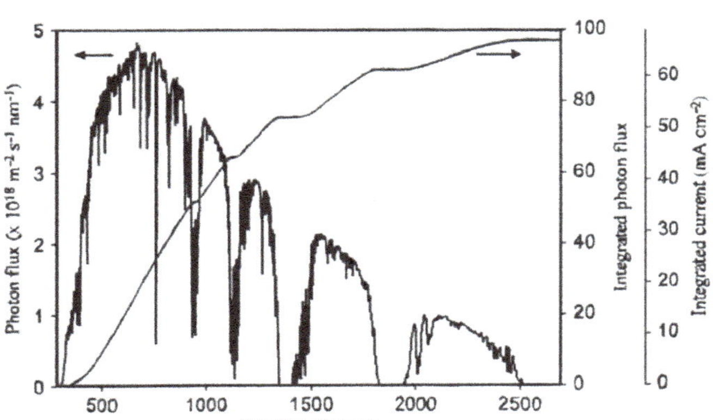

Figure 2. Solar photon flux at the earth's surface *vs.* wavelength and integrated current density obtainable from an ideal PV cell (adapted from [72] with permission of Elsevier).

Furthermore, not all the energy in these photons is available for chemical energy storage after absorption due to intraband*thermalization*, *i.e.*, ultra-fast relaxation from excited vibrational states of the first electronic excited state (or conduction band in a semiconductor).

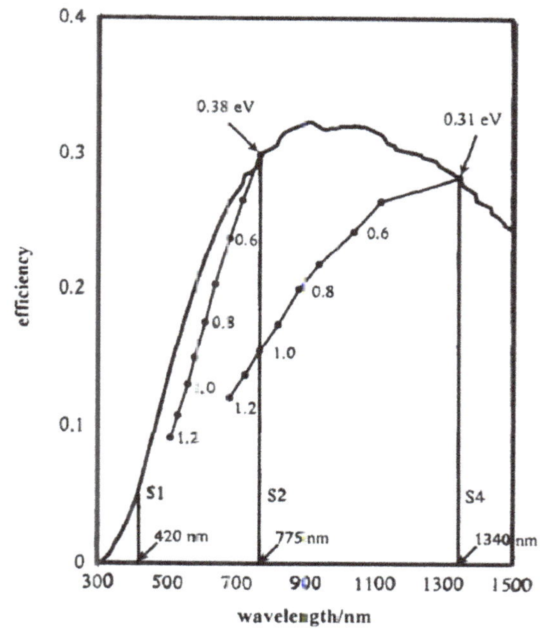

Figure 3. Solar photon-to-chemical energy conversion efficiency in a single bandgap current device (adapted from [71] with permission of Elsevier).

To compensate for this energy loss the wave-length threshold must shift to 775 nm in this so-called S2 process, as shown in Figure 3. When entropic losses (non-equivalence of internal energy and Gibbs energy) are also factored in, the maximum efficiency (η_{STH}) of a device for solar-to-hydrogen (STH) energy conversion based on a single photo-system is around 30%. Also shown in Figure 3 are plots with more realistic entropic losses (>0.4 eV) and their progressive erosion of efficiency. In principle, a 4 e^- (S4) process extends the useful spectral range to 1340 nm based on successive absorption of two low energy photons to drive a single electron event like proton discharge ($H^+ + e^- \rightarrow H\cdot$). For a semiconductor, this is a futuristic concept because it would require hypothetical long-lived mid-gap states to be populated in the first photon absorption event, as shown in Figure 4. This will require a major advance in "bandgap engineering", but may yield so-called 3rd generation photocatalysts (*vide infra*) [73,74].

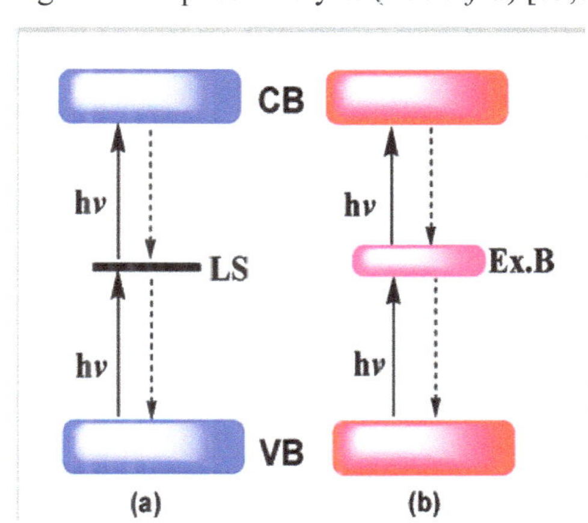

Figure 4. Successive two-photon excitation (**a**) via localized dopant states (LS) (**b**) via delocalized states in an extrinsic band (Ex.B) (reproduced from [74] with permission of ACS.

On the other hand, coupling of two absorbers of complementary bandgap and band-edge position (type II -see Figure 5) each absorbing two photons in a dual (D4) or tandem device is a viable water splitting configuration and offers better matching to the solar spectrum [75,76]. In practice, the mechanistic complexity (kinetic barrier) in the water oxidation half-cell reaction has kept conversion efficiencies below 2% until recently [60,61,62,77,78]. Efficiencies in the more challenging process of CO_2 photo-reduction are still below 0.5% [79]. The minimum workable efficiency for implementation of any solar-to-hydrogen (STH) process is taken as 10%, leading to a H_2 price approaching the DOE target of $4.00 per kg [60]. Immediately obvious from Figure 2 and Figure 3 is that pristine TiO_2, the prototypical semiconductor photocatalyst [80,81] with a band-gap of ~3 eV ($\lambda \leq 400$ nm), absorbs just 4% of solar light. However, since this oxide is cheap and non-toxic and has otherwise excellent material properties, suitable energetics (band edge positions) to drive both proton reduction and water oxidation, stability in aqueous environment, *etc.*, ways of *sensitizing* TiO_2 to visible light are being studied intensively. At the same time, other non-TiO_2-based systems with intrinsic visible absorption are urgently sought.

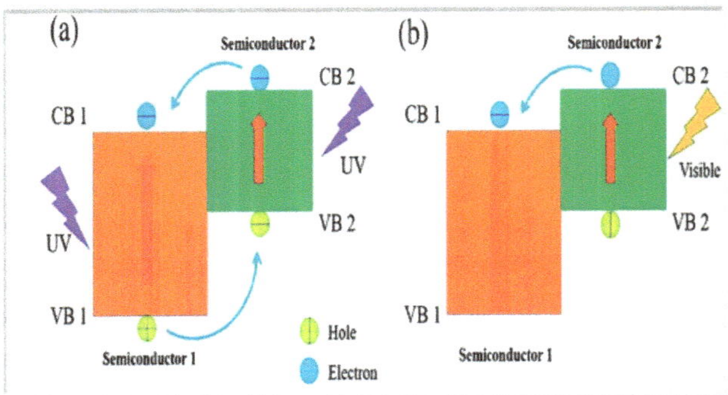

Figure 5. Dual absorbers with complementary bandgap/band edge positions (type II heterojunction) under (**a**) UV; and (**b**) Visible excitation (reproduced from [82] with permission of PCCP Owner Societies).

Incorporation of materials that perform efficient solar light harvesting is a fundamental (thought not the only) pre-condition for an effective energy conversion device. While it is generally assumed that absorption of one photon of energy exceeding the bandgap creates one electron/hole pair (exciton) with 100% efficiency, the fate of these excited states is less clear. One advantage of screening photocatalysts in "wired" mode, e.g., as photoanodes in an electrochemical cell, is that it promotes better charge separation via "band-bending", viz., the space-charge (depletion) layer formed spontaneously at the semiconductor/liquid junction. It also facilitates measurement of the combined efficiency of the two key initiatory processes (light absorption and charge collection) in terms of the resulting photocurrent. The incident photon-to-current conversion efficiency (IPCE) is a figure of merit for any photovoltaic (PV) cell. The photocurrent density limit shown in Figure 2 (~65 $mA \cdot cm^{-2}$) is the ideal value achievable under so-called AM1.5G (1 Sun) illumination, corresponding to an incident optical power close to 1000 $W \cdot m^{-2}$ or 100 $mW \cdot cm^{-2}$ [72,83]. In solar fuel generation, these photo-generated charges are required instead to drive surface redox chemistry, in which the presence of suitable co-catalysts and rapid turnover of the substrate (diffusion) are crucial to performance. Due to these kinetic limitations, STH efficiencies lie below the PV (IPCE) value.

ADVANCES IN ABSORBER MATERIALS WITH IMPROVED SOLAR SPECTRAL MATCHING

Modified TiO_2

The greatest success has been achieved in the so-called mesoscopic Dye-Sensitized Solar Cell (DSSC), or Grätzel cell as named after its inventor [84]. The basic principle is illustrated in Figure 6. Upon photo-excitation of the chemically-anchored dye monolayer, electrons are injected to the conduction band of the mesoporous TiO_2 substrate indirectly via an excited state (S*). Back electron transfer to the oxidized dye is prevented by rapid dye regeneration (reduction) by a donor species (I^-) present at high concentration in the electrolyte. To complete the circuit, the oxidized form of the mediator (I_3^-) is discharged by electron flow through the external circuit to the Pt cathode. The energy difference between the TiO_2conduction band and the redox level of the mediator (I^-/I_3^-) determines the maximum open circuit voltage ($V_{oc} \approx 0.8$ V), while the IPCE generally exceeds 70% up to 700 nm, leading to photocurrent densities ≈ 15 $mA \cdot cm^{-2}$. With a typical "fill factor" (non-ideality in the power curve) of around 0.75, the resulting overall efficiency is ~10%.

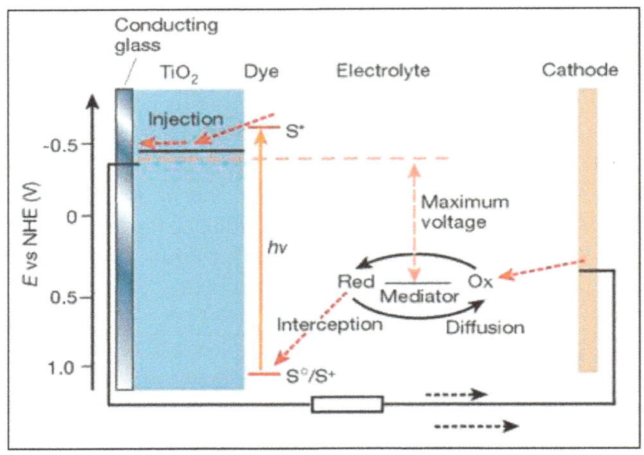

Figure 6. Working principle of dye-sensitized solar cell (DSSC) (reproduced from [76] with permission of Nature Publishing Group).

Most recently, this has been extended to 13% while at the same time reducing costs, substituting the previous Ru-based panchromatic (black) dyes with more absorptive (Zn-based) donor-π-acceptor porphyrins and the slightly corrosive I^-/I_3^- redox mediator by Co^{2+}/Co^{3+}-based electrolytes [85]. Although the DSSC is strictly speaking a photovoltaic (PV) device rivaling conventional Si, their recent coupling into water splitting PEC cells has led to remarkably high efficiencies of H_2generation (*vide infra*) [61,62]. Future advances can be expected from co-sensitization utilizing mixed dye systems, one of which absorbs in the near infrared region [86,87,88,89,90,91]. Alternatively, by varying its surface binding configuration, a single dye may achieve charge injection by both *direct* (type II—as exemplified by catechol, EDTA, *etc.* (see Figure 7b [92]) and *indirect* (type I—see Figure 7a) modes, thereby rendering it more panchromatic in response [93,94]. Although it has not yet come to fruition, pure type II sensitization should increase DSSC efficiency by eliminating the electron injection overpotential, *i.e.*, the energy loss due to thermalization from the excited state of the dye (S*) in the conventional (type I) process [93]. The general prospects for exploiting direct ligand-to-metal (Ti^{4+}) charge transfer (LMCT) absorption in photocatalysis over TiO_2 have been reviewed [95]. It can be recognized by the appearance of a new absorption band absent in either of the free components. One serendipitous example is self-activation of hydrogen peroxide (which forms the yellow peroxotitanate complex on adsorption) for visible-driven environmental applications of TiO_2 [95,96]. The most exciting recent discovery in the DSSC field is the advantage of utilizing panchromatic semiconductor alkyl-ammonium lead (tin) trihalide perovskites as solid-state (layer) sensitizers [97,98,99,100], simultaneously replacing both the conventional dye and redox mediator. This technology-disruptive advance should ease fabrication costs and accelerate the development of a cheaper alternative solid-state PV cell of similar durability and efficiency to Si.

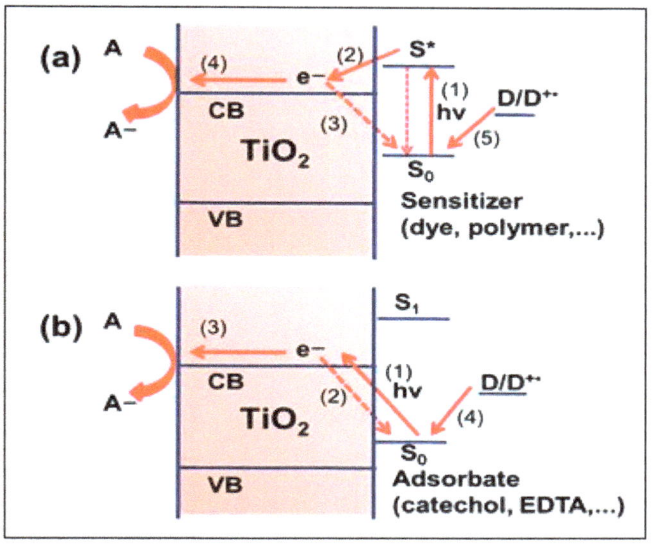

Figure 7. Scheme of charge transfer modes from sensitizer (S) to the TiO_2 conduction band (CB): (**a**) indirect (type I); (**b**) direct (type II or LMCT) (reproduced from [95] with permission of the Royal Society of Chemistry).

In particulate systems, bulk doping of TiO_2 with inorganic (metallic or non-metallic) elements has been a major strategy for visible sensitization in the last decade. Although this approach offers better long-term photostability (as compared to sensitization by organic dyes), the complexities of the solid-state chemistry have been decidedly more challenging. This is linked primarily to restricted solubility of dopants when introduced individually (1–2 at%), leading to insufficient visible light absorption. Furthermore, the concomitant introduction of defects that act as recombination centres has often led to efficiency losses. These are commonly O-vacancies but the dopant site itself can act deleteriously if present above a certain threshold concentration [101]. It is inadequate merely to impart colour to TiO_2, e.g., from transition metal ions (TMI) with a d^1–d^9 electronic configuration. Such optical transitions, being mostly localized (d ↔ d) type, do not involve charge transfer and merely act as parasitic absorbers competing with genuine (delocalized) charge injection. In a few cases, intervalence charge transfer may be effective provided the energy state of the photo-reduced acceptor lies above the bottom of the conduction band, e.g., Ti^{3+}, V^{2+}, or Fe^{2+} [102]. Early interest in TMI doping in TiO_2 was intended to improve the efficiency of charge separation and/or to extend the lifetime of surface-trapped carriers for photochemical action [103,104]. Any advantage to be gained by applying the TMI strategy to visible-light sensitization remains debatable as attested by more recent literature [105,106]. Indeed, it is increasingly recognized that incorporation of TMIs with empty (d^0) or filled (d^{10}) d-shells gets better results [107,108]. The recent flurry of excitement over the discovery of "black" TiO_2, obtained by high-pressure hydrogenation [109], is fading since it was confirmed that little or no visible photoactivity is generated [110], even though UV activity can be dramatically increased. Nevertheless, it has led to theoretical modeling [111] and renewed interest in defect engineering in pure TiO_2 [82,101,112].

The advent of what are now considered "2nd generation" photocatalysts [82,113] was triggered by independent reports in 2001 and 2002 that doping with electron acceptors, *i.e.*, elements forming anionic species such as N [114] and C [115], was the most effective way to impart visible-light sensitization. This was soon corroborated [116,117,118,119], and studies were extended to include F [120], S [121], and P [122]. Awareness of the benefits (synergies) of anionic *co-doping* gradually followed [123,124,125]. Thanks to good underpinning by DFT (calculational) modeling [117,126,127], most studies have focused on modified N-TiO_2 [128,129,130]. However, the visible sensitization effect is limited. N-TiO_2 generally appears pale yellow due to the low level of dopant achievable (<2 at % N), conferring only weak absorption in the blue-green region (seeFigure 8).

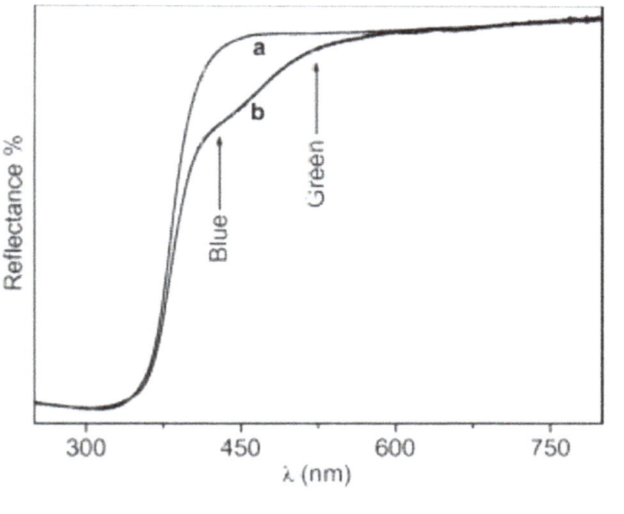

Figure 8. UV-visible reflectance spectra of (a) pristine TiO_2 (b) N-doped TiO_2(reproduced from [126] with permission of Elsevier).

Furthermore, a number of studies have found a loss in oxidative power upon illumination within this visible band, and have attributed this mainly to faster charge carrier recombination [131,132,133]. Di Valentin *et al.* [126] have shown that the most stable doping configuration depends on the chemical potential of ambient oxygen during preparation. Since the most common reagent, NH_3, has reducing properties O-poor conditions prevail, favouring combination of two substitutional nitrogen sites compensated by one O-vacancy ($2N_s + V_o$). The implied diamagnetic material of formula $TiO_{(2-3x)}N_{2x}$ has been affirmed as most likely by more recent modeling [127]. In an O-rich environment, a species consisting of an interstitial nitrogen atom associated with O (N_i-O) may also be stable, interacting with lattice Ti atoms through its π-bonding states, as shown in Figure 9. In either case, new (N_{2p}) energy states predicted to lie slightly above the valence band could be responsible for the observed visible absorption band around 450 nm. EPR spectroscopy has shown that irradiation within this band transfers an electron from the bulk diamagnetic N centre to the TiO_2 surface leading to formation of superoxide species ($O_2^{\bullet -}$), a key activation process in environmental photocatalysis.

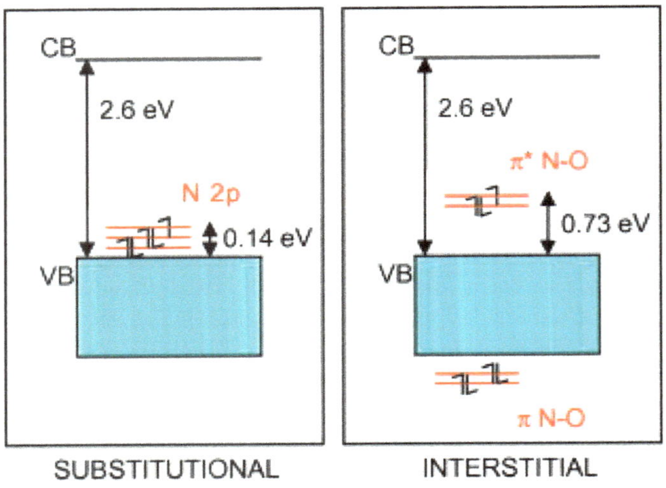

Figure 9. New energy states introduced into TiO_2 by substitutional and interstitial nitrogen. (reproduced from [126] with permission of Elsevier).

The current theoretical viewpoint on bulk doping is to take a donor/acceptor cooperative approach. This can be electro-neutral, e.g., "B + N" [134] or "Mo + 2N" [135], combinations that form intra-gap states extending from the bottom of the conduction band and top of the valence band, respectively. Charge compensation reduces the risk of introducing detrimental structural defects (interstitials, vacancies, *etc.*), while enabling the incorporation of higher dopant levels. Alternatively, even non-compensated p,n- type co-doping has been proposed. Using an excess of donor, e.g., Cr > N, the creation of states of intermediate energy results in a quasi-continuum visible absorption and an apparent bandgap energy of 1.5 eV [136]. The extrinsic states responsible for enhanced visible photoresponse often involve paramagnetic centres that can be explored by EPR [136,137].

One caveat on bulk doping should be mentioned. While re-affirming that N-TiO_2 has visible activity in formic acid mineralization, the same study claimed IR spectroscopic evidence for defective Ti≡N bonds and correlated this with weaker UV photoactivity due to related loss of crystallinity [138]. It is not clear if this trade-off in performance is inevitable [130]. Finally, it should be recognized that the "band narrowing" strategy vis-à-vis solar fuel generation may ultimately be constrained by high overpotentials associated with key redox processes, e.g., water oxidation (*vide infra*). The corollary is that *co-catalysts* will have a more vital role to play in lowering kinetic barriers in visible-active semiconductors, materials of intrinsically lower redox power than pristine TiO_2 (see Section 3.3).

A new class of more intensely coloured "multilayer-sensitized" titanias related to N-doped TiO_2 has emerged recently. These are obtained via mild preparative routes like sol-gel, hydrothermal, *etc.*, where the N-source is usually organic instead of ammonia, the preferred reagent for bulk doping. Starting from urea, the organic moiety transforms stepwise during calcination into the yellow-brown melon structure based on tri-*s*-triazine (heptazine) rings [139,140,141], a process catalyzed by acidic (H)-titanates [142]. Insofar as melon is structurally related to the more-condensed (fully dehydrogenated) graphitic carbon nitride, g-C_3N_4, a visible-absorbing semiconductor *per se* [143], these N-modified materials resemble nanocomposites. Figure 10 shows examples of these structural tectons (building blocks) and their inter-relationship. Starting from amines or alkyl-ammonium salts, the material appears more intensely coloured (brown) already below 200 °C due to a strong absorption tail extending across most of the visible region [96,144,145,146,147]. However, unlike the case of melon, the exact identity of the chromophore is uncertain and it is thermally labile. Calcination weakens both visible absorption and photoactivity [96,148]. Representative UV-Vis spectra of various C,N-based sensitizers loaded onto biphasic anatase/titanates (A/T) are shown in Figure 11.

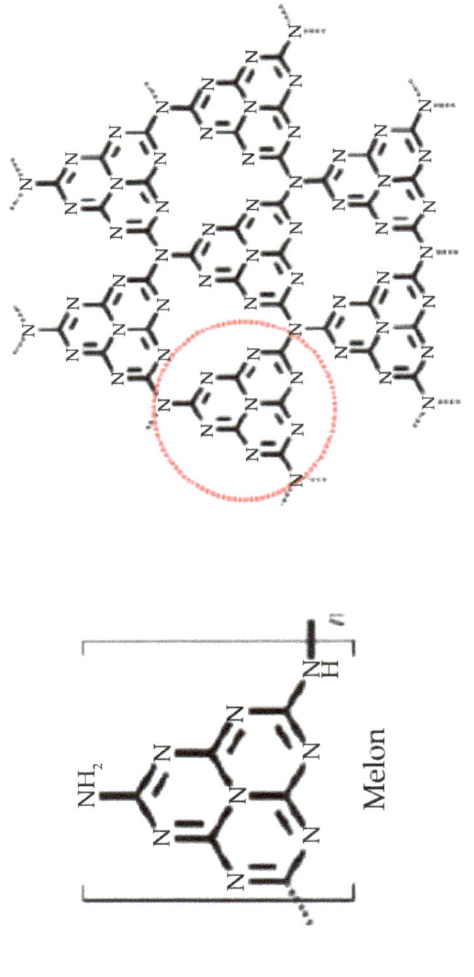

Figure 10. Melon and related tri-*s*-triazine unit (ringed) as building block for g-C_3N_4(adapted from [143] with permission of Wiley).

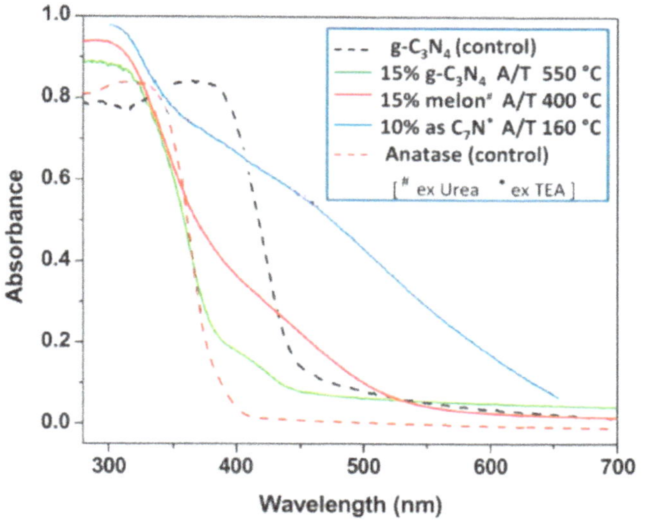

Figure 11. UV-Vis spectra of C/N-based sensitizers on biphasic anatase/titanates (adapted from [96,142] with permission).

Visible sensitization of TiO_2 via "plasmonics" is another rapidly intensifying field that may yield 3rd generation photoactive materials [73,74]. Localized Surface Plasmon Resonance (LSPR) is responsible for the now familiar intense coloration of mono-dispersed colloidal noble metals like Au and Ag, in which the absorption band may be "tuned" by varying the particle size and shape. The first convincing report that visible-light-induced metal-to-semiconductor electron transfer can be induced in Au/TiO_2 appeared in 2005 [149]. The action spectrum (IPCE) for photo-oxidation of ethanol was found to match the Au optical absorption peaking at ~550 nm (see Figure 12).

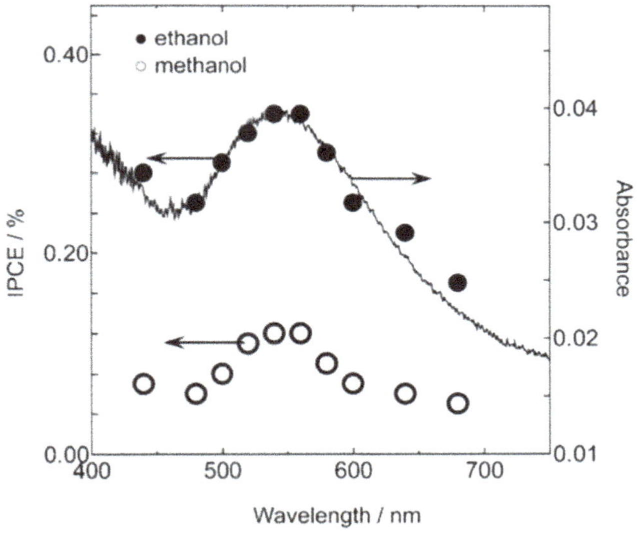

Figure 12. IPCE action spectrum in ethanol (methanol) photo-oxidation *vs.* LSPR (visible absorption) spectrum of gold in Au/TiO_2 (adapted from [149]; copyright (2005) American Chemical Society).

In addition, the characteristic spectrum of self-trapped electrons in TiO_2 (λ_{max} ≈ 680 nm) developed under N_2. As shown in Figure 13, this was quenched in the presence of O_2 as electron acceptor. More extensive studies on Au/TiO_2 and Ag/TiO_2have reached similar conclusions but efficiencies still need improvement [150,151,152,153,154].

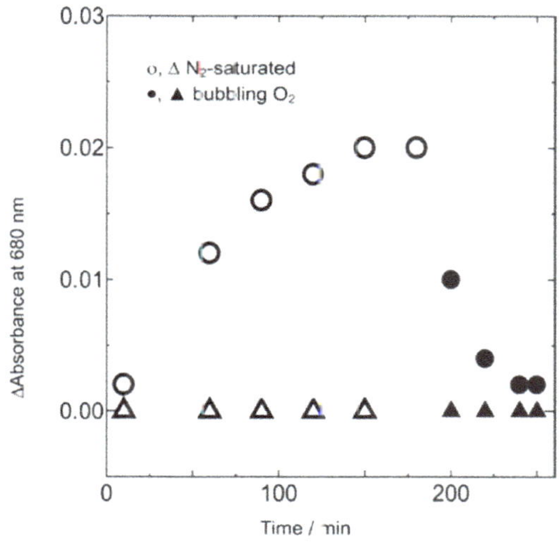

Figure 13. "Electron" absorption in TiO_2 @ 680 nm induced by visible illumination (λ > 500 nm) of Au deposits and its quenching by ambient O_2 (adapted from [149]; copyright (2005) American Chemical Society).

The theoretical basis of plasmonics is developing rapidly, but two distinct types of interaction have already been identified [155,156,157,158,159]. The first is "hot electron" transfer from the metal to the semiconductor, which is the LSPR-induced analogue of indirect charge injection from a sensitizer dye in the DSSC, as described above. A second mechanism is also operative if their respective absorption bands overlap. This is termed plasmon resonant energy transfer (PRET), but its directionality may be reversed under UV irradiation (semiconductor → metal) via Forster resonant energy transfer (FRET). These "near-field" effects do not even require electrical contact at the interface. Notable examples are long-lived hot electron injection from Au to TiO_2 [160], PRET from insulated Ag nanocubes to N-doped TiO_2 [161]; or PRET from Au to α-Fe_2O_3 nanoplatelets [162]. Despite the promise of plasmonic sensitization, the long term economic outlook dictates a shift towards more earth abundant (but inevitably less stable) elements such as Cu [163,164,165], Al [166] and doped oxides [167].

Individual Alternatives to TiO_2

Recent reviews affirm that applied photocatalysis research is still largely (>80%) based on TiO_2, albeit in increasingly sophisticated (modified) forms [168,169,170]. Being amenable to nano-architecturing [170], and providing multi-phase heterojunctions for improved charge separation [171,172], this benchmark material has close to ideal properties as a photocatalyst [173,174] excepting its poor solar light response. One overdue task in TiO_2 research is a more quantitative evaluation of the importance of trapping states that do not lead to fast recombination but, on the contrary, extend charge carrier lifetime into the seconds or minutes time domain [175,176,177,178,179,180,181]. Recent modeling studies on photoexcited anatase show that the energetics (site stability—surface *vs.* bulk) favour surface-trapping of both the hole and the electron [182], with beneficial implications for surface redox chemistry. The search for visible-light active semiconductors that also satisfy other key criteria for practical photocatalysis on a large scale is a difficult task While various alternative semiconductors exist with suitable bandgap (E_g = 1.5–3.0 eV), most are inferior to TiO_2 in other respects, e.g., in having lower majority carrier conductivity, shorter minority carrier diffusion length (faster recombination), a less positive valence band edge (lower oxidizing power), instability under illumination (photo-corrosion), toxicity and/or high cost [183]. For these reasons, emphasis is now shifting to the development of type II (staggered bandgap) composites or tandem arrangements that perform complementary functions, coupled by directional electron transfer at the common heterojunction to "close the photochemical circuit" (*vide infra*) [111,183].

One notable exception that has emerged recently is graphitic carbon nitride (g-C_3N_4), which has a similar bandgap to N-TiO_2 ($E_g \approx 2.7$ eV—see also Figure 11), suitable energetics (band edge positions) for water splitting [142], and can be doped (bandgap tuned) and nano-textured to promote efficient charge migration [184,185,186]. However, a recent modeling study has identified a major kinetic constraint (large overpotential) linked to oxidative dissociation of water [187], helping to rationalize why co-catalysts are urgently needed for O_2 evolution [188]. In contrast, modeling studies have shown that N-doping of anatase TiO_2 may actually promote water dissociation [189]. Elsewhere, research into prospects for hematite (α-Fe_2O_3) has undergone a strong revival in the last decade, mainly due to efforts by the EPFL (Lausanne, Switzerland) group [162,183,190,191]. Pristine hematite is a cheap and stable indirect n-type semiconductor that absorbs visible light up to ~600 nm (E_g = 1.9–2.2 eV), offering a maximum photocurrent density of 12.6 mA cm^{-2}, or a solar-to-hydrogen (STH) conversion efficiency (η_{STH}) of ~16%. One limitation is its relatively low majority carrier (electron) conductivity, but this is readily overcome by incorporation of suitable dopants, e.g., Ti^{4+}, Sn^{4+}, Nb^{5+}, *etc.* [192]. However, it also suffers from several more challenging (deleterious) properties as a photo(electro)catalyst. Having a conduction band edge too low in energy for proton reduction and a large overpotential for O_2 evolution means water splitting over α-Fe_2O_3 will only work under external bias [183,191,193]. Furthermore, its high optical absorption depth (~400 nm [194]), coupled with a very short minority carrier (hole) diffusion length (~4 nm [195]) translates into a very low quantum efficiency for charge collection (IPCE). Nevertheless, structuring highly-crystalline deposits on the 20–30 nm scale has already raised the IPCE to over 30% with photocurrent densities exceeding 3 mA·cm^{-2} [190,191,196,197]. Reinforcing the contention that long-lived charge carriers are of key importance (*vide ultra*), application of transient optical and electrochemical techniques on α-Fe_2O_3 photo-electrodes held under positive (anodic) bias has shown a quantitative correlation between accumulated surface-trapped holes and photocurrent (electron) density [198]. The hole lifetime (τ = 0.1–1 s) is sufficient for photo-oxidation of water, which has a rate constant in the range 0.1–10 s^{-1} [199,200]. The holes are reported to be of two distinct types, O_{2p} (O^-) and Fe_{3d} [Fe^{n+} ($n > 3$)], but both have similar activity [201].

Tandem (D4) Photoelectrochemical Cells, Composites, and the Role of Co-catalysts in Water Splitting

It is 30 years ago now that visible light-driven electron transfer from CdS to TiO_2 in an aqueous suspension of aggregated nanoparticles was first demonstrated [202]. The STH efficiency for the composite was very low but better than either of the pure components due to effective spatial separation and localization of electrons into TiO_2 (for proton reduction), the holes remaining on CdS (for H_2S oxidation) due to the relative energetics of the respective band edges (type II—see Figure 5 [82]). Here, the particles act as complementary self-biasing "photoelectrochemical diodes" in an efficient S2 mechanism (2 photons per H_2 molecule). Since that time, remarkable progress has been made in the visible-driven reduction half-reaction of water splitting:

$$2\,H^+ + 2\,e^- \rightarrow H_2\ [E^\circ\,(V) = 0.00 - (0.059 \times pH)] \qquad (5)$$

A quantum yield of 93% was recorded over Pt/PdS/CdS at 420 nm, evolving 9 mmol/h H_2 with sulphide/sulfite as sacrificial donors [203]. Conventional wisdom has it that the low levels of Pt (0.3%) and PdS (0.13%) act as cocatalysts [51,53,203,204]. PdS is believed to promote oxidation of S^{2-} and SO_3^{2-} and transfer electrons to CdS. However, recent literature suggests that the combination of PdS and CdS may also be classified as an optical tandem system. PdS is an n-type semiconductor with a bandgap of ~1.6 eV and under investigation as a photovoltaic material *per se* [205,206]. It confers extra absorption in CdS composites that extends into the near IR region [207,208]. Unfortunately, the rarity of Pt and Pd, the tendency of sulphides to photocorrode, and the toxicity of cadmium ion, all militate against their use on a large scale. As shown in Figure 14, Pt as cocatalyst traps electrons from the semi- conductor to discharge protons, forming Pt-H bonds of ideal (intermediate) strength for H-H combination and desorption as molecular H_2 (Sabatier Principle). While often black in appearance, co-catalysts are strictly not photoactive and promote only dark elementary steps in the reaction. As shown inFigure 15, there is an optimum amount due to competing (beneficial and deleterious) effects. In practice loadings fall below 1%, fortuitously mitigating costs (many are precious metal-based, e.g., IrO_2, RuO_2, *etc.*) while minimizing parasitic light absorption. Prospects for alternative earth-abundant cocatalysts in photocatalytic water splitting have been reviewed [209]. Promising substitutes for Pt in H_2 evolution under neutral or alkaline conditions are Ni nanoclusters [210], Ni/Mo alloy [211,212], and $Cu(OH)_2$ [213].

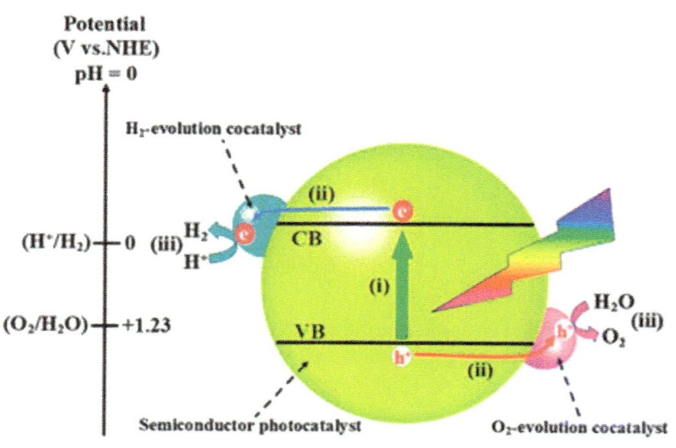

Figure 14. Photocatalytic water splitting over a visible-absorbing semiconductor loaded with H_2- & O_2-evolution co-catalysts (reproduced from [209] with permission of the Royal Society of Chemistry).

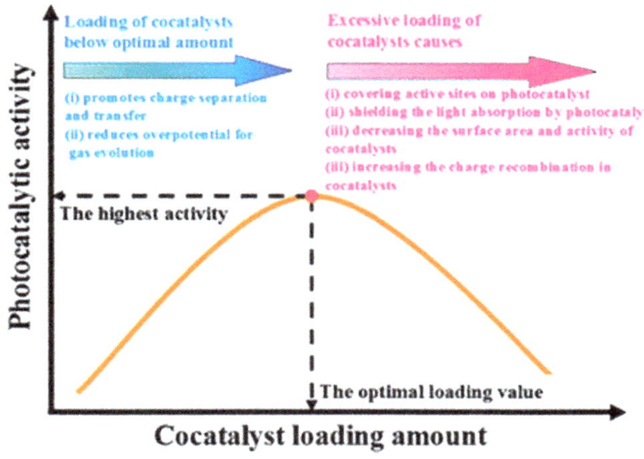

Figure 15. Principle of optimum loading of cocatalyst on a visible-absorbing semiconductor photocatalyst (reproduced from [209] with permission of the Royal Society of Chemistry).

The water oxidation half reaction [or oxygen evolution reaction (OER)]:

$$2\ H_2O \rightarrow O_2 + 4\ H^+ + 4\ e^-\ [E^\circ\ (V) = +1.23 - (0.059 \times pH)] \quad (6)$$

is the main obstacle to efficient water splitting as it suffers from a large activation energy barrier (overpotential > 0.4V) due to the necessary transfer of 4 charges per O_2 molecule in a complex proton-coupled electron transfer mechanism [46,214]. Nature's catalyst in photosynthetic water oxidation is the $CaMn_4O_x$ cubane-related molecular complex [215]. Traditionally, simple though expensive oxides like RuO_2 and IrO_2 have been used at the anode of PEM (acid) water electrolyzers [216], while mixed oxide Ni and Co spinels or perovskites (with inclusion of Cu) are favoured in alkali electrolyzers [217]. In heterogeneous particulate systems, the earth-abundant oxide CoO_x loaded onto $LaTiO_2N$ had an OER quantum efficiency (Φ_{O2}) of 27% at 440 nm [218], while CoO_x or MnO_x on TiO_2 nanosheets achieved $\Phi_{O2} \approx 15\%$ at 365 nm [219]. In overall water splitting, only composite (tandem) absorbers, each optimized (with cocatalysts) for a single half-cell reaction have achieved quantum efficiencies greater than 5% (*vide infra*). Most progress has been made with oxysulphides or oxynitrides of d° or d^{10} metal cations [107,220]. It should also be noted that quantum efficiency (Φ) values are not to be mistaken for η_{STH}. The last is theoretically ~10% for an absorber with a 500 nm absorption cut-off even at 100% quantum efficiency ($\Phi = 1$). In reality, despite intensive efforts over the last three decades η_{STH} values for particulate systems have yet to exceed 1% [220,221].

Prospects for PEC cells look promising due to the efficient separation, collection and transport of photo-separated charges in a wired system. For most photoanodes, e.g., α-Fe_2O_3, WO_3, or other materials of more suitable bandgap (E_g = 1.4–2.0 eV), the conduction band energy is so positioned that any injected electrons can only thermalize into the valence band of the photocathode, e.g., TiO_2, where they effectively neutralize holes created by direct photo-excitation (of TiO_2). By analogy with photosynthesis, such a configuration is generally referred to as a Z-scheme as originally proposed for spatially separated photoelectrodes [75]. Since most electrons reaching the photocathode conduction band are from the photoanode sensitizer and undergo two successive excitation steps, the mechanism is said to be of type D4 (4 photons per H_2), As shown in Figure 16 for a visible-absorbing WO_3 photoanode, if bare TiO_2 is replaced by a photoactive cathode (or a solar cell whose cathode is configured to evolve H_2) the theoretical combined efficiency can rise substantially due to wider light harvesting. This is most notably so (η_{STH} > 40%) in a series arrangement with an "in- front" photoanode (E_g ~1.8 eV) that absorbs visible wavelengths to evolve O_2 from water. The transmitted near-IR light is incident on the solar cell (E_g ~0.95 eV), which provides a voltage bias for H_2 evolution at the cathode (see Figure 17). The Z scheme principle has been extended to particulate systems but with limited success [50,107]. An added complexity here is the need to promote interparticle electron transfer using a redox mediator, e.g., IO_3^-/I^-, in solution. However, this suffers from "chemical short-circuiting", *i.e.*, competitive reactions between water and the mediator, and especially reaction of its oxidized form, necessarily present in excess, with product H_2. Nevertheless, Maeda *et al.* [222] have reported a respectable overall quantum efficiency of 6.3% at 420 nm for a Pt-doped ZrO_2-protected TaON "cathode" (for H_2 evolution) suspended with PtO_x-loaded WO_3 as "anode" (for O_2 evolution). Better prospects may lie in elimination of mediators and the development of "all-solid-state" Z-scheme analogues, *i.e.*, composite particles with heterojunctions [223,224,225].

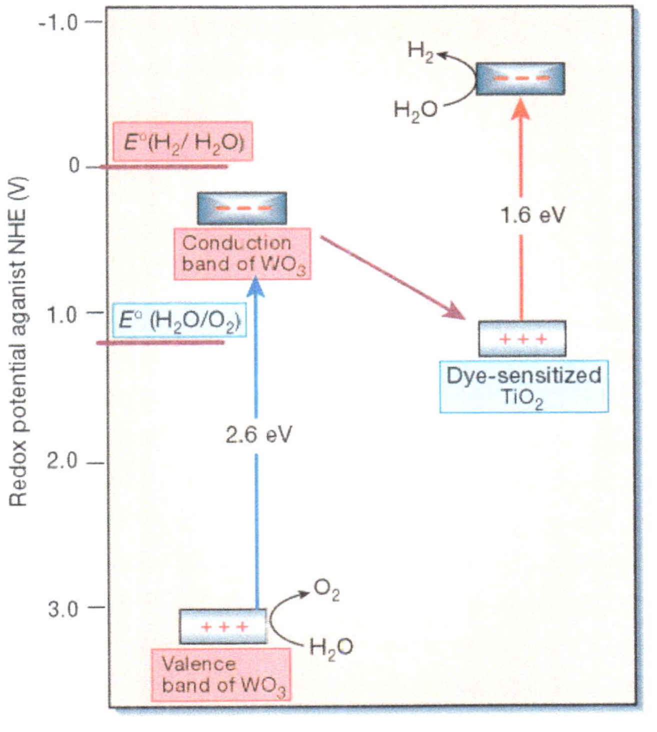

Figure 16. Example of Z-scheme (4D): WO_3 photoanode ($E_{cb} < E°_{H^+}/H_2$) coupled to dye-sensitized TiO_2 (reproduced from [76] with permission of the Nature Publishing Group).

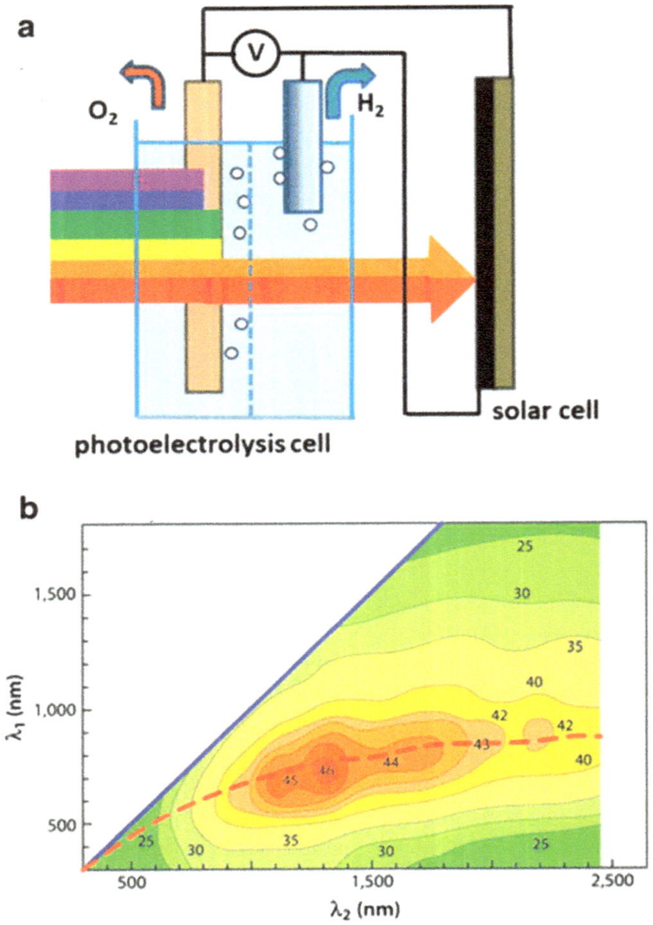

Figure 17. (**a**) Water splitting tandem cell: photoanode passes NIR light to solar cell, giving cathodic bias for H_2 evolution; (**b**) Dual absorber efficiency curve (η_{STH} > 40% at $\lambda_a \leq$ 750nm (1.6 eV), λ_{sc} = 750–1300 nm (0.95 eV) (reproduced from [200] with permission from Springer).

The choice of materials for practical tandem PEC cells is restricted. Figure 18 shows the bandgap and bend edge position of representative semiconductors. They must be cheap (earth-abundant) and stable ideally in strongly acidic and/or alkaline conditions for good electrolyte conductance.

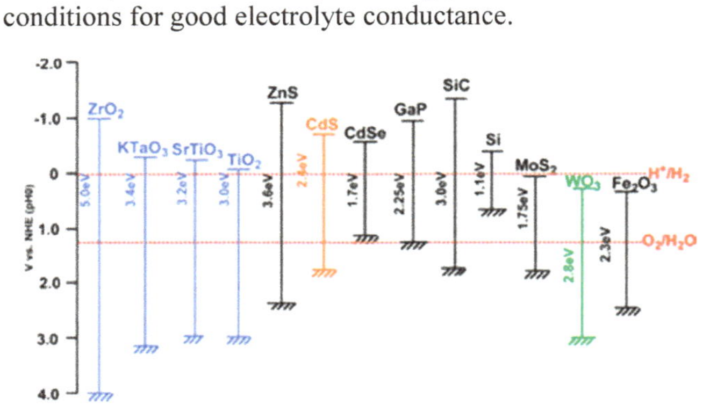

Figure 18. Bandgaps and band edge positions of representative semiconductors in relation to the redox potentials for water splitting at pH = 0. (Reproduced from [220] with permission of the Royal Society of Chemistry).

Bandgaps must also be higher to compensate for inevitable voltage losses, e.g., device series (ohmic) resistance, and electrode overpotentials. Nevertheless, the efficiencies of these tandem devices are expected to exceed 25% [226,227]. Major advances towards this goal were reported by Nocera *et al.* (η_{STH} = 4.7%, 2.5% wireless) [62] and Grätzel et al (η_{STH} = 3.1%) [61], but with a quite different system approach. Nocera's design consisted of a triple-junction (T6 or 6-photon) amorphous Si absorber loaded with cobalt phosphate (for the oxygen evolution reaction—OER) [228]) and Ni/Mo alloy (for H_2evolution [211,212]) on a Ni mesh in phosphate or borate/nitrate buffer. The wireless layout suffered from an added resistance loss due to the longer migratory path imposed on proton transport from the front (anode) to the rear (cathode) of the cell. Grätzel's design was based on a single WO_3 (Fe_2O_3) photoanode coupled to a single DSSC, as shown in Figure 16and Figure 17. It is not clear if only Fe_2O_3 or both were loaded with IrO_2 co-catalyst. Continuing their exploration of a buried junction PV configuration with loaded electrocatalysts (EC), in which the Si absorber is protected from the electrolyte, the Nocera team most recently achieved η_{STH}= 10% with these so-called PV-EC tandem devices [78]. It consists of 4 single-junction crystalline Si solar cells connected in series to a NiMo cathode and a nickel borate anode, all immersed in a borate buffer (pH 9.2). The Ni-based anode is comparable in performance and cheaper than cobalt phosphate, but needs prior anodization (after deposition) to create the mixed-oxidation $Ni^{III/IV}$ state responsible for OE activity [229]. However, both Ni and Co salts can be electrodeposited, conveniently forming the anode *in-situ* from divalent ions in the appropriate buffer. The latest efficiency advance reported by the Grätzel team is η_{STH} = 12.3% in an analogous PV-EC device [77]. This was achieved with two DSSCs connected in series, each providing a short-circuit photocurrent density of 21.3 $mA{\cdot}cm^{-2}$, open-circuit voltage (V_{oc}) = 1.06 V, and a fill factor of 0.76. A combined solar-to-electric power conversion efficiency of 15.7% was attained with superior light harvesting by lead iodide perovskite ($CH_3NH_3PbI_3$) sensitizers, prepared using a simple two-step spin-coating method at 100 °C. A cheap Ni-foam supported Ni/Fe layered double hydroxide, obtained by one-step hydrothermal growth, served as both cathode and anode in 1M NaOH electrolyte. It should be noted that a comparable performance may be obtained in PV-driven ("brute force") electrolysis, *i.e.*, by combining state-of-the-art PV modules and electrolyzers optimized independently. For example, $\eta_{STH} \approx$ 12% has been reported in standalone systems where the voltage was maintained at ~1.7 V per cell (in a 20 cell PEM electrolyzer stack) with a DC-DC converter [230,231]. The PEC cell is a less expensive single integrated unit, provides a higher open circuit photovoltage, and reduces potential loss channels. However, it is susceptible to electrolyte resistance and polarization losses [227,232], especially under neutral conditions needed for operational stability of many earth-abundant co-catalysts. In view of such complications and exciting results with hybrid (PV-EC) devices, co-development of PV-electrolyzers and PEC water splitting cells may offer the best prospects [233,234].

HYDROGEN PEROXIDE AS SOLAR FUEL AND SUSTAINABLE CHEMICAL

Water splitting to H_2 and O_2 has been considered the "Holy Grail" of chemists working in the energy field. However, the co-production of H_2 and H_2O_2 is arguably a yet more valuable process:

$$2\,H_2O \rightarrow H_2 + H_2O_2\ [\Delta G° = +342\ \text{kJ/mol}\ H_2O_2] \quad (7)$$

and may actually be easier, *i.e.*, the kinetic barrier may be lower, because it is just a 2 e^- process:

$$2\,H_2O \rightarrow 2\,H^+ + 2\,e^- + H_2O_2\ [E° = -1.77\ V] \quad (8)$$

This requires two photons at 171 kJ/mol (photons), corresponding to wavelengths below ~690 nm, which comprises ~50% of the solar power spectrum. Hydrogen peroxide is a valuable commodity chemical serving as a green oxidant in environmental clean-up, pulp bleaching, detergents, *etc.* [235]. It is now made largely by the *Anthraquinone Process* but research has intensified in recent years into direct catalytic synthesis from H_2 and O_2:

$$H_2 + O_2 \rightarrow H_2O_2 \quad (9)$$

which is exothermic ($\Delta H° = -136$ kJ/mol) and a competitive option for small-scale on-site production (< 10^4 t/y) [236,237]. However, it deals with potentially explosive mixtures and only works efficiently over expensive rare metal (Pd or Pd/Au) catalysts. It is also an example of a highly selective partial oxidation reaction in which reaction with a second H_2 molecule:

$$H_2O_2 + H_2 \rightarrow 2\,H_2O_{(g)} \quad (10)$$

is even more exothermic (Equation (10) is the reverse of Equation (7)) and must be kinetically inhibited [238]. A $PtHg_4$/C electrocatalyst was shown to be active and highly selective for H_2O_2 synthesis, as predicted by DFT modeling [239]. Alloying leaves an isolated surface Pt atom for hydroperoxide (HOO*) stabilization in the on-top position while eliminating the hollow adsorption sites preferred by activated oxygen (O*) species, thereby inhibiting water as product. Unfortunately, it is still a rare-metal based formulation. The cathodic half-cell reacton for peroxide synthesis from water (Equation (8)) can be written in two ways depending on the electron acceptor.

For co-production of H_2 (Equation (7) overall) this is:

$$2\,H^+ + 2\,e^- \rightarrow H_2\ [E° = 0.00\ V] \quad (11)$$

However, peroxide can also be synthesized by O_2 reduction:

$$O_2 + 2\,H^+ + 2\,e^- \rightarrow H_2O_2\ [E° = 0.68\ V] \quad (12)$$

Summing the two half reactions (8) and (12), each yielding one peroxide molecule, gives Equation (13):

$$2\,H_2O + O_2 \rightarrow 2\,H_2O_2\ [E° = -1.09\ V] \quad (13)$$

This is still a potentially visible-driven endergonic process so that H_2O_2 alone can be considered as an energy carrier derivable from cheap reactants. Supplied commercially as 30% aqueous solution (~9M) it is already in an energy dense form, unlike H_2 gas, and this underlines recent interest in peroxide as a solar fuel. Having acceptor and donor properties (reverse of Equations (8) and (12), respectively), the theoretical voltage of a "direct H_2O_2" fuel cell based on its own dismutation (reverse of Equation (13)) is 1.09 V. Although it enables a simplified fuel cell design (use of a single compartment is possible), the current densities are not yet of technical interest [240,241,242,243,244,245,246]. Alternatively, as a more powerful oxidant (than O_2) supplied to the cathode, it increases the operating voltage in cells based on H_2 [247] or liquid fuels such as aqueous $NaBH_4$ [248,249] and ethanol [250]. However, its powerful oxidizing properties and susceptibility to decomposition by traces of metal (ions) and redox-active surfaces can also lead to under-performance (mixed potentials) and introduces more stringent material compatibility issues, especially concerning long-term stability of polymer membranes and possibly carbon as an electrocatalyst support. Potential cathode materials explored to date include $PbSO_4$ [240], Au on Vulcan [249], and $LaNiO_3$ on N-doped graphene, which promotes the ORR [251].

In energetic terms, the photosynthetic route involving co-generation of hydrogen (Equation (7)) would be preferred as long as the H_2O_2/H_2 mixture remains stable. This has been reported for platinized calcium niobate [252] and Pt/TiO_2 [253]. Both showed relatively low quantum efficiencies (Φ_{uv} < 1%) but the 1:1 product stoichometry of Equation (7) was confirmed. A liquid water environ-ment would probably impede further reaction because any H_2O_2 dissolves selectively. However, little peroxide was found in solution, most remaining associated with the catalyst in a form seemingly immune to any back-reaction with H_2. This is consistent with previous literature showing that, in the absence of O_2, a stable form of peroxide (O_2^{2-}, HO_2^-, *etc.*) builds up and deactivates the photo-process in a few hours, typically affording only micromoles of products. Although the intermediate(s) can be decomposed easily to yield O_2 (the "missing" product in early studies of water splitting [254,255]), a way must be found to displace it as intact H_2O_2, e.g., by exploring more weakly-adsorbing (non-oxide) semiconductors and/or loading an oxidation co-catalyst. A viable system needs to sustain *milli* moles per hour productivity indicative of a broad spectral response and quantum yields exceeding 10%. This may be achievable but not easily recognized in practice because the peroxide, once formed, can decompose adventitiously, e.g., in the presence of trace metal ions, or excited by the UV component of a solar simulator, *etc.* Such H_2O_2 production rates (Φ_{uv} = 1%–30%) have been reported over quantum-sized ZnO and TiO_2 particles but only under oxygenated conditions, implying that photo-reduction of O_2 (Equation (12)) is the main source of peroxide [256]. Fluorination of TiO_2 improved dramatically the yield by weakening the surface adsorption of peroxide, a precursor step in self-decomposition [257].

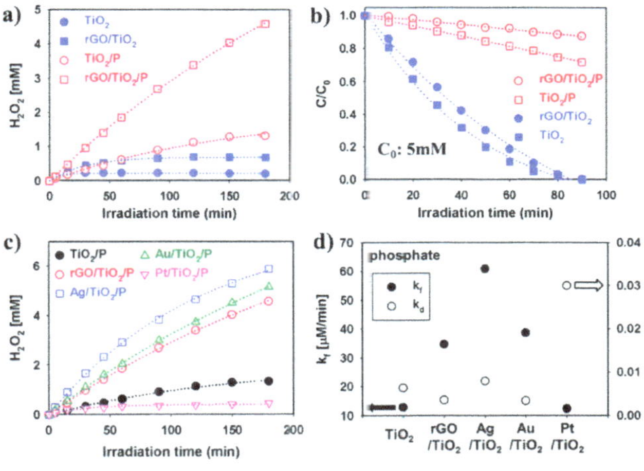

Figure 19. (**a**) Effect of pre-treating TiO_2 with phosphate on H_2O_2 photosynthesis (λ > 320 nm), and (**b**) 5 mM H_2O_2 decomposition over 6 wt % rGO/TiO_2 (0.5 g/L) in O_2-saturated aqueous buffer (pH 3) containing 5 vol % 2-propanol as hole scavenger; (**c**) Comparison of H_2O_2 synthesis rates over rGO/TiO_2(F) and 1 wt % metalized TiO_2(P) samples; (**d**) Rate constants of H_2O_2 synthesis (k_f) and decomposition (k_d) over TiO_2(P)-supported composites (reproduced from [258] with permission of the Royal Society of Chemistry).

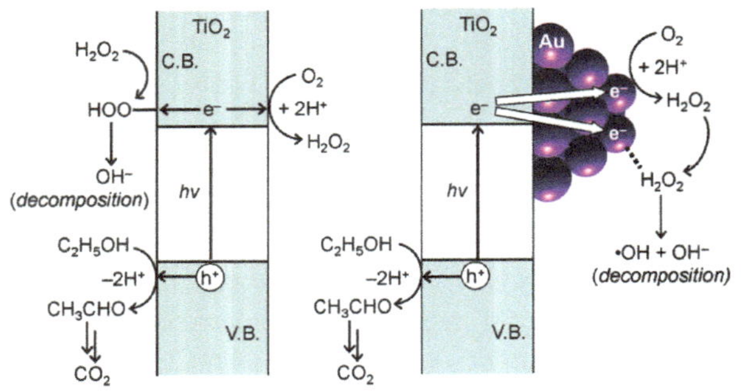

Figure 20. Scheme of synthesis and decomposition of H_2O_2 on (**a**) TiO_2; and (**b**) Au/TiO_2 photocatalysts. (Reproduced from [259]. Copyright (2012) American Chemical Society).

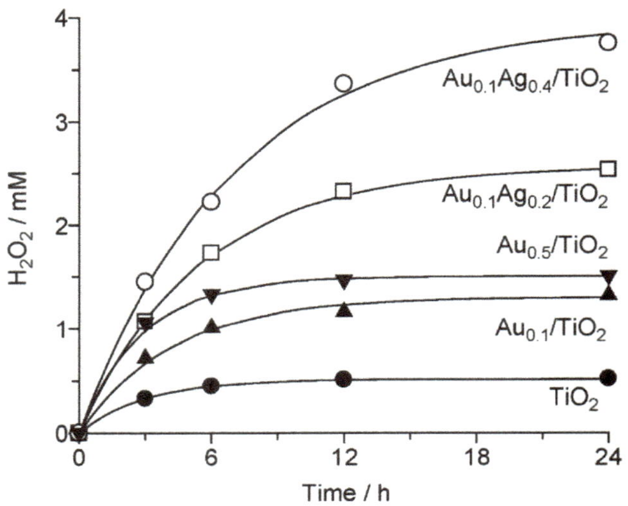

Figure 21. Build-up of photostationary levels of H_2O_2 on Au-Ag/TiO_2 catalysts (5 mg) in 5 mL aerated 4% aqueous ethanol ($\lambda > 280$ nm, $I_{uv} \approx 14$ mW)). (Reproduced from [259]. Copyright (2012) American Chemical Society).

As shown in Figure 19, the same effect was seen in a reduced graphene oxide (rGO)/TiO_2 composite in which the TiO_2surface was phosphated to prevent simultaneous degradation, which is otherwise responsible for the attainment of photostationary product levels [258]. Alternatively, as shown in Figure 20 and Figure 21, alloying Au (on TiO_2) with Ag suppresses selectively the intrinsic tendency of Au to simultaneously decompose its own product, thereby raising photostationary yields [259]. In contrast, it has been reported that TiO_2 *per se* is a poor catalyst in H_2O_2 decomposition due to its low affinity for the OH• radical [260,261]. This powerful but non-selective oxidant:

$$HO^{\cdot} + H^+ + e^- \rightarrow H_2O\ [E° = 2.80\ V] \quad (14)$$

is produced from hydrogen peroxide by an electron donor:

$$H_2O_2 + H^+ + e^- \rightarrow HO^{\cdot} + H_2O\ [E° = 0.72\ V] \quad (15)$$

such as Fe^{2+} in the Photo-Fenton process, one of a variety of advanced oxidation processes (AOP) [13,96,262]. Unfortunately, it also promotes the autocatalytic decomposition of peroxide [263]:

$$H_2O_2 + OH^{\cdot} \rightarrow H_2O + HO_2^{\cdot}\ [E° = 1.40\ V] \quad (16)$$

$$HO_2^{\cdot} \rightarrow O_2 + H^+ + e^-\ [E° = -0.05\ V] \quad (17)$$

Hole scavengers (H·donors) always increase decomposition rates, suggesting that the 2 e^- water oxidation half-reaction (Equation (8)) is rate-determining in the synthesis. Under these conditions, *i.e.*, with little H_2 or H_2O_2 produced from water, the process is of no interest as an energy conversion scheme. Nevertheless, the peroxide can still be considered a value-added green reagent obtained efficiently and cheaply by photo-oxidation of organic wastes. Evidence has just been reported for visible-driven H_2O_2 photosynthesis from oxygenated ethanol over pristine g-C_3N_4 [264], whose conduction band minimum ($E° \approx -1.3$ V [265]) exceeds the reduction potential for the O_2/O_2^- couple ($E° \approx -0.3$ V [266]). A 1,4-endoperoxide intermediate stabilized by the g-C_3N_4 surface was identified by Raman spectroscopy. Addition of Pt had a deleterious effect on yield due to its tendency to break the O–O bond [239].

PHOTOREFORMING OF BIO-OXYGENATES

In the renewable energy scheme under consideration (*cf.* Figure 1), bio-oxygenates like sugars, alcohols, and polyols, all serve as CO_2-neutral energy carriers provided that their incipient H_2 can be extracted efficiently by catalytic reforming with steam [31,32,55,56]:

$$C_nH_mO_k + (2n - k)\ H_2O \rightarrow n\ CO_2 + (2n + m_{/2} - k)\ H_2 \quad (18)$$

Despite being highly endothermic, Equation (18) is favoured thermodynamically above a threshold temperature due to the large volume expansion (entropy factor). Input thermal energy is converted into chemical energy (H_2) and represents a significant gain in exergy (20%–30%), as can be seen by comparing the heats of combustion of reactant and product. This is known as "chemical recuperation" [267,268], and pre-reforming of natural gas is likely to be incorporated into future gas turbine technologies [269]. Methanol is not currently made on a large scale from biomass or renewable H_2 and can be readily reformed by conventional (thermal) catalysis [23,24,25,28]. In contrast, ethanol comprises 90% of biofuel production and, due to its high growth forecast [270], is now being considered as a renewable platform chemical, e.g., for butadiene synthesis [271]. Ethanol is also a good model oxygenate as it is one of the simplest compounds containing C–C, C–H, and C–O bonds. However, its catalytic conversion in high activity and selectivity poses a major challenge [272,273].

Bio-ethanol obtained by fermentation of glucose:

$$C_6H_{12}O_6 \rightarrow 2\ CO_2 + 2\ C_2H_5OH \quad (19)$$

is an excellent energy carrier since almost the entire heating value of the original sugar (~2800 kJ/mol) is retained in the product (two moles liquid ethanol at 1365 kJ/mol). Ethanol steam reforming (ESR):

$$C_2H_5OH\ (g) + 3\ H_2O\ (g) \rightarrow 2\ CO_2 + 6\ H_2\ [\Delta H° = +174\ kJ/mol] \quad (20)$$

raises the fuel value substantially (six moles H_2 at 286 kJ/mol). In addition, Equation (20) has a crossover ($\Delta G° \leq 0$) temperature as low as 210 °C [19], suggesting it could be driven by "low-quality" heat provided a suitable catalyst can be found. This explains the major interest in ESR in recent years, e.g., as an on-board source of H_2 for PEM fuel cell (electric) vehicular propulsion [274,275]. However, the low rates encountered over many oxide-supported transition metals (Pt, Ni, Co, Rh, Ru) necessitate working above 400 °C, where rapid deactivation by coking ensues, possibly linked to acetic acid intermediate [30,276,277,278]. DFT modeling supports experimental data showing that the rate-determining step in ESR is initial dehydrogenation to acetaldehyde [279]:

$$C_2H_5OH_{liq} \rightarrow CH_3CHO + H_2\ [\Delta H° = +85\ kJ/mol] \quad (21)$$

a modestly endothermic reaction but with an apparent activation energy (E_{app}) as high as +150 kJ/mol [280,281]. This is typically followed by decarbonylation of the aldehyde [281,282]:

$$CH_3CHO \rightarrow CO + CH_4\ [\Delta H° = +7\ kJ/mol] \quad (22)$$

an almost thermoneutral process giving an undesirable alkane product. Photocatalysis may circumvent these activity/selectivity issues because it works by an alternate mechanism (lowering E_{app}) and at low temperature where deleterious side reactions are inhibited. The pioneering work of Pichat *et al.* [283,284] established that photo-dehydrogenation (PDH) of alcohols (see Equation (21)) proceeds in high quantum efficiency ($\Phi_{uv} \geq 0.1$) over Pt/TiO_2. More recently, photo-assisted water-gas shift (WGS) [285], photoreforming of methanol [286,287], and combined photo-/thermal reforming of methanol (or glycerol) to H_2/CO_2 co-products over Pt/TiO_2 [288] (or Pd/TiO_2 [289]) have been studied. As shown in Figure 22 and Figure 23, mild heating is a useful adjunct in photocatalysis when dark processes are rate-controlling. However, despite the sharp rise (×2–×5) in quantum efficiency ($\Phi_{uv} \approx 7\%$ for CH_3OH at 65 °C [288]) and better reaction stability (due to more complete product recovery [288,290]), this synergism has still not been widely exploited.

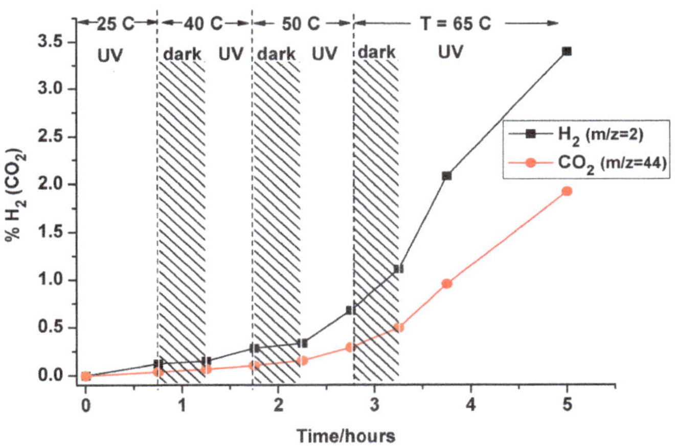

Figure 22. MS response (H_2/CO_2) showing effect of temperature on rate of vapour-phase CH_3OH photoreforming over Pt/TiO_2 (reproduced from [288] with permission of RSC).

Photoreforming *per se* has been extended to ethanol [290,291,292,293,294], various alcohol mixtures [35,294,295,296,297,298], glycerol [33,55,56,289,299], sugars [300,301], and acetic acid [302]. In the last case, no CH_4 product was seen, an encouraging result in view of previous claims of a novel Photo-Kolbé process over a similar catalyst [303]. All these studies used TiO_2-supported precious metals responding only to UV light at high efficiency.

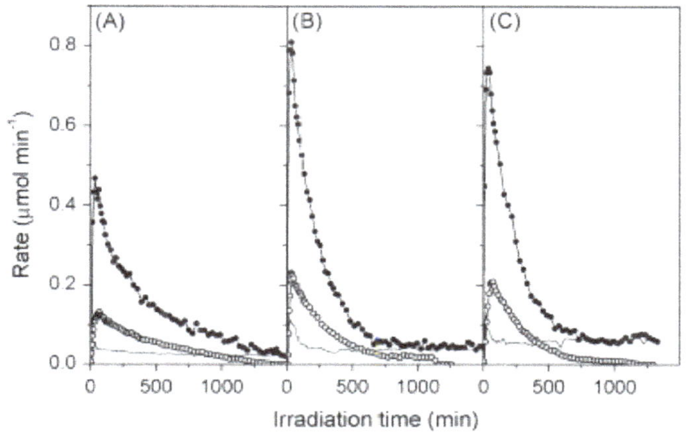

Figure 23. H_2 (full circles) & CO_2 (closed circles) evolution during liquid-phase (0.37 mM) aqueous glycerol reforming at (**A**) 40 °C; (**B**) 60 °C; (**C**) 80 °C (reproduced from [291] with permission of Elsevier).

For 80% aqueous ethanol over a well-dispersed Pt/TiO_2 film at low light intensities ($I_{uv} \approx 0.8$ mW/cm², or 0.2 suns), a remarkable quantum efficiency ($\Phi_{uv} \approx 74\%$) was estimated [294]. Close to solar intensities, these were rather lower (Φ_{uv} = 10%–30%) for ethanol [290,295,297], but several groups have reported mass-specific H_2 evolution rates of technical interest (>2 mmol/h/g_{cat}) [33,291,293]. As shown in Figure 24 and Figure 25, bio-oxygenates are reformed at comparable rates [297,298], while quantum efficiencies can be raised significantly (to $\Phi \geq 30\%$) due to the superior illumination geometry available in an optical fibre honeycomb reactor [297].

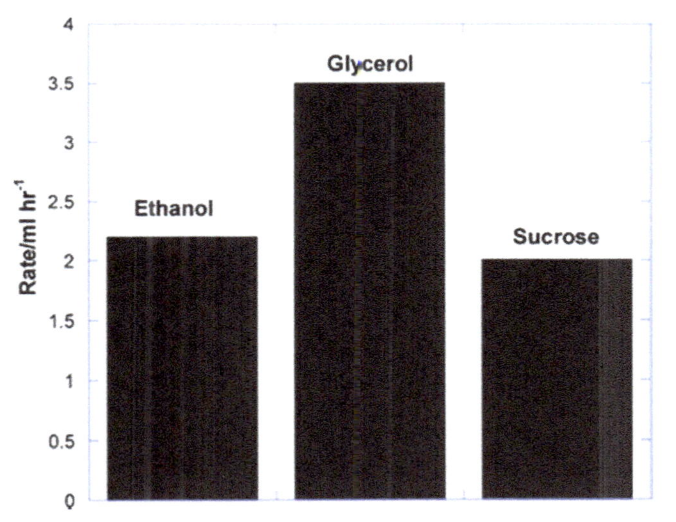

Figure 24. Rate of H_2 evolution by photo reforming of 3 important bio-oxygenates (1% in H_2O) over 0.3 wt % Pd/TiO_2 (P25) (reproduced from [298] with permission of RSC).

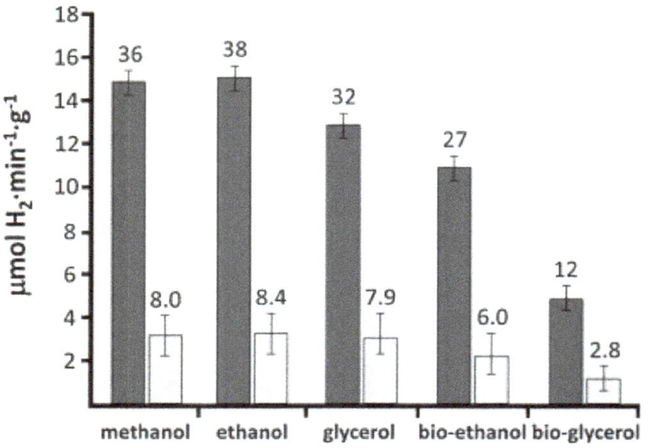

Figure 25. H_2 evolution rates and quantum efficiencies of bio-alcohols (1:1 H_2O) in a catalytic wall reactor (dark) or a slurry reactor (light) over 1 wt % Au/TiO_2. (reproduced from [297] with permission of Elsevier).

Early investigations of photo-reforming over visible-active catalysts include SPR metals on TiO_2 [304], $Cd_{1-x}Zn_xS$/ZnO [305], CdS/TiO_2 [306], ε-Fe_2O_3 [307,308], and a potentially self-degrading glucose-TiO_2 charge transfer complex [309]. However, only the chalcogenide-based system gave a respectable quantum efficiency, $\Phi_{>420nm}$ = 9.6% [305]. Glycerol photoreforming over earth-abundant co-catalysts CuO_x/TiO_2 [310,311] and NiO_x/TiO_2 [312] has also been reported.

PHOTOREDUCTION OF CARBON DIOXIDE: ARTIFICIAL PHOTOSYNTHESIS

As mentioned in the *Introduction*, a valuable (solar) energy-storing artificial photosynthetic process which couples two stages in the energy scheme is methanol synthesis from carbon dioxide and water:

$$CO_2 + 2\,H_2O \leftrightarrow CH_3OH + 1.5\,O_2 \qquad (23)$$

although ethanol synthesis:

$$2\,CO_2 + 3\,H_2O \leftrightarrow C_2H_5OH + 3\,O_2 \qquad (24)$$

would be even better due to its higher energy density, lower toxicity and volatility. However, Equation (24) is probably unrealistic due to its mechanistic complexity and associated selectivity issues, at least as reported in CO_2 hydrogenation [26,27]. Methanol synthesis *per se* (Equation (23)) is already complex as it subsumes the water splitting process (see Equation (6)) but is 50% more demanding energetically (ΔH = +727 kJ/mol) due to the evolution of proportionately more O_2, as shown by the relevant half-cell reactions:

$$CO_2 + 6\,H^+ + 6\,e^- \rightarrow CH_3OH + H_2O\ [E° = -0.32\ V] \qquad (25)$$

$$3\,H_2O \rightarrow 1.5\,O_2 + 6\,H^+ + 6\,e^-\ [E° = +1.23\ V] \qquad (26)$$

and comparison of Equations (6) and (26). While this 6 e^- process can be driven by near-IR photons, it is mechanistically more complex than water splitting as two elementary steps (proton reduction and H atom coupling) at the photocathode are replaced by activation of a stable gas molecule of low aqueous solubility, its multi-step reductive de-oxygenation, and progressive hydrogenation (Equation (25)). It perhaps comes as no surprise that carbon-based solar-to-fuel conversion efficiencies lag far behind those of water splitting at < 1%, with turnover rates (10–100 µmol/h/g_{cat}) over pristine semiconductors still too low for technical exploitation [43,45,79]. Achieving high selectivity to methanol is also a challenge insofar as CH_4 is preferentially obtained over hydrated anatase TiO_2 [313,314], unless the surface Ti^{4+} (electron trap) centre is highly-dispersed or isolated [315].

Nonetheless, as shown in Figure 26, there is general agreement that the photocatalyst plays an indispensible role in activating CO_2 (the probable rate-determining step) by electron transfer and stabilizing the highly energetic $CO_2{\cdot}^-$ radical ion in coordinated form(s) [43,316,317]. Possible sequences of proton-coupled electron transfer in CO_2 conversion to formic acid are shown in Figure 27. DFT-modeled energy barriers favor the green route via bidentate coordination mode B1 (0.87 eV) and the red route via the linear monodentate mode A1 (0.82 eV) although this is less likely as it requires a simultaneous two e^- transfer. Routes (black) via carbonato-type complexes A2 and B2 proceeding through carboxyl (COOH) intermediate have much higher energy barriers, 2.25 eV and 1.73 eV, respectively.

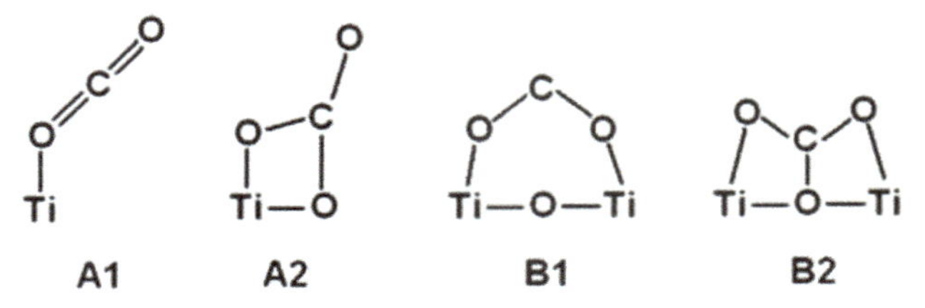

Figure 26. DFT-modeled states of neutral/anionic CO_2 adsorbed on TiO_2 anatase (101) (reproduced from [316] with permission of the Royal Society of Chemistry).

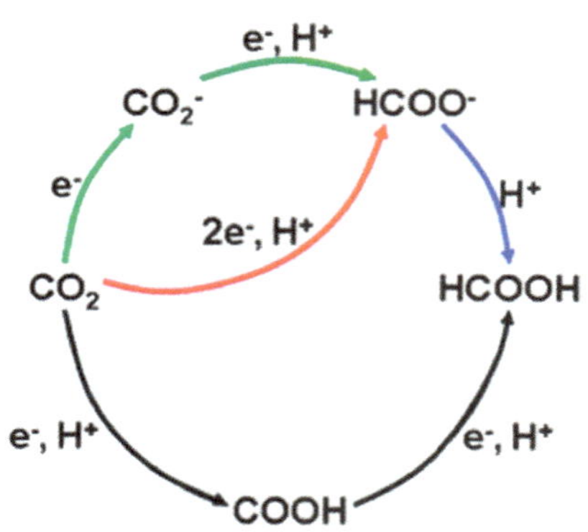

Figure 27. Pathways of $2e^-/2H^+$ photoreduction of CO_2 to formic acid. DFT energy barriers (<1 eV) favour the green route (via B1) and the red route (via A1) (reproduced from [316] with permission of the Royal Society of Chemistry).

It is evident from the literature [24,43,317] that earth-abundant Cu is a ubiquitous co-catalyst in CO_2 reduction, yielding mainly methanol in gas-solid photocatalysis or methane by electrocatalysis. This selectivity effect has been rationalized [318] as due to reaction of the surface-bound methoxy (CH_3O-Cu) intermediate either with a co-adsorbed H atom (favouring CH_3OH by a lateral surface mechanism) as in the industrial synthesis [319], or with a proton from aqueous solution (favouring CH_4 via attack on the protruding CH_3 moiety). The prevalence of Cu (or CuO_x) in composite photocatalysts [41,43,44,45,320,321,322,323] and electrodes [319,324,325,326] suggests that the dark mechanism "post-formate" is still operative but in which water photo-oxidation provides the electron/proton pairs in Equation (25), *i.e.*, the half-cell equivalents of H atoms from H_2 dissociation. If this is the case, the thermal mechanism remains important [19] and an approach toward industrial synthesis conditions is worthy of study, e.g., mild heating/pressurization of CO_2/H_2O vapour under illumination. Publications have recently proliferated on the role of nanocarbons and graphenes as cocatalysts in composite photocatalysts [327,328,329], including photo- [40,330,331,332] and electro-reduction of CO_2 [333]. This is linked mainly to their effectiveness in promoting charge separation [327,328,329,334], although there are tentative claims for visible sensitization, possibly due to adventitious C doping [40,327,328,329]. There is also growing evidence that graphene oxide can act as a photocatalyst *per se* [335] although CH_3OH synthesis activity is improved by addition of Cu [336] and/or molecular sensitizers [337]. Otherwise, visible response has been conferred to TiO_2 nanocomposites by incorporating CdS/Bi_2S_3 [338], CdSe quantum dots [339], or plasmonic metal deposits [340]. While promising advances have been made in "self-biasing" particulates, the tendency of methanol to act as an efficient hole scavenger may ultimately militate against a gas-phase photocatalytic approach in favour of a PEC (membrane-separated photo-electrode) arrangement, as shown inFigure 28. This is despite the known limitations of the latter, viz., CO_2 solubility/mass transfer issues in the liquid phase [79], and a higher probability of obtaining CH_4 on a photoelectrode [318,326].

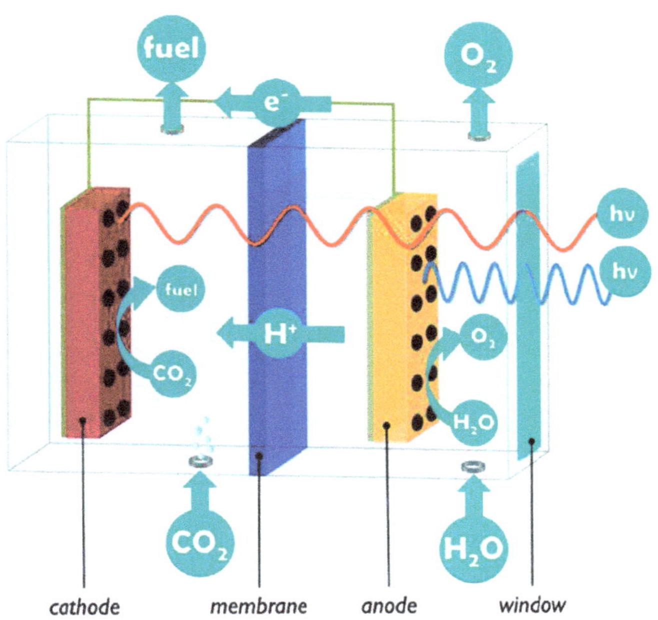

Figure 28. Liquid-phase 2-compartment PEC cell for CO_2 photo-reduction. (reproduced from [79] with permission of the Royal Society of Chemistry).

As a potential solution, and borrowing concepts from PEM fuel cell technology, gas-phase electrocatalytic studies using a gas diffusion membrane electrode (GDM) configuration have given improved CO_2 conversion rates (Faradaic efficiencies) and a selectivity shift towards oxygenates over carbon nanotube-supported Pt and Fe [333,341,342], as shown in Figure 29. The beneficial effect of mild heating is also evident.

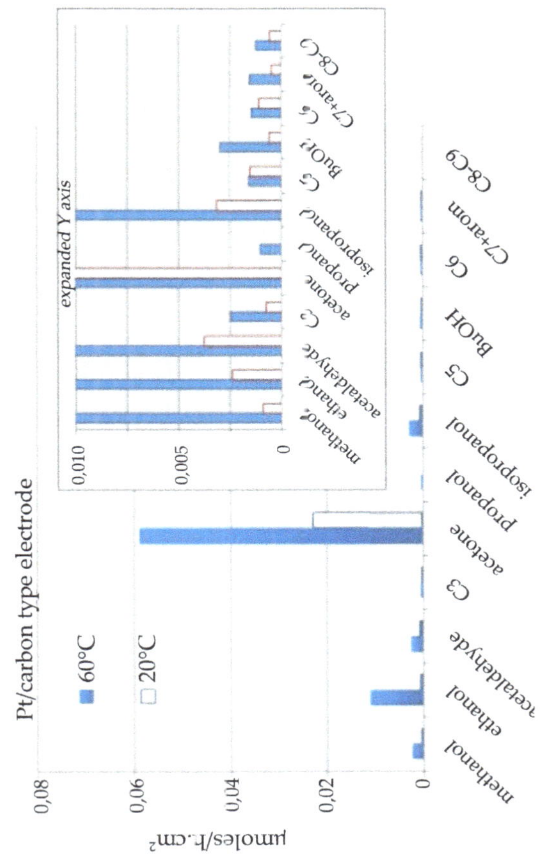

Figure 29. Temperature effect in gas phase CO_2 electro-reduction on 10% Pt/Carbon/ Nafion117 GDM electrode (reproduced from [333] with permission of Elsevier).

Regarding the dark mechanism in industrial methanol synthesis (from CO_2/CO) over supported Cu, a recent microkinetic/DFT modeling paper [343] casts doubt on the formate (HCOO) → dioxomethylene (H_2COO) route in favour of carboxyl (COOH) → formic acid [HC(OH)O] → hydroxymethoxy [$H_2C(OH)O$] → formaldehyde (hydroxyl) [CH_2O (OH)] → methoxy (CH_3O). The hydroxymethoxy species is rightly considered the key intermediate but its identity as the conjugate base of methanediol or methylene glycol (formaldehyde hydrate) has seemingly not been recognized [343,344]. This species is implicated in the Cannizzaro disproportionation involving a hydride or proton-coupled 2 e^- transfer reaction [345,346,347,348]:

$$2\ H_2C(OH)O \rightarrow CH_3O + HCOO + H_2O \qquad (27)$$

thereby providing a simpler (direct) path to methoxy.

SUMMARY AND OUTLOOK

From this brief overview of a burgeoning field of research over the last two to three decades, it is somewhat puzzling and disappointing to admit that a "solar fuels" industry is still no more than a futuristic concept [349] with no clear indication as to when to expect its realization [58], or at least the emergence of (pre-commercial) demonstration systems. A recent review of the proliferating patent literature reveals that they are almost exclusively based on incremental advances at the fundamental level, many being filed by academic scientists [42]. However, looked at *sub specie aeternatitis,* one must recognize the enormity of the challenge facing Mankind, viz., a paradigm shift towards a sustainable global economy based on an entirely new foundation, or what could be termed "renewable petrochemistry" [350]. The expanding role of industrial catalysis in such a future will require a thorough evaluation of scalability issues, not least in global elemental resources [68] for a sector established largely on the exploitation of rare metals. Photocatalysis and the maturation of commercial photoreactor design [351] in environmental detoxification is certainly gaining an industrial foothold [13], but this reviewer has only come across one pertinent reference to H_2 photogeneration on a pilot scale, and even this did not involve water splitting (O_2 co-generation) but photoreforming of aqueous organic contaminants over Pt/N-TiO_2 and Pt/CdS/ZnS [352]. Otherwise, nanoparticulate suspensions have demonstrated very low efficiencies to date ($\leq$1%) and offer merely simplicity and convenience in operation. Disadvantages include unwieldy reactor size/catalyst charge and an associated explosion hazard in the absence of a H_2/O_2 separation stage. In contrast, implementation of a PEC-membrane-integrated tandem system that yields pure H_2 at STH efficiencies already exceeding 10% in the laboratory (mandated to meet the US DOE cost target of USD 2.00–4.00 per kg H_2 [353]), is hampered by complexities associated with device design and scale-up, especially geometric factors and their role in loss (optical and overvoltage) minimization [354,355,356]. Indeed, the question as to whether an integrated (PV-PEC) design will ultimately outperform coupled PV-electrolyzers (with independently optimized components), or if their co-development has advantages [230,233,234,357], is still open to debate [58,59,226,358,359]. Certainly the former raises more materials compatibility issues since electrolyzers work best in acid or alkaline environments. According to McKone *et al.* [58], priority needs in fundamental work include: (a) Higher efficiency electrolysis in buffered pH-neutral electrolyte (for greater durability of earth-abundant absorbers and catalysts); (b) More conductive anion (alkaline) exchange membranes (thinner separators lead to higher efficiency electrolysis and faster pH equilibration); (c) More earth-abundant OER catalysts stable in acidic media (a major expansion in PEM-based devices will otherwise be limited by their present dependence on precious metal catalysts); (d) Optical transparency in lower mass-specific activity co-catalysts (minimal parasitic light absorption); and (e) A wider range of stable (acid/alkali -resistant) visible light absorbers and transparent (ultra-thin protective) layers for Si.

Two excellent articles have appeared recently that address the technoeconomic feasibility of centralized facilities for solar hydrogen [60] and the more complex case of solar methanol, which includes the problem of CO_2 sourcing [231]. Pinaud *et al.*[60] considered four types of reactor systems of increasing complexity designed to generate 1 ton per day of pure H_2 at 20 bar from 0.1 M KOH solution/electrolyte. Type 1 was an array of 18 shallow plastic "baggies" (323 × 12 × 0.1 m) housing a single-bed photocatalyst particle suspension working at an assumed STH efficiency (η) of 10% (theoretical $\eta_{max} \approx 23\%$). Type 2 was similar but had two (types of photocatalyst) beds for separate generation of H_2 and O_2 coupled by redox mediators (Z-scheme)—$\eta = 5\%$ ($\eta_{max} \approx 15\%$). Type 3 was a fixed panel array of ~27,000 monolithic tandem absorber PEC cells—$\eta = 10\%$ ($\eta_{max} \approx 30\%$), while Type 4 was similar but with a drastically reduced number of cells (~2,000, generating equivalent power) due to coupling with a tracking concentrator (×10) assembly—$\eta = 15\%$. The estimated cost of H_2 (per kg) for the four cases was \$1.60, \$3.20, \$10.40, and \$4.00, respectively. These are encouraging figures but it should be recognized that the efficiencies assumed for the particulate systems were "target" values several times higher than the current state-of-the-art. Clearly, the most promising and realistic system for early demonstration would be Type 4. However, the inclusion of low-power concentrators adds further complexity [360] and raises the question as to how PEC-based systems actually respond to light intensification and to what extent it is influenced by device configuration. While it is well-known that efficiencies in PV arrays are maintained (or even improved) by concentration of sunlight [361,362], the progressive drop in photochemical quantum yields with increasing incident power in suspended particulate systems is notorious [363], explaining the virtual absence of concentrating optics in advanced oxidation photoreactors, excepting perhaps the compound parabolic reflector for improved collection of diffuse sunlight [351]. The problem is generally attributed to kinetic limitations in surface redox processes and/or O_2 supply due to low aqueous solubility. Some of these "chemical" constraints are shared by PEC cells, such that a strong case can be made for a systematic evaluation of the sensitivity of water splitting efficiency to solar light concentration. The only positive evidence in the open literature is the NREL claim of a record-breaking 12% efficiency (in 1998) using a prototype hybrid PV/PEC system comprising a $GaInP_2$ photocathode voltage-biased by an underlying GaAs (PV) absorber under 11 Suns illumination [364].

Herron *et al.* [231] have taken a broader approach in designing a "transitional" solar refinery that produces H_2 as an intermediate in the generation of liquid fuels but with sub-systems still dependent to some degree on fossil fuel energy input, as represented schematically in Figure 30. First, a feasibility study was made by assessing the energy balance for the*indirect* route based on existing (sub-system) efficiencies. The CAMERE process [365] was selected and modeled, viz., methanol synthesis (at 1 kg/s ≡ 22.7 MW_{HHV}) from CO_2 (captured from a fossil fuel power station) with H_2 (produced increasingly from solar energy), and defining the solar or primary "energy incorporation efficiency" (EIE) as being positive (viable) only when the methanol energy content exceeded the sum of all fossil energy inputs. This rather severe criterion was not satisfied even when all the H_2 was derived from solar energy. Thus, there is a need for diversification and greater implementation of solar technologies in early demonstration systems. A good candidate (energy-intensive) process for fossil fuel substitution is solar heat-driven CO_2 recovery in amine scrubbers and, ultimately, atmospheric trapping devices [20,21].

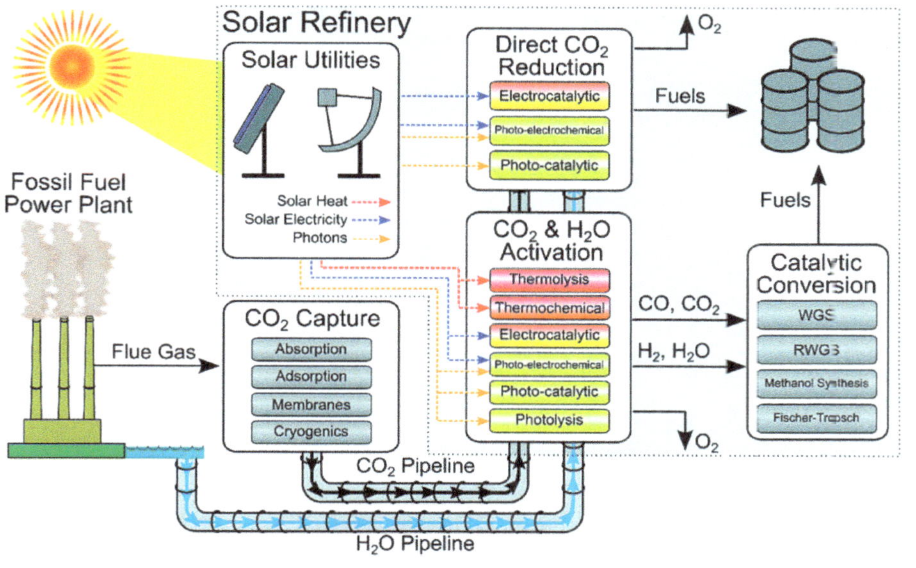

Figure 30. Scheme of a solar refinery based on CO_2 reduction to methanol *indirectly* by renewable H_2 (ex solar photon- or heat-driven electrolysis) or *directly* ("one pot") with H_2O using solar-electricity (PEC) or solar photons (photocatalysis) (reproduced from [231] with permission of the Royal Society of Chemistry).

A case study estimating the threshold CO_2 single-pass conversion for positive EIE based on a *direct* (photocatalytic) route was also made, assuming a reaction selectivity of 40% as below:

$$CO_2 + 2\,H_2O \rightarrow 0.6\,CH_4 + 0.4\,CH_3OH + 1.8\,O_2 \qquad (28)$$

and a catalyst mass-specific rate of ~1 μmol/g_{cat}/h (Φ_{uv} = 0.28% over Ti-containing porous SiO_2 [366]). As shown in Figure 31, it is primarily the CO_2 capture stage (from dilute flue gas) that impedes the energy efficiency break-even point (EIE = 0), such that present costs of around 5.5 MJ/kg_{CO2} must be further reduced. If the enthalpy from burning the renewable CH_4co-product is valorized, positive energy incorporation is feasible below 50% conversion. Better still, if a more selective catalyst can be developed, it would simplify the process and have a dramatic effect on minimum one-pass conversion (EIE > 0), as shown in Figure 32. However, by the authors' own admission, this case is quite impractical due to the low activity of the photocatalyst, roughly 3 orders of magnitude below that required as discussed in Section 4 and Section 5. Even at 1 mmol/g_{cat}/h, for an overall production rate of 1 kg/s, it would need 100 metric tons of catalyst and a reactor volume of 100 m^3 severely constrained in one dimension by optical factors.

In closing, the author feels obliged to point out that the earlier review [19] was written on the premise that future market penetration by the renewable energy sector would be substantial and already impacting (slowing the rise of, if not stabilizing) global CO_2 levels in the atmosphere. Unfortunately, governments have not provided sufficient incentives or regulatory measures to curb our dependence on fossil fuels and the oil industry has conducted very much a "business-as-usual" policy. In the last 20 years, the rate of emissions has increased by 50% and a cumulative level of 400 ppm CO_2 has been reached, *i.e.*, 50 ppm in excess of the threshold considered necessary to avoid a mean global temperature rise of more than 2 °C with probable catastrophic effects. While the penetration of CO_2-neutral energy systems will help to ease the burden on natural sinks for CO_2, it now appears essential to augment these with artificial (man-made) disposal methods. The author's directions in research have broadened accordingly in recent years, favouring mineralization as the only technology that will guarantee CO_2 sequestration on the requisite (geological) time-scale [367,368,369,370,371]

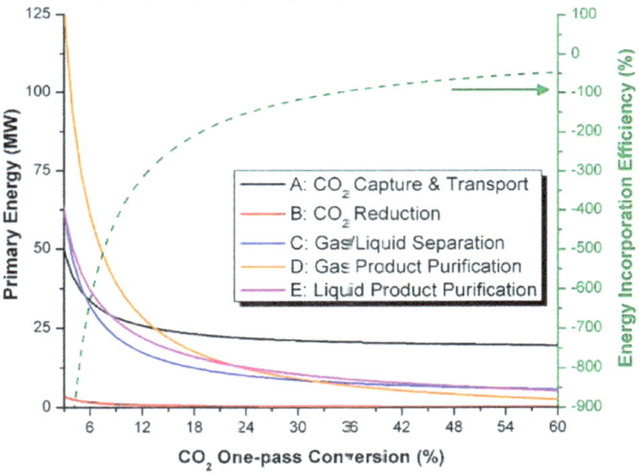

Figure 31. Dominance of CO_2 capture stage on primary energy costs in photocatalytic CH_3OH synthesis from CO_2/H_2O (reproduced from [232] with permission of the Royal Society of Chemistry).

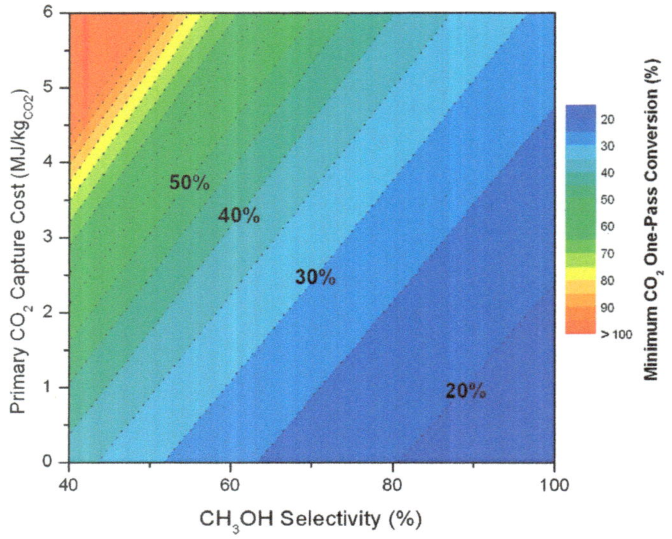

Figure 32. Effect of photocatalyst selectivity and CO_2 capture cost on minimum one-pass conversion for positive energy incorporation efficiency (EIE > 0) (reproduced from [231] with permission of the Royal Society of Chemistry).

Acknowledgments

The author is grateful to Eric Claude, Innovation Director at *Axane* (Air Liquide, Grenoble, France), for valuable advice on the hydrogen peroxide fuel cell.

Conflicts of Interest

The author declares no conflict of interest.

REFERENCES

1. Smalley, R.E. Future Global Energy Prosperity: The Terawatt Challenge. *MRS Bull.* **2005**, *30*, 412.
2. Bell, A.T. The Impact of Nanoscience on Heterogeneous Catalysis. *Science* **2003**, *299*, 1688–1691.
3. Rolison, D.R. Catalytic Nanoarchitectures—The Importance of Nothing and the Unimportance of Periodicity. *Science***2003**, *299*, 1698–1701.
4. Nørskov, J.K.; Bligaard, T.; Rossmeisl, J.; Christensen, C.H. Towards the Computational Design of Solid Catalysts. *Nat. Chem.* **2009**, *1*, 37–46.
5. Semagina, N.; Kiwi-Minsker, L. Recent Advances in the liquid-phase synthesis of metal nanostructures with controlled shape and size for catalysis. *Catal. Rev. Sci. Eng.* **2009**, *51*, 147–217.
6. Cuenya, B.R. Synthesis and Catalytic Properties of Metal Nanoparticles: Size, Shape, Support, Composition, and Oxidation State Effects. *Thin Solid Films* **2010**, *518*, 3127–3150.
7. Yuan, Y.; Yan, N.; Dyson, P.J. Advances in the Rational Design of Rhodium Nanoparticle Catalysts: Control via Manipulation of the Nanoparticle Core and Stabilizer. *ACS Catal.* **2012**, *2*, 1057–1069.
8. Guo, Z.; Liu, B.; Zhang, Q.H.; Deng, W.P.; Wang, Y.; Yang, Y.H. Recent Advances in Heterogeneous Selective Oxidation Catalysis for Sustainable Chemistry. *Chem. Soc. Rev.* **2014**, *43*, 3480–3524.
9. Schlögl, R.; Abd Hamid, S.B. Nanocatalysis: Mature Science Revisited or Something Really New? *Angew. Chem. Int. Ed.* **2004**, *43*, 1628–1637.
10. Chen, X.; Li, C.; Grätzel, M.; Kostecki, R.; Mao, S.S. Nanomaterials for Renewable Energy Production and Storage.*Chem. Soc. Rev.* **2012**, *41*, 7909–7937.
11. Thomas, J.M. Heterogeneous Catalysis and the Challenges of Powering the Planet, Securing Chemicals for Civilised Life, and Clean Efficient Utilization of Renewable Feedstocks. *ChemSusChem* **2014**, *7*, 1801–1832.
12. Kondratenko, E.V.; Mul, G.; Baltrusaitis, J.; Larrazabal, G.O.; Perez-Ramirez, J. Status and Perspectives of CO_2Conversion into Fuels and Chemicals by Catalytic, Photocatalytic and Electrocatalytic Processes. *Energy Environ. Sci.***2013**, *6*, 3112–3135.
13. Lu, M. *Photocatalysis and Water Purification: From Fundamentals to Recent Applications*; Pichat, P., Ed.; Wiley-VCH Verlag: Weinheim, Germany, 2013.
14. Ohtani, B. Photocatalysis A to Z—What We Know and What We Do Not Know in a Scientific Sense. *J. Photochem. Photobiol. C Photochem. Rev.* **2010**, *11*, 157–178.
15. Kalyanasundaram, K.; Grätzel, M. Artificial Photosynthesis: Biomimetic Approaches to Solar Energy Conversion and Storage. *Curr. Opin. Biotechnol.* **2010**, *21*, 298–310.
16. Wang, Y.; Chen, K.S.; Mishler, J.; Cho, S.C.; Adroher, X.C. A review of polymer electrolyte membrane fuel cells: Technology, applications, and needs on fundamental research. *Appl. Energy* **2011**, *88*, 981–1007.
17. Durbin, D.J.; Malardier-Jugroot, C. Review of Hydrogen Storage Techniques for On-Board Vehicle Applications. *Int. J. Hydrogen Energy* **2013**, *38*, 14595–14617.
18. Muradov, N.Z.; Veziroglu, T.N. "Green" Path from Fossil-Based to Hydrogen Economy: An Overview of Carbon-Neutral Technologies. *Int. J. Hydrogen Energy* **2008**, *33*, 6804–6839.
19. Highfield, J.G. The Central Role of Catalysis in a Future Energy Cycle based on Renewable Hydrogen and Carbon Dioxide as Reactive Liquefier. *Trends Phys. Chem.* **1995**, *5*, 91–156.
20. Goeppert, A.; Czaun, M.; Prakash, G.K.S.; Olah, G.A. Air as the Renewable Carbon Source of the Future: An Overview of CO_2 Capture from the Atmosphere. *Energy Environ. Sci.* **2012**, *5*, 7833–7853.
21. Lackner, K.S. The Thermodynamics of Direct Air Capture of Carbon Dioxide. *Energy* **2013**, *50*, 38–46.
22. Alonso, D.M.; Bond, J.Q.; Dumesic, J.A. Catalytic Conversion of Biomass to Biofuels. *Green Chem.* **2010**, *12*, 1493–1513.
23. Olah, G.A.; Goeppert, A.; Prakash, G.K.S. *Beyond Oil and Gas: The Methanol Economy*, 1st ed.; Wiley-VCH: Weinheim, Germany, 2006.
24. Goeppert, A.; Czaun, M.; Jones, J.P.; Prakash, G.K.S.; Olah, G.A. Recycling of Carbon Dioxide to Methanol

and Derived Products—Closing the Loop. *Chem. Soc. Rev.* **2014**, *43*, 7995–8048.

25. Galindo Cifre, P.; Badr, O. Renewable Hydrogen Utilisation for the Production of Methanol. *Energy Convers. Manag.***2007**, *48*, 519–527.
26. Wang, W.; Wang, S.; Ma, X.; Gong, J. Recent Advances in Catalytic Hydrogenation of Carbon Dioxide. *Chem. Soc. Rev.***2011**, *40*, 3703–3727.
27. Velu, S.; Gangwal, S.K. A Review of Recent Literature to Search for an Efficient Catalytic Process for the Conversion of Syngas to Ethanol. *Energy Fuels* **2008**, *22*, 814–839.
28. Palo, D.R.; Dagle, R.A.; Holladay, J.D. Methanol Steam Reforming for Hydrogen Production. *Chem. Rev.* **2007**, *107*, 3992–4021.
29. Huang, L.; Choong, C.; Chen, L.W.; Wang, Z.; Zhong, Z.Y.; Campos-Cuerva, C.; Lin, J.Y. Monometallic Carbonyl-Derived CeO_2-Supported Rh and Co Bicomponent Catalysts for CO-Free, High-Yield H_2 Generation from Low-Temperature Ethanol Steam-Reforming. *ChemCatChem* **2013**, *5*, 220–234.
30. Mattos, L.V.; Jacobs, G.; Davis, B.H.; Noronha, F.B. Production of Hydrogen from Ethanol: Review of Reaction Mechanism and Catalyst Deactivation. *Chem. Rev.* **2012**, *112*, 4094–4123.
31. De la Piscina, P.R.; Homs, N. The Use of Biofuels to Produce Hydrogen (Reformation Processes). *Chem. Soc. Rev.* **2008**,*37*, 2459–2467.
32. De Rogatis, L.; Fornasiero, P.; Catalyst Design for Reforming of Oxygenates. *Catalysis for Sustainable Energy Production*; Barbaro, P., Bianchini, C., Eds.; Wiley-VCH Verlag GmbH: Weinheim, Germany, 2009; Chapter 6; pp. 173–232.
33. Tran, N.H.; Kannangara, G.S.K. Conversion of Glycerol to Hydrogen Rich Gas. *Chem. Soc. Rev.* **2013**, *42*, 9454–9479.
34. Chheda, J.N.; Huber, G.W.; Dumesic, J.A. Liquid-Phase Catalytic Processing of Biomass-Derived Oxygenated Hydrocarbons to Fuels and Chemicals. *Angew. Chem. Int. Ed.* **2007**, *46*, 7164–7183.
35. Bowker, M. Photocatalytic Hydrogen Production and Oxygenate Photoreforming. *Catal. Lett.* **2012**, *142*, 923–929.
36. Jiang, Z.; Xiao, T.; Kuznetsov, V.L.; Edwards, P.P. Turning Carbon Dioxide into Fuel. *Philos. Trans. R. Soc. A* **2010**, *368*, 3343–3364.
37. Li, C.; Domen, K.; Hutchings, G.J. Special Issue on Photocatalysis and Photoelectrolysis. *J. Catal.* **2014**, *310*, 1–108.
38. Dasgupta, S.; Brunschwig, B.S.; Winkler, J.R.; Gray, H.B. Solar Fuels Editorial. *Chem. Soc. Rev.* **2013**, *42*, 2205–2472.
39. Qu, Y.; Duan, X. Progress, Challenge and Perspective of Heterogeneous Photocatalysts. *Chem. Soc. Rev.* **2013**, *42*, 2568–2580.
40. Sun, H.; Wang, S. Research Advances in the Synthesis of Nanocarbon-based Photocatalysts and Their Applications for Photocatalytic Conversion of Carbon Dioxide to Hydrocarbon Fuels. *Energy Fuels* **2014**, *28*, 22–36.
41. Li, K.; An, X.; Park, K.H.; Khraisheh, M.; Tang, J. A Critical Review of CO_2 Photoconversion: Catalysts and Reactors.*Catal. Today* **2014** *224*, 3–12.
42. Protti, S.; Albini, A.; Serpone, N. Photocatalytic Generation of Solar Fuels from the Reduction of H_2O and CO_2: A Look at the Patent Literature. *Phys. Chem. Chem. Phys.* **2014**, *16*, 19790–19827.
43. Habisreutinger, S.N.; Schmidt-Mende, L.; Stolarczyk, J.K. Photocatalytic Reduction of CO_2 on TiO_2 and Other Semiconductors. *Angew. Chem. Int. Ed.* **2013**, *52*, 7372–7408.
44. Izumi, Y. Recent Advances in the Photocatalytic Conversion of Carbon Dioxide to Fuels with Water and/or Hydrogen Using Solar Energy and Beyond. *Coord. Chem. Rev.* **2013**, *257*, 171–186.
45. Dhakshinamoorthy, A.; Navalon, S.; Corma, A.; Garcia, H. Photocatalytic CO_2 Reduction by TiO_2 and Related Titanium Containing Solids. *Energy Environ. Sci.* **2012**, *5*, 9217–9233.
46. Nocera, D.G. The Artificial Leaf. *Acc. Chem. Res.* **2012**, *45*, 767–776.
47. Hoffmann, M.R.; Moss, J.A.; Baum, M.M. Artificial Photosynthesis: Semiconductor Photocatalytic Fixation of CO_2 to Afford Higher Organic Compounds. *Dalton Trans.* **2011**, *40*, 5151–5153.
48. Ismail, A.A.; Bahnemann, D.W. Photochemical Splitting of Water for Hydrogen Production by Photocatalysis: A Review. *Sol. Energy Mater. Sol. Cells* **2014**, *128*, 85–101.
49. Hisatomi, T.; Kubota, J.; Domen, K. Recent Advances in Semiconductors for Photocatalytic and Photoelectrochemical Water Splitting. *Chem. Soc. Rev.* **2014**, *43*, 7520–7535.
50. Maeda, K. Z-Scheme Water Splitting Using Two Different Semiconductor Photocatalysts. *ACS Catal.* **2013**, *3*, 1486–1503.
51. Yang, J.; Wang, D.; Han, H.; Li, C. Roles of Cocatalysts in Photocatalysis and Photoelectrocatalysis. *Acc. Chem. Res.***2013**, *46*, 1900–1909.
52. Horiuchi, Y.; Toyao, T.; Takeuchi, M.; Matsuoka, M.; Anpo, M. Recent Advances in Visible-Light-Responsive Photocatalysts for Hydrogen Production and Solar Energy Conversion—From Semiconducting TiO_2 to MOF/PCP Photocatalysts. *Phys. Chem. Chem. Phys.* **2013**, *15*, 13243–13253.
53. Yang, J.; Yan, H.; Zong, X.; Wen, F.; Liu, M.; Li, C. Roles of Cocatalysts in Semiconductor-Based Photocatalytic Hydrogen Production. *Philos. Trans. R. Soc. A* **2013**, *371*, 20110430.
54. Cargnello, M.; Diroll, B.T. Tailoring Photocatalytic Nanostructures for Sustainable Hydrogen Production. *Nanoscale***2014**, *6*, 97–105.
55. Cargnello, M.; Gasparotto, A.; Gombac, V.; Montini, T.; Barreca, D.; Fornasiero, P. Photocatalytic H_2 and Added-Value By-Products—The Role of Metal Oxide Systems in Their Synthesis from Oxygenates. *Eur. J. Inorg. Chem.* **2011**, *28*, 4309–4323.

56. Navarro, R.M.; Sánchez-Sánchez, M.C.; Alvarez-Galvan, M.C.; del Valle, F.; Fierro, J.L.G. Hydrogen Production from Renewable Sources: Biomass and Photocatalytic Opportunities. *Energy Environ. Sci.* **2009**, *2*, 35–54.
57. Smestad, G.P.; Steinfeld, A. Review: Photochemical and Thermochemical Production of Solar Fuels from H_2O and CO_2 using Metal Oxide Catalysts. *Ind. Eng. Chem. Res.* **2012**, *51*, 11828–11840.
58. McKone, J.R.; Lewis, N.S.; Gray, H.B. Will Solar-Driven Water-Splitting Devices See the Light of Day? *Chem. Mater.***2013**, *26*, 407–414.
59. Nielander, A.C.; Shaner, M.R.; Papadantonakis, K.M.; Francis, S.A.; Lewis, N.S. A Taxonomy for Solar Fuels Generators. *Energy Environ. Sci.* **2015**, *8*, 16–25.
60. Pinaud, B.A.; Benck, J.D.; Seitz, L.C.; Forman, A.J.; Chen, Z.; Deutsch, T.G.; James, B.D.; Baum, G.N.; Ardo, S.; Wang, H.; *et al.* Technical and Economic Feasibility of Centralized Facilities for Solar Hydrogen Production via Photocatalysis and Photoelectrochemistry. *Energy Environ. Sci.* **2013**, *6*, 1983–2002.
61. Brillet, J.; Yum, J.H.; Cornuz, M.; Hisatomi, T.; Solarska, R.; Augustynski, J.; Graetzel, M.; Sivula, K. Highly Efficient Water Splitting by a Dual-Absorber Tandem Cell. *Nat. Photon.* **2012**, *6*, 824–828.
62. Reece, S.Y.; Hamel, J.A.; Sung, K.; Jarvi, T.D.; Esswein, A.J.; Pijpers, J.J.H.; Nocera, D.G. Wireless Solar Water Splitting Using Silicon-Based Semiconductors and Earth-Abundant Catalysts. *Science* **2011**, *334*, 645–648.
63. McKone, J.R.; Marinescu, S.C.; Brunschwig, B.S.; Winkler, J.R.; Gray, H.B. Earth-Abundant Hydrogen Evolution Electrocatalysts. *Chem. Sci.* **2014**, *5*, 865–878.
64. Bozic-Weber, B.; Constable, E.C.; Housecroft, C.E. Light Harvesting with Earth-Abundant d-Block Metals: Development of Sensitizers in Dye-Sensitized Solar Cells (DSCs). *Coord. Chem. Rev.* **2013**, *257*, 3089–3106.
65. Thoi, V.S.; Sun, Y.J.; Long, J.R.; Chang, C.J. Complexes of Earth-Abundant Metals for Catalytic Electrochemical Hydrogen Generation under Aqueous Conditions. *Chem. Soc. Rev.* **2013**, *42*, 2388–2400.
66. Tran, P.D.; Pramana, S.S.; Kale, V.S.; Nguyen, M.; Chiam, S.Y.; Batabyal, S.K.; Wong, L.H.; Barber, J.; Loo, J. Novel Assembly of an MoS_2 Electrocatalyst onto a Silicon Nanowire Array Electrode to Construct a Photocathode Composed of Elements Abundant on the Earth for Hydrogen Generation. *Chem. Eur. J.* **2012**, *18*, 13994–13999.
67. Du, P.W.; Eisenberg, R. Catalysts Made of Earth-Abundant Elements (Co, Ni, Fe) for Water Splitting: Recent Progress and Future Challenges. *Energy Environ. Sci.* **2012**, *5*, 6012–6021.
68. Vesborg, P.C.K.; Jaramillo, T.F. Addressing the Terawatt Challenge: Scalability in the Supply of Chemical Elements for Renewable Energy. *RSC Adv.* **2012**, *2*, 7933–7947.
69. Bolton, J.R.; Strickler, S.J; Connolly, J.S. Limiting and Realizable Efficiencies of Solar Photolysis of Water. *Nature* **1985**,*316*, 495–500.
70. Archer, M.D; Bolton, J.R. Requirements for Ideal Performance of Photochemical and Photovoltaic Solar Energy Converters. *J. Phys. Chem.* **1990**, *94*, 8028–8036.
71. Bolton, J.R. Solar Photoproduction of Hydrogen: A Review. *Sol. Energy* **1996**, *57*, 37–50.
72. Smestad, G.P; Krebs, F.C; Lampert, C.M; Granqvist, C.G.; Chopra, K.L.; Mathew, X.; Takakura, H. Reporting Solar Cell Efficiencies in Solar Energy Materials and Solar Cells. *Sol. Energy Mater. Sol. Cells* **2008**, *92*, 371–373.
73. Emeline, A.V.; Kuznetsov, V.N.; Ryabchuk, V.K.; Serpone, N. On the Way to the Creation of Next Generation Photoactive Materials. *Environ. Sci. Pollut. Res.* **2012**, *19*, 3666–3675.
74. Serpone, N.; Emeline, A.V. Semiconductor Photocatalysis—Past, Present, and Future Outlook. *J. Phys. Chem. Lett.* **2012**,*3*, 673–677.
75. Bard, A.J. Photoelectrochemistry and Heterogeneous Photocatalysis at Semiconductors. *J. Photochem.* **1979**, *10*, 59–75.
76. Grätzel, M. Photoelectrochemical Cells. *Nature* **2001**, *414*, 338–344.
77. Luo, J.; Im, J.H.; Mayer, M.T.; Schrieier, M.; Nazeeruddin, K.; Park, N.G.; Tilley, S.D.; Fan, H.J.; Grätzel, M. Water Photolysis at 12.3% Efficiency via Perovskite Photovoltaics and Earth-Abundant Catalysts. *Science* **2014**, *345*, 1593–1596.
78. Cox, C.R.; Lee, J.Z.; Nocera, D.G.; Buonassisi, T. Ten-Percent Solar-to-Fuel Conversion with Nonprecious Materials.*Proc. Natl. Acad. Sci. USA* **2014**, *111*, 14057–14061.
79. Rongé, J.; Bosserez, T.; Martel, D.; Nervi, C.; Boarino, L.; Taulelle, F.; Decher, G.; Bordiga, S.; Martens, J.A. Monolithic Cells for Solar Fuels. *Chem. Soc. Rev.* **2014**, *43*, 7963–7981.
80. Fujishima, A.K.; Honda, K. Electrochemical Photolysis of Water at a Semiconductor Electrode. *Nature* **1972**, *238*, 37–38.
81. Borgarello, E.; Kiwi, J.; Pelizzetti, E.; Visca, M.; Graetzel, M. Photochemical Cleavage of Water by Photocatalysis.*Nature* **1981**, *289*, 158–160.
82. Wang, Z.; Liu, Y.; Huang, B.; Dai, Y.; Lou, Z.; Wang, G.; Zhang, X.; Qin, X. Progress on Extending the Light Absorption Spectra of Photocatalysts. *Phys. Chem. Chem. Phys.* **2014**, *16*, 2758–2774.
83. Varghese, O.K.; Grimes, C.A. Appropriate Strategies for Determining the Photoconversion Efficiency of Water Photoelectrolysis Cells: A Review with Examples using Titania Nanotube Array Photoanodes. *Sol. Energy Mater. Sol. Cells* **2008**, *92*, 374–384.
84. Grätzel, M. Conversion of Sunlight to Electric Power by Nanocrystalline Dye-Sensitized Solar Cells. *J. Photochem. Photobiol. A* **2004**, *164*, 3–14.
85. Mathew, S.; Yella, A.; Gao, P.; Humphry-Baker, R.; Churchod, B.F.E.; Ashari-Astani, N.; Tavernelli, I.; Rothlisberger, U.; Nazeeruddin, M.K.; Grätzel, M. Dye-Sensitized Solar Cells with 13% Efficiency Achieved through the Molecular Engineering of Porphyrin Sensitizers. *Nat. Chem.* **2014**, *6*, 242–247.

86. Hardin, B.E.; Snaith, H.J.; McGehee, M.D. The Renaissance of Dye-Sensitized Solar Cells. *Nat. Photon.* **2012**, *6*, 162–169.

87. Li, L.L.; Diau, E.W. Porphyrin-Sensitized Solar Cells. *Chem. Soc. Rev.* **2013**, *42*, 291–304.

88. Balasingam, S.K.; Lee, M.; Kang, M.G.; Jun, Y. Improvement of Dye-Sensitized Solar Cells Toward the Broader Light Harvesting of the Solar Spectrum. *Chem. Commun.* **2013**, *49*, 1471–1487.

89. Macor, L.; Fungo, F.; Tempesti, T.; Durantini, E.N.; Otero, L.; Barea, E.M.; Fabregat-Santiago, F.; Bisquert, J. Near-IR Sensitization of Wide Band Gap Oxide Semiconductor by Axially Anchored Si-Naphthalocyanines. *Energy Environ. Sci.* **2009**, *2*, 529–534.

90. Ono, T.; Yamaguchi, T.; Arakawa, H. Study on Dye-Sensitized Solar Cell using Novel Infrared Dye. *Sol. Energy Mater. Sol. Cells* **2009**, *93*, 831–835.

91. Altobello, S.; Argazzi, R.; Caramori, S.; Contado, C.; da Fré, S.; Rubino, P.; Chon, C.; Larramona, G.; Bignozzi, C.A. Sensitization of Nanocrystalline TiO_2 with Black Absorbers based on Os and Ru Polypyridine Complexes. *J. Am. Chem. Soc.* **2005**, *127*, 15342–15343.

92. Persson, P.; Bergstrom, R.; Lunell, S. Quantum Chemical Study of Photoinjection Processes in Dye-sensitized TiO_2Nanoparticles. *J. Phys. Chem. B* **2000**, *104*, 10348–10351.

93. Ooyama, Y.; Yamada, T.; Fujita, T.; Harima, Y.; Ohshita, J. Development of D-π-Cat Fluorescent Dyes with a Catechol Group for Dye-Sensitized Solar Cells based on Dye-to-TiO_2 Charge Transfer. *J. Mater. Chem. A* **2014**, *2*, 8500–8511.

94. Pratik, S.M.; Datta, A. Computational Design of Concomitant Type-I and Type-II Porphyrin Sensitized Solar Cells.*Phys. Chem. Chem. Phys.* **2013**, *15*, 18471–18481.

95. Zhang, G.; Kim, G.; Choi, W. Visible-light driven photocatalysis mediated via ligand-to-metal charge transfer (LMCT): An alternative approach to solar activation of Titania. *Energy Environ. Sci.* **2014**, *7*, 954–966.

96. Cheng, Y.H.; Subramaniam, V.P.; Gong, D.G.; Tang, Y.; Highfield, J.; Pehkonen, S.O.; Pichat, P.; Schreyer, M.K.; Chen, Z. Nitrogen-Sensitized Dual Phase Titanate/Titania for Visible-Light Driven Phenol Degradation. *J. Solid State Chem.***2012**, *196*, 518–527.

97. Grätzel, M. The Light and Shade of Perovskite Solar Cells. *Nat. Mater.* **2014**, *13*, 838–842.

98. Green, M.A.; Ho-Baillie, A.; Snaith, H.J. The Emergence of Perovskite Solar Cells. *Nat. Photon.* **2014**, *8*, 506–514.

99. Gao, P.; Grätzel, M.; Nazeeruddin, M.K. Organohalide lead perovskites for photovoltaic applications. *Energy Environ. Sci.* **2014**, *7*, 2448–2463.

100. Kazim, S.; Nazeeruddin, M.K.; Grätzel, M.; Ahmad, S. Perovskite as Light Harvester: A Game Changer in Photovoltaics. *Angew. Chem. Int. Ed.* **2014**, *53*, 2812–2824.

101. Nowotny, M.K.; Sheppard, L.R.; Bak, T.; Nowotny, J. Defect Chemistry of Titanium Dioxide. Application of Defect Engineering in Processing of TiO_2-Based Photocatalysts. *J. Phys. Chem. C* **2008**, *112*, 5275–5300.

102. Mizushima, K.; Tanaka, M.; Asai, A.; Iiga, S.; Goodenough, J.B. Impurity Levels of Iron-Group Ions in TiO_2 (II). *J. Phys. Chem. Solids* **1979**, *40*, 1129–1140.

103. Choi, W.; Termin, A.; Hoffmann, M.R. The Role of Metal Ion Dopants in Quantum-Sized TiO_2: Correlation between Photoreactivity and Charge Carrier Recombination Dynamics. *J. Phys. Chem.* **1994**, *98*, 13669–13679.

104. Moser, J.; Grätzel, M.; Gallay, R. Inhibition of Electron-Hole Recombination in Substitution-ally Doped Colloidal Semiconductor Crystallites. *Helv. Chim. Acta* **1987**, *70*, 1596–1604.

105. Choi, J.; Park, H.; Hoffmann, M.R. Effects of Single Metal-Ion Doping on the Visible-Light Photoreactivity of TiO_2. *J. Phys. Chem. C* **2010**, *114*, 783–792.

106. Kernazhitsky, L.; Shymanovska, V.; Gavrilko, T.; Naumov, V.; Kshnyakin, V.; Khalyavka, T. A Comparative Study of Optical Absorption and Photocatalytic Properties of Nanocrystalline Single-Phase Anatase and Rutile TiO_2 Doped with Transition Metal Cations. *J. Solid State Chem.* **2013**, *198* 511–519.

107. Maeda, K. (Oxy)nitrides with d°-Electronic Configuration as Photocatalysts and Photoanodes that Operate under a Wide Range of Visible Light for Overall Water Splitting. *Phys. Chem. Chem. Phys.* **2013**, *15*, 10537–10548.

108. Teoh, W.Y.; Scott, J.A.; Amal, R. Progress in Heterogeneous Photocatalysis: From Classical Radical Chemistry to Engineering Nanomaterials and Solar Reactors. *J. Phys. Chem. Lett.* **2012**, *3*, 629–639.

109. Chen, X.B.; Liu, L.; Yu, P.Y.; Mao, S.S. Increasing Solar Absorption for Photocatalysis with Black Hydrogenated Titanium Dioxide Nanocrystals. *Science* **2011**, *331*, 746–750.

110. Pesci, F.M.; Wang, G.; Klug, D.R.; Li, Y.; Cowan, A.J. Efficient Suppression of Electron-Hole Recombination in Oxygen-Deficient Hydrogen-Treated TiO_2 Nanowires for Photoelectro-chemical Water Splitting. *J. Phys. Chem. C* **2013**, *117*, 25837–25844.

111. Liu, L.; Yu, P.Y.; Chen, X.; Mao, S.S.; Shen, D.Z. Hydrogenation and Disorder in Engineered Black TiO_2. *Phys. Rev. Lett.* **2013**, *111*, 065505-(1–5).

112. Su, J.; Zou, X.; Chen, J.S. Self-Modification of Titanium Dioxide Materials by Ti^{3+} and/or Oxygen Vacancies: New Insights into Defect Chemistry of Metal Oxides. *RSC Adv.* **2014**, *4*, 13979–13988.

113. Fresno, F.; Portela, R.; Suárez, S.; Coronado, J.M. Photocatalytic Materials: Recent Achievements and Near Future Trends. *J. Mater. Chem. A* **2014**, *2*, 2863–2884.

114. Asahi, R.; Morikawa, T.; Ohwaki, T.; Aoki, K.; Taga, Y. Visible-Light Photocatalysis in Nitrogen-Doped Titanium Oxides. *Science* **2001**, *293*, 269–271.

115. Khan, S.U.M.; al-Shahry, M.; Ingler, W.B. Efficient Photochemical Water Splitting by a Chemically Modified n-TiO_2.*Science* **2002**, *297*, 2243–2245.

116. Diwald, O.; Thompson, T.L.; Goralski, E.G.; Walck,

S.D.; Yates, J.T. The Effect of Nitrogen Ion Implantation on the Photoactivity of TiO_2 Rutile Single Crystals. *J. Phys. Chem. B* **2004**, *108*, 52–57.

117. Di Valentin, C.; Pacchioni, G.; Selloni, A. Origin of the Different Photoactivity of N-Doped Anatase and Rutile TiO_2.*Phys. Rev. B* **2004**, *70*, 085116.

118. Nakamura, R.; Tanaka, T.; Nakato, Y. Mechanism for Visible Light Responses in Anodic Photocurrents at N-Doped TiO_2 Film Electrodes. *J. Phys. Chem. B* **2004**, *108*, 10617–10620.

119. Choi, Y.; Umebayashi, T.; Yoshikawa, M. Fabrication and Characterization of C-Doped Anatase TiO_2 Photocatalysts. *J. Mater. Sci.* **2004**, *39*, 1837–1839.

120. Yu, J.C.; Yu, J.G.; Ho, W.K.; Jiang, Z.T.; Zhang, L.Z. Effects of F-Doping on the Photocatalytic Activity and Microstructures of Nanocrystalline TiO_2 Powders. *Chem. Mater.* **2002**, *14*, 3808–3816.

121. Ohno, T.; Akiyoshi, M.; Umebayashi, T.; Asai, K.; Mitsui, T.; Matsumura, M. Preparation of S-Doped TiO_2Photocatalysts and Their Photocatalytic Activities under Visible Light. *Appl. Catal. A Gen.* **2004**, *265*, 115–121.

122. Lin, L.; Lin, W.; Zhu, Y.X.; Zhao, B.Y.; Xie, Y.C. Phosphor-Doped Titania—A Novel Photocatalyst Active in Visible Light. *Chem. Lett.* **2005**, *34*, 284–285.

123. Yin, S.; Ihara, K.; Aita, Y.; Komatsu, M.; Sato, T. Visible-Light Induced Photocatalytic Activity of $TiO_{(2-x)}A(_y)$ (A = N, S) Prepared by Precipitation Route. *J. Photochem. Photobiol. A Chem.* **2006**, *179*, 105–114.

124. Chen, D.M.; Jiang, Z.Y.; Geng, J.Q.; Wang, Q.; Yang, D. Carbon and Nitrogen Co-Doped TiO_2 with Enhanced Visible-Light Photocatalytic Activity. *Ind. Eng. Chem. Res.* **2007**, *46*, 2741–2746.

125. In, S.; Orlov, A.; Berg, R.; Garcia, F.; Pedrosa-Jimenez, S.; Tikhov, M.S.; Wright, D.S.; Lambert, R.M. Effective Visible Light-Activated B-Doped and B,N-Co-Doped TiO_2 Photocatalysts. *J. Am. Chem. Soc.* **2007**, *129*, 13790–13791.

126. Di Valentin, C.; Finazzi, E.; Pacchioni, G.; Selloni, A.; Livraghi, S.; Paganini, M.C.; Giamello, E. N-Doped TiO_2: Theory and Experiment. *Chem. Phys.* **2007**, *339*, 44–56.

127. Harb, M.; Sautet, P.; Raybaud, P. Origin of the Enhanced Visible-Light Absorption in N-Doped Bulk Anatase $TiO_2$from First-Principles Calculations. *J. Phys. Chem. C* **2011**, *115*, 19394–19404.

128. Marschall, R.; Wang, L.Z. Non-Metal Doping of Transition Metal Oxides for Visible-Light Photocatalysis. *Catal. Today***2014**, *225*, 111–135.

129. Gomathi Devi, L.; Kavitha, R. Review on Modified N-TiO_2 for Green Energy Applications under UV/Visible Light: Selected Results and Reaction Mechanisms. *RSC Adv.* **2014**, *4*, 28265–28299.

130. Asahi, R.; Morikawa, T.; Irie, H.; Ohwaki, T. Nitrogen-Doped Titanium Dioxide as Visible-Light-Sensitive Photocatalyst: Designs, Developments, and Prospects. *Chem. Rev.* **2014**, *114*, 9824–9852.

131. Mrowetz, M.; Balcerski, W.; Colussi, A.J.; Hoffmann, M.R. Oxidative Power of Nitrogen-Doped TiO_2 Photocatalysts under Visible Illumination. *J. Phys. Chem. B* **2004**, *108*, 17269–17273.

132. Lindgren, T.; Lu, J.; Hoel, A.; Granqvist, C.G.; Torres, G.R.; Lindquist, S.E. Photoelectrochemical Study of Sputtered Nitrogen-Doped Titanium Dioxide Thin Films in Aqueous Electrolyte. *Sol. Energy Mater. Sol. Cells* **2004**, *84*, 145–157.

133. Tang, J.; Cowan, A.J.; Durrant, J.R.; Klug, D.R. Mechanism of O_2 Production from Water Splitting: Nature of Charge Carriers in Nitrogen Doped Nanocrystalline TiO_2 Films and Factors Limiting O_2 Production. *J. Phys. Chem. C* **2011**,*115*, 3143–3150.

134. Di Valentin, C.; Pacchioni, G. Trends in Non-Metal Doping of Anatase TiO_2: B, C, N and F. *Catal. Today* **2013**, *206*, 12–18.

135. Yin, W.J.; Tang, H.W.; Wei, S.H.; Al-Jassim, M.M.; Turner, J.; Yan, Y.F. Band Structure Engineering of Semiconductors for Enhanced Photoelectrochemical Water Splitting: The Case of TiO_2. *Phys. Rev. B* **2010**, *82*, 045106-(1–6).

136. Zhu, W.; Qiu, X.; Iancu, V.; Chen, X.Q.; Pan, H.; Wang, W.; Dimitrijevic, N.M.; Rajh, T.; Meyer, H.M., III; Paranthaman, M.P.; *et al.* Band Gap Narrowing of Titanium Oxide Semiconductors by Non-Compensated Anion-Cation Codoping for Enhanced Visible Light Photoactivity. *Phys. Rev. Lett.* **2009**, *103*, 226401:1–226401:4.

137. Czoska, A.M.; Livraghi, S.; Paganini, M.C.; Giamello, E.; di Valentin, C.; Pacchioni, G. The Nitrogen-Boron Paramagnetic Center in Visible Light-Sensitized N-B-Codoped TiO_2. Experimental and Theoretical Characterization.*Phys. Chem. Chem. Phys.* **2011**, *13*, 136–143.

138. Balcerski, W.; Ryu, S.Y.; Hoffmann, M.R. Visible-Light Photoactivity of Nitrogen-Doped TiO_2: Photo-Oxidation of HCO_2H to CO_2 and H_2O. *J. Phys. Chem. C* **2007**, *111*, 15357–15362.

139. Mitoraj, D.; Kisch, H. The Nature of Nitrogen-Modified Titanium Dioxide Photocatalysts Active in Visible Light. *Angew. Chem. Int. Ed.* **2008**, *47*, 9975–9978.

140. Mitoraj, D.; Kisch, H. On the Mechanism of Urea-Induced Titania Modification. *Chem. Eur. J.* **2010**, *16*, 261–269.

141. Ang, T.P.; Chan, Y.M. Comparison of the Melon Nanocomposites in Structural Properties and Photocatalytic Activities. *J. Phys. Chem. C* **2011**, *115*, 15965–15972.

142. Gong, D.; Highfield, J.G.; Zhong, E.N.; Tang, Y.; Ho, W.C.J.; Tay, Q.; Chen, Z. Poly Tri-s-triazines as Visible Light Sensitizers in Titania-Based Composite Photocatalysts: Promotion of Melon Development from Urea over Acid Titanates. *ACS Sustain. Chem. Eng.* **2014**, *2*, 149–157.

143. Wang, Y.; Wang, X.; Antonietti, M. Polymeric Graphitic Carbon Nitride as a Heterogeneous Organocatalyst: From Photochemistry to Multipurpose Catalysis to Sustainable Chemistry. *Angew. Chem. Int. Ed.* **2012**, *51*, 68–89.

144. Gole, J.L.; Stout, J.D.; Burda, C.; Lou, Y.; Chen, X. Highly Efficient Formation of Visible-Light Tunable $TiO_{2-x}N_x$Photocatalysts and Their Transformation at the Nanoscale. *J. Phys. Chem. B* **2004**, *108*, 1230–1240.
145. Cantau, C.; Pigot, T.; Dupin, J.C.; Lacombe, S. N-Doped TiO_2 by Low Temperature Synthesis: Stability, Photo-Reactivity, and Singlet Oxygen Formation in the Visible Range. *J. Photochem. Photobiol. A Chem.* **2010**, *216*, 201–208.
146. Spadavecchia, F.; Cappelletti, G.; Ardizzone, S.; Bianchi, C.L.; Cappelli, S.; Oliva, C.; Scardi, P.; Leoni, M.; Fermo, P. Solar Photoactivity of Nano-N-TiO_2 from Tertiary Amine: Role of Defects and Paramagnetic Species. *Appl. Catal. B: Environ.* **2010**, *96*, 314–322.
147. Cheng, Y.H.; Huang, Y.; Kanhere, P.D.; Subramaniam, V.P.; Gong, D.; Zhang, S.; Highfield, J.; Schreyer, M.K.; Chen, Z. Dual-Phase Titanate/Anatase with Nitrogen Doping for Enhanced Degradation of Organic Dye under Visible Light.*Chem. Eur. J.* **2011**, *17*, 2575–2578.
148. Zhao, Y.; Qiu, X.; Burda, C. The Effects of Sintering on the Photocatalytic Activity of N-doped TiO_2 Nanoparticles. *Chem. Mater.* **2008**, *20*, 2629–2636.
149. Tian, Y.; Tatsuma, T. Mechanisms and Applications of Plasmon-Induced Charge Separation at TiO_2 Films Loaded with Gold Nanoparticles. *J. Am. Chem. Soc.* **2005**, *127*, 7632–7637.
150. Kowalska, E.; Mahaney, O.O.P.; Abe, R.; Ohtani, B. Visible-Light-Induced Photocatalysis through Surface Plasmon Excitation of Gold on Titania Surfaces. *Phys. Chem. Chem. Phys.* **2010**, *12*, 2344–2355.
151. Gong, D.; Ho, W.C.J.; Tang, Y.; Tay, Q.; Lai, Y.; Highfield, J.G.; Chen, Z. Silver-Decorated Titanate/Titania Nanostructures for Efficient Solar Driven Photocatalysis. *J. Solid State Chem.* **2012**, *189*, 117–122.
152. Sellapan, R.; Nielsen, M.G.; Gonzalez-Posada, F.; Vesborg, P.C.K.; Chorkendorff, I.; Chakarov, D. Effects of Plasmon Excitation on Photocatalytic Activity of Ag/TiO_2 and Au/TiO_2 Nanocomposites. *J. Catal.* **2013**, *307*, 214–221.
153. Mubeen, S.; Lee, J.; Singh, N.; Krämer, S.; Stucky, G.D.; Moskovits, M. An autonomous photosynthetic device in which all charge carriers derive from surface Plasmons. *Nat. Nanotechnol.* **2013**, *8*, 247–251.
154. Ma, X.; Dai, Y.; Yu, L.; Huang, B. New Basic Insights into the Low Hot Electron Injection Efficiency of Gold-Nanoparticle-Photosensitized Titanium Dioxide. *ACS Appl. Mater. Interfaces* **2014**, *6*, 12388–12394.
155. Clavero, C. Plasmon-Induced Hot-Electron Generation at Nanoparticle/Metal-Oxide Interfaces for Photovoltaic and Photocatalytic Devices. *Nat. Photon.* **2014**, *8*, 95–103.
156. Xu, H.; Ouyang, S.; Liu, L.; Reunchan, P.; Umezawa, N.; Ye, J. Recent Advances in TiO_2-based Photocatalysis. *J. Mater. Chem. A* **2014**, *2*, 12642–12661.
157. Warren, S.C.; Thimsen, E. Plasmonic Solar Water Splitting. *Energy Environ. Sci.* **2012**, *5*, 5133–5146.
158. Linic, S.; Christopher, P.; Ingram, D.B. Plasmonic-Metal Nanostructures for Efficient Conversion of Solar to Chemical Energy. *Nat. Mater.* **2011**, *10*, 911–921.
159. Zhou, X.M.; Liu, G.; Yu, J.G.; Fan, W.H. Surface Plasmon Resonance-Mediated Photocatalysis by Noble Metal-Based Composites under Visible Light. *J. Mater. Chem.* **2012**, *22*, 21337–21354.
160. DuChene, J.S.; Sweeny, B.C.; Johnston-Peck, A.C.; Su, D.; Stach, E.A.; Wei, W.D. Prolonged Hot Electron Dynamics in Plasmonic-Metal/Semiconductor Heterostructures with Implications for Solar Photocatalysis. *Angew. Chem. Int. Ed.***2014**, *53*, 7887–7891.
161. Ingram, D.B.; Christopher, P.; Bauer, J.L.; Linic, S. Predictive Model for the Design of Plasmonic Metal/Semiconductor Composite Photocatalysts. *ACS Catal.* **2011**, *1*, 1441–1447.
162. Thimsen, E.; le Formal, F.; Grätzel, M.; Warren, S.C. Influence of Plasmonic Au Nanoparticles on the Photoactivity of Fe_2O_3 Electrodes for Water Splitting. *Nano Lett.* **2011**, *11*, 35–43.
163. Logar, M.; Bračko, I.; Potočnik, A.; Jančar, B. Cu and CuO/Titanate Nanobelt Based Network Assemblies for Enhanced Visible Light Photocatalysis. *Langmuir* **2014**, *30*, 4852–4862.
164. Nogawa, T.; Isobe, T.; Matsushita, S.; Nakajima, A. Preparation and Visible-Light Photo-catalytic Activity of Au- and Cu-modified TiO_2 Powders. *Mater. Lett.* **2012**, *82*, 174–177.
165. Kazuma, E.; Yamaguchi, T.; Sakai, N.; Tatsuma, T. Growth Behaviour and Plasmon Resonance Properties of Photocatalytically Deposited Cu Nanoparticles. *Nanoscale* **2011**, *3*, 3641–3645.
166. Knight, M.W.; King, N.S.; Liu, L.; Everitt, H.O.; Nordlander, P.; Halas, N.J. Aluminum for Plasmonics. *ACS Nano* **2014**,*8*, 834–840.
167. Lounis, S.D.; Runnerstrom, E.L.; Llordés; Milliron, D.J. Defect Chemistry and Plasmon Physics of Colloidal Metal Oxide Nanocrystals. *J. Phys. Chem. Lett.* **2014**, *5*, 1564–1574.
168. Henderson, M.A. A Surface Science Perspective on TiO_2 Photocatalysis. *Surf. Sci. Rep.* **2011**, *66*, 185–297.
169. Hernández-Alonso,M.D.;Fresno,F.;Suárez,S.;Coronado, J.M. Development of Alternative Photocatalysts to TiO_2: Challenges and Opportunities. *Energy Environ. Sci.* **2009**, *2*, 1231–1257.
170. Chen, X.; Mao, S.S. Titanium Dioxide Nanomaterials: Synthesis, Properties, Modifications, and Applications. *Chem. Rev.* **2007**, *107*, 2891–2959.
171. Zhang, J.; Xu, Q.; Li, M; Li, C. Importance of the Relationship between Surface Phases and Photocatalytic Activity of TiO_2. *Angew. Chem. Int. Ed.* **2008**, *47*, 1766–1769.
172. Li, G.; Gray, K.A. The Solid-Solid Interface: Explaining the High and Unique Photocatalytic Reactivity of TiO_2-Based Nanocomposite Materials. *Chem. Phys.* **2007**, *339*, 173–187.
173. Chen, X.; Selloni, A. Introduction: Titanium Dioxide (TiO_2) Nanomaterials. *Chem. Rev.* **2014**, *114*, 9281–9282.

174. Kapilashrami, M.; Zhang, Y.; Liu, Y.S.; Hagfeldt, A.; Guo, J. Probing the Optical Property and Electronic Structure of TiO_2 Nanomaterials for Renewable Energy Applications. *Chem. Rev.* **2014**, *114*, 9662–9707.

175. Schneider, J.; Matsuoka, M.; Takeuchi, M.; Zhang, J.; Horiuchi, Y.; Anpo, M.; Bahnemann, D.W. Understanding TiO_2Photocatalysis: Mechanisms and Materials. *Chem. Rev.* **2014**, *114*, 9919–9986.

176. Di Iorio, Y.; Aguirre, M.E.; Brusam, M.A.; Grela, M.A. Surface Chemistry Determines Electron Storage Capabilities in Alcoholic Sols of Titanium Dioxide Nanoparticles. A Combined FTIR and Room Temperature EPR Investigation. *J. Phys. Chem. C* **2012**, *116*, 9646–9652.

177. Murakami, N.; Mahaney, O.O.P.; Abe, R.; Torimoto, T.; Ohtani, B. Double-Beam Photo- acoustic Spectroscopic Studies on Transient Absorption of Titanium(IV) Oxide Photocatalyst Powders. *J. Phys. Chem. C* **2007**, *111*, 11927–11935.

178. Chen, T.; Feng, Z.; Wu, G.; Shi, J.; Ma, G.; Ying, P.; Li, C. Mechanistic Studies of Photocatalytic Reaction of Methanol for Hydrogen Production on Pt/TiO_2 by In-Situ Fourier-Transform IR and Time-Resolved IR Spectroscopy. *J. Phys. Chem. C* **2007**, *111*, 8005–8014.

179. Yamakata, A.; Ishibashi, T.; Onishi, H. Effects of Water Addition on the Methanol Oxidation on Pt/ TiO_2 Photocatalyst Studied by Time-Resolved Infrared Absorption Spectroscopy. *J. Phys. Chem. B* **2003**, *107*, 9820–9823.

180. Szczepankiewicz, S.H.; Moss, J.A.; Hoffmann, M.R. Slow Surface Charge Trapping Kinetics on Irradiated TiO_2. *J. Phys. Chem. B* **2002**, *106*, 2922–2927.

181. Highfield, J.G.; Grätzel, M. Discovery of Reversible Photochromism in Titanium Dioxide using Photoacoustic Spectroscopy: Implications for the Investigation of Light-Induced Charge Separation and Surface Redox Processes in Titanium Dioxide. *J. Phys. Chem.* **1988**, *92*, 464–467.

182. Di Valentin, C.; Selloni, A. Bulk and Surface Polarons in Photoexcited Anatase. *J. Phys. Chem. Lett.* **2011**, *2*, 2223–2228.

183. Peter, L.M.; Wijayantha, K.G.U. Photoelectrochemical Water Splitting at Semiconductor Electrodes: Fundamental Problems and New Perspectives. *ChemPhysChem* **2014**, *15*, 1983–1995.

184. Wang, X.C.; Biechert, S.; Antonietti, M. Polymeric Graphitic Carbon Nitride for Heterogeneous Photocatalysis. *ACS Catal.* **2012**, *2*, 1596–1606.

185. Martin, D.J.; Qiu, K.P.; Shevlin, S.A.; Handoko, A.D.; Chen, X.W.; Guo, Z.X.; Tang, J.W. Highly Efficient Photocatalytic H_2 Evolution from Water Using Visible Light and Structure-Controlled Graphitic Carbon Nitride. *Angew. Chem. Int. Ed.* **2014**, *53*, 9240–9245.

186. Hollmann, D.; Karnahl, M.; Tschierlei, S.; Kailasam, K.; Schneider, M.; Radnik, J.; Grabow, K.; Bentrup, U.; Junge, H.; Beller, M.; *et al.* Structure-Activity Relationships in Bulk Polymeric and Sol-Gel Derived Carbon Nitrides During Photocatalytic Hydrogen Production. *Chem. Mater.* **2014**, *26*, 1727–1733.

187. Wirth, J.; Neumann, R.; Antonietti, M.; Saalfrank, P. Adsorption and Photocatalytic Splitting of Water on Graphitic Carbon Nitride: A First Principles and Semi-Empirical Study. *Phys. Chem. Chem. Phys.* **2014**, *16*, 15917–15926.

188. Maeda, K.; Wang, X.; Nishihara, Y.; Lu, D.; Antonietti, M.; Domen, K. Photocatalytic Activities of Graphitic Carbon Nitride Powder for Water Oxidation and Reduction under Visible Light. *J. Phys. Chem. C* **2009**, *113*, 4940–4947.

189. Wang, X.; Pehkonen, S.O.; Rämö, J.; Väänänen, M.; Highfield, J.G.; Laasonen, K. Experimental and Computational Studies of Nitrogen-Doped Degussa P25 TiO_2: Application to Visible-Light Driven Photo-Oxidation of As (III). *Catal. Sci. Technol.* **2012**, *2*, 784–793.

190. Sivula, K.; le Formal, F.; Grätzel, M. Solar Water Splitting: Progress Using Hematite (α-Fe_2O_3) Photoelectrodes. *ChemSusChem* **2011**, *4*, 432–449.

191. Warren, S.C.; Voitchovsky, K.; Dotan, H.; Leroy, C.M.; Cornuz, M.; Stellacci, F.; Hébert, C.; Rothschild, A.; Grätzel, M. Identifying Champion Nanostructures for Solar Water-Splitting. *Nat. Mater.* **2013**, *12*, 842–849.

192. Shinar, R.; Kennedy, J.H. Photoactivity of Doped α-Fe_2O_3 Electrodes. *Sol. Energy Mater.* **1982**, *6*, 323–335.

193. Dare-Edwards, M.P.; Goodenough, J.B.; Hamnett, A.; Trevellick, P.R. Electrochemistry and Photoelectrochemistry of Iron(III) Oxide. *J. Chem. Soc. Faraday Trans. I* **1983**, *79*, 2027–2041.

194. Itoh, K.; Bockris, J.O. Thin Film Photoelectrochemistry: Iron Oxide. *J. Electrochem. Soc.* **1984**, *131*, 1266–1271.

195. Kennedy, J.H.; Frese, K.W. Photooxidation of Water at α-Fe_2O_3 Electrodes. *J. Electrochem. Soc.* **1978**, *125*, 709–714.

196. Wang, K.X.Z.; Wu, Z.F.; Liu, V.; Brongersma, M.L.; Jaramillo, T.F.; Fan, S.H. Nearly Total Solar Absorption in Ultrathin Nanostructured Iron Oxide for Efficient Photoelectrochemical Water Splitting. *ACS Photon.* **2014**, *1*, 235–240.

197. Wheeler, D.A.; Wang, G.M.; Ling, Y.C.; Li, Y.; Zhang, J.Z. Nanostructured hematite: Synthesis, characterization, charge carrier dynamics, and photoelectrochemical properties. *Energy Environ. Sci.* **2012**, *5*, 6682–6702.

198. Pendlebury, S.R.; Cowan, A.J.; Barroso, M.; Sivula, K.; Ye, J.; Grätzel, M.; Klug, D.R.; Tang, J.; Durrant, J.R. Correlating Long-Lived Photogenerated Hole Populations with Photocurrent Densities in Hematite Water Oxidation Photoanodes. *Energy Environ. Sci.* **2012**, *5*, 6304–6312.

199. Le Formal, F.; Pendlebury, S.R.; Cornuz, M.; Tilley, S.D.; Grätzel, M.; Durrant, J.R. Back Electron-Hole Recombination in Hematite Photoanodes for Water Splitting. *J. Am. Chem. Soc.* **2014**, *136*, 2564–2574.

200. Peter, L.M. Energetics and Kinetics of Light-Driven Oxygen Evolution at Semiconductor Electrodes: The Example of Hematite. *J. Solid State Electrochem.* **2013**, *17*, 315–326.

201. Braun, A.; Sivula, K.; Bora, D.K.; Zhu, J.; Zhang, L.; Grätzel, M.; Guo, J.; Constable, E.C. Direct Observation of Two Electron Holes in a Hematite Photoanode During Photoelectro-chemical Water Splitting. *J. Phys. Chem. C* **2012**, *116*, 16870–16875.

202. Serpone, N.; Borgarello, E.; Grätzel, M. Visible Light Induced Generation of H_2 from H_2S in Mixed Semiconductor Dispersions: Improved Efficiency through Interparticle Electron Transfer. *J. Chem. Soc. Chem. Commun.* **1984**, 342–344.

203. Yan, H.; Yang, J.; Ma, G.; Wu, G.; Lei, Z.; Shi, J.; Li, C. Visible-Light-Driven Hydrogen Production with Extremely High Quantum Efficiency on Pt-PdS/CdS Photocatalyst. *J. Catal.* **2009**, *266*, 165–168.

204. Yang, J.; Yan, H.; Wang, X.; Wen, F.; Wang, Z.; Fan, D.; Shi, J.; Li, C. Roles of Cocatalysts in Pt-PdS/CdS with Exceptionally High Quantum Efficiency for Photocatalytic Hydrogen Production. *J. Catal.* **2012**, *290*, 151–157.

205. Ehsan, M.A.; Ming, H.N.; McKee, V.; Peiris, T.A.N.; Gamage, U.W.K.; Arifin, Z.; Mazhar, M. Vysotskite Structured Photoactive Palladium Sulphide Thin Films from Dithiocarbamate. Derivatives. *New J. Chem.* **2014**, *38*, 4083–4091.

206. Ferrer, I.J.; Díaz-Chao, P.; Sánchez, C. An Investigation on Palladium Sulphide (PdS) Thin Films as a Photovoltaic Material. *Thin Solid Films* **2007**, *515*, 5783–5786.

207. Meng, J.; Yu, Z.; Li, Y.; Li, Y. PdS-Modified CdS/NiS Composite as an Efficient Photocatalyst for H_2 evolution in Visible Light. *Catal. Today* **2014**, *225*, 136–141.

208. Zhang, S.; Chen, Q.; Jing, D.; Wang, Y.; Guo, L. Visible Photoactivity and Anticorrosion Performance of PdS-CdS Photocatalysts Modified by Polyaniline. *Int. J. Hydrog. Energy* **2012**, *37*, 791–796.

209. Ran, J.R.; Zhang, J.; Yu, J.G.; Jaroniec, M.; Qiao, S.Z. Earth-Abundant Cocatalysts for Semiconductor-Based Photocatalytic Water Splitting. *Chem. Soc. Rev.* **2014**, *43*, 7787–7812.

210. Dinh, C.; Pham, M.; Kleitz, F.; Do, T. Design of Water-Soluble CdS-Titanate-Nickel Nanocomposites for Photocatalytic Hydrogen production Under Sunlight. *J. Mater. Chem. A* **2013**, *1*, 13308–13313.

211. McKone, J.R.; Warren, E.L.; Bierman, M.J.; Boettcher, S.W.; Brunscwig, B.S.; Lewis, N.S.; Gray, H.B. Evaluation of Pt, Ni, and Ni-Mo Electrocatalysts for Hydrogen Evolution on Crystalline Si Electrodes. *Energy Environ. Sci.* **2011**, *4*, 3573–3583.

212. Highfield, J.G.; Claude, E.; Oguro, K. Electrocatalytic Synergism in Ni/Mo Cathodes for Hydrogen Evolution in Acid Medium: A New Model. *Electrochim. Acta* **1999**, *44*, 2805–2814.

213. Yu, J.G.; Ran, J.R. Facile Preparation and Enhanced Photocatalytic H_2-Production Activity of $Cu(OH)_2$ Cluster Modified TiO_2. *Energy Environ. Sci.* **2011**, *4*, 1364–1371.

214. Gagliardi, C.J.; Vannucci, A.K.; Concepcion, J.J.; Chen, Z.; Mayer, T.J. The Role of Proton Coupled Electron Transfer in Water Oxidation. *Energy Environ. Sci.* **2012**, *5*, 7704–7717.

215. Armstrong, F.A. Why Did Nature Choose Manganese to Make Oxygen? *Philos. Trans. R. Soc. B* **2008**, *363*, 1263–1270.

216. Carmo, M.; Fritz, D.L.; Mergel, J.; Stolten, D. A Comprehensive Review on PEM Water Electrolysis. *Int. J. Hydrog. Energy* **2013**, *38*, 4901–4934.

217. Fabbri, E.; Habereder, A.; Waltar, K.; Kötz, R.; Schmidt, T.J. Developments and Perspectives of Oxide-Based Catalysts for the Oxygen Evolution Reaction. *Catal. Sci. Technol.* **2014**, *4*, 3800–3821.

218. Zhang, F.; Yamakata, A.; Maeda, K.; Moriya, Y.; Takata, T.; Kubota, J.; Teshima, K.; Oishi, S.; Domen, K. Cobalt-Modified Porous Single-Crystalline $LaTiO_2N$ for Highly Efficient Water Oxidation Under Visible Light. *J. Am. Chem. Soc.* **2012**, *134*, 8348–8351.

219. Liu, L.C.; Li, Z.Y.; Zou, W.X.; Gu, X.R.; Deng, Y.; Gao, F.; Tang, C.J.; Dong, L. *In Situ* Loading Transition Metal Oxide Clusters on TiO_2 Nanosheets as Co-Catalysts for Exceptional High Photoactivity. *ACS Catal.* **2013**, *3*, 2052–2061.

220. Kudo, A.; Miseki, Y. Heterogeneous Photocatalytic Materials for Water Splitting. *Chem. Soc. Rev.* **2009**, *38*, 253–278.

221. Maeda, K.; Domen, K. Photocatalytic Water Splitting: Recent Progress and Future Challenges. *J. Phys. Chem. Lett.* **2010**, *1*, 2655–2661.

222. Maeda, K.; Higashi, M.; Lu, D.; Abe, R.; Domen, K. Efficient Non-Sacrificial Water Splitting through Two-Step Photoexcitation by Visible Light Using a Modified Oxynitride as a Hydrogen Evolution Photocatalyst. *J. Am. Chem. Soc.* **2010**, *132*, 5858–5868.

223. Sasaki, Y.; Nemoto, H.; Saito, K.; Kudo, A. Solar Water Splitting Using Powdered Photocatalysts Driven by Z-Schematic Interparticle Electron Transfer without an Electron Mediator. *J. Phys. Chem. C* **2009**, *113*, 17536–17542.

224. Zhou, P.; Yu, J.; Jaroniec, M. All-Solid-State Z-Scheme Photocatalytic Systems. *Adv. Mater.* **2014**, *26*, 4920–4935.

225. Wang, Y.; Wang, Q.; Zhan, X.; Wang, F.; Safdar, M.; He, J. Visible Light Driven Type II Heterostructures and Their Enhanced Photocatalysis Properties: A Review. *Nanoscale* **2013**, *5*, 8326–8339.

226. Prévot, M.S.; Sivula, K. Photoelectrochemical Tandem Cells for Solar Water Splitting. *J. Phys. Chem. C* **2013**, *117*, 17879–17893.

227. Hu, S.; Xiang, C.; Haussener, S.; Berger, A.D.; Lewis, N.S. An Analysis of the Optimal Band Gaps of Light Absorbers in Integrated Tandem Photoelectrochemical Water-Splitting Systems. *Energy Environ. Sci.* **2013**, *6*, 2984–2993.

228. Kanan, M.W.; Nocera, D.G. *In Situ* Formation of an Oxygen-Evolving Catalyst in Neutral Water Containing Phosphate and Co^{2+}. *Science* **2008**, *321*, 1072–1075.

229. Bediako, D.K.; Lasalle-Kaiser, B.; Surendranath, Y.; Yano, J.; Yachandra, V.K.; Nocera, D.G. Structure-Activity Correlations in a Nickel-Borate Oxygen

Evolution Catalyst. *J. Am. Chem. Soc.* **2012**, *134*, 6801–6809.

230. Gibson, T.L.; Kelly, N.A. Optimization of Solar Powered Hydrogen Production Using Photovoltaic Electrolysis Devices. *Int. J. Hydrog. Energy* **2008**, *33*, 5931–5940.

231. Herron, J.A.; Kim, J.; Upadhye, A.A.; Huber, G.W.; Maravelias, C.T. A General Framework for the Assessment of Solar Fuel Technologies. *Energy Environ. Sci.* **2015**, *8*, 126–157.

232. Hernández-Pagán, E.A.; Vargas-Barbosa, N.M.; Wang, T.H.; Zhao, Y.; Smotkin, E.S. Resistance and Polarization Losses in Aqueous Buffer-Membrane Electrolytes for Water-Splitting Photoelectrochemical Cells. *Energy Environ. Sci.* **2012**, *5*, 7582–7589.

233. Jacobsson, T.J.; Fjällstrom, V.; Edoff, M.; Edvinsson, T. Sustainable Solar Hydrogen Production: From Photoelectrochemical cells to PV-Electrolyzers and Back Again. *Energy Environ. Sci.* **2014**, *7*, 2056–2070.

234. Tributsch, H. Photovoltaic Hydrogen Generation. *Int. J. Hydrog. Energy* **2008**, *33*, 5911–5930.

235. Campos-Martin, J.M.; Blanco-Brieva, G.; Fierro, J.L.G. Hydrogen Peroxide Synthesis: An Outlook beyond the Anthraquinone Process. *Angew. Chem. Int. Ed.* **2006**, *45*, 6962–6984.

236. García-Serna, J.; Moreno, T.; Biasi, P.; Cocero, M.J.; Mikkola, J.P.; Salmi, T.O. Engineering in Direct Synthesis of Hydrogen Peroxide: Targets, Reactors and Guidelines for Operational Conditions. *Green Chem.* **2014**, *16*, 2320–2345.

237. Samanta, C. Direct Synthesis of Hydrogen Peroxide from Hydrogen and Oxygen: An Overview of Recent Developments in the Process. *Appl. Catal. A Gen.* **2008**, *350*, 133–149.

238. Edwards, J.K.; Solsona, B.; Ntainjua, E.; Carley, A.F.; Herzing, A.A.; Kiely, C.J.; Hutchings, G.J. Switching Off Hydrogen Peroxide Hydrogenation in the Direct Synthesis Process. *Science* **2009**, *323*, 1037–1041.

239. Siahrostami, S.; Verdaguer-Casadevall, A.; Karamad, M.; Deiana, D.; Malacrida, P.; Wickman, B.; Escudero-Escribano, M.; Paoli, E.A.; Frydendal, R.; Hansen, T.W.; *et al.* Enabling Direct H_2O_2 Production Through Rational Electrocatalyst Design. *Nat. Mater.* **2013**, *12*, 1137–1143.

240. Sanli, A.E. A Possible Future Fuel Cell: the Peroxide/Peroxide Fuel Cell. *Int. J. Energy Res.* **2013**, *37*, 1488–1497.

241. Kato, S.; Jung, J.; Suenobu, T.; Fukuzumi, S. Production of Hydrogen Peroxide as a Sustainable Solar Fuel from Water and Dioxygen. *Energy Environ. Sci.* **2013**, *6*, 3756–3764.

242. Fukuzumi, S.; Yamada, Y.; Karlin, K.D. Hydrogen Peroxide as a Sustainable Energy Carrier: Electrocatalytic Production of Hydrogen Peroxide and the Fuel Cell. *Electrochim. Acta* **2012**, *82*, 493–511.

243. Sanli, A.E.; Aytaç, A. Response to Disselkamp: Direct Peroxide/Peroxide Fuel Cell as a Novel Type Fuel Cell. *Int. J. Hydrog. Energy* **2011**, *36*, 869–875.

244. Disselkamp, R.S. Can Aqueous Hydrogen Peroxide be used as a Standalone Energy Source? *Int. J. Hydrogen Energy***2010**, *35*, 1049–1053.

245. Yamazaki, S.; Siroma, Z.; Senoh, H.; Iroi, T.; Fujiwara, N.; Yasuda, K. A Fuel Cell with Selective Electrocatalysts Using Hydrogen Peroxide as Both an Electron Acceptor and a Fuel. *J. Power Sources* **2008**, *178*, 20–25.

246. Disselkamp, R.S. Energy Storage Using Aqueous Hydrogen Peroxide. *Energy Fuels* **2008**, *22*, 2771–2774.

247. Bussayajarn, N.; Harrington, D.A.; Therdthianwong, S.; Therdthianwong, A.; Djilali, N. The Cathodic Polarization Prediction of H_2/H_2O_2 Fuel Cells using EIS Spectra. In Paper A-012, Proceedings of the 2nd Joint International Conference on Sustainable Energy & Environment, Bankok, Thailand, 21–23 November 2006.

248. Ponce de León; Walsh, F.C.; Rose, A.; Lakeman, J.B.; Browning, D.J.; Reeve, R.W. A Direct Borohydride—Acid Peroxide Fuel Cell. *J. Power Sources* **2007**, *164*, 441–448.

249. Miley, G.H.; Luo, N.; Mather, J.; Burton, R.; Hawkins, G.; Gu, L.; Byrd, E.; Gimlin, R.; Shrestha, P.J.; Benavides, G.; *et al.* Direct $NaBH_4/H_2O_2$ Fuel Cells. *J. Power Sources* **2007**, *165*, 509–516.

250. An, L.; Zhao, T.S.; Zhou, X.L.; Wei, L.; Yan, X.H. A High Performance Ethanol—Hydrogen Peroxide Fuel Cell. *RSC Adv.* **2014**, *4*, 65031–65034.

251. Amirfakhri, S.J.; Meunier, J.L.; Berk, D. Electrocatalytic Activity of $LaNiO_3$ toward H_2O_2 Reduction Reaction: Minimization of Oxygen Evolution. *J. Power Sources* **2014**, *272*, 248–258.

252. Compton, O.C.; Osterloh, F.E. Niobate Nanosheets as Catalysts for Photochemical Water Splitting into Hydrogen and Hydrogen Peroxide. *J. Phys. Chem. C* **2009**, *113*, 479–485.

253. Daskalaki, V.M.; Panagiotopoulou, P.; Kondarides, D.I. Production of Peroxide Species in Pt/TiO_2 Suspensions under Conditions of Photocatalytic Water Splitting and Glycerol Photo-Reforming. *Chem. Eng. J.* **2011**, *170*, 433–439.

254. Arakawa, H. Water Photolysis by TiO_2 Particles—Significant Effect of Na_2CO_3 Addition on Water Splitting. In*Photocatalysis Science and Technology*; Kaneko, M., Okura, I., Eds.; Springer: New York, NY, USA, 2002; p. 235.

255. Yesodharan, E.; Yesodharan, S.; Grätzel, M. Photolysis of Water with Supported Noble Metal Cluster. The Fate of Oxygen in Titania Based Water Cleavage Systems. *Sol. Energy Mater.* **1984**, *10*, 287–302.

256. Hoffman, A.J.; Carraway, E.R.; Hoffmann, M.R. Photocatalytic Production of H_2O_2 and Organic Peroxides on Quantum-Sized Semiconductor Colloids. *Environ. Sci. Technol.* **1994**, *28*, 776–785.

257. Maurino, V.; Minero, C.; Mariella, G.; Pelizzetti, E. Sustained Production of H_2O_2 on Irradiated TiO_2-Fluoride Systems. *Chem. Commun.* **2005**, 2627–2629.

258. Moon, G.M.; Kim, W.; Bokare, A.; Sung, N.; Choi, W. Solar Production of H_2O_2 on Reduced Graphene Oxide-TiO_2Hybrid Photocatalysts Consisting of Earth-

Abundant Elements Only. *Energy Environ. Sci.* **2014**, *7*, 4023–4028.

259. Tsukamoto, D.; Shiro, A.; Shiraishi, Y.; Sugano, Y.; Ichikawa, S.; Tanaka, S.; Hirai, T. Photocatalytic H_2O_2 Production from Ethanol/O_2 System Using TiO_2 Loaded with Au-Ag Bimetallic Alloy Nanoparticles. *ACS Catal.* **2012**, *2*, 599–603.

260. Diesen, V.; Jonsson, M. Formation of H_2O_2 in TiO_2 Photocatalysis of Oxygenated and Deoxygenated Aqueous Systems: A Probe for Photocatalytically Produced Hydroxyl Radicals. *J. Phys. Chem. C* **2014**, *118*, 10083–10087.

261. Lousada, C.M.; Johansson, A.J.; Brinck, T.; Jonsson, M. Mechanism of H_2O_2 Decomposition on Transition Metal Oxide Surfaces. *J. Phys. Chem. C* **2012**, *116*, 9533–9543.

262. Bokare, A.D.; Choi, W. Review of Iron-Free Fenton-like Systems for Activating H_2O_2 in Advanced Oxidation Processes. *J. Hazard. Mater.* **2014**, *275*, 121–135.

263. Koppenol, W.H. The Haber-Weiss Cycle—70 Years Later. *Redox Rep.* **2001**, *6*, 229–234.

264. Shiraishi, Y.; Kanazawa, S.; Sugano, Y.; Tsukamoto, D.; Sakamoto, H.; Ichikawa, S.; Hirai, T. Highly Selective Production of Hydrogen Peroxide on Graphitic Carbon Nitride (g-C_3N_4) Photocatalyst Activated by Visible Light.*ACS Catal.* **2014**, *4*, 774–780.

265. Zhang, J.; Chen, X.; Takanabe, K.; Maeda, K.; Domen, K.; Epping, J.D.; Fu, X.; Antonietti, M.; Wang, X. Synthesis of a Carbon Nitride Structure for Visible-Light Catalysis by Copolymerization. *Angew. Chem. Int. Ed.* **2010**, *49*, 441–444.

266. Abe, R.; Takami, H.; Murakami, N.; Ohtani, B. Pristine Simple Oxides as Visible Light Driven Photocatalysts: Highly Efficient Decomposition of Organic Compounds over Platinum-Loaded Tungsten Oxide. *J. Am. Chem. Soc.* **2008**, *130*, 7780–7781.

267. Rostrup-Nielsen, J.R. Steam Reforming and Chemical Recuperation. *Catal. Today* **2009**, *145*, 72–75.

268. Hodoshima, S.; Shono, A.; Saito, Y. Chemical Recuperation of Low-Quality Waste Heats by Catalytic Dehydrogenation of Organic Chemical Hydrides and Its Exergy Analysis. *Energy Fuels* **2008**, *22*, 2559–2569.

269. Poullikkas, A. An Overview of Current and Future Sustainable Gas Turbine Technologies. *Renew. Sustain. Energy Rev.***2005**, *9*, 409–443.

270. Sun, J.; Wang, Y. Recent Advances in Catalytic Conversion of Ethanol to Chemicals. *ACS Catal.* **2014**, *4*, 1078–1090.

271. Makshina, E.V.; Dusselier, M.; Janssens, W.; Degrève, J.; Jacobs, P.A.; Sels, B.F. Review of Old Chemistry and New Catalytic Advances in the On-Purpose Synthesis of Butadiene. *Chem. Soc. Rev.* **2014**, *43*, 7917–7953.

272. Alcalá, R.; Mavrikakis, M.; Dumesic, J.A. DFT Studies for Cleavage of C−C and C−O Bonds in Surface Species Derived from Ethanol on Pt(111). *J. Catal.* **2003**, *218*, 178–190.

273. Ferrin, P.; Simonetti, D.; Kandoi, S.; Kunkes, E.; Dumesic, J.A.; Nørskov, J.K.; Mavrikakis, M. Modeling Ethanol Decomposition on Transition Metals: A Combined Application of Scaling and Brønsted-Evans-Polanyi Relations. *J. Am. Chem. Soc.* **2009**, *131*, 5809–5815.

274. Salemme, L.; Menna, L.; Simeone, M. Thermodynamic Analysis of Ethanol Processors—PEM Fuel Cell Systems. *Int. J. Hydrog. Energy* **2010**, *35*, 3480–3489.

275. Ni, M.; Leung, D.Y.C.; Leung, M.K.H. A Review on Reforming Bio-Ethanol for Hydrogen Production. *Int. J. Hydrog. Energy* **2007**, *32*, 3238–3247.

276. Xu, W.Q.; Liu, Z.Y.; Johnston-Peck, A.C.; Senanayake, S.D.; Zhou, G.; Stacchiloa, D.; Stach, E.A.; Rodriguez, J.A. Steam Reforming of Ethanol on Ni/CeO_2: Reaction Pathway and Interaction between Ni and the CeO_2 Support. *ACS Catal.* **2013**, *3*, 975–984.

277. Sanchez-Sanchez, M.C.; Yerga, R.M.N.; Kondarides, D.I.; Verykios, X.E.; Fierro, J.L.G. Mechanistic Aspects of the Ethanol Steam Reforming Reaction for Hydrogen Production on Pt, Ni, and Pt/Ni Catalysts Supported on γ-Al_2O_3. *J. Phys. Chem. A* **2010**, *114*, 3873–3882.

278. Highfield, J.; Geiger, F.; Uenala, E.; Schucan, T. Hydrogen Release by Steam-Reforming of Ethanol for Efficient & Clean Fuel Applications. In Proceedings of the 10th World Hydrogen Energy Conference, Cocoa Beach, FL, USA, 20–24 June 1994; Hydrogen Energy Progress X. Block, D.L., Veziroglu, T.N., Eds.; Volume 2, p. 1039.

279. Sutton, J.E.; Panagiotopoulou, P.; Verykios, X.E.; Vlachos, D.G. Combined DFT, Microkinetic, and Experimental Study of Ethanol Steam Reforming on Pt. *J. Phys. Chem. C* **2013**, *117*, 4691–4706.

280. Maihom, T.; Probst, M.; Limtrakul, J. Density Functional Theory Study of the Dehydrogenation of Ethanol to Acetaldehyde over the Au-Exchanged ZSM-5 Zeolite: Effect of Surface Oxygen. *J. Phys. Chem. C* **2014**, *118*, 18564–18572.

281. Morgenstern, D.A.; Fornango, J.P. Low-Temperature Reforming of Ethanol over Copper-Plated Raney Nickel: A New Route to Sustainable Hydrogen for Transportation. *Energy Fuels* **2005**, *19*, 1708–1716.

282. Bowker, M.; Holroyd, R.; Perkins, N.; Bhantoo, J.; Counsell, J.; Carley, A.; Morgan, C. Acetaldehyde Adsorption and Catalytic Decomposition on Pd (110) and the Dissolution of Carbon. *Surf. Sci.* **2007**, *601*, 3651–3660.

283. Pichat, P.; Mozzanega, M.N.; Disdier, J.; Herrmann, J.M. Pt Content and Temperature Effects on the Photocatalytic H_2production from Aliphatic Alcohols over Pt/TiO_2. *New J. Chem. (Nouv. J. Chim.)* **1982**, *6*, 559–564.

284. Pichat, P.; Hermann, J.M.; Disdier, J.; Courbon, H.; Mozzanega, M.N. Photocatalytic Hydrogen Production from Aliphatic Alcohols over a Bifunctional Platinum on Titanium Dioxide Catalyst. *New J. Chem. (Nouv. J. Chim.)* **1981**, *5*, 627–636.

285. Millard, L.; Bowker, M. Photocatalytic Water-Gas Shift Reaction at Ambient Temperature. *J. Photochem. Photobiol. A Chem.* **2002**, *148*, 91–95.

286. Bowker, M.; James, D.; Stone, P.; Bennett, R.; Perkins, N.; Millard, L.; Greaves, J.; Dickinson, A. Catalysis

at the Metal-Support Interface: Exemplified by the Photocatalytic Reforming of Methanol on Pd/TiO_2. *J. Catal.* **2004**, *217*, 427–433.

287. Al-Mazroai, L.S.; Bowker, M.; Davies, P.; Dickinson, A.; Greaves, J.; James, D.; Millard, L. The Photocatalytic Reforming of Methanol. *Catal. Today* **2007**, *122*, 46–50.

288. Highfield, J.G.; Chen, M.H.; Nguyen, P.T.; Chen, Z. Mechanistic Investigations of Photo-Driven Processes over $TiO_2$by in-situ DRIFTS-MS: Part 1: Platinization and Methanol Reforming. *Energy Environ. Sci.* **2009**, *2*, 991–1002.

289. Daskalaki, V.M.; Kondarides, D.I. Efficient Production of Hydrogen by Photo-Induced Reforming of Glycerol at Ambient Conditions. *Catal. Today* **2009**, *144*, 75–80.

290. Taboada, E.; Angurell, I.; Llorca, J. Dynamic Photocatalytic Hydrogen Production from Ethanol—Water Mixtures in an Optical Fiber Honeycomb Reactor Loaded with Au/TiO_2. *J. Catal.* **2014**, *309*, 460–467.

291. Puga, A.V.; Forneli, A.; García, H.; Corma, A. Production of H_2 by Ethanol Photoreforming on Au/TiO_2. *Adv. Funct. Mater.* **2014**, *24*, 241–248.

292. Murdoch, M.; Waterhouse, G.I.N.; Nadeem, M.A.; Metson, J.B.; Keane, M.A.; Howe, R.F.; Llorca, J.; Idriss, H. The Effect of Gold Loading and Particle Size on the Photocatalytic Hydrogen Production from Ethanol on Au/TiO_2Nanoparticles. *Nat. Chem.* **2011**, *3*, 489–492.

293. Gallo, A.; Marelli, M.; Psaro, R.; Gombac, V.; Montini, T.; Fornasiero, P.; Pievo, R.; Dal Santo, V. Bimetallic Au-Pt/TiO_2Photocatalysts Active under UV-A and Simulated Sunlight for H_2 Production from Ethanol. *Green Chem.* **2012**, *14*, 330–333.

294. Strataki, N.; Bekiari, V.; Kondarides, D.I.; Lianos, P. Hydrogen production by photocatalytic alcohol reforming employing highly efficient Nanocrystalline Titania films. *Appl. Catal. B Environ.* **2007**, *77*, 184–189.

295. Languer, M.P.; Scheffer, F.R.; Feil, A.F.; Baptista, D.L.; Migowski, P.; Macahdo, G.J.; de Moraes, D.P.; Dupont, J.; Teixeira, S.R.; Weibel, D.E. Photo-Induced Reforming of Alcohols with Improved Hydrogen Apparent Quantum Yield on TiO_2 Nanotubes Loaded with Ultra-Small Pt Nanoparticles. *Int. J. Hydrog. Energy* **2013**, *38*, 14440–14450.

296. Bowker, M.; Morton, C.; Kennedy, J.; Bahruji, H.; Greves, J.; Jones, W.; Davies, P.R.; Brookes, C.; Wells, P.P.; Dimitratos, N. Hydrogen Production by Photoreforming of Biofuels using Au, Pd and Au-Pd/TiO_2 Photocatalysts. *J. Catal.* **2014**, *310*, 10–15.

297. Taboada, E.; Angurell, I.; Llorca, J. Hydrogen Photoproduction from Bio-derived Alcohols in an Optical Fiber Honeycomb Reactor Loaded with Au/TiO_2. *J. Photochem. Photobiol. A Chem.* **2014**, *281*, 35–39.

298. Bowker, M. Sustainable Hydrogen Production by the Application of Ambient Temperature Photocatalysis. *Green Chem.* **2011**, *13*, 2235–2246.

299. Bowker, M.; Davies, P.R.; al-Mazroai, L.S. Photocatalytic Reforming of Glycerol over Gold and Palladium as an Alternative Fuel Source. *Catal. Lett.* **2009**, *128*, 253–255.

300. Luo, N.; Jiang, Z.; Shi, H.; Cao, F.; Xiao, T.; Edwards, P.P. Photo-Catalytic Conversion of Oxygenated Hydrocarbons to Hydrogen over Heteroatom-Doped TiO_2 Catalysts. *Int. J. Hydrog. Energy* **2009**, *34*, 125–129.

301. Fu, X.; Long, J.; Wang, X.; Leung, D.Y.C.; Ding, Z.; Wu, L.; Zhang, Z.; Li, Z.; Fu, X. Photocatalytic Reforming of Biomass: A Systematic Study of Hydrogen Evolution from Glucose Solution. *Int. J. Hydrog. Energy* **2008**, *33*, 6484–6491.

302. Zheng, X.J.; Wei, L.F.; Zhang, Z.H.; Jiang, Q.J.; Wei, Y.J.; Xie, B.; Wei, M.B. Research on Photocatalytic H_2 Production from Acetic Acid Solution by Pt/TiO_2 Nanoparticles under UV Irradiation. *Int. J. Hydrog. Energy* **2009**, *34*, 9033–9041.

303. Kraeutler, B.; Bard, A.J. Heterogeneous Photocatalytic Synthesis of Methane from Acetic Acid—New Kolbe Reaction Pathway. *J. Am. Chem. Soc.* **1978**, *100*, 2239–2240.

304. Tanaka, A.; Sakaguchi, S.; Hashimoto, K.; Kominami, H. Preparation of Au/TiO_2 with Metal Cocatalysts Exhibiting Strong Surface Plasmon Resonance Effective for Photoinduced Hydrogen Formation under Irradiation of Visible Light. *ACS Catal.* **2013**, *3*, 79–85.

305. Lyubina, T.P.; Markovskaya, D.V.; Kozlova, E.A.; Parmon, V.N. Photocatalytic Hydrogen Evolution from Aqueous Solutions of Glycerol under Visible Light Irradiation. *Int. J. Hydrog. Energy* **2013**, *38*, 14172–14179.

306. Melo, M.D.; Silva, L.A. Visible-Light-Induced Hydrogen Production from Glycerol Aqueous Solution on Hybrid Pt-CdS-TiO_2 Photocatalysts. *J. Photochem. Photobiol. A Chem.* **2011**, *226*, 36–41.

307. Carraro, G.; Gasparotto, A.; Maccato, C.; Gombac, V.; Rossi, F.; Montini, T.; Peeters, D.; Bontempi, E.; Sada, C.; Barreca, D.; *et al.* Solar H_2 Generation via Ethanol Photoreforming on ε-Fe_2O_3 Nanorod Arrays Activated by Ag and Au Nanoparticles. *RSC Adv.* **2014**, *4*, 32174–32179.

308. Carraro, G.; Maccato, C.; Gasparotto, A.; Montini, T.; Turner, S.; Lebedev, O.I.; Gombac, V.; Adami, G.; van Tendeloo, G.; Barreca, D.; *et al.* Enhanced Hydrogen Production by Photoreforming of Renewable Oxygenates through Nanostructured Fe_2O_3 Polymorphs. *Adv. Funct. Mater.* **2014**, *24*, 372–378.

309. Kim, G.; Lee, S.H.; Choi, W. Glucose-TiO_2 Charge Transfer Complex-Mediated Photocatalysis under Visible Light.*Appl. Catal. B Environ.* **2015**, *162*, 463–469.

310. Clarizia, L.; Spasiano, D.; di Somma, I.; Marotta, R.; Andreozzi, R.; Dionysiou, D.D. Copper Modified-TiO_2 Catalysts for Hydrogen Generation through Photoreforming of Organics. A Short Review. *Int. J. Hydrog. Energy* **2014**, *39*, 16812–16831.

311. Ampelli, C.; Passalacqua, R.; Genovese, C.; Perathoner, S.; Centi, G.; Montini, T.; Gombac, V.; Jaen, J.J.D.; Fornasiero, P. H_2 Production by Selective Photo-Dehydrogenation of Ethanol in Gas and Liquid Phase on

CuO_x/TiO_2Nanocomposites. *RSC Adv.* **2013**, *3*, 21776–21788.

312. Liu, R.X.; Yoshida, H.; Fujita, S.; Arai, M. Photocatalytic Hydrogen Production from Glycerol and Water with NiO_x/TiO_2 Catalysts. *Appl. Catal. B Environ.* **2014**, *144*, 41–45.

313. Shkrob, I.A.; Dimitrijevic, N.M.; Marin, T.W.; He, H.; Zapol, P. Heteroatom-Transfer Coupled Photoreduction and Carbon Dioxide Fixation on Metal Oxides. *J. Phys. Chem. C* **2012**, *116*, 9461–9471.

314. Shkrob, I.A.; Marin, T.W.; He, H.; Zapol, P. Photoredox Reactions and the Catalytic Cycle for Carbon Dioxide Fixation and Methanogenesis on Metal Oxides. *J. Phys. Chem. C* **2012**, *116*, 9450–9460.

315. Mori, K.; Yamashita, H.; Anpo, M. Photocatalytic Reduction of CO_2 with H_2O on Various Titanium Oxide Photocatalysts. *RSC Adv.* **2012**, *2*, 3165–3172.

316. He, H.; Zapol, P.; Curtiss, L.A. Computational Screening of Dopants for Photocatalytic Two-Electron Reduction of CO_2 on Anatase (101) Surfaces. *Energy Environ. Sci.* **2012**, *5*, 6196–6205.

317. Tu, W.; Zhou, Y.; Zou, Z. Photocatalytic Conversion of CO_2 into Renewable Hydrocarbon Fuels: State-of-the-Art Accomplishment, Challenges, and Prospects. *Adv. Mater.* **2014**, *26*, 4607–4626.

318. Peterson, A.A.; Abild-Pedersen, F.; Studt, F.; Rossmeisl, J.; Nørskov, J.K. How Copper Catalyzes the Electroreduction of Carbon Dioxide into Hydrocarbon Fuels. *Energy Environ. Sci.* **2010**, *3*, 1311–1315.

319. Behrens, M.; Studt, F.; Kasatkin, I.; Kühl, S.; Hävecker, M.; Abild-Pedersen, F.; Zander, S.; Girgsdies, F.; Kurr, P.; Kniep, B.L.; *et al.* The Active Site of Methanol Synthesis over $Cu/ZnO/Al_2O_3$ Industrial Catalysts. *Science* **2012**, *336*, 893–897.

320. Ola, O.; Maroto-Valer, M.M. Copper-based TiO_2 Honeycomb Monoliths for CO_2 Photoreduction. *Catal. Sci. Technol.***2014**, *4*, 1631–1637.

321. Nunez, J.; O'Shea, V.A.D.; Jana, P.; Coronado, J.M.; Serrano, D.P. Effect of Copper on the Performance of ZnO and $ZnO_{1-x}N_x$ Oxides as CO_2 Photoreduction Catalysts. *Catal. Today* **2013**, *209*, 21–27.

322. Mao, J.; Li, K.; Peng, T.Y. Recent Advances in the Photocatalytic CO_2 Reduction over Semiconductors. *Catal. Sci. Technol.* **2013**, *3*, 2481–2498.

323. Richardson, P.L.; Perdigoto, M.L.N.; Wang, W.; Lopes, R.J.G. Manganese- and Copper-Doped Titania Nanocomposites for the Photocatalytic Reduction of Carbon Dioxide into Methanol. *Appl. Catal. B Environ.* **2012**, *126*, 200–207.

324. Kuhl, K.P.; Hatsukade, T.; Cave, E.R.; Abram, D.N.; Kibsgaard, J.; Jaramillo, T.F. Electrocatalytic Conversion of Carbon Dioxide to Methane and Methanol on Transition Metal Surfaces. *J. Am. Chem. Soc.* **2014**, *136*, 14107–14113.

325. Constentin, C.; Robert, M.; Saveant, J.M. Catalysis of the Electrochemical Reduction of Carbon Dioxide. *Chem. Soc. Rev.* **2013**, *42*, 2423–2436.

326. Kuhl, K.P.; Cave, E.R.; Abram, D.N.; Jaramillo, T.F. New Insights into the Electrochemical Reduction of Carbon Dioxide on Metallic Copper Surfaces. *Energy Environ. Sci.* **2012**, *5*, 7050–7059.

327. Tu, W.; Zhou, Y.; Zou, Z. Versatile Graphene-Promoting Photocatalytic Performance of Semiconductors: Basic Principles, Synthesis, Solar Energy Conversion, and Environmental Applications. *Adv. Funct. Mater.* **2013**, *23*, 4996–5008.

328. Fan, W.; Zhang, Q.; Wang, Y. Semiconductor-Based Nanocomposites for Photocatalytic H_2 Production and CO_2Conversion. *Phys. Chem. Chem. Phys.* **2013**, *15*, 2632–2649.

329. Leary, R.; Westwood, A. Carbonaceous Nanomaterials for the Enhancement of TiO_2 Photocatalysis. *Carbon* **2011**, *49*, 741–772.

330. Liang, Y.T.; Vijayan, B.K.; Gray, K.A.; Hersam, M.C. Minimizing Graphene Defects Enhances Titania Nanocomposite-Based Photocatalytic Reduction of CO_2 for Improved Solar Fuel Production. *Nano Lett.* **2011**, *11*, 2865–2870.

331. Tu, W.G.; Zhou, Y.; Liu, Q.; Tian, Z.P.; Gao, J.; Chen, X.Y.; Zhang, H.T.; Liu, J.G.; Zou, Z.G. Robust Hollow Spheres Consisting of Alternating Titania Nanosheets and Graphene Nanosheets with High Photocatalytic Activity for CO_2Conversion into Renewable Fuels. *Adv. Funct. Mater.* **2012**, *22*, 1215–1221.

332. Xia, X.H.; Jia, Z.H.; Yu, Y.; Liang, Y.; Wang, Z.; Ma, L.L. Preparation of Multi-Walled Carbon Nanotube Supported TiO_2 and Its Photocatalytic Activity in the Reduction of CO_2 with H_2O. *Carbon* **2007**, *45*, 717–721.

333. Centi, G.; Perathoner, S. Opportunities and Prospects in the Chemical Recycling of Carbon Dioxide to Fuels. *Catal. Today* **2009**, *148*, 191–205.

334. Yao, Y.; Li, G.; Ciston, S.; Lueptow, R.M.; Gray, K.A. Photoreactive TiO_2/Carbon Nanotube Composites: Synthesis and Reactivity. *Environ. Sci. Technol.* **2008**, *42*, 4952–4957.

335. Hsu, H.C.; Shown, I.; Wei, H.Y.; Chang, Y.C.; Du, H.Y.; Lin, Y.G.; Tseng, C.A.; Wang, C.H.; Chen, L.C.; Lin, Y.C.; *et al.* Graphene Oxide as a Promising Photocatalyst for CO_2 to Methanol Conversion. *Nanoscale* **2013**, *5*, 262–268.

336. Shown, I.; Hsu, H.C.; Chang, Y.C.; Lin, C.H.; Roy, P.K.; Ganguly, A.; Wang, C.H.; Chang, J.K.; Wu, C.I.; Chen, L.C.; *et al.* Highly Efficient Visible Light Photocatalytic Reduction of CO_2 to Hydrocarbon Fuels by Cu-Nanoparticle Decorated Graphene Oxide. *Nano Lett.* **2014**, *14*, 6097–6103.

337. Kumar, P.; Kumar, A.; Sreedhar, B.; Sain, B.; Ray, S.S.; Jain, S.L. Cobalt Phthalocyanine Immobilized on Graphene Oxide: An Efficient Visible-Active Catalyst for the Photoreduction of Carbon Dioxide. *Chem. Eur. J.* **2014**, *20*, 6154–6161.

338. Li, X.; Liu, H.; Luo, D.; Li, J.; Huang, Y.; Li, H.; Fang, Y.; Xu, Y.; Zhu, L. Adsorption of CO_2 on Heterostructure $CdS(Bi_2S_3)/TiO_2$ Nanotube Photocatalysts and Their

Photocatalytic Activities in the Reduction of CO_2 to Methanol under Visible Light Irradiation. *Chem. Eng. J.* **2012**, *180*, 151–158.

339. Wang, C.; Thompson, R.L.; Baltrus, J. Matranga, Visible Light Photoreduction of CO_2 Using CdSe/ Pt/TiO_2Heterostructured Catalysts. *J. Phys. Chem. Lett.* **2010**, *1*, 48–53.

340. Hou, W.B.; Hung, W.H.; Pavaskar, P.; Goeppert, A.; Aykol, M.; Cronin, S.B. Photocatalytic Conversion of CO_2 to Hydrocarbon Fuels via Plasmon-Enhanced Absorption and Metallic Interband Transitions. *ACS Catal.* **2011**, *1*, 929–936.

341. Centi, G.; Perathoner, S. Catalysis: Role and Challenges for a Sustainable Energy. *Top. Catal.* **2009**, *52*, 948–961.

342. Gangeri, M.; Perathoner, S.; Caudo, S.; Centi, G.; Amadou, J.; Bégon, D.; Pham-Huu, C.; Ledoux, M.J.; Tessonier, J.P.; Su, D.S.; *et al.* Fe and Pt Carbon Nanotubes for the Electrocatalytic Conversion of Carbon Dioxide to Oxygenates.*Catal. Today* **2009**, *143*, 57–63.

343. Grabow, L.C.; Mavrikakis, M. Mechanism of Methanol Synthesis on Cu through CO_2 and CO Hydrogenation. *ACS Catal.* **2011**, *1*, 365–384.

344. Cheng, D.; Negreiros, F.R.; Aprà, E.; Fortunelli, A. Computational Approaches to the Chemical Conversion of Carbon Dioxide. *ChemSusChem* **2013**, *6*, 944–965.

345. Lazo, N.D.; Murray, D.K.; Kieke, M.L.; Haw, J.F. *In-Situ* C-13 Solid-State NMR Study of the Cu/ZnO/ Al_2O_3 Methanol Synthesis Catalyst. *J. Am. Chem. Soc.* **1992**, *114*, 8552–8559.

346. Idriss, H.; Kim, K.S.; Barteau, M.A. Surface-Dependent Pathways for Formaldehyde Oxidation and Reduction on TiO_2(001). *Surf. Sci.* **1992**, *262*, 113–127.

347. Busca, G.; Lamotte, J.; Lavalley, J.C.; Lorenzelli, V. FTIR Study of the Adsorption and Transformation of Formaldehyde on Oxide Surfaces. *J. Am. Chem. Soc.* **1987**, *109*, 5197–5202.

348. Rzepa, H.S.; Miller, J. An MNDO SCF-MO Study of the Mechanism of the Cannizzaro Reaction. *J. Chem. Soc. Perkin Trans. II* **1985**, 717–723. CrossRef

349. Harriman, A. Prospects for Conversion of Solar Energy into Chemical Fuels: The Concept of a Solar Fuels Industry.*Philos. Trans. R. Soc. A* **2013**, *371*, 20110415.

350. Thomas, J.M. Reflections on the Topic of Solar Fuels. *Energy Environ. Sci.* **2014**, *7*, 19–20.

351. Malato, S.; Fernández-Ibáñez, P.; Maldonado, M.I.; Oller, I.; Polo-López, M.I. Solar Photocatalytic Pilot Plants: Commercially Available Reactors. In *Photocatalysis and Water Purification: From Fundamentals to Recent Applications*; Pichat, P., Ed.; Wiley-VCH Verlag: Weinheim, Germany, 2013; Chapter 15.

352. Villa, K.; Domènech, X.; Malato, S.; Maldonado, M.I.; Peral, J. Heterogeneous Photocatalytic Hydrogen Generation in a Solar Pilot Plant. *Int. J. Hydrog. Energy* **2013**, *38*, 12718–12724.

353. James, B.D.; Baum, G.N.; Perez, J.; Baum, K.N. *Technoeconomic Analysis of Photoelectrochemical (PEC) Hydrogen Production*; Directed Technologies Inc.: Arlington, VA, USA, 2009; US DOE Contract no. GS-10F-009.

354. Döscher, H.; Geisz, J.F.; Deutsch, T.G.; Turner, J.A. Sunlight Absorption in Water—Efficiency and Design Implications for Photoelectrochemical Devices. *Energy Environ. Sci.* **2014**, *7*, 2951–2956.

355. Seitz, L.C.; Chen, Z.; Forman, A.J.; Pinaud, B.A.; Benck, J.D.; Jaramillo, T.F. Modeling Practical Performance Limits of Photoelectrochemical Water Splitting Based on the Current State of Materials Research. *ChemSusChem* **2014**, *7*, 1372–1385.

356. Haussener, S.; Xiang, C.; Spurgeon, J.M.; Ardo, S.; Lewis, N.S.; Weber, A.Z. Modeling, Simulation, and Design Criteria for Photoelectrochemical Water-Splitting Systems. *Energy Environ. Sci.* **2012**, *5*, 9922–9935.

357. Rodriguez, C.A.; Modestino, M.A.; Psaltis, D.; Moser, C. Design and Cost Considerations for Practical Solar-Hydrogen Generators. *Energy Environ. Sci.* **2014**, *7*, 3828–3835.

358. Chandross, E.A. Shining a Light on Solar Water Splitting. *Science* **2014**, *344*, 469.

359. Turner, J.A. Response to Chandross. *Science* **2014**, *344*, 469–470.

360. Chen, Y.; Xiang, C.; Hu, C.; Lewis, N.S. Modeling the Performance of an Integrated Photoelectrolysis System with 10× Solar Concentrators. *J. Electrochem. Soc.* **2014**, *161*, F1101–F1110.

361. Khamooshi, M.; Salati, H.; Egelioglu, F.; Faghiri, A.H.; Tarabishi, J.; Babadi, S. A Review of Solar Photovoltaic Concentrators. *Int. J. Photoenergy* **2014**.

362. Swanson, R.M. The Promise of Concentrators. *Prog. Photovolt. Res. Appl.* **2000**, *8*, 93–111.

363. Hoffmann, M.R.; Martin, S.T.; Choi, W.; Bahnemann, D.W. Environmental Applications of Semiconductor Photocatalysis. *Chem. Rev.* **1995**, *95*, 69–96.

364. Khaselev, O.; Turner, J.A. A Monolithic Photovoltaic-Photoelectrochemical Device for Hydrogen Production via Water Splitting. *Science* **1998**, *280*, 425–427.

365. Joo, O.S.; Jung, K.D.; Jung, Y.S. CAMERE Process for Methanol Synthesis from CO_2 Hydrogenation. In *Carbon Dioxide Utilization for Global Sustainability*; Park, S.E., Chang, J.S., Lee, K.W., Eds.; Elsevier: Amsterdam, The Netherlands, 2004; Volume 153, pp. 67–72.

366. Ikeue, K.; Nozaki, S.; Ogawa, M.; Anpo, M. Photocatalytic Reduction of CO_2 with H_2O on Ti-Containing Porous Silica Thin-Film Photocatalysts. *Catal. Lett.* **2002**, *80*, 111–114.

367. Lackner, K.S. A Guide to CO_2 Sequestration. *Science* **2003**, *300*, 1677–1678.

368. Gerdemann, S.J.; O'Connor, W.K.; Dahlin, D.C.; Penner, L.R.; Rush, H. Ex Situ Aqueous Mineral Carbonation. *Environ. Sci. Technol.* **2007**, *41*, 2587–2593.

369. Khoo, H.H.; Sharratt, P.N.; Bu, J.; Yeo, T.Y.; Borgna, A.; Highfield, J.G.; Björklöf, T.G.; Zevenhoven, R. Carbon Capture and Mineralization in Singapore: Preliminary Environmental Impacts and Costs via LCA. *Ind. Eng. Chem. Res.* **2011**,*50*, 11350–11357.

370. Zevenhoven, R.; Fagerlund, J.; Nduagu, E.; Romão, I.; Bu, J.; Highfield, J. Carbon Storage by Mineralisation (CSM): Serpentinite Rock Carbonation via $Mg(OH)_2$ Reaction Intermediate without CO_2 Pre-Separation GHGT-11. *Energy Proced.* **2013**, *37*, 5945–5954.

371. Sanna, A.; Uibu, M.; Caramanna, G.; Kuusik, R.; Maroto-Valer, M.M. A Review of Mineral Carbonation Technologies to Sequester CO_2. *Chem. Soc. Rev.* **2014**, *43*, 8049–8080.

CHAPTER 14

IMPACT OF THE SPATIAL DISTRIBUTION OF MORPHOLOGICAL PATTERNS ON THE EFFICIENCY OF ELECTROCATALYTIC GAS EVOLVING REACTIONS

Aleksandar R. Žerađanin

Keywords: Energy efficiency, electrocatalysis, morphology, SECM, activity, stability

Due to their technical importance, electrocatalytic gas-evolving reactions (hydrogen, chlorine and oxygen evolution) have been the subject of investigation for decades.1–3 From well-established large scale industrial processes, such as chlor-alkali technology towards water electrolysis and electrochemical CO_2 reduction, the efficiency of electrocatalytic gas evolving reactions (GER) has been a major challenge due to the large amount of electricity necessary to conduct electrode reactions at a desirable rate.4 Energy efficiency in one electrochemical reactor is usually monitored by the voltage corresponding to the predefined turnover of product controlled by the imposed current intensity. An essential task is to minimize the overvoltage, which represents the difference between the actual voltage and the voltage in the state of equilibrium (reversible voltage).

The energy efficiency during one electrocatalytic process strongly depends on the design of the electrochemical interface. The functionality of an electrochemical interface is largely influenced by the nature of the material of the chosen electrode that belongs to the scope of electrocatalysis.2 Electrocatalysis essentially deals with the impact of the electrode material on the rate of an electrode reaction. The predictive bases on which electrocatalysis rests are empirical or theoretical models known as "volcano" plots, in which the kinetics of the electrode reaction is expressed as a function of some catalytic descriptor, which is in fact some physicochemical property of the material (binding energy of intermediates, number of outer electrons, etc.).5 The apex of "volcano" responds to the maximal kinetics, which is essentially determined by the optimal value of a catalytic descriptor (Sabatier Principle). For example, a noticeable fact is that RuO_2 is at the apex of the "volcano" plot for the oxygen evolution reaction (OER) and the chlorine evolution reaction (CER).6,7 If a material whose activity responds to the maximum of the "volcano" plot is identified, the question that would naturally arise is: how to improve further the efficiency of the given reaction?

A high intrinsic activity of the chosen electrocatalyst, reflected by the high turnover per active site, however, goes hand in hand with a certain self-inhibiting effect. Namely, the phenomenon of supersaturation with the product is actually a prerequisite for the gas formation.8,9 The formed gas-bubbles cover a fraction of the active surface area and induce additional overpotential. It is important to mention that the gas bubble effect was detected to be the major source of overpotential in a case of chlor-alkali technology.2 Consequently, this issue requires serious investigations, which could be beneficial for all gas-evolving reactions, even for those that are not driven by the electrode potential. At the same time, electrode materials, especially those used as catalysts for anodic electrocatalytic reactions, are subjected to intensive corrosion.10 The electrode materials used to catalyze the mentioned electrode reactions usually originate from the family of noble metal oxides. Considering that the price of noble metals is subject to unpredictable fluctuations (for example, the price of Ru jumped in 2007/2008 by 1000 %11), it is of major importance to find strategies to 1) stabilize electrocatalytic materials, 2) reduce the loading with noble metals and/or 3) replace noble metals with cheaper and more abundant materials.

Citation: Aleksandar R. Žerađanin, Impact of the spatial distribution of morphological patterns on the efficiency of electrocatalytic gas evolving reactions, doi: 10.2298/JSC131002106Z.

A promising approach to address the above-mentioned issues is a rational design of the surface morphology. Namely, morphology has strong impact on the gas evolution (for example, it influences the contact angle at the triple phase boundary) and influences the utilization of the active surface area by determining the accessibility of active sites.12 Additionally, expecting that high local overpotentials and current densities could enhance degradation of electrocatalysts,13 it is essential to investigate to what extent the stability of electrocatalysts is influenced by the morphological pattern. State of the art anodes for the electrocatalytic gas evolution are DSA (dimensionally stable anodes)-type electrodes. Usually, "valve" metal-like titanium is coated with a mixed transition metal oxide. Catalytically, the active component is a noble metal-based oxide (for example ruthenium oxide), the long-term stability of which is assured by mixing with a "valve" metal-based oxide (for example titanium oxide). An important pre-condition for mixed oxide formation is that the parameters of the crystal lattice match (according to the Hume Rothery Rules).14 A typical morphological pattern of DSAs, the so-called "mud-crack" structure, contains two interesting features, namely "islands" and "channels", which could be potentially beneficial. The edges of the islands can allow high local activities, while the channels can contribute to mass transport by allowing certain internal hydraulic regimes via microconvection.

Two studies provide particularly interesting insights into the behavior of DSAs. Trasatti and co-workers derived a method to distinguish easily accessible "outer" surface area from the hardly accessible "inner" surface area (originating in nanopores, etc.).15 In the mentioned work, the authors stated that only the outer surface is functional and accessible for the reaction. Simultaneously, Evdokimov reported that at very high current densities (up to 4 A cm^{-2}), the inner surface area starts to participate in the reaction.16,17 The results of Evdokimov could be understood in terms of an increase of the interfacial overpressure, which allows solvent penetration towards the nanopores (according to the Young–Laplace Equation for capillary forces).18 While Evdokimov was increasing the imposed intensity of the process (rate of reaction) at the analyzed electrode, the question was whether the opposite was possible. Namely, whether it is possible for the pre-defined intensity of the process to gradually change the morphology and cause activation of the inner surface area or at least increase its utility. To achieve better utility of surface area, it is important to bear in mind that gas evolution is an oscillatory phenomenon.19 Thermal oscillations of the gas-bubbles cause periodic displacement of the surrounding liquid, which in return influences the root of the surrounding gas-bubbles and ultimately cause their detachment. The smaller the gas-bubbles are, the more frequently they oscillate and finally detach. Consequently, given the oscillatory behavior, a key parameter of the macrokinetics of a gas-evolving electrode is the frequency of gas-bubble detachment. In this sense, the goal was to find a way to generate gas-bubbles with small radius at electrode/electrolyte interface. Knowing that gas bubble evolution consists of three processes: 1) nucleation, 2) growth and coalescence and 3) detachment, it seems rational to design electrodes with adequate thickness of the catalyst layer and size of the "channels" so the gas bubbles can nucleate inside the "channels" that would limit their growth. In this respect, the thickness of the coating and size of the "channels" should be similar to the critical radius of gas-bubble nucleation.20 To achieve the targeted pattern, the synthesis was based on controlling the tensile stress in the catalyst layer. The tensile stress delivered to the coating could be controlled by adequate thermal treatment (calcination temperature, rate of heating/cooling...) and by creating mismatch in the coefficients of thermal expansion between support and catalyst layer. In this sense, the "breakage" of the catalyst layer is under certain measures of control as is the size of the "channels". Besides the design of the morphological pattern, it was important to ensure that the desirable surface features are uniformly distributed over the surface.21 While DSA coatings were usually synthesized using the thermal decomposition procedure, important progress was made through the introduction of sol–gel route, as shown by Nikolić and co-workers.22 The sol–gel procedure allowed coatings to be obtain with significantly reduced surface inhomogeneities in the morphology and/or composition. Reducing surface inhomogeneities is already recognized as an important aspect of electrode design and as one of the potential strategies for further improvement of DSA performance.23 Although, the impact of the spatial distribution of activity on electrode performance is still not adequately understood, it has been proven to be of major importance for efficient gas evolution.21

Besides the design of the features of the electrode surface, of major importance is simultaneously to have a realistic insight into the catalytic performance of a particular material, which further strongly depends on the employment of adequate analytical tools and procedures. Particularly important was to characterize and adequately interpret gas-bubble evolution. The frequency of gas bubble detachment at various potentials was monitored using scanning electrochemical microscopy (SECM).24 This is the first work in which an SECM setup was adopted for the characterization of GER by recording potential dependent frequency spectra. From the recorded spectra, the point of transition from pseudo-periodical to periodical behavior could be detected, which was related with a drop in the efficiency. All the tested DSA coatings for which a characteristic frequency of the gas-bubble detachment could not be estimated exhibited relatively low activity. This could be related to the spatial distribution of the activity, which can also be visualized in situ using SECM.21,25 Finally SECM in combination with cyclic voltammetry (CV) was used for the most challenging task, namely to develop a methodology that quantifies what fraction of the catalyst layer (what percentage of overall active sites) really participate in the reaction in dependence on the applied potential. Finally, as a part of an analysis of catalyst stability, time resolved dissolution profiles were recorded using a recently developed setup based on the coupling of electrochemistry with mass spectrometry (SFC––ICPMS).26,27 From the dissolution profiles, it was clear that the more active sample, due to the morphological pattern that promotes efficient gas evolution and assures a larger effective surface area, at the same time could be more stable, which is of major importance on a technical scale.

REFERENCES

1. J. O'M. Bockris, J. Electrochem. Soc. 131 (1984) 290
2. S. Trasatti, Electrochim. Acta 45 (2000) 2377
3. J. Suntivich, K. J. May, H. A. Gasteiger, J. B.

Goodenough, Y. Shao-Horn, Science 334 (2011) 1383
4. N. S. Lewis, D. G. Nocera, Proc. Natl. Acad. Sci. 103 (2006) 15729
5. A. R. Zeradjanin, N. Menzel, P. Strasser, W. Schuhmann, ChemSusChem 5 (2012) 1897
6. S. Trasatti, Electrochim. Acta 29 (1984) 1503
7. I. C. Man, H.-Y. Su, F. Calle-Vallejo, H. A. Hansen, J. I. Martínez, N. G. Inoglu, J. Kitchin, T. F. Jaramillo, J. K. Nørskov, J. Rossmeisl, ChemCatChem 3 (2011) 1159
8. H. Vogt, J. Appl. Electrochem. 23 (1993) 1323
9. N. P. Brandon, G. H. Kelsall, J. Appl. Electrochem. 15 (1985) 475
10. G. N. Martelli, R. Ornelas, G. Faita, Electrochim. Acta 39 (1994) 1551
11. J. Pérez-Ramírez, C. Mondelli, T. Schmidt, O. F.-K. Schlüter, A. Wolf, L. Mleczko, T. Dreier, Energy Environ. Sci. 4 (2011) 4786
12. S. Ardizzone, J. Electrochem. Soc. 129 (1982) 1689
13. B. V. Tilak, V. I. Birss, J. Wang, C.-P. Chen, S. K. Rangarajan, J. Electrochem. Soc. 148 (2001) D112
14. R. D. Shannon, Solid State Commun. 6 (1968) 139
15. S. Ardizzone, G. Fregonara, S. Trasatti, Electrochim. Acta 35 (1990) 263
16. S. V. Evdokimov, Russ. J. Electrochem. 36 (2000) 236
17. S. V. Evdokimov, Russ. J. Electrochem. 36 (2000) 489
18. H. Wendt, Electrochim. Acta 39 (1994) 1749
19. I. G. Malenkov, J. Eng. Phys. 20 (1971) 704
20. A. R. Zeradjanin, F. La Mantia, J. Masa, W. Schuhmann, Electrochim. Acta 82 (2012) 408
21. R. Chen, V. Trieu, A. R. Zeradjanin, H. Natter, D. Teschner, J. Kintrup, A. Bulan, W. Schuhmann, R. Hempelmann, Phys. Chem. Chem. Phys. 14 (2012) 7392
22. V. V. Panić, A. Dekanski, S. K. Milonjić, R. T. Atanasoski, B. Ž. Nikolić, Colloids Surfaces, A 157 (1999) 269
23. H. Over, Electrochim. Acta 93 (2013) 314
24. A. R. Zeradjanin, E. Ventosa, A. S. Bondarenko, W. Schuhmann, ChemSusChem 5 (2012) 1905
25. A. R. Zeradjanin, T. Schilling, S. Seisel, M. Bron, W. Schuhmann, Anal. Chem. 83 (2011) 7645
26. A. A. Topalov, I. Katsounaros, M. Auinger, S. Cherevko, J. C. Meier, S. O. Klemm, K. J. J. Mayrhofer, Angew. Chem. Int. Ed. 51 (2012) 12613
27. S. Cherevko, A. A. Topalov, A. R. Zeradjanin, I. Katsounaros, K. J. J. Mayrhofer, RSC Adv. 3 (2013) 16516.

CHAPTER 15

POLYANILINE-DERIVED ORDERED MESOPOROUS CARBON AS AN EFFICIENT ELECTROCATALYST FOR OXYGEN REDUCTION REACTION

Kai Wan, Zhi-Peng Yu and Zhen-Xing Liang

Keywords: Ammonia-activation, electrocatalysis, fuel cells, nitrogen-doped mesoporous carbon, oxygen reduction reaction

INTRODUCTION

The oxygen reduction reaction (ORR) is one key electrochemical process for the energy conversion devices, like fuel cells and metal-air batteries. Pt-based materials have been so far acknowledged to be the most effective catalysts for the ORR at low temperatures [1,2]; however, the source scarcity and high cost pose great challenges to the large-scale applications to fuel cells [3,4,5]. Hence, enormous effort has been devoted to search for greater efficiency, durability and less cost [6,7].

In recent years, nanostructured carbon materials have attracted increasing attention as the Pt-alternative electrocatalysts. Dai [8] synthesized 1D nitrogen-doped carbon nanotubes by the chemical vapor deposition (CVD) method with iron (II) phthalocyanine as the precursor, which featured high charge transfer and, thus, facilitated the ORR. Feng [9] synthesized 2D graphene-based carbon nitride nanosheets, of which the high specific surface favored the dense assembling of the active sites on the surface. Lu [10] synthesized 3D hierarchically porous nitrogen-doped carbons with a hieratical porous structure, which enabled low mass transfer resistance and improved accessibility of catalytic sites for the ORR.

Among them, nitrogen-doped ordered mesoporous carbon (NOMC) features high specific surface area and a uniform pore structure, which respectively facilitate the reaction kinetics and mass transfer to the electrode [11,12]. The hard-template method is a universal strategy to synthesize such materials, and SBA-15 is one of the most often used templates to synthesize NOMCs. Asefa [13] synthesized NOMC catalysts by pyrolyzing polyaniline in the framework of SBA-15. It was found that the NOMC catalyst pyrolyzed at 800 °C showed the best ORR performance. Guo [14] synthesized a series of NOMC catalysts with honey as the precursor, which showed a high specific surface area ranging from 1050 to 1273 $m^2 \cdot g^{-1}$. Mullen [15] synthesized high-performance NOMC catalysts by pyrolyzing ionic liquid within SBA-15. All of the above-mentioned catalysts showed a decent electrocatalytic activity to the ORR in alkaline media. Other mesoporous silicas have been also used for the synthesis. For example, Popov [16] synthesized nitrogen-doped ordered porous carbon with the aid of SBA-12, which also showed a superior activity for the ORR. Joo [17] synthesized a series of carbon catalysts by using various templates, like SBA-15, MSU-F, KIT-6 and fumed Carb-O-sil M-5. It was found that the template SBA-15 yielded an extremely high surface area of 1500 $m^2 \cdot g^{-1}$ and the best electrocatalytic activity.

Beside the template, the carbon precursor has a considerable effect on the microstructure and composition of the final carbon materials. A variety of nitrogen-containing organic chemicals, like phthalocyanine [18], porphyrin [19,20] and ionic liquid [21,22], have been used as the carbon precursor to synthesize the nitrogen-doped carbon catalyst. Polyaniline represents an aromatic ring connected via nitrogen-containing groups and, thus, facilitates the incorporation of nitrogen-containing active sites into the carbon matrix during the heat treatment [23]. Zelenay [23] found that the polyaniline-derived carbon electrocatalyst showed a superior performance to the

Citation: Kai Wan, Zhi-Peng Yu and Zhen-Xing Liang, Polyaniline-Derived Ordered Mesoporous Carbon as an Efficient Electrocatalyst for Oxygen Reduction Reaction, doi:10.3390/catal5031034.

ORR, which exhibited the highest maximum power density of 0.55 $W \cdot cm^{-2}$ at 0.38 V.

It has been well acknowledged that the template, carbon precursor and pyrolysis conditions have significant and complicated effects on the composition, structure and electrocatalytic activity of the ORR. In our previous work, we developed a method to synthesize the nitrogen-doped ordered mesoporous carbon featuring a high specific surface area [24,25]. Additionally, the active sites for the ORR were claimed to be the nitrogen-activated carbon atoms, on which the ORR proceeded by a surface-confined redox-mediation mechanism in both acid and alkaline media [24,26]. In this work, we will use polyaniline as the carbon source to synthesize the NOMC catalysts and to optimize the pyrolysis conditions. In order to improve the electrocatalytic activity, the as-prepared catalyst is further subjected to the NH_3-activation at high temperatures. Then, the nitrogen sorption isotherm, electron microscopy and X-ray photoelectron spectroscopy are used to study the microstructure and composition. The cyclic voltammetry and rotating-ring-disk electrode methods are used to investigate the electrochemical behavior for the ORR.

RESULTS AND DISCUSSION

Figure 1 shows the nitrogen sorption isotherm of the as-synthesized carbon materials. It is seen that all curves show a similar shape of the typical Type-IV isotherm, which indicates the mesoporous nature of the synthesized carbon (see Figure S1). Pore parameters are then extracted from the isotherms and listed in Table 1. It is seen that the specific surface area increases from 470 to 629 $m^2 \cdot g^{-1}$ with increasing the pyrolysis temperature from 800 to 1000 °C, which can be ascribed to the deepened decomposition in this course. Then, a further increase in the pyrolysis temperature to 1100 °C results in a decrease in the specific surface area, which may be linked to the collapse of the carbon framework. A similar trend is also seen in the specific pore volume, which reaches the highest value of 0.81 $cm^{-3} \cdot g^{-1}$ at 1000 °C. It is noted that the post-treatment in NH_3yields a dramatic increase in the specific surface area (1312 $m^2 \cdot g^{-1}$), which can be attributed to the gasification of amorphous carbon and the consequent formation of micro-/meso-pores in the active atmosphere [27,28,29].

Table 1. Pore features of the synthesized carbon materials.

Samples	$A_{BET}/m^2 \cdot g^{-1}$	$A_{MP}/m^2 \cdot g^{-1}$	D_{BJH}/nm	$V/cm^3 \cdot g^{-1}$
C-PA-800	470	67	5.4	0.58
C-PA-900	569	41	5.9	0.80
C-PA-1000	629	132	6.1	0.81
C-PA-1100	517	61	5.9	0.67
C-PA-1000-NH₃	1312	229	5.9	1.73

The surface composition is characterized by XPS, for which the survey spectra are shown in Figure S2. The content of the main elements are qualified and listed in Table 2. It is found that both the pyrolysis temperature and NH_3-activation yield a significant effect on the nitrogen content. First, the nitrogen content shows a monotonic decrease from 5.07 to 1.25 at. % with increasing the pyrolysis temperature from 800 to 1100 °C, which should be attributed to the enhanced decomposition of the nitrogen-containing functional groups at higher temperatures. Second, the nitrogen content is 2.20 at. % for C-PA-1000 and 3.15 at. % for C-PA-1000-NH_3, indicating that the nitrogen doping can be enhanced by the pyrolysis in the nitrogen-containing active gases. The change in the surface composition is expected to yield effects on the electrocatalysis, as discussed below.

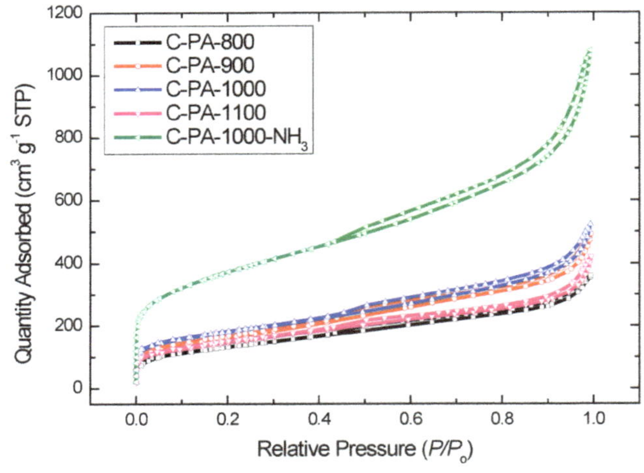

Figure 1. Nitrogen sorption isotherms of the synthesized carbon materials (STP: standard temperature and pressure).

Table 2. Elemental composition (at. %) of the synthesized carbon materials.

Samples	C	N	O	N:C
C-PA-800	86.63	5.07	5.33	0.057
C-PA-900	92.03	3.13	4.40	0.034
C-PA-1000	93.45	2.20	3.91	0.024
C-PA-1100	94.88	1.25	3.57	0.013
C-PA-1000-NH3	94.32	3.15	2.33	0.033

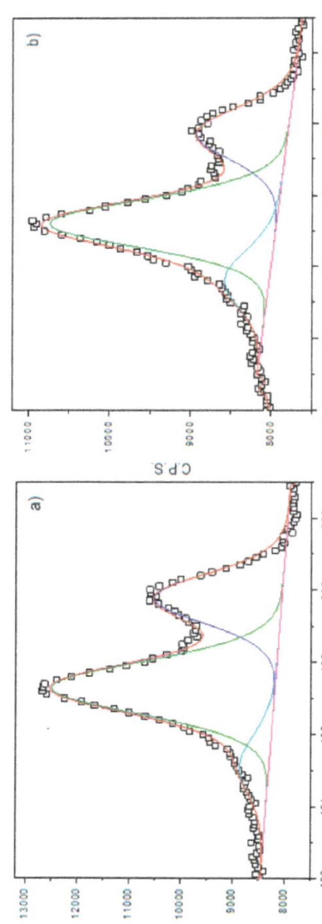

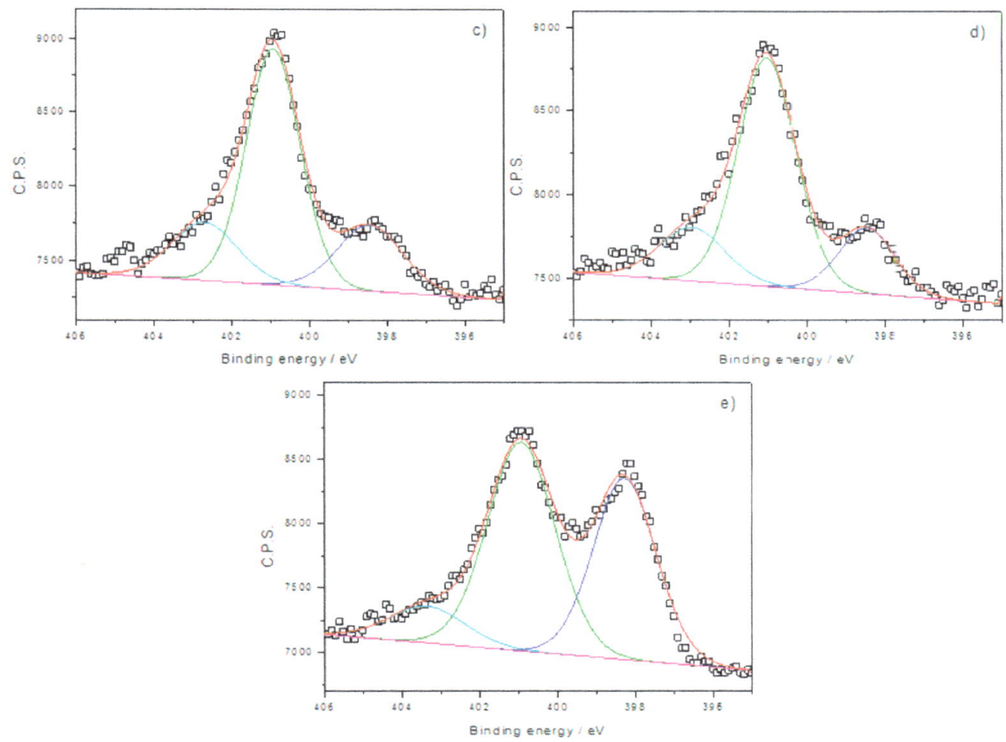

Figure 2. N 1s peak and the peak fitting results of the following materials: (**a**) C-PA-800; (**b**) C-PA-900; (**c**) C-PA-1000; (**d**) C-PA-1100; and (e) C-PA-1000-NH_3. The unit of *y*-axis unit is counts per second (C.P.S.).

Figure 2 shows the high-resolution XPS spectra of N 1s. These spectra can be deconvoluted to three peaks based on the binding energy: 398.4 ± 0.2, 401.0 ± 0.1 and 401.5–404 eV, which respectively correspond to pyridinic-nitrogen, graphitic-nitrogen and nitrogen-oxide [30,31]. Then, the content of each species is seen in Table 3. It is seen that the graphitic nitrogen is the predominant component among the three species, of which the content slightly increases with the pyrolysis temperature. In comparison, the content of the pyridinic-nitrogen decreases from 31.88% to 17.90% with increasing the pyrolysis temperature from 800 to 1100 °C. These results are consistent with the previous findings that the graphitic nitrogen is the most stable nitrogen species at high temperatures [32,33]. Finally, C-PA-1000-NH_3 shows an extraordinarily high content of pyridinic-nitrogen (40.31%), which should be attributed to the nitrogen doping at the edge of the graphite plane during the NH_3-etching process. It seems that the content of the graphitic nitrogen decreases to 49.88% in the etching process; however, it should be noted that the "absolute" content of this species remains unchanged, as compared with the un-etched one, by considering the total nitrogen content (see Table 2).

In our previous work, the relationship has been well established between the content of the nitrogen-activated carbon and the electrocatalytic activity for ORR [24]. In line with this understanding, the curve fitting of high-resolution C1s peak is performed (see Figure S3), and the results are listed in Table 4. It is found that the content of the nitrogen-activated carbon slightly increases with increasing pyrolysis temperature from 800 to 1000 °C and then decreases at 1100 °C. Additionally, C-PA-1000-NH_3 shows the highest content of nitrogen-activated carbon among all of the carbon materials. The ORR performance is expected to follow this change in content, as seen below.

Table 3. Content of each nitrogen component (%) of the synthesized carbon materials.

Samples	Pyridinic-N	Graphitic-N	O-N
C-PA-800	31.88	59.10	9.02
C-PA-900	26.42	59.55	14.04
C-PA-1000	20.90	60.62	18.48
C-PA-1100	17.90	64.72	17.38
C-PA-1000-NH_3	40.31	49.88	9.81

Table 4. Content of each carbon component (%) of the synthesized carbon materials.

Samples	C–C=C	C–N	C–O/C=N	C=O	COOH
C-PA-800	65.58	13.19	14.39	5.52	1.33
C-PA-900	68.29	15.60	8.81	5.46	1.84
C-PA-1000	68.84	18.51	7.53	3.85	.28
C-PA-1100	73.60	15.11	6.10	3.71	.48
C-PA-1000-NH_3	67.98	20.49	7.84	1.65	2.04

Figure 3 shows the CV curves in Ar-saturated 0.10 M KOH. It is seen that all of the curves are similar in shape with large capacitance currents, and broad symmetrical redox peaks are found in the potential range of 0 to 0.9 V. It is understandable that a large capacitance current should result from the high specific surface area, and the pseudocapacitance current is associated with the chemical adsorption of OH^- onto the enriched redox couples. Basically, the capacitance current shows a volcano-shape relationship with the pyrolysis temperature, which first increases and then dramatically decreases at temperatures of 1100 °C. Notably, C-PA-1000-NH_3 shows the largest capacitance current. Such a change can be rationalized as a result of the specific surface area (*vide supra*). In comparison, the pseudocapacitance current shows a monotonic decrease with increasing pyrolysis temperature. Additionally, this result should be associated with the deepened decomposition of the electrochemically-active functional groups (like nitrogen-containing species) on the surface at higher temperatures. In line with the above analysis, C-PA-1000-NH_3 should yield the largest pseudocapacitance/capacitance currents due to its extraordinarily high specific surface area and nitrogen content. However, the increase in the capacitance current is not that large, and the pseudocapacitance current does not increase. This seeming contradiction may be understandable by considering the pore structure. The NH_3 post-treatment can effectively gasify/etch the amorphous carbon, leaving enriched micropores in the bulk. Additionally, these pores may not be fully utilized due to the lack of contact with the liquid electrolyte.

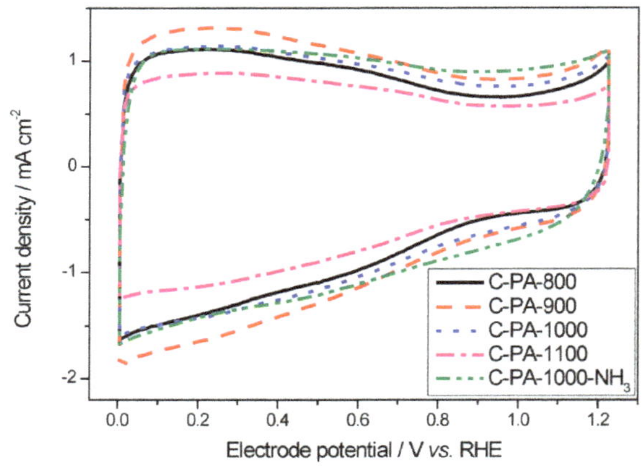

Figure 3. Cyclic voltammograms of the carbon materials in Ar-saturated 0.10 M KOH solution. RHE, reversible hydrogen electrode.

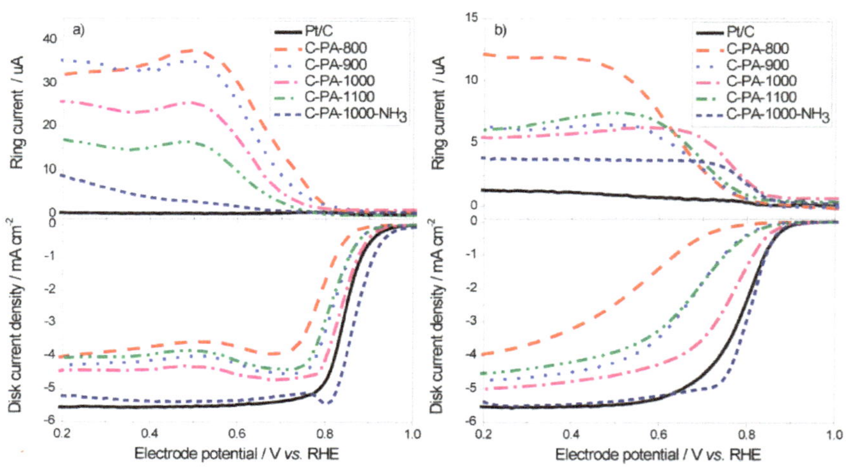

Figure 4. Ring (top) and disk (bottom) currents for the RRDE tests of the carbon materials: (**a**) O_2-saturated 0.10 M KOH solution; (**b**) O_2-saturated 0.10 M $HClO_4$solution.

Figure 4 shows the polarization curves of the carbon materials in O_2-saturated 0.10 M KOH and 0.10 M $HClO_4$ solutions.Figure 4a shows that in alkaline media, the electrocatalytic activity shows a volcano-shape relationship, which increases with the pyrolysis temperature from 800 to 1000 °C and then decreases at 1100 °C. The same trend is further seen in the change of the yield of hydrogen peroxide and electron transfer number (see Figure S4a). Increasing the pyrolysis temperature lowers the yield of hydrogen peroxide, and thereby, selectively favors the 4-e reduction of oxygen. Such a trend in the electrocatalytic activity and selectivity correlates well with the change in the content of nitrogen-activated carbon and the specific surface area (*vide supra*). These findings further confirm that the active sites for the ORR should be the nitrogen-activated carbon atoms [24,26]. In acid media, the change in the electrocatalytic activity is similar to that in alkaline media (Figure 4b), revealing that the active site should remain substantially the same in a wide range of pH. Finally, it is noted that for C-PA-1000-NH_3, both the activity and the selectivity are considerably improved upon NH_3-activation. This result quantitatively agrees with the change in the specific surface area and the content of nitrogen-activated carbon, as discussed above. However, it seems irrational to directly correlate the small increase in the nitrogen-activated carbon (Table 4) with the extremely high activity and low H_2O_2 yield. The mechanism has not been fully understood yet, but the reason may be associated with high content of the pyridinic-nitrogen after the NH_3 treatment.

EXPERIMENTAL SECTION

Materials Preparation

Nitrogen-doped ordered mesoporous carbon (NOMC) was synthesized by a modified nanocasting method [24]. The process is briefly described as follows. (i) Synthesis of the template SBA-15 [34]: An aqueous mixture, consisting of Pluronic P123, HCl and tetraethoxysilane, was stirred for 20 h at 35 °C and then hydrothermally treated at 100 °C for 24 h. The resultant powders were calcined in air at 550 °C for 6 h, and SBA-15 was finally obtained. (ii) Impregnation of the carbon precursor: Aniline was impregnated into SBA-15 by the vaporization-capillary condensation method [24]. Then, the monomer was polymerized upon adding $FeCl_3$ to form polyaniline (PA). (iii) Pyrolysis and template removal: The resultant powders were then subjected to the pyrolysis at high temperatures (800, 900, 1000, 1100 °C) for 3 h in argon, respectively. Finally, the carbon catalyst was obtained by removing the silicate template by boiling in 10 M NaOH solution for 24 h. The samples were referred to as C-PA-X. Here, X refers to the pyrolysis temperature, *viz.* 800, 900, 1000 and 1100. In addition, the sample C-PA-1000 was heat-treated again at 1000 °C in ammonia for 30 min, which was labeled as C-PA-1000-NH_3.

Physical Characterizations

X-ray photoelectron spectroscopy (XPS, Physical Electronics PHI 5600, Chanhassen, MN, USA) measurement was carried out with a multi-technique system using an Al monochromatic X-ray at a power of 350 W. Transmission electron microscopy (TEM) was performed on an FEI Tecnai G2 F20 S-TWIN (Hillsboro, OR, USA) operated at 200 kV. Nitrogen adsorption/desorption isotherms were measured at 77 K using a Micromeritics TriStar II 3020 analyzer (Norcross, GA, USA). The total surface area was analyzed with the well-established Brunauer–Emmett–Teller (BET) method; the microporous surface area was obtained with the MP (micropore) method (*t*-plot method); and the pore size distribution was analyzed by the Barrett–Joyner–Halenda (BJH) method.

Electrochemical Characterization

The electrochemical behavior of the catalyst was characterized by the cyclic voltammetry (CV) and linear sweeping voltammetry (LSV) using a three-electrode cell with an electrochemical work station Zennium (Zahner, Germany) at room temperature (25 °C). A platinum wire and a double-junction Ag/AgCl reference electrode were used as the counter and reference electrodes, respectively. The working electrode was a rotating ring-disk electrode (RRDE, glassy carbon disk: 5.0 mm in diameter; platinum ring: 6.5 mm inner diameter and 7.5 mm outer diameter). The thin-film electrode on the disk was prepared as follows. Ten milligrams of the catalyst were dispersed in 1.0 mL Nafion/ethanol (0.84 wt. % Nafion) by sonication for 120 min. Then, 10 μL of the dispersion were transferred onto the glassy carbon disk by using a pipette, yielding the catalyst loading of 0.50 mg·cm^{-2}. The ORR activity of the Pt/C catalyst (HiSPEC4000, Johnson Matthey, London, UK) with the metal loading of 20 μg·cm^{-2} was collected for comparison.

The electrolyte solution, 0.10 M KOH, was first bubbled with argon for 60 min. Then, a CV test was conducted at 20 mV·s^{-1} in the potential range between 0 and 1.23 V (*vs.* reversible hydrogen electrode, RHE) for 20 cycles. If not specified, the LSV curve was collected by scanning the disk potential from 1.2 down to 0 V at 5 mV·s^{-1} in the oxygen-saturated electrolyte solution under 1600 rpm, from which the ORR polarization curve was extracted by subtracting the capacitive current. During the collection, the potential of the ring was set to be 0.5 V (*vs.* RHE) to determine the yield of hydrogen peroxide.

The electron-transfer number (n) and hydrogen peroxide yield (H_2O_2%) in the ORR were calculated from the following equations:

$$n = \frac{4|i_d|}{|i_d| + i_r/N} \quad (1)$$

$$H_2O_2(\%) = \frac{2i_r/N}{|i_d| + i_r/N} \times 100 \quad (2)$$

where i_d is the disk current, i_r is the ring current and N is the collection efficiency (=20.50%).

CONCLUSIONS

Nitrogen-doped ordered mesoporous carbon was synthesized by the modified nanocasting method with high electrocatalytic activities to the ORR in both acid and alkaline media. The results revealed that both the pyrolysis temperature and the NH_3-activation yielded significant effects on the specific surface area, nitrogen doping and, thus, the electrocatalytic activity, as well. First, the pyrolysis temperature yielded a volcano-shape relationship with the specific surface area and the content of the nitrogen-activated carbon. Additionally, it was found that such a change could be correlated with the electrocatalytic activity to the ORR, revealing the importance of the specific surface area and the chemical nature of the active sites. Second, the post-treatment in NH_3 could further increase the specific surface and enhance the nitrogen doping, which thereby improved the electrocatalytic activity and selectivity to the ORR. Additionally, the C-PA-1000-NH_3 catalyst outperformed the Pt/C one in both acid and alkaline media, which make it promising to be applied in fuel cells.

Acknowledgments

The work described in this paper was jointly supported by the National Natural Science Foundation of China (No. 21476087), the Pearl River S&T Nova Program of Guangzhou (No. 2013J2200041), the Science & Technology Research Project of Guangdong Province (No. 2014A010105041), the Guangdong Natural Science Foundation (No. S2013010012469) and the Innovation Project of Guangdong Department of Education (No. 2014KTSCX016).

Conflicts of Interest

The authors declare no conflicts of interest.

REFERENCES

1. Job, N.; Lambert, S.; Zubiaur, A.; Cao, C.; Pirard, J.P. Design of Pt/carbon xerogel catalysts for pem fuel cells. *Catalysts***2015**, *5*, 40–57.
2. Jia, Q.; Liang, W.; Bates, M.K.; Mani, P.; Lee, W.; Mukerjee, S. Activity descriptor identification for oxygen reduction on platinum-based bimetallic nanoparticles: *In situ* observation of the linear composition-strain-activity relationship.*ACS Nano* **2015**, *9*, 387–400.
3. Ye, T.N.; Lv, L.B.; Li, X.H.; Xu, M.; Chen, J.S. Strongly veined carbon nanoleaves as a highly efficient metal-free electrocatalyst. *Angew. Chem. Int. Ed.* **2014**, *53*, 6905–6909.
4. Wu, G.; Zelenay, P. Nanostructured nonprecious metal catalysts for oxygen reduction reaction. *Acc. Chem. Res.* **2013**,*46*, 1878–1889.
5. Zhang, Y.; Zhuang, X.D.; Su, Y.Z.; Zhang, F.; Feng, X.L. Polyaniline nanosheet derived B/N co-doped carbon nanosheets as efficient metal-free catalysts for oxygen reduction reaction. *J. Mater. Chem. A* **2014**, *2*, 7742–7746.
6. Su, D.S.; Perathoner, S.; Centi, G. Nanocarbons for the development of advanced catalysts. *Chem. Rev.* **2013**, *113*, 5782–5816.
7. Wang, H.L.; Dai, H.J. Strongly coupled inorganic-nano-carbon hybrid materials for energy storage. *Chem. Soc. Rev.***2013**, *42*, 3088–3113.
8. Gong, K.; Du, F.; Xia, Z.; Durstock, M.; Dai, L. Nitrogen-doped carbon nanotube arrays with high electrocatalytic activity for oxygen reduction. *Science* **2009**, *323*, 760–764.
9. Yang, S.; Feng, X.; Wang, X.; Müllen, K. Graphene-based carbon nitride nanosheets as efficient metal-free electrocatalysts for oxygen reduction reactions. *Angew. Chem. Int. Ed.* **2011**, *50*, 5339–5343.
10. He, W.; Jiang, C.; Wang, J.; Lu, L. High-rate oxygen electroreduction over graphitic-n species exposed on 3D hierarchically porous nitrogen-doped carbons. *Angew. Chem. Int. Ed.* **2014**, *53*, 9503–9507.
11. Zhu, Y.; Zhang, B.; Liu, X.; Wang, D.W.; Su, D.S. Unravelling the structure of electrocatalytically active Fe–N complexes in carbon for the oxygen reduction reaction. *Angew. Chem. Int. Ed.* **2014**, *53*, 10673–10677.
12. Sun, X.; Song, P.; Zhang, Y.; Liu, C.; Xu, W.; Xing, W. A class of high performance metal-free oxygen reduction electrocatalysts based on cheap carbon blacks. *Sci. Rep.* **2013**, *3*, 2505.
13. Silva, R.; Voiry, D.; Chhowalla, M.; Asefa, T. Efficient metal-free electrocatalysts for oxygen reduction: Polyaniline-derived N- and O-doped mesoporous carbons. *J. Am. Chem. Soc.* **2013**, *135*, 7823–7826.
14. Ramaswamy, N.; Tylus, U.; Jia, Q.; Mukerjee, S. Activity descriptor identification for oxygen reduction on nonprecious electrocatalysts: Linking surface science to coordination chemistry. *J. Am. Chem. Soc.* **2013**, *135*, 15443–15449.
15. Liang, H.W.; Wei, W.; Wu, Z.S.; Feng, X.; Mullen, K. Mesoporous metal-nitrogen-doped carbon electrocatalysts for highly efficient oxygen reduction reaction. *J. Am. Chem. Soc.* **2013**, *135*, 16002–16005.
16. Liu, G.; Li, X.G.; Ganesan, P.; Popov, B.N. Development of non-precious metal oxygen-reduction catalysts for pem fuel cells based on N-doped ordered porous carbon. *Appl. Catal. B* **2009**, *93*, 156–165.
17. Cheon, J.Y.; Kim, T.; Choi, Y.; Jeong, H.Y.; Joo, S.H. Ordered mesoporous porphyrinic carbons with very high electrocatalytic activity for the oxygen reduction reaction. *Sci. Rep.* **2013**, *3*, 2715.
18. Yin, H.; Zhang, C.; Liu, F.; Hou, Y. Hybrid of iron nitride and nitrogen-doped graphene aerogel as synergistic catalyst for oxygen reduction reaction. *Adv. Funct. Mater.* **2014**, *24*, 2930–2937.
19. Xi, P.B.; Liang, Z.X.; Liao, S.J. Stability of hemin/C electrocatalyst for oxygen reduction reaction. *Int. J. Hydrogen Energy* **2012**, *37*, 4606–4611.
20. Liang, Z.X.; Song, H.Y.; Liao, S.J. Hemin: A highly effective electrocatalyst mediating the oxygen reduction reaction. *J. Phys. Chem. C* **2011**, *115*, 2604–2610.
21. Fellinger, T.P.; Hasche, F.; Strasser, P.; Antonietti, M. Mesoporous nitrogen-doped carbon for the electrocatalytic synthesis of hydrogen peroxide. *J. Am. Chem. Soc.* **2012**, *134*, 4072–4075.
22. Sa, Y.J.; Park, C.; Jeong, H.Y.; Park, S.H.; Lee, Z.; Kim, K.T.; Park, G.G.; Joo, S.H. Carbon nanotubes/heteroatom-doped carbon core-sheath nanostructures as highly active, metal-free oxygen reduction electrocatalysts for alkaline fuel cells.*Angew. Chem. Int. Ed.* **2014**, *53*, 4102–4106.
23. Wu, G.; More, K.L.; Johnston, C.M.; Zelenay, P. High-performance electrocatalysts for oxygen reduction derived from polyaniline, iron, and cobalt. *Science* **2011**, *332*, 443–447.
24. Wan, K.; Long, G.F.; Liu, M.Y.; Du, L.; Liang, Z.X.; Tsiakaras, P. Nitrogen-doped ordered mesoporous carbon: Synthesis and active sites for electrocatalysis of oxygen reduction reaction. *Appl. Catal. B* **2015**, *165*, 566–571.
25. Long, G.F.; Wan, K.; Liu, M.Y.; Li, X.H.; Liang, Z.X.; Piao, J.H. Effect of pyrolysis conditions on nitrogen-doped ordered mesoporous carbon electrocatalysts. *Chin. J. Catal.* **2015**.
26. Wan, K.; Yu, Z.P.; Li, X.H.; Liu, M.Y.; Yang, G.; Piao, J.H.; Liang, Z.X. PH effect on electrochemistry of nitrogen-doped carbon catalyst for oxygen reduction reaction. *ACS Catal.* **2015**, *5*, 4325–4332.
27. Meng, H.; Larouche, N.; Lefèvre, M.; Jaouen, F.; Stansfield, B.; Dodelet, J.-P. Iron porphyrin-based cathode catalysts for polymer electrolyte membrane fuel cells: Effect of NH_3 and Ar mixtures as pyrolysis gases on catalytic activity and stability. *Electrochim. Acta* **2010**, *55*, 6450–6461.
28. Zhao, Y.; Watanabe, K.; Hashimoto, K. Self-supporting

oxygen reduction electrocatalysts made from a nitrogen-rich network polymer. *J. Am. Chem. Soc.* **2012**, *134*, 19528–19531.

29. Liang, H.W.; Zhuang, X.; Bruller, S.; Feng, X.; Mullen, K. Hierarchically porous carbons with optimized nitrogen doping as highly active electrocatalysts for oxygen reduction. *Nat. Commun.* **2014**, *5*, 4973.

30. Chen, S.; Bi, J.Y.; Zhao, Y.; Yang, L.J.; Zhang, C.; Ma, Y.W.; Wu, Q.; Wang, X.Z.; Hu, Z. Nitrogen-doped carbon nanocages as efficient metal-free electrocatalysts for oxygen reduction reaction. *Adv. Mater.* **2012**, *24*, 5593–5597.

31. Hu, Y.; Jensen, J.O.; Zhang, W.; Cleemann, L.N.; Xing, W.; Bjerrum, N.J.; Li, Q. Hollow spheres of iron carbide nanoparticles encased in graphitic layers as oxygen reduction catalysts. *Angew. Chem. Int. Ed.* **2014**, *53*, 3675–3679.

32. Geng, D.; Chen, Y.; Chen, Y.; Li, Y.; Li, R.; Sun, X.; Ye, S.; Knights, S. High oxygen-reduction activity and durability of nitrogen-doped graphene. *Energ. Environ. Sci.* **2011**, *4*, 760–764.

33. Favaro, M.; Perini, L.; Agnoli, S.; Durante, C.; Granozzi, G.; Gennaro, A. Electrochemical behavior of n and ar implanted highly oriented pyrolytic graphite substrates and activity toward oxygen reduction reaction. *Electrochim. Acta* **2013**, *88*, 477–487.

34. Zhao, D.Y.; Feng, J.L.; Huo, Q.S.; Melosh, N.; Fredrickson, G.H.; Chmelka, B.F.; Stucky, G.D. Triblock copolymer syntheses of mesoporous silica with periodic 50 to 300 angstrom pores. *Science* **1998**, *279*, 548–552.

CHAPTER 16

THE ROLE OF OXYGEN AND WATER ON MOLYBDENUM NANOCLUSTERS FOR ELECTRO CATALYTIC AMMONIA PRODUCTION

Jakob G. Howalt, and Tejs Vegge

INTRODUCTION

Molybdenum nanoclusters have been identified as a prime candidate for electrochemical ammonia production with seemingly low Faradaic losses to hydrogen evolution [1,2]. To produce ammonia electrochemically, one can either use a liquid or a solid electrolyte, but these effectively require wet conditions to obtain sufficient protonic conduction [3,4]. The presence of water may give rise to oxygen or hydroxide adsorption on the surface, which can occupy or block important surfaces sites. The adsorbed oxygen species can either decrease or totally hinder other chemical reactions taking place at that site. Oxygen poisoning of the surface is indeed a main inhibitor for ammonia production [5,6]. In this paper, the presence of oxygen species, e.g., resulting from a dehydrogenation reaction of residual water from a solid electrolyte or an aqueous electrolyte, will be investigated to understand the implications it has on the catalytic properties for electrochemical ammonia production. In addition, the blocking of active sites by oxygen species has been explored; together with a determination of reduction pathways to electrochemically reduce the blocking oxygen off the surface.

COMPUTATIONAL METHOD

DFT Calculations

The calculations were carried out with density functional theory (DFT) calculations [7,8] using the RPBE exchange correlation functional [9] along with the projector augmented wave method [10,11] as implemented in the GPAW code [12-14]. A grid of (3,3) for the finite difference stencils have been used together with a grid spacing of 0.18 Å and a minimum of 20 free bands above the Fermi level. Periodic boundary conditions with a Monkhorst–Pack [15] k-point sampling of 2 × 2 × 2 were used to perform a methodological consistent comparison with previously obtained results on stepped and flat surfaces [1,16]. A 7 Å vacuum layer on all sides of the nanocluster is applied, giving a separation of 14 Å between the clusters. When solving the electronic density self-consistently, the convergence criteria have been chosen such that the changes were $\leq 10^{-5}$ eV for the energy and 10^{-4} electrons per valence electron for the density. In all calculations, a Fermi smearing of 10^{-4} eV has been used. The atomic simulation environment ASE [17] was used to set up the atomic structure of these systems. All calculations were performed without spin polarization in order to enable full structural (and atomic) relaxation of the full Mo13 nanocluster together with the adsorbates (N, H, O, NH, etc.); these were carried out using the BFGS and FIRE [18] optimizers within ASE.

Electrochemical Modelling

The procedure for electrochemical reduction of nitrogen molecules through the associative mechanism is outlined in [2,16]. It was shown that the associative mechanism is the preferred route for electrochemical ammonia production, where the protonation of the nitrogen molecules give rise to a weakened N–N bond and subsequent splitting at the third or fourth protonation step. The dissociative mechanism, the other route for electrochemical ammonia production, is due to high dissociation barriers (1.8 eV) of N2 not taking place at the surface. For the purposes of analyzing the reduction of oxygen, a twostep electron-transfer process was assumed and

Citation: Jakob G. Howalt and Tejs Vegge, The role of oxygen and water on molybdenum nanoclusters for electro catalytic ammonia production, doi:10.3762/bjnano.5.11.

simulated using the Heyrovsky-type [19] reaction. In an acidic environment, the reaction comprises of these elementary reaction steps:

$$O^* + 2(H^+ + e^-) \rightleftharpoons OH^* + (H^+ + e^-)$$

$$OH^* + (H^+ + e^-) \rightleftharpoons H_2O^*$$

$$H_2O^* \rightleftharpoons H_2O(g) + ^*,$$

where e^- is an electron in the electrode, H^+ is a proton in the electrolyte and * is a surface site. The reference potential is set to that of the standard hydrogen electrode, and through the computational hydrogen electrode approach the electrons and protons are introduced into the analysis, where $1/2\ H_2 \rightleftharpoons H^+ +$. Hereby, a description of the effects of an external applied potential U on the electrons and the concentration of protons in the electrolyte [20-24] are implemented.

The adsorption energy of O* under electrochemical reaction conditions under an applied potential U are calculated as $\Delta E_0(U) = E_{surface/O} - E_{surface} - E_{H_2O(g)} + 2E_{H^+ + e^-}(U)$, where $E_{surface/O}$ and $E_{surface}$ are the total energies of the molybdenum nanocluster with and without the specific oxygen atom adsorbed, and $E_{H_2O(g)}$ is the gas phase energy of water and $E_{H^+ + e^-}(U)$ are the energy of protons and electrons with an applied potential defined by the computational hydrogen electrode approach. Molecular O_2 is not expected to be present under operating conditions and the sole contributor of O atoms is water. At standard ambient temperature and pressure the adsorption energies are corrected with zero-point energy, E_{ZPE}, and entropy, TS, contributions at room temperatures.

$$\Delta G = \Delta E + \Delta E_{ZPE} - T\Delta S - neU + kT \cdot \ln[H^+]$$
$$= \Delta E + \Delta E_{ZPE} - T\Delta S - neU + 2.3 \cdot kT \cdot pH \quad (1)$$

In this paper the pH value is set to 0 and the values of the free energy corrections for all the oxygen containing species were found in literature [25], were the $\Delta E_{ZPE} - T\Delta S$ corrections for O is 0.05 eV, for OH is 0.35 eV and for H_2O is 0.67 eV. A pH of, e.g., 7 would change the energy by 0.41 eV for all the reduction mechanisms, leaving the relative activity unchanged.

RESULTS AND DISCUSSION

Oxygen Adsorbed in Nitrogen Vacancies

We have previously shown that partially reduced nitrogen covered molybdenum nanoclusters are promising catalysts for electrochemical production of ammonia [2]. N_2 adsorption is preferred over H in nitrogen vacancy sites at lower nitrogen coverages at the potential of U = −0.6 V needed for electrochemical ammonia production through the associative mechanism [2]. Adsorption of oxygen atoms at nitrogen vacancy sites is presented in Figure 1, where they have been adsorbed at the vacant bridge site.

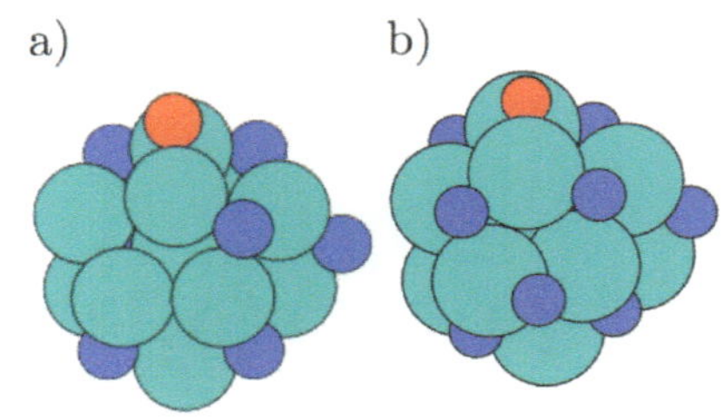

Figure 1: (a) Adsorption of oxygen at a nitrogen vacancy site on $Mo_{13}N_{10}$, and (b) adsorption of oxygen at a nitrogen vacancy site on $Mo_{13}N_{12}$. Oxygen is adsorbed in the bridge site in both cases.

The free energies for oxygen absorption are shown in Table 1 for neutral bias and the potential needed for ammonia production, and listed together with the energies for nitrogen and hydrogen (from [2]). At neutral bias, oxygen is the preferred adsorbate with adsorption energies of −1.58 eV increasing to −2.15 eV. Under these conditions, the vacancy sites will be filled with oxygen and block nitrogen molecules from adsorbing to the active sites and stop the electrochemical formation of ammonia.

At the potentials needed for ammonia production (−0.6 V), the adsorption of oxygen is weakened by 1.2 eV, due to the change in energy from oxidation of water at the applied potential, i.e., −2U. Under these conditions, the adsorption of oxygen has a comparable stability to nitrogen and hydrogen.

The strong preference for oxidation at neutral bias means that the particles will often be (partially) oxidized during synthesis or sample transfer [24]. To free the active sites for N2 adsorption and subsequent electrochemical ammonia production, the adsorbed oxygen atoms needs to be removed from the surface sites. One way of re-activating the molybdenum nanocluster for electrochemical ammonia production is therefore to reduce the oxygen electrochemically to water, which will be less strongly bound to the active site and make it possible to bind N_2 preferentially.

In the case of oxygen adsorption on the more nitrogen rich molybdenum nanoclusters, Table 2 shows the required potentials for the two-proton transfer steps in the reduction of oxygen.

Table 2: Reduction of oxygen adsorbed on the $Mo_{13}Nx$ nanocluster.

	$U_{O \rightarrow OH}$ [V]	$U_{OH \rightarrow H2O}$ [V]
$Mo_{13}N_{12}O$	-0.35	-1.62
$Mo_{13}N_{11}O$	-0.97	-1.25
$Mo_{13}N_{10}O$	-1.18	-1.28
$Mo_{13}N_{9}O$	-0.72	-1.41

The potentials range from −1.28 to −1.62 V for the second protonation step, where the formation of H_2O is hindered by the adsorbed nitrogen atoms in the three fold hollow sites, surrounding the adsorption site of O, OH and H_2O. The high potential needed for the formation of H_2O is due to the repositioning of the adsorption site of OH (bridge site) to the adsorption site of H_2O (ontop site), which is structurally hindered by the nearby adsorbed nitrogen atoms. The removal of oxygen from the partially nitrogen covered molybdenum surface and oxygen will therefore constitute a strong blocking of the active sites and subsequently limit the ammonia production rate through the associative mechanism on partially oxidized nitrogen covered molybdenum nanoclusters.

Table 1: Stability of nitrogen, hydrogen and oxygen with (U = −0.6 V) and without (U = 0 V) an applied potential at nitrogen vacancy sites on Mo13Nx nanoclusters. The potential required to produce ammonia electrochemically at partially nitrogen covered molybdenum nanoclusters was shown to be U = −0.6 V [2]. The energies are given with respect to H_2O (g), H_2 (g) and N_2 (g). (The values for ΔGN_2, ΔGH (U = 0 V) and ΔGH (U = −0.6 V) are taken from [2]).

	ΔG_{N2} [eV]	ΔG_H [eV] (U = 0 V)	ΔG_O [eV] (U = 0 V)	ΔG_H [eV] (U = −0.6 V)	ΔG_O [eV] (U = −0.6 V)
$Mo_{13}N_{12}$	−0.06	−0.66	−1.58	−1.26	−0.38
$Mo_{13}N_{11}$	−0.43	−0.71	−2.24	−1.31	−1.04
$Mo_{13}N_{10}$	−1.2	−0.73	−2.33	−1.33	−1.13
$Mo_{13}N_9$	−1.13	−0.59	−2.15	−1.19	−0.95

Direct reduction of the residual nitrogen skin will, however, still be possible and the potential will not be influenced by the presence of oxygen and the nitrogen skin will be reduced electrochemically at −0.6 V as shown in [2].

An Oxygen Skin

The electrochemical production of ammonia will not only occur on nitrogen covered molybdenum clusters, but could also take place at very low or no nitrogen coverage [2]. A clean molybdenum nanocluster in contact with nitrogen, hydrogen and water at neutral bias will also adsorb oxygen from water on the surface, see Figure 2. The figure shows the total adsorption free energies of oxygen, nitrogen and hydrogen as the coverage evolves and until saturation of oxygen and nitrogen is obtained on the surface. This oxygen skin is approximate 1–2 eV more energetically favoured than a nitrogen skin at low coverage. At higher coverage, the oxygen skin becomes even more energetically favoured. However, an applied potential of U = −0.6 V, destabilizes oxygen (dashed blue line) significantly with respect to nitrogen and hydrogen (dashed red line), such that an overlayer of oxygen arising from electrochemical splitting of water should not be expected.

A cyclic pattern is seen in the binding energy for both nitrogen and oxygen adatoms, and the distance between the cycles correspond roughly to an addition of 10 electrons, corresponding to the filling of a d-shell. Figure S1 in Supporting Information File 1 shows the d-band of the molybdenum nanocluster as the oxygen coverage increases and it is observed that the d-band broadens and the energetically lower lying d-orbitals are filled additionally as the coverage of oxygen is increased. The preferred adsorption sites for the oxygen atoms are the three fold hollow sites, see Figure 3a. When all the three fold sites are filled, the oxygen atoms will adsorb in the four fold hollow sites, see Figure 3c, where additional adsorption of a few oxygen atoms greatly distorts the surface; this restructuring allows a higher filling of oxygen on the surface, see Figure 3d. The maximum filling of the surface increases to 16 oxygen atoms and further additions of oxygen atoms are energetically unfavourable. Close to a full overlayer, the binding sites of oxygen becomes asymmetrically and the oxygen atoms now binds in a mix of a three fold hollow site and a bridge site.

Reduction of the Oxygen Overlayer

For a fully oxygen covered molybdenum nanocluster, nitrogen molecules cannot adsorb to the surface. Therefore, a full oxygen skin eliminates electrochemical ammonia production. It is therefore necessary to reduce the surface, in order to create active sites for nitrogen adsorption. To investigate the reduction of oxygen on a partially or fully oxygen covered molybdenum nanocluster, the reduction of all adsorbed oxygen atoms should, in principle, be analysed. This is, however, not computationally feasible and instead up to three representative adsorbed oxygen atoms are reduced at each investigated oxygen coverage. Each of the selected oxygen atoms represents a unique oxygen adsorption position and the coverage ranges from 16 oxygen atoms to only one oxygen atom. The potentials needed for the two proton transfer processes are presented in Table 3. For other proton transfer processes (not shown), the local geometries around the adsorption site can hinder both protonation steps, but most often the second protonation is hindered. One such hindrance can be the movements from the adsorption site of the OH species to an on top site, where the H_2O species is energetically most stable. Such movement requires a restructuring of the local environment, and hence the protonation step can become strongly endothermic. The largest variation is therefore seen for the second protonation step. For the presented reduction steps of OH to H_2O, the required potentials are in the range of −0.2 to −0.7 V. The formation of OH from O will require a negative potential to stabilize the OH specie compared to the O atom, and in most cases the potential required are in the range of −0.3 V to −0.7 V.

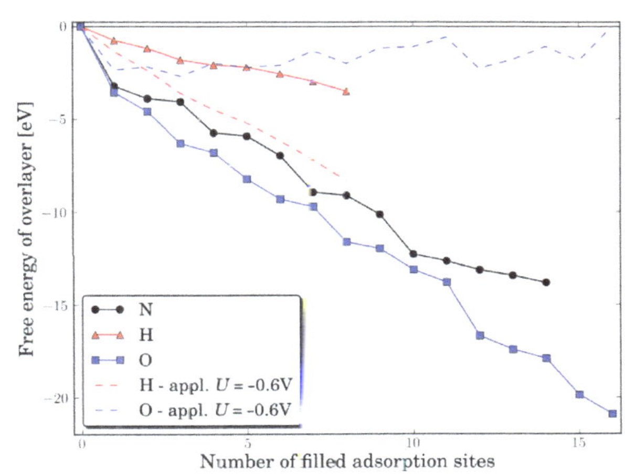

Figure 2: The total free energy for covering the Mo13 nanocluster with nitrogen, oxygen or hydrogen. The filled black line shows the filling of the nitrogen skin, while the filled blue line shows the evolution of oxygen coverage at neutral bias. The red coloured line shows the adsorption of hydrogen, while the dashed red coloured line shows

adsorption of hydrogen and the dashed blue line shows the adsorption of oxygen at an external applied potential of −0.6 V, which is the potential needed for electrochemically ammonia production previously determined for the clean and nitrogen covered molybdenum nanocluster [2].

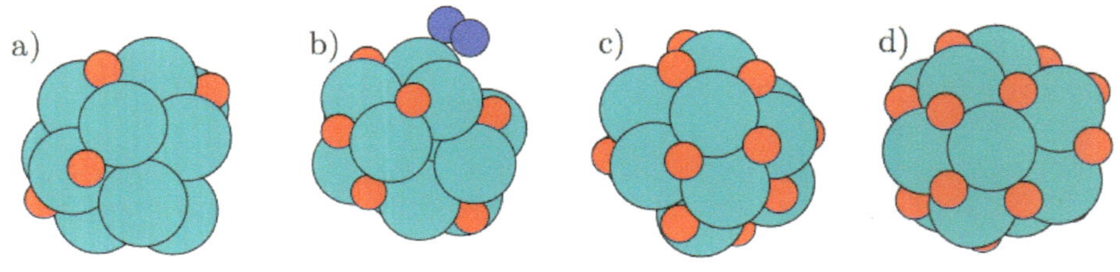

Figure 3: (a) The $Mo_{13}O_6$ nanocluster, (b) the $Mo_{13}O_9$ nanocluster with N_2 adsorbed, (c) the $Mo_{13}O_{12}$ with an almost filled oxygen skin, (d) the $Mo_{13}O_{16}$ with a filled oxygen skin. At the $Mo_{13}O_{16}$, the oxygen adsorption sites are a mix of three fold hollow sites and bridge sites.

For most of the studied coverages, the reaction free energies for desorption of H_2O off the surface are either low or exergonic. In general, desorption energies are lower than 0.4 eV, but for a single adsorbed H_2O molecule, it is as high as 1.2 eV. Water should therefore desorb thermally from the surface, except at very low oxygen coverages.

Table 3: Reduction of an oxygen overlayer on molybdenum nanoparticles. Multiple adsorbed oxygen atoms have been reduced at the studied oxygen coverage. This is done to probe the reduction of all the unique adsorption sites at the studied oxygen coverage.

	$U_{O \to OH}$ [V]	$U_{OH \to H2O}$ [V]
$Mo_{13}O_{16}$	−0.49	−0.44
$Mo_{13}O_{14}$	0.04	−0.47
$Mo_{13}O_{12}$	−0.64	−0.66
$Mo_{13}O_9$	−0.27	−0.67
$Mo_{13}O_6$	−0.29	−0.24
$Mo_{13}O_2$	−0.44	−0.31
$Mo_{13}O$	−0.67	−0.72

The study of oxygen reduction produces adsorption energies for O, OH, H_2O at different coverages. No apparent correlation is found between the adsorption energies of EO, EOH and EH_2O on the cluster, see Figure S_2 and Figure S3 in Supporting Information File 1, in contrast to what has previously been observed on metal surfaces, where scaling relations are applicable for OH species on close-packed and stepped surfaces for low coverages [26]. The close packed and stepped surfaces typically have less restructuring during the adsorption of O, OH and H_2O. On the molybdenum nanocluster surface, larger local restructuring takes place when either O is added or removed and when either O or OH is protonated. The restructuring of the molybdenum cluster involves all atoms in the nanocluster. This is an effect of the small size of the nanocluster, where the local impact from an adsorption or a reduction step influences the whole nanocluster.

The lowest required potential for water formation at the different coverages is in the range of −0.3 V to −0.7 V. These values are lower or comparable to the required potential needed to form ammonia on the molybdenum nanocluster [2]. Overall, it seems that it is possible to reduce oxygen of the surface at moderate potential of −0.72 V. The potential required for reduction of a surface oxygen atom indicates that the nanoclusters can be reactivated after exposure to water from the electrolyte or the fabrication process.

Formation of Ammonia at Relative High Oxygen Coverage

Nitrogen molecules are not able to adsorb to neither an oxygen vacancy site nor ontop of an oxygen atom at high oxygen coverages. Once the oxygen skin has been partially reduced, the nitrogen molecules adsorb onto the surface at the oxygen vacancy sites, see Figure 3b. The adsorption of nitrogen become stable at a coverage of less than 10 oxygen atoms, see Table 4. The adsorption energies of N_2 range from −1.0 eV to −1.8 eV depending on the oxygen coverage. The corresponding hydrogen adsorption energies are lower, ranging from −0.65 eV to −0.86 eV, and nitrogen molecules are therefore preferred over hydrogen on the surface at neutral bias. At an applied potential of −0.6 V, which is the potential shown to electrochemically produce ammonia on the clean molybdenum nanocluster, the reaction free energy of adsorbing either a hydrogen atom (coming from H^+ and e^-) or a nitrogen molecule will at certain coverages be in favour of H ($Mo_{13}O_8$ and $Mo_{13}O_6$) and others of N_2 ($Mo_{13}O_9$ and $Mo_{13}O_7$).

For nitrogen molecules adsorbed at a vacancy site on either $Mo_{13}O_9$ or $Mo_{13}O_6$, the potential for driving the electrochemical production of ammonia has been determined. These nanoclusters were selected to describe the two regimes with either strong or weak N_2 adsorption compared to hydrogen adsorption, respectively. On the $Mo_{13}O_9$ nanocluster, N_2 is bound most strongly at an applied potential of −0.6 V, while hydrogen is bound stronger on the $Mo_{13}O_6$ nanocluster. Investigations of the associative pathway on the $Mo_{13}O_9$ nanocluster, shows the first protonation step to require an onset potential of −0.7 V similar to that needed to reduce oxygen (see Table 3). The onset potentials are shown in Figure 4 for the following electrochemical reaction steps:

Table 4: The adsorption free energies of nitrogen and hydrogen with and without an applied potential on a partly oxygen covered Mo_{13} nanocluster

	ΔG_{N2} [eV]	ΔG_H [eV] (U = 0 V)	ΔG_H [eV] (U = −0.6 V)
$Mo_{13}O_{12}$	No binding	−0.23	−0.83
$Mo_{13}O_{10}$	0.33	−0.84	−1.44
$Mo_{13}O_9$	−1.82	−0.72	−1.32
$Mo_{13}O_8$	−1.02	−0.75	−1.35
$Mo_{13}O_7$	−1.77	−0.86	−1.46
$Mo_{13}O_6$	−0.93	−0.65	−1.25

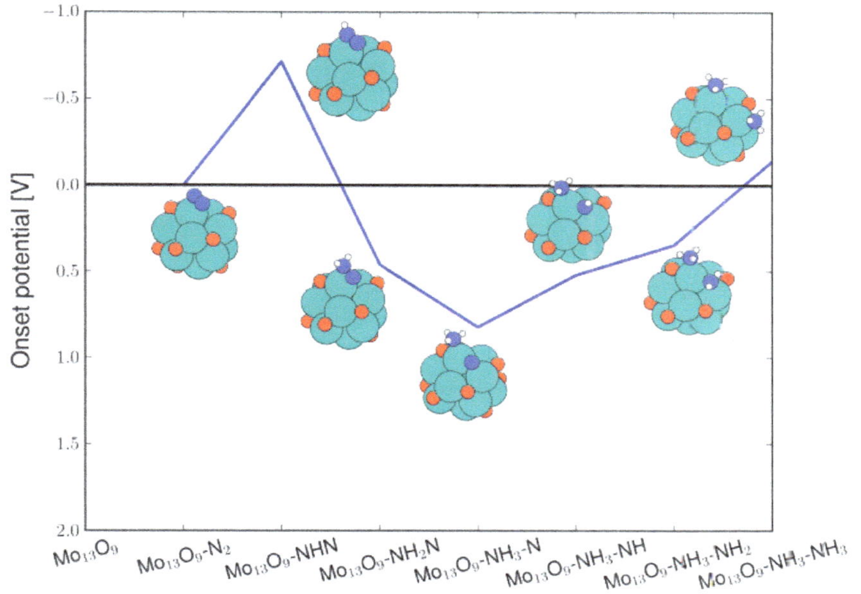

Figure 4: Diagram of the required applied potential to make each reaction step exergonic for electrochemical ammonia production on the $Mo_{13}O_9$ through the associative mechanism. For all five studied reaction pathways, the highest required potential is −0.7 V for the first protonation step.

$$
\begin{aligned}
&NN^* + 6(H^+ + e^-) \rightarrow \\
&NHN^* + 5(H^+ + e^-) \rightarrow \\
&NH_2N^* + 4(H^+ + e^-) \rightarrow \\
&NH_3^* + N^* + 3(H^+ + e^-) \rightarrow \\
&NH_3^* + NH^* + 2(H^+ + e^-) \rightarrow \\
&NH_3^* + NH_2^* + (H^+ + e^-) \rightarrow \\
&NH_3^* + NH_3^*
\end{aligned}
$$

The x-axis show the adsorbed reaction intermediate for each protonation step and NH_2N denotes a doubly protonated nitrogen molecule with the N–N backbone intact, while NH_3–NH denotes NH_3 and NH adsorbed on the surface and the N–N backbone has been dissociated. Five pathways were studied, and all of them had the first initial protonation step to be the potential limiting step with identical required onset potentials.

For a more reduced oxygen skin, e.g., the $Mo_{13}O_6$ nanocluster, the energetics of the electrochemical production of ammonia are much more diverse. Here, two routes are very favourably, demanding only potentials of around −0.45 V, to drive the electrochemical production of ammonia, see Figure 5. The routes not shown require potentials of −0.85 V to −1.3 V. The reaction intermediates for the preferred ammonia formation route have been used as the bottom x-axis. The blue filled line on Figure 5 shows that the rate-limiting step is the last protonation; while for the alternative route, marked with the green dashed line and the corresponding reaction intermediates have been used as the top x-axis in Figure 5, the rate-limiting steps are found to be both the third and the last protonation step. The limiting step for this reaction path is the formation of the NH_2–NH intermediate on the surface. The reaction is then followed by a very exergonic reaction step, where the N–N bond breaks. In both pathways, the N–N bond breaking in the associative process is very exergonic and no apparent activation barrier is observed for the N–N cleavage at either the third or fourth protonation step. The onset potential presented here for electrochemical ammonia production is similar to those obtained in previous studies on both clean and nitrogen covered molybdenum nanoclusters [2].

Instead of protonating the adsorbed nitrogen molecule, the potential could drive the reduction of oxygen. In Table 5, the potentials for driving the oxygen reduction processes are presented. Here, a potential of −0.77 V is required for the high oxygen coverage case, while the last reduction step for the lower oxygen coverage requires significantly larger potentials, i.e., −1.3 V to −1.75 V, to make the reaction exergonic, H_2O is found to be unstable on $Mo_{13}O_6$, while $Mo_{13}O_9$ has a stable adsorption of water.

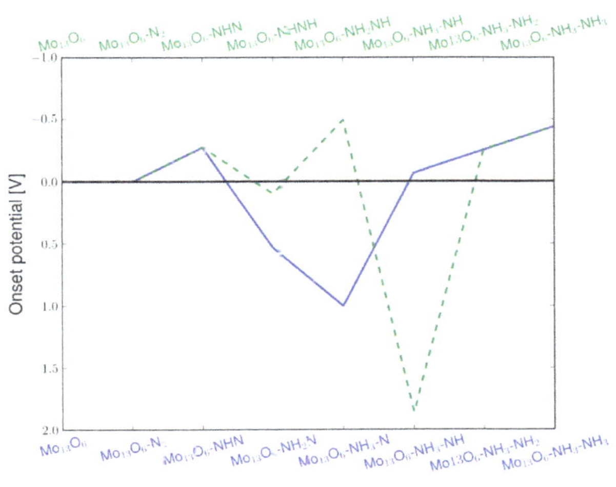

Figure 5: Diagram of the required applied potential to make each reaction step exergonic for electrochemical ammonia production on the $Mo_{13}O_6$ through the associative mechanism. The limiting reaction is the last protonation to form the second ammonia on the surface requiring a potential of −0.45 V to become exergonic.

Table 5: The required potentials for reducing oxygen at two oxygen coverages, when nitrogen molecules are present on the partly covered oxygen surfaces.

	$U_{O \to OH}$ [V]	$U_{OH \to H2O}$ [V]
$Mo_{13}O_9$-N_2	−0.53	−0.77
$Mo_{13}O_6$-N_2	−0.26	−1.30

COMPETING REACTION MECHANISMS

Here, a possible reaction pathway is presented for electrochemical ammonia production on the $Mo_{13}O_9$ nanocluster with adsorbed N_2 at an applied potential U = −0.7 V. Three possible mechanisms must be considered: the reduction of N_2 to ammonia, hydrogen adsorption and evolution, and last the reduction of oxygen and subsequently formation of water. Here, we assume that the adsorption of H with N_2 present on the surface will be equal to adsorption of H on a surface without N_2 present.

The reaction free energies of forming H, OH, and N_2H at U = −0.7 V are: $\Delta G[N_2^* + (H^+ + e^-) \rightarrow N_2H^*] = 0$ eV, $\Delta G[(H^+ + e^-) \rightarrow H^*] = -1.42$ eV, $\Delta G[O^* + (H_+ + e_-) \rightarrow OH_*] = -0.17$ eV, respectively. The reaction step with the lowest reaction free energy is hydrogen adsorption on the surface. For adsorbed hydrogen, the next step will be formation of H_2(g) with a reaction free energies of $\Delta G[H^* + (H^+ + e^-) \rightarrow H_2(g)] = 0.02$ eV. The next reaction to occur on the surface will therefore be the reduction of O to OH at −0.17 V. For this reaction intermediate, the next step will produce H_2O with a reaction free energy of $\Delta G[OH_* + (H^+ + e^-) \rightarrow H_2O^*] = 0.08$ eV. Small energy differences exist for the next protonation step, but N_2H will be formed on the surface.

The subsequent reaction intermediates for ammonia formation are exergonic at U = −0.7 V. The initial production of an overlayer of H is followed by reduction of O into OH, until the electrochemical reduction of N_2 becomes possible. When the ammonia molecules have desorbed from the surface, adsorption of N_2 is 0.4 eV more stable than H, which would make the adsorption of nitrogen at a vacancy site energetically favourable. At higher pH-values this difference would increase in favour of N_2 adsorption because a pH relatively destabilizes the adsorption of H, see Equation 1. Because of the very small difference between the formation of N_2H, H_2O and H_2 (g) a Faradaic loss due to hydrogen evolution should be expected and a reduction of the oxygen skin on the surface would also be expected at this oxygen coverage and potential.

For the $Mo_{13}O_6$ nanocluster with N_2 adsorbed at an applied potential of U = −0.45 V, the preferred mechanism is the following. First hydrogen is bound to the surface, then O is reduced to OH and then finally the reduction of N_2 can take place. Here, the reaction free energy for production of gas phase hydrogen is $\Delta G[H^* + (H^+ + e^-) \rightarrow H_2(g)] = 0.20$ eV and for water formation is $\Delta G[OH + (H^+ + e^-) \rightarrow H_2O^*] = 0.59$ eV, while ΔG for all electrochemical reaction steps for ammonia production, see Figure 5, are zero or lower as in the previous example on the $Mo_{13}O_9$ nanocluster.

In the discussion of the preferential reactions on the 9 and $Mo_{13}O_6$ nanoclusters, all possible adsorbate–adsorbate interactions are not included due to the large computational cost associated with a comprehensive investigation. This is a potentially significant assumption, since the adsorption energies may decrease with higher coverage of certain species [27]. For hydrogen evolution on the $Mo_{13}O_6$ nanocluster, this could, e.g., have the consequence that the free energy barrier for producing gas phase hydrogen molecules is lowered. The results for ammonia production on a partially oxidized molybdenum nanocluster indicates that the formation of ammonia on the molybdenum nanocluster is possible at a low onset potential, but with a low Faradaic efficiency due to the parasitic formation of both adsorbed H or OH on $Mo_{13}O_9$ and $Mo_{13}O_6$. Adsorption of nitrogen seems to hinder the further reduction of OH at lower oxygen coverages. This indicates that the nanocluster should be fully reduced, i.e., no oxygen present on the surface, before nitrogen is able to reach the catalyst surface. Small amounts of oxygen present at the molybdenum nanocluster surface, arising from either the electrolyte or from the preparation of the electrocatalyst, should therefore not affect the electrochemical production of ammonia.

CONCLUSION

Density functional theory calculations have been employed to investigate the adsorption and reduction of oxygen on molybdenum nanoclusters. The computational hydrogen electrode was used to determine potentials for reduction of nitrogen and oxygen and the hydrogen evolution reaction.

First, a partially nitrogen covered molybdenum nanocluster was exposed to gaseous water, showing preferential adsorption of oxygen atoms compared to both hydrogen atoms and nitrogen molecules at neutral bias (U = 0 V), while the adsorption energies are comparable at the potential needed to produce ammonia (U = −0.6 V). The consequence is that the presence of water will lead to preferential oxygen adsorption at the nitrogen vacancy sites unless a negative bias is applied. The reduction of the oxygen atoms at the nitrogen rich molybdenum nanocluster was studied, and potentials more negative than −1.25 V is required to reduce the oxygen atoms into water. The main challenge is the second protonation step, where the reaction step is very dependent on the local environment. The oxygen atom will bind to the vacancy site blocking the adsorption of nitrogen molecules and thereby greatly reduce the efficiency of the nitrogen rich molybdenum nanocluster as an electro catalyst for ammonia production through the associative mechanism. A direct reduction of the nitrogen skin will, however, still be possible at U = −0.6 V [2].

A clean molybdenum nanocluster in contact with oxygen, hydrogen and nitrogen will preferentially form an oxygen skin at neutral bias and nitrogen adsorption on an oxygen covered molybdenum nanocluster is found to be impossible. The reduction of an oxygen overlayer was therefore studied and it was found that the reduction requires potentials of −0.29 V to −0.72 V to successfully produce water from adsorbed oxygen atoms. For oxygen coverages of nine or less oxygen atoms, adsorption of nitrogen and hydrogen becomes possible. The electrochemical production of ammonia for adsorbed nitrogen molecules at partial oxygen coverage will only require potentials of −0.45 V to −0.7 V to make the reaction exergonic. These onset potentials are similar to values reported in earlier studies of electrochemical ammonia production on molybdenum nanocluster with or without a nitrogen skin [2]. At the potentials needed to make the ammonia production exergonic, hydrogen is found to be present on the surface, and a reduction of oxygen to OH is observed, before electrochemical production of ammonia is possible.

On the basis of the work presented here, we propose that (partially) oxidized and nitrided molybdenum nanoclusters can be re-activated for ammonia production by electrochemical reduction of the adsorbed oxygen resulting from, e.g., the catalyst synthesis procedure or the presence of water in the electrolyte, at a potential of approximate −0.7 V. This makes molybdenum nanoclusters a highly promising catalyst for electrochemical ammonia production via the associative mechanism.

SUPPORTING INFORMATION

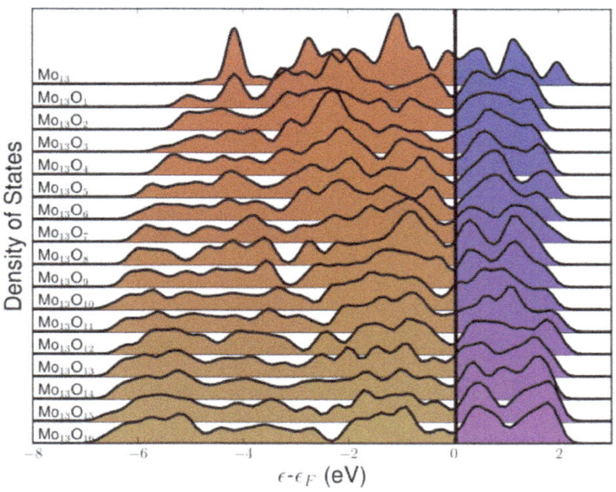

Figure S1: The d-band of molybdenum nanoclusters as the oxygen coverage evolves.

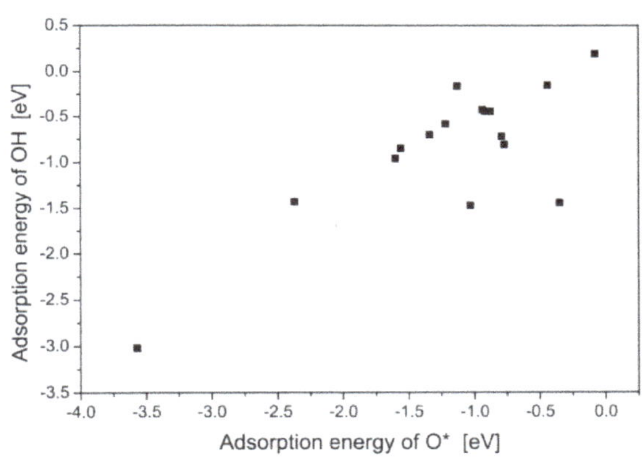

Figure S2: The adsorption energy of OH plotted as a function of adsorption energy of O for the oxygen covered molybdenum nanoclusters.

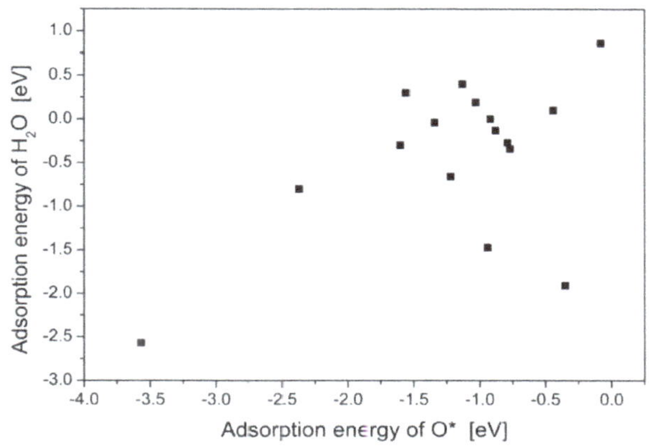

Figure S3: The adsorption energy of H_2O plotted as a function of adsorption energy of O for the oxygen covered molybdenum nanoclusters.

REFERENCES

1. Howalt, J. G.; Bligaard, T.; Rossmeisl, J.; Vegge, T. Phys. Chem. Chem. Phys. 2013, 15, 7785–7795. doi:10.1039/c3cp44641g
2. Howalt, J. G.; Vegge, T. Phys. Chem. Chem. Phys. 2013, 15, 20957–20965. doi:10.1039/c3cp53160k
3. Sone, Y.; Ekdunge, P.; Simonsson, D. J. Electrochem. Soc. 1996, 143, 1254–1259. doi:10.1149/1.1836625
4. Kreuer, K.-D.; Paddison, S.; Spohr, E.; Schuster, M. Chem. Rev. 2004, 104, 4637–4678. doi:10.1021/cr020715f
5. Fastrup, B.; Nygård Nielsen, H. Catal. Lett. 1992, 14, 233–239. doi:10.1007/BF00765236
6. Kuchaev, V. L.; Shapatina, E. N.; Avetisov, A. K. Russ. J. Electrochem. 2009, 45, 983–995. doi:10.1134/S1023193509090031
7. Hohenberg, P.; Kohn, W. Phys. Rev. 1964, 136, B864–B871. doi:10.1103/PhysRev.136.B864
8. Kohn, W.; Sham, L. J. Phys. Rev. 1965, 140, A1133–A1138. doi:10.1103/PhysRev.140.A1133
9. Hammer, B.; Hansen, L. B.; Nørskov, J. K. Phys. Rev. B 1999, 59, 7413–7421. doi:10.1103/PhysRevB.59.7413
10. Blöchl, P. E. Phys. Rev. B 1994, 50, 17953–17979. doi:10.1103/PhysRevB.50.17953
11. Blöchl, P. E.; Först, C. J.; Schimpl, J. Bull. Mater. Sci. 2003, 26, 33–41. doi:10.1007/BF02712785
12. Mortensen, J. J.; Hansen, L. B.; Jacobsen, K. W. Phys. Rev. B 2005, 71, 035109. doi 10.1103/PhysRevB.71.035109
13. Enkovaara, J.; Rostgaard, C.; Mortensen, J. J.; Chen, J.; Dułak, M.; Ferrighi, L.; Gavnholt, J.; Glinsvad, C.; Haikola, V.; Hansen, H. A.; Kristoffersen, H. H.; Kuisma, M.; Larsen, A. H.; Lehtovaara, L.; Ljungberg, M.; Lopez-Acevedo, O.; Moses, P. G.; Ojanen, J.; Olsen, T.; Petzold, V.; Romero, N. A.; Stausholm-Møller, J.; Strange, M.; Tritsaris, G. A.; Vanin, M.; Walter, M.; Hammer, B.; Häkkinen, H.; Madsen, G. K. H.; Nieminen, R. M.; Nørskov, J. K.; Puska, M.; Rantala, T. T.; Schiøtz, J.; Thygesen, K. S.; Jacobsen, K. W. J. Phys.: Condens. Matter 2010, 22, 253202. doi:10.1088/0953-

8984/22/25/253202

14. The GPAW code is available as a part of the CAMPOS software: http://wiki.fysik.dtu.dk/gpaw/
15. Monkhorst, H.; Pack, J. D. Phys. Rev. B 1976, 13, 5188–5192. doi:10.1103/PhysRevB.13.5188
16. Skúlason, E.; Bligaard, T.; Gudmundsdóttir, S.; Studt, F.; Rossmeisl, J.; Abild-Pedersen, F.; Vegge, T.; Jónsson, H.; Nørskov, J. K. Phys. Chem. Chem. Phys. 2012, 14, 1235–1245. doi:10.1039/c1cp22271f
17. The ASE code is available as a part of the CAMPOS software: http://wiki.fysik.dtu.dk/ase/
18. Bitzek, E.; Koskinen, P.; Gähler, F.; Moseler, M.; Gumbsch, P. Phys. Rev. Lett. 2006, 97, 170201. doi:10.1103/PhysRevLett.97.170201
19. Heyrovský, J. Recl. Trav. Chim. Pays-Bas 1927, 46, 582–585. doi:10.1002/recl.19270460805
20. Rossmeisl, J.; Logadottir, A.; Nørskov, J. K. J. Chem. Phys. 2005, 319, 178–184. doi:10.1016/j.chemphys.2005.05.038
21. Rossmeisl, J.; Qu, Z.-W.; Kroes, G.-J.; Nørskov, J. K. J. Electroanal. Chem. 2007, 607, 83–89. doi:10.1016/j.jelechem.2006.11.008
22. Hummelshøj, J. S.; Blomquist, J.; Datta, S.; Vegge, T.; Rossmeisl, J.; Thygesen, K. S.; Luntz, A. C.; Jacobsen, K. W.; Nørskov, J. K. J. Chem. Phys. 2010, 132, 071101. doi:10.1063/1.3298994
23. Reuter, K.; Scheffler, M. Phys. Rev. B 2001, 65, 035406. doi:10.1103/PhysRevB.65.035406
24. Rossmeisl, J.; Nørskov, J. K.; Taylor, C. D.; Janik, M. J.; Neurock, M. J. Phys. Chem. B 2006, 110, 21833–21839. doi:10.1021/jp0631735
25. Nørskov, J. K.; Rossmeisl, J.; Logadottir, A.; Lindqvist, L.; Kitchin, J. R.; Bligaard, T.; Jónsson, H. J. Phys. Chem. B 2004, 108, 17886–17892. doi:10.1021/jp047349j
26. Abild-Petersen, F.; Greeley, J.; Studt, F.; Rossmeisl, J.; Munter, T. R.; Moses, P. G.; Skúlason, E.; Bligaard, T.; Nørskov, J. K. Phys. Rev. Lett. 2007, 99, 016105. doi:10.1103/PhysRevLett.99.016105
27. Getman, R. B.; Schneider, W. F. ChemCatChem 2010, 2, 1450–1460. doi:10.1002/cctc.201000146

INDEX